PLANT LIFE IN AQUATIC AND AMPHIBIOUS HABITATS

DAVID SPENCE 1925–1985

TO THE MEMORY OF
DAVID HUGH NEVEN SPENCE

Foundation Professor of Botany
The University of Makere
1962–1966

Professor of Botany
The University of St Andrews
1971–1985

PLANT LIFE IN AQUATIC AND AMPHIBIOUS HABITATS

SPECIAL PUBLICATION NUMBER 5 OF THE
BRITISH ECOLOGICAL SOCIETY

PRODUCED AS A TRIBUTE TO

D. H. N. SPENCE

EDITED BY

R. M. M. CRAWFORD

Department of Plant Biology and Ecology
The University of St Andrews
St Andrews, Fife, KY16 9TH
Scotland

BLACKWELL SCIENTIFIC PUBLICATIONS
OXFORD LONDON EDINBURGH
BOSTON PALO ALTO MELBOURNE
1987

© 1987 by The British Ecological Society
and published for them by
Blackwell Scientific Publications
Osney Mead, Oxford, OX2 0EL
8 John Street, London, WC1N 2ES
23 Ainslie Place, Edinburgh, EH3 6AJ
52 Beacon Street, Boston
  Massachusetts 02108, U.S.A.
667 Lytton Avenue, Palo Alto
  California 94301, U.S.A.
107 Barry Street, Carlton
  Victoria 3053, Australia

First published 1987

Set by Spire Print Services Ltd, Salisbury
Wilts.
Printed and bound in Great Britain by
Butler & Tanner Ltd, Frome and London

DISTRIBUTORS

U.S.A. and Canada
  Blackwell Scientific Publications Inc
  PO Box 50009, Palo Alto
  California 94303

Australia
  Blackwell Scientific Book Publications
  (Australia) Pty Ltd.
  107 Barry Street
  Carlton, Victoria 3053

British Library
Cataloguing in Publication Data

Plant life in aquatic and amphibious habitats:
  produced as a tribute to D. H. N Spence.—
  (Special publication of the British
  Ecological Society, ISSN 0262-7027; no. 5)
  1. Plants, Effects of floods on
  I. Crawford, R. M. M.   II. Spence, D. H. N.
  III. Series
  581.5′263   QK870

  ISBN 0-632-01628-0

Library of Congress
Cataloging-in-Publication Data

Plant life in aquatic and amphibious habitats.
  (Special publication number 5 of the British
Ecological Society)
  "Produced as a tribute to D. H. N. Spence."
  Bibliography: p.
  Includes index.
  1. Aquatic plants. 2. Aquatic ecology. 3. Plants,
Effect of floods on. 4. Wetland flora. 5. Wetland
ecology. I. Crawford, R. M. M.   II. Spence, David Hugh
Neven, 1925–1985. III. Series: Special publication . . .
of the British Ecological Society; no. 5.
QK930.P57 1987   581.5′263   86-20752
ISBN 0-632-01628-0

Contents

List of Contributors and Addresses   viii

R. M. M. Crawford: Introduction and tribute to the late Professor D. H. N. Spence   xi

PART I: AQUATIC PLANTS

A: The Aquatic Environment

M. G. Holmes & W. H. Klein. The light and temperature environments   3

M. J. Dring. Light climate in intertidal and subtidal zones in relation to photosynthesis and growth of benthic algae: a theoretical model   23

G. Russell. Salinity and seaweed vegetation   35

I. Ridge. Ethylene and growth control in amphibious plants   53

B: Photosynthesis under Water

G. Bowes. Aquatic plant photosynthesis: strategies that enhance carbon gain   79

K. Sand-Jensen. Environmental control of bicarbonate use among freshwater and marine macrophytes   99

J. E. Keeley. The adaptive radiation of photosynthetic modes in the genus *Isoetes* (Isoetaceae)   113

J. A. Raven, H. Griffiths & J. J. Macfarlane. The application of carbon isotope discrimination techniques   129

C: Growth, Development and Dispersal in Aquatic Plants

D. H. N. Spence, M. R. Bartley & R. Child. Photomorphogenic processes in freshwater angiosperms   153

B. Frankland, M. R. Bartley & D. H. N. Spence. Germination under water   167

C. D. K. Cook. Dispersion in aquatic and amphibious vascular plants   179

Contents

PART II: AMPHIBIOUS PLANTS AND FLOODING TOLERANCE

A: Oxygen Stress in Seeds and Seedlings

R. A. Kennedy, M. E. Rumpho & T. C. Fox. Germination physiology of rice and rice weeds: metabolic adaptations to anoxia   193

B. B. Vartapetian, H. H. Snkhchian & I. P. Generozova. Mitochondrial fine structure in imbibing seeds and seedlings of *Zea mays* L. under anoxia   205

B: Root Physiology under Oxygen Stress

T. ap Rees, L. E. T. Jenkin, A. M. Smith & P. M. Wilson. The metabolism of flood-tolerant plants   227

I. A. Mendelssohn & K. L. McKee. Root metabolic response of *Spartina alterniflora* to hypoxia   239

A. Bertani, I. Brambilla & R. Reggiani. Effect of exogenous nitrate on anaerobic root metabolism   255

D. D. Davies, P. Kenworthy, B. Mocquot & K. Roberts. The effects of anoxia on the ultrastructure of pea roots   265

C: Whole Plant Responses to Oxygen Stress

D. D. Hook & S. Denslow. Metabolic responses of four families of loblolly pine to two flood regimes   281

C. Studer & R. Braendle. Ethanol, acetaldehyde, ethylene release and ACC concentration of rhizomes from marsh plants under normoxia, hypoxia and anoxia   293

T. J. Gaynard & W. Armstrong. Some aspects of internal plant aeration in amphibious habitats   303

M. C. Drew. Mechanisms of acclimation to flooding and oxygen shortage in non-wetland species   321

G. R. Stewart & M. Popp. The ecophysiology of mangroves   333

√ **D. A. Rugg & T. A. Norton.** *Pelvetia canaliculata*, a high-shore seaweed that shuns the sea   347

√ **C. A. Maggs & M. D. Guiry.** Environmental control of macroalgal phenology   359

R. M. M. Crawford, L. S. Monk & Z. M. Zochowski. Enhancement of anoxia tolerance by removal of volatile products of anaerobiosis   375

D. R. Drakeford & D. M. Reid. Some rapid responses of sunflower to flooding 385

R. Braendle & R. M. M. Crawford. Rhizome anoxia tolerance and habitat specialization in wetland plants 397

T. L. Setter, I. Waters, B. J. Atwell, T. Kapanchanakul & H. Greenway. Carbohydrate status of terrestrial plants during flooding 411

Author Index 435

Subject Index 444

List of Contributors
and addresses

W. Armstrong, *Department of Plant Biology and Genetics, The University, Hull, HU6 7RX*

B. J. Atwell, *Dryland Crops and Soils Research Program, CSIRO, Wembley, W.A. 6014, Australia*

M. R. Bartley, *I.C.I. plc Plant Protection Division, Jealott's Hill Research Station, Bracknell, Berkshire, RG12 6EY*

A. Bertani, *Istituto Biosintesi Vegetale, Via Bassini 15, 20133 Milano, Italy*

G. Bowes, *Department of Botany, University of Florida, Gainesville, Florida 32611, U.S.A.*

R. Braendle, *Institut für Pflanzenphysiologie, Universität Bern, Altenbergrain 21, Bern, CH 3013, Switzerland*

I. Brambilla, *Istituto Biosintesi Vegetale, Via Bassini 15, 20133 Milano, Italy*

R. Child, *Department of Plant Biology and Ecology, The University, St Andrews, Fife, KY16 9TH*

C. D. K. Cook, *Institut für Systematische Botanik, Universität Zurich, Zollikerstrasse 107, Zurich, CH 8008, Switzerland*

R. M. M. Crawford, *Department of Plant Biology and Ecology, The University, St Andrews, Fife, KY16 9TH*

D. D. Davies, *School of Biological Sciences, University of East Anglia, Norwich, NR4 7TJ*

S. Denslow, *Department of Forestry, Clemson University, Clemson, South Carolina 29407, U.S.A.*

D. R. Drakeford, *Plant Physiology Research Group, Department of Biology, University of Calgary, Calgary, Alberta, T2N 1N4, Canada*

M. C. Drew, *Department of Horticultural Sciences, Texas A & M University, College Station, Texas 77843-2133, U.S.A.*

M. J. Dring, *Department of Botany, Queen's University, Belfast, BT7 1NN, Northern Ireland*

T. C. Fox, *Department of Horticulture, The Ohio State University, 2001 Fyffe Court, Columbus, Ohio 43210, U.S.A.*

B. Frankland, *School of Biological Sciences, Queen Mary College, Mile End Road, London, E1 4NS*

T. J. Gaynard, *c/o Dr W. Armstrong, Department of Plant Biology and Genetics, The University, Hull, HU6 7RX*

I. P. Generozova, *Timiriazev Institute of Plant Physiology, Academy of Sciences, Botanicheskaya yl.35, Moscow 127276, U.S.S.R.*

H. Greenway, *School of Agriculture, University of Western Australia, Nedlands, W.A. 6009, Australia*

H. Griffiths, *Department of Plant Biology, The University, Newcastle upon Tyne, NE1 7RU*

M. D. Guiry, *Department of Botany, University College, Galway, Ireland*

M. G. Holmes, *Botany School, Downing Street, Cambridge, CB2 3EA*

D. D. Hook, *Department of Forestry, Clemson University, Clemson, South Carolina 29407, U.S.A.*

L. E. T. Jenkin, *c/o Dr T. ap Rees, Botany School, Downing Street, Cambridge, CB2 3EA*

T. Kapanchanakul, *Huntra Rice Experiment Station, Ayutthaya, Thailand*

J. E. Keeley, *Department of Biology, Occidental College, Los Angeles, California 90041, U.S.A.*

R. A. Kennedy, *Department of Horticulture, The Ohio State University, 2001 Fyffe Court, Columbus, Ohio 43210 U.S.A.*

P. Kenworthy, *School of Biological Sciences, University of East Anglia, Norwich, NR4 7TJ*

W. H. Klein, *Plant Science Laboratories, University of Reading, Whiteknights, P.O. Box 221, Reading, Berkshire*

J. J. Macfarlane, *Department of Biological Sciences, The University, Dundee, DD1 4HN*

K. L. McKee, *Laboratory for Wetland Soils and Sediments, Center for Wetland Resources, Louisiana State University, Baton Rouge, Louisiana 70803, U.S.A.*

C. A. Maggs, *Department of Botany, Queen's University, Belfast, BT7 1NN, Northern Ireland*

I. A. Mendelssohn, *Laboratory for Wetland Soils and Sediments, Center for Wetland Resources, Louisiana State University, Baton Rouge, Louisiana 70803, U.S.A.*

B. Mocquot, *Station de Physiologie Végétale, INRA, Pont-de-la-Maye, Bordeaux, 33140, France*

L. S. Monk, *Department of Plant Biology and Ecology, The University, St Andrews, Fife, KY16 9TH*

T. A. Norton, *Department of Marine Biology, University of Liverpool, Port Erin, Isle of Man*

M. Popp, *Institut für Pflanzenphysiologie, Universität Wien, Althanstrasse 14, Postfach 285, Vienna, A 1091, Austria*

J. A. Raven, *Department of Biological Sciences, The University, Dundee, DD1 4HN*

T. ap Rees, *Botany School, Downing Street, Cambridge, CB2 3EA*

R. Reggiani, *Istituto Biosintesi Vegetale, Via Bassini 15, 20133 Milano, Italy*

D. M. Reid, *Plant Physiology Research Group, Department of Biology, University of Calgary, Calgary, Alberta, T2N 1N4, Canada*

I. Ridge, *Department of Biology, The Open University, Walton Hall, Milton Keynes, MK7 6AA*

K. Roberts, *John Innes Institute, Colney Lane, Norwich, NR4 7UH*

D. A. Rugg, *c/o Dr T. A. Norton, Department of Marine Biology, University of Liverpool, Port Erin, Isle of Man*

M. E. Rumpho, *Department of Horticulture, The Ohio State University, 2001 Fyffe Court, Columbus, Ohio 43210, U.S.A.*

G. Russell, *Department of Botany, The University, P.O. Box 147, Liverpool, L69 3BX*

K. Sand-Jensen, *Freshwater Biological Laboratory, Helsingørsgade 51, DK3400, Hillerød, Denmark*

T. L. Setter, *School of Agriculture, University of Western Australia, Nedlands, W.A. 6009, Australia*

A. M. Smith, *John Innes Institute, Colney Lane, Norwich, NR4 7UH*

H. H. Snkhchian, *Timiriazev Institute of Plant Physiology, Academy of Sciences, Botanicheskaya yl.35, Moscow 127276, U.S.S.R.*

The late D. H. N. Spence, *Department of Plant Biology and Ecology, The University, St Andrews, Fife, KY16 9TH*

G. R. Stewart, *Department of Botany, Birkbeck College, University of London, Malet Street, London, WC1E 7HX*

C. Studer, *Institut für Pflanzenphysiologie, Universität Bern, Altenbergrain 21, Bern, CH 3013, Switzerland*

B. B. Vartapetian, *Timiriazev Institute of Plant Physiology, Academy of Sciences, Botanicheskaya yl.35, Moscow 127276, U.S.S.R.*

I. Waters, *School of Agriculture, University of Western Australia, Nedlands, W.A. 6009, Australia*

P. M. Wilson, *c/o Dr T. ap Rees, Botany School, Downing Street, Cambridge, CB2 3EA*

Z. M. Zochowski, *Department of Plant Biology and Ecology, The University, St Andrews, Fife, KY16 9TH*

Introduction and tribute to the late Professor D. H. N. Spence

R. M. M. CRAWFORD

Department of Plant Biology and Ecology, University of St Andrews

The physiological ecology of amphibious and intertidal plants was the subject of a special symposium held by the British Ecological Society in March 1985. This was the first time that this particular aspect of plant life had been the subject of a symposium and, as can be seen from the list of contributors, the meeting brought together a large number of researchers from many parts of the world. 'Amphibious' is not a word that is usually associated with plants yet there are many species of higher plants that have to adjust suddenly from dry-land to an aquatic existence. When soils flood or the tide comes in over a salt marsh, air is displaced from the soil and microbial activity removes any remaining dissolved oxygen in the soil water in less than 24 h. If the leaves are also submerged this can deprive the plants of light and carbon dioxide for photosynthesis. The problems encountered by both crop and wild plants under these conditions formed the subject of this international meeting which was organized by the Department of Plant Biology and Ecology of the University of St Andrews.

It is perhaps not unexpected that Scotland with its reputation for rain and bogs should be a venue for those interested in the effects of flooding. However, what is surprising at first sight, is the extent of interest in flooding that now exists in arid countries. The ever-increasing use of irrigation creates problems of temporary waterlogging for crops that would normally never expect to be flooded. Furthermore, in arid countries irrigation leads to the accumulation of salt in upper soil levels. Thus modern agriculture has brought the problems of amphibious and intertidal plant life into the cultivation of crops in arid zones. In an attempt to gain a better understanding of the adaptations that allow plants to withstand the stresses of amphibious life the symposium examined every aspect of plant development, from the germination of seeds under water to the dispersion of seeds and fruits. All phases of plant growth can be affected by flooding while even the survival of reputedly flood-tolerant plants such as rice can be adversely affected by flooding at certain stages in its growth.

It is a curious paradox that in many cultivated plants it is sufficient for flooding to last only a few days for the eventual harvest to be much reduced, while the vegetation of wetland soils and marshes can be among the most productive types of natural plant communities. There is a growing climate of opinion that instead of overcoming the problems of flooding by drainage, more attention should be paid to adapting crops to withstand the dangers of flooding. Crop breeding has produced varieties of plants to suit varying lengths of growing season and temperature regimes and it would be just as feasible to do this also in relation to flooding tolerance, and thus replace the habit of adapting the environment to suit the crop with the development of crops to suit the environment. This approach, if adopted, would not only tap the enormous production potential of wetlands but would do much to save our remaining natural wetlands, which due to the extent and popularity of current agricultural improvement schemes are dwindling just as rapidly as our forests. For this to be possible, however, our knowledge of plant responses to flooding in tolerant and intolerant species needs to be greatly extended. As the papers in this volume will show, the inherent variation that exists in plants ensures that although species may resemble each other in being tolerant or intolerant of flooding, the means by which they survive or the reasons for which they succumb can be highly variable.

The organization of this symposium was a direct consequence of Professor David Spence's active interest for many years in the ecology of aquatic plants. His sudden death only three weeks after this symposium came as an extraordinary shock to all his many friends and colleagues. To all who knew David's infectious vitality and warmth of spirit it seemed unreal that he should have passed away so rapidly. His tireless enthusiasm for his subject, with hours of underwater recording at all seasons of the year combined with painstaking laboratory studies, earned him an international reputation on the ecology of aquatic vegetation. This volume contains his last two papers which illustrate his continuing flair for combining field work with perceptive experimentation. Even in the last few months of his life when his illness was unknowingly undermining his physical strength he was diving in remote Scottish lochs in the depth of winter. It is a small consolation that his scientific life concluded with the happy occasion of this symposium which not only brought to St Andrews so many of his friends but gives us the opportunity to dedicate this volume to his memory.

R.M.M.C.

PART I
AQUATIC PLANTS

A. The Aquatic Environment

The light and temperature environments

M. G. HOLMES AND W. H. KLEIN

Smithsonian Environmental Research Center, 12441 Parklawn Drive, Rockville, MD 20852, USA

SUMMARY

The light environment experienced by aquatic plants is reviewed in terms of both the energy input from radiation and the information about the environment which an aquatic organism can derive from the quantity and quality of the radiation. Account is taken of the spectral attenuation of natural radiation by surrounding objects such as vegetation, water and suspended particles. The additional influence of screening by tissue surrounding the photoreceptor is incorporated into the analysis. The implications for photoresponses under water are then derived from knowledge about the spectral and quantum operating ranges of plant photoreceptors. Conclusions about photosynthetic limitations, photomorphogenesis, and photoperiodism under water are discussed in relation to the restrictions imposed by the underwater temperature environment.

INTRODUCTION

The survival of an organism depends on its ability to sense and exploit its environment. Of the many variables which amphibious and intertidal plants experience, both the light and temperature environments play a major role in determining the suitability of a habitat for a particular species. The fundamental restriction on phototrophic organisms is the availability of sufficient radiation to provide for the synthesis of their organic constituents. If this criterion is fulfilled, and temperature is not a limiting factor, the radiation environment can be utilized by photomorphogenetic pigments to modulate growth and development. We must therefore regard the radiation environment not just as a source of energy, but also as a source of information.

The radiation reaching a photoreceptor has been attenuated both quantitatively and qualitatively by the earth's atmosphere and by the tissue surrounding the photoreceptor. The same radiation may be additionally modified by surrounding vegetation and, in aquatic habitats, by water and suspended particles. Clearly, therefore, there is an interaction of many factors in the determination of the radiation available to potential photoreceptors. We have attempted here to reach a compromise between the desire to provide a precise and accurate description of the photoreceptor's radiation environment in a wide variety of habitats, and the need to quantify radiation in general terms which can be related to the biological response. The approach we have taken is to consider variations in the radiation environment in amphibious habitats which are within the known spectral and quantum operating range of plant photoreceptors. As plants exploit radiation both as a source of energy and as a source of information, it is convenient to consider radiation in terms of the availability of energy for photosynthesis, and also in terms of the information which can be derived from the radiation environment.

Photoreceptor spectral requirements

Photosynthetically active radiation (PAR) has been variously defined as being the energy flux within the 400–700 nm waveband (Gabrielsen 1940), the energy flux within the 380–710 nm waveband (Nichiporovich 1960), and the quantum (photon) flux within the 400–700 nm waveband (Federer & Tanner 1966). The most extensive comparative study was made by McCree (1972a), who concluded that measurement of the incident photon flux in the 400–700 nm waveband provided an acceptable description of PAR (McCree 1972b), and because most published measurements have been made in this way, we have used this method. It should, however, be borne in mind continuously that terrestrial plants do use radiation outside this waveband for photosynthesis (McCree 1972a, b; Inada 1978). Indeed, there is evidence that 28% or more of the energy trapped for photosynthesis in far-red rich environments is derived from wavelengths longer than 700 nm (Holmes, Sager & Klein 1986).

The radiation which plants use to obtain information on their environment (photomorphogenetically active radiation) is more difficult to define. The two main wavebands which are commonly acknowledged as being photomorphogenetically active are the ultraviolet/blue (UV/B) waveband and the red/far red (R/FR) waveband. Responses to UV/B fall largely into two categories: those which can be induced by both the UV and the B waveband, and those which are induced solely by the B waveband. For our consideration of the natural radiation environment, we consider UV to constitute wavelengths below 400 nm, and the B waveband radiation between 400 and 500 nm.

Although the photomorphogenetic photoreceptor phytochrome absorbs throughout the entire 300–800 nm waveband (Butler, Hendricks & Siegelman 1960; Vierstra & Quail 1983), the relative quantum efficiency of photoconversion is much greater in the R and FR wavebands than at other wavelengths (Pratt & Briggs 1966). We therefore consider the R (600–700 nm) and FR (700–800 nm) wavebands as being the most significant for phytochrome-modulated responses. The approach can be refined further by referring to the R:FR ratio which is the relative photon flux ratio in 10 nm bandwidths centred at 660 and 730 nm, respectively (Holmes & Smith 1975). As will be seen later, knowledge of the R:FR ratio provides useful information on the relative extent of photoconversion of the two main component forms (Pr and Pfr) of phytochrome (Smith & Holmes 1977).

Photoreceptor quantum requirements

The lowest photon flux of PAR at which net photosynthetic carbon gain can occur is typically in the range 10^{-1} to 10^{0} μmol m^{-2} s^{-1} (e.g. Kain 1966; Lüning & Dring

1979), although underwater zonation studies (Lüning & Dring 1979) and direct measurements in terrestrial plants indicate that values up to c. $3 \times 10^{-1} \mu$mol m^{-2} s^{-1} may be more applicable to plants in amphibious environments. Clearly the exact value will vary from species to species, and will ultimately determine the suitability of a habitat. Whatever value applies in an individual case, the range of photon fluxes determining the CO_2 compensation point is much narrower than the apparent range of irradiances required by the information-providing photoreceptors.

The threshold quantum requirements for B-induced photomorphogenetic and phototropic responses typically lie within the range 10^{-4} to 10^{-3} μmol m^{-2}, although saturation may require as much as 10^1, or even 10^2 μmol m^{-2} (e.g. Zimmermann & Briggs 1963; Everett 1974; Hartmann & Schmidt 1980). Further examples can be found in Senger (1980). Threshold requirements in the R waveband for phytochrome-induced responses have been reported to be as low as c. 5×10^{-4} μmol m^{-2} (Withrow 1959), but c. 5×10^{-2} μmol m^{-2} may be a more conservative figure (Klein et al. 1957). Reference to Holmes & Wagner (1980; their Fig. 2) shows that most phytochrome-induced responses—and measurable phytochrome photoconversion—occur following irradiation with between 10^{-1} and 10^3 μmol m^{-2} R light.

Interpretation of radiation measurements

The radiation environment is usually expressed in terms of the radiation arriving at the plant surface. The physiologically effective radiation, however, is that which arrives at, and is absorbed by, the photoreceptor itself. External radiation measurements are perfectly adequate for interpreting the potential effectiveness of natural radiation in terms of photosynthesis because laboratory or field measurements of photosynthesis are also related to external radiation measurements.

External radiation measurements are also suitable for most purposes of relating the natural light environment to the quantum requirement for B-induced photoresponses because these are also quantified in the laboratory by external radiation measurements. The important point to bear in mind is that the quantum requirements for a B-induced response in etiolated (chlorophyll-free) material may not be comparable with that for green material because the chlorophyll reduces the flux reaching the photoreceptor. Although other factors can be involved, green tissue can exhibit an apparent decrease in sensitivity to monochromatic radiation relative to its chlorophyll-free counterpart (e.g. Everett 1974; Beggs et al. 1980).

THE AERIAL RADIATION ENVIRONMENT

Although the quantity and quality of solar radiation reaching the earth's atmosphere are fairly constant, many factors interact to modify this radiation before it reaches plant photoreceptors. Fundamental properties of the radiation environment, such as day length and rate of change in day length are strongly dependent on latitude. Day length is constant in non-mountainous regions at the equator, but with increasing distance from the equator, days become longer in summer and shorter in winter. The rate of day length change also depends on latitude, but is substantially modified by season, being slow near the winter and summer solstices and fastest in spring and autumn. These important properties of the radiation environment and the additional influence of tidal effects in amphibious habitats will be covered by M. J. Dring in a later chapter and will therefore only be referred to in the following pages in passing.

The daily changes in energy flux caused by the earth's rotation are attended by only small changes in the spectral quality of natural daylight. At solar elevations of less than 10–15°, however, the air mass through which direct solar radiation has to pass is greatly increased. Differential attenuation of the direct radiation by scattering and absorption in the atmosphere results in a decrease in the R:FR ratio of the direct (solar) beam. Twilight ratios of 600–700:700–800 nm radiation as low as 0·125 have been observed in direct beam radiation (Goldberg & Klein 1971, 1977; Shropshire 1973). However, plants receive both direct beam and scattered sky radiation simultaneously, and the largest observed drop in R:RF ratio in this global radiation is to *c.* 0·45, which corresponds to a decrease in Pft/Ptot to around 0·40 (Holmes & Smith 1977a). A decrease in R:FR ratio from around 1·15 at higher solar elevations to *c.* 0·70 is more typical (Holmes & Smith 1977a).

EFFECT OF VEGETATION CANOPIES

The most dramatic changes in both the spectral quality and the quantity of radiation received by terrestrial plants are those caused by shading from overlying vegetation. Radiation impinging on a green leaf is selectively absorbed, reflected, and transmitted. Chlorophyll and accessory pigments within the leaves absorb most of the radiation in the 400–700 nm waveband, but most of the radiation between 700 and 800 nm is transmitted. The result of this selective attenuation by individual leaves is a marked change in the spectral characteristics of radiation within plant canopies compared to the incident radiation above (Fig. 1).

At least three of the spectral changes depicted in Fig. 1 can be of physiological significance. First, there is a large reduction in the availability of PAR below the canopy (Table 1), with obvious consequences for the potential rate of dry mass accumulation. Second, the photon flux in the B waveband is drastically reduced, although the quantity of radiation available will rapidly fulfil the quantum requirements of B-induced responses.

A third spectral change which has been demonstrated to be of major ecological significance in terrestrial plants is the combined strong depletion of the R waveband and weak depletion of the FR waveband. The reductions in R:FR ratio found below vegetation canopies have been demonstrated empirically to cause reductions in phytochrome photoequilibrium *in vivo*. When etiolated tissue is exposed to a wide variety of natural and artificial broadband spectra (Holmes &

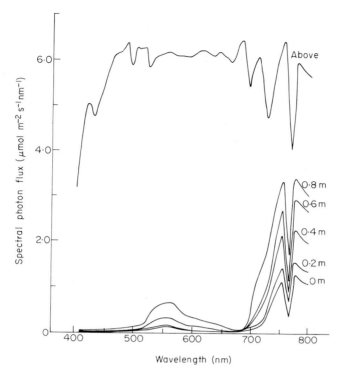

FIG. 1. Spectral quality of radiation at various heights above ground level within a wheat (*Triticum aestivum* L.) canopy. The spectra were measured on 7 June, 1973 with a crop height of 0·90–0·95 m under a clear sky (0/8–2/8 cloud cover, solar disc not obscured). Further characteristics of these spectra are presented in Table 1. (After Holmes & Smith 1977a.)

TABLE 1. Characteristics of radiation at various heights above ground level within a wheat (*Triticum aestivum* L.) canopy as presented in Fig. 1

Height above ground (m)	R:FR ratio (ζ)	PAR as %	B as %
1·5	1·18	100	100
0·8	0.07	3.4	3.1
0·6	0·06	2·1	1·6
0·4	0·18	3·9	1·8
0·2	0·10	1·1	1·4
0·0	0·10	0·9	1·1

The crop height = 0·90–0·95 m.
PAR at 1·5 m above ground (i.e. above the canopy) = 1700 μmol m^{-2} s^{-1}
B (400–500 nm) at 1·5 m above ground = 490 μmol m^{-2}s^{-1}
(After Holmes, 1975 and Holmes & Smith, 1977b).

Smith 1977a, b), a hyperbolic relation exists between the R:FR ratio of the radiation source and the photoequilibrium established in the tissues (Holmes & McCartney 1976; Smith & Holmes 1977). A notable feature of this relation is that phytochrome is most sensitive to changes in R:FR ratio which are characteristic of those found within natural vegetation canopies, i.e. between approximately 1·15 in natural daylight and 0·05 in dense canopy shade (Holmes & Smith 1977b).

Although there is now extensive circumstantial evidence that the photochromic properties of phytochrome are used to perceive shading by other plants, it is emphasized that at least three factors may lead to deviations from this relationship. First, phytochrome exists as intermediate forms between Pr and Pfr at the high photon fluxes typical of unattenuated daylight (Kendrick & Spruit 1972) with the possible result that formation of Pfr from Pr may be limited owing to the accumulation of the rate-limiting intermediate meta-Rb (Kendrick & Spruit 1976). Second, photoequilibrium is affected by the molecular environment, and both anaerobiosis and dehydration affect reaction rates *in vivo*. Finally, although phytochrome photoequilibria cannot be measured directly in green tissue, calculations of photostationary state in green *Phaseolus vulgaris* leaves predict that the screening effect of chlorophyll on the internal light environment will result in a lower photostationary state than is measured in chlorophyll-free etiolated tissue (Holmes & Fukshansky 1979). Although the photostationary state will be lower in green tissue, the systematic relation between R:FR ratio and photoequilibrium will not be affected (Morgan & Smith 1978) and the relevance of this prediction will be expanded on in the following section.

THE AQUATIC RADIATION ENVIRONMENT

Aquatic radiation environments are determined by (a) the quantity and quality of the radiation incident on the water surface, (b) the absorptive properties of the water and any dissolved substances, and (c) the absorptive, reflective and scattering properties of particulate material within the water. As mentioned in the previous section, there is a diurnal change in the photon flux of radiation reaching the water surface. This contrasts with the spectral quality of daylight which remains fairly constant throughout the day, but exhibits major changes during twilight. The final point in respect of radiation arriving at the water surface is that overlying vegetation such as that found on the banks of many estuaries will cause a reduction in the quantity of the entire PAR waveband, and also a reduction in the R:FR ratio.

Attenuation by water and soluble inorganic material

Pure water is highly transparent to long wavelength UV and to PAR. The relative absorbance increases significantly at wavelengths longer than *c.* 700 nm (Curcio & Petty 1951). This results in an increase in R:FR ratio with increasing depth. This

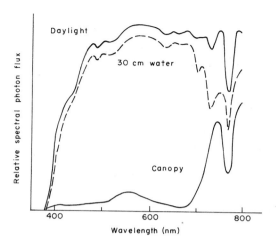

FIG. 2. Spectral attenuation of natural daylight by 30 cm of pure water. Spectral attenuation of the same daylight spectrum by an aerial vegetation canopy. (After Smith 1980.)

contrasts with the aerial environment where only decreases in R:FR ratio have been observed (Fig. 2).

Spectral changes due to scattering of radiation by water molecules becomes significant in deep (several metres) water. A typical example of attenuation in relatively pure water has been measured in Crater Lake, U.S.A. (Tyler & Smith 1967) in which green wavelengths penetrate furthest because increasingly shorter wavelengths are increasingly scattered by water and increasingly longer wavelengths are increasingly absorbed by water.

A typical characteristic of dissolved inorganic material, in particular fertilizer-derived sulphates, is a more marked attenuation of longer wavelengths—resulting in a further increase in R:FR ratio—and some attenuation of the shorter wavelength UV and B wavebands (Pfund 1939; Kasha 1948; Withrow & Price 1953).

A factor which deserves mention but has received surprisingly little attention is the influence of solar elevation on the attenuation of radiation by water bodies. Under clear skies, direct solar radiation contributes approximately 80% of the total global radiation, the remaining 20% coming from diffuse sky radiation (Rozenberg 1966). As solar elevation declines from around 25° to 0°, the contribution of direct radiation decreases from around 75% to zero. Whereas the photon flux change during twilight is gradual in aerial environments (ignoring the influence of atmospheric variables such as cloud cover), a more dramatic change in photon flux may be expected below placid water bodies as the critical angle for total external reflectance of direct radiation is approached. With an optically smooth surface, total exclusion of the direct beam is predicted at solar elevations of less than 41·4° (Weinberg 1976). This contrasts with a typical loss of direct radiation of less than 10% at higher solar elevations (Spence 1975).

Attenuation by turbid water

In turbid water containing insoluble material, scattering begins to play an important role in the attenuation of radiation. Suspended particles in turbid water influence the attenuation of radiation by both scattering and absorption. Many objects, including silts, clays, phytoplankton, and larger plant surfaces, contribute to scattering in a form which is largely independent of wavelength and may therefore be regarded as neutral density filtration. Spectrally selective scattering is caused primarily by small inorganic particles (<700 nm diameter; Spence 1981). The degree of scattering increases with decreasing wavelength and can severely reduce the penetration of the UV and B wavebands in turbid water containing a large quantity of small inorganic particles.

In addition to selective absorption of FR by water itself, other wavelengths are absorbed in turbid water. The two main absorbers are 'Gelbstoff', or yellow substance, and living suspended organic material. Gelbstoff consists of dissolved and colloidal substances of organic origin. It represents about 95% of total organic carbon (Spence 1981) and is considered to be contained by all natural waters (Hutchinson 1957; Jerlov 1968). Its major influence is to absorb the UV and B wavebands, although increased attenuation of longer wavelengths is usually evident. Suspended organic material, such as phytoplankton, absorbs differently in that it absorbs throughout the PAR waveband but transmits in the FR.

The attenuation of radiation by turbid water is most accurately described by measuring the transmission properties of the water. An example is shown in Fig. 3 which depicts the transmissivity of the relatively turbid water of Rhode River, a sub-estuary of Chesapeake Bay, Md, U.S.A. The data were obtained by measuring

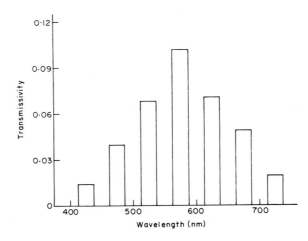

FIG. 3. Transmissivity of radiation as a function of wavelength in 50 cm of turbid water (see text for details). Measurements were made in May, 1981 in the Rhode River estuary, Md., U.S.A. by W. H. Klein, B. Goldberg, D. L. Correll and J. W. Pierce (Smithsonian Institution).

the transmission of radiation from a stabilized halogen lamp source through 500 m of water which was cylindrically surrounded by a series of black, closely-spaced baffles which limited the amount of scattered and natural light reaching the detector. It can be seen that the combined effects of the factors described above are: (a) strong attenuation of the entire PAR waveband, (b) marked depletion of the B waveband, and (c) strong attenuation of the R waveband, but proportionally greater depletion of FR. The influence of these factors in determining the spectral quality of radiation 0·8 m below the surface of Rhode River is seen in Fig. 4.

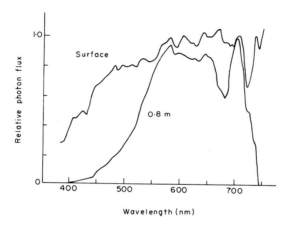

FIG. 4. Relative spectral photon flux (normalized to 710 nm) of the radiation incident at the surface and at a depth of 0·8 m in the Rhode River estuary. (After Seliger & Loftus 1974.)

Spence and colleagues have made an extensive study of the radiation environment within a variety of Scottish lochs. Spence (1975, 1981) recognizes two broad categories, primarily on the basis of the presence or absence of more than trace amounts of yellow substances and phytoplankton. Loch Borralie (Fig. 5) falls into the first category, having very low levels of yellow substances and phytoplankton. The characteristics of this type of lake are low attenuation of both the B and R/FR wavebands and accordingly high levels of PAR at depth. Black Loch falls into the second category, owing to high levels of yellow substances (Fig. 5). The result is a marked increase in attenuation of all wavelengths, but in particular in the B waveband. The radiation environment in Loch Leven (Fig. 5) is a modification of the situation seen in Black Loch. In Loch Leven a substantial quantity of phytoplankton resulted in increased absorption of the PAR waveband, but had negligible effect on FR radiation with the result that a markedly higher R:FR ratio prevailed than in the low phytoplankton water of Black Loch.

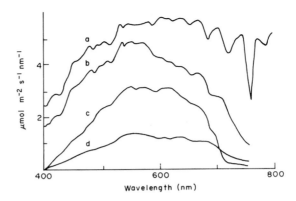

FIG. 5. Comparison of the spectral properties of the incident daylight (a), with the spectral distribution of radiation at 1·0 m depth in three Scottish lakes of differing turbidity. (b) Loch Borralie, (c) Black Loch, (d) Loch Leven. The aquatic data were calculated from the measurements of Spence (1975, 1981) by Smith (1980).

Photosynthesis under water

Radiation-dependent limitations on photosynthesis underwater are best defined by determining the depth of the euphotic zone and then determining the quantity of PAR available. This approach has been used by Lüning & Dring (1979) for various organic boundaries of the subtidal zone near Helgoland (Table 2) and they observed species-dependent euphotic limits of between *c.* 0·7 and 1·4 $\times 10^2$ μmol m^{-2} s^{-1} PAR, based on the monthly mean of instantaneous values. As the authors

TABLE 2. Underwater radiation climate during July at various depths near Helgoland in 1975

Water depth (m)	Biological boundary	Monthly mean of instantaneous values (July)			Yearly integral (400–700 nm) (mol m^{-2} year^{-1})
		Blue 400–500 nm	Green 500–600 nm	Red 600–700 nm	
2	Deepest *Laminaria digitata*	37·1	73·7	31·8	1037
4	Deepest *L. hyperborea* forest	10·7	34·9	10·0	388
8	Deepest *L. hyperborea*	1·1	8·2	1·0	71
10	Deepest red bushy algae	0·4	4·0	0·3	33
15	Deepest macroalgae (crusts)	0·0	0·7	0·0	6

The photon flux is expressed in units of μmol m^{-2} s^{-1}. (From Lüning & Dring 1979).

Table 3. 1% and 0·05% depths for total visible radiation (350–700 nm) in different water types

Water type		1% depth (m)	0·05% depth (m)
Oceanic	I	105	175
	II	55	95
	III	32	55
Coastal	1	27	48
	3	17·5	31·5
	5	11·5	20
	7	8·0	14
	9	6·0	10·5

(From Dring & Lüning (1983) based on data from Jerlov (1976) and Lüning & Dring (1979)).

pointed out, these photon flux values depend on water type and the aerial radiation climate. Dring & Lüning (1983) described limits of the euphotic zone in terms of the depth at which radiation between 350 and 700 nm radiation is reduced to give a percentage of its surface value (Table 3). Broadly speaking, the euphotic limit for both large brown algae (seaweeds) and phytoplankton occurs at about the 1% depth whereas the limit for small multicellular algae, in particular red algae, is at about 0·05% depth (Dring & Lüning 1983).

Photomorphogenetic considerations

The attenuation of radiation with increasing depth can be described approximately using the equation (Spence 1981):

$$K = \frac{\ln I_o - \ln I_z}{z} \tag{1}$$

where K is the vertical diffuse attenuation coefficient (m^{-1}), I_0 is the photon flux at subsurface, I_z is the photon flux at depth z, where z is in metres. If it is assumed that I_0 approximates to the photon flux incident at the surface, a value of $10^3 \ \mu mol \ m^{-2} \ s^{-1}$ can be used to approximate 'typical' values around noon in temperate regions. Using the data of Bodkin (1979) and Spence & Allen (in Spence 1981) r ·K, the depth in metres at which $\ln I_z$ will become limiting for photosynthesis can be calculated. Using values of $20 \ \mu mol \ m^{-2} \ s^{-1}$ and $1·0 \ \mu mol \ m^{-2} \ s^{-1}$ (PAR) for plausible values for the compensation point of aquatic angiosperms and phytoplankton respectively, the approximate photosynthetic depth limitations in the three lakes described in Fig. 5 can be calculated (Table 4). Using the same equation, but the specific spectral attenuation coefficients for each waveband (in Spence 1981), we can then calculate the photon flux available (at the various photosynthetically limiting depths) for induction of photomorphogenetic responses controlled by the B and R wavebands (Table 4). Although many variables are not accounted for in detail, two generalizations may be made about the availability of radiation for controlling photomorphogenetic responses at the lower levels of photic zones. First although B-induced responses may be rapidly

TABLE 4(a). Depth in metres at which photosynthesis becomes a limiting factor for growth in the lakes depicted in Fig. 5, assuming compensation points of either 20 μmol m^{-2} s^{-1} or 1·0 μmol m^{-2} s^{-1} and using 10^3 μmol m^{-2} s^{-1} as the photon flux incident at the water surface (all values are for the 400–700 nm waveband). (b) Time in seconds required to achieve a B- or R-induced response at the depths described in Table 4(a), assuming a quantum required of 10^2 μmol m^{-2} for both the B and R wavebands

(a)

μmol m^{-2}s^{-1} (400–700 nm)	Depth (m)		
	Loch Leven	Black Loch	Loch Borralie
20·0	2·11	3·80	8·50
1·0	3·73	6·71	15·00

(b)

Waveband	PAR	Time required for 10^2 μmol m^{-2} (s)		
		Loch Leven	Black Loch	Loch Borralie
B	20·0	35	39	2
B	1·0	1250	1470	9
R	20·0	8	4	8
R	1·0	111	36	115

saturated in clear lakes (Loch Borralie), the time taken to saturate the responses in lakes containing high levels of yellow substances (Loch Leven, Black Loch) may be too long to provide useful information to an organism under any but the highest aerial photon fluxes. Second, the quantum requirements of phytochrome in the R waveband are saturated rapidly in comparison to B, and there is no major dependence on water type for efficient function.

Potential difficulties arise when we wish to assess the significance of variations in the external radiation environment in terms of phytochrome function. The most complex—and the most striking—effects of spectral attenuation by plant tissues occur when we compare the relative distribution of the R and FR wavebands. A substantial amount of empirical information is available on the spectral attenuation of radiation within plant leaves and buds. The general trend of the results of these studies is that wavelengths below *c*. 700 nm and wavelengths above *c*. 700 nm are distributed differently within plant organs. The highest levels of radiation below 700 nm are situated immediately below the irradiated surface and decrease markedly with increasing distance from that surface (Fig. 6). Whereas the gradient of radiation above 700 nm also decreases with increasing distance from the irradiated surface (Fig. 6), internal reflectance of these poorly absorbed wavelengths can result in a substantially higher photon flux near the irradiated surface than in the incident radiation itself. Studies with cotyledons of *Cucurbita pepo* (Seyfried & Fukshanksy 1983) and with leaves of *Crassula falcata* (Vogelmann & Björn 1984) indicate that internal reflectance can result in an approximately threefold greater level of FR (730–750 nm) immediately below the irradiated surface than that contained in the incident radiation. Some error is inevitable with these technically difficult measurements, in particular from shading by the fibre-optic probe and from rupture of tissues by the probe. Also, greater loss

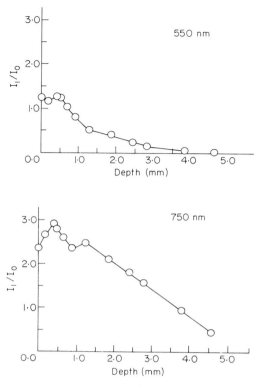

FIG. 6. Energy flux gradient at 550 nm (upper) and 750 nm (lower) across a leaf of *Crassula falcata*. The gradient is a result of both collimated and scattered radiation. Radiation quantity is expressed as the ratio of the radiation inside (I_1) to the radiation outside (I_o). (From Vogelmann & Björn 1984.)

of FR may be expected in submerged organs owing to greater similarity of the refractive index of tissue and water than tissue and air. Nevertheless, we can apply the information described above to determine any gross effects of internal attenuance of radiation in submerged plant tissue.

If we consider the attenuation of 660 nm radiation to be fivefold greater than that of 550 nm radiation (Fig. 7) and the attenuance of 730 nm radiation to be the same as 750 nm radiation (Fig. 7), we can calculate the attenuance of 660 nm and 730 nm radiation at different depths within a *C. falcata* leaf in air. This information provides estimates of the 660:730 nm (R:FR) ratio at different depths. Estimates of P_{fr}/P_{tot} at those depths can then be read off directly from Fig. 2 in Smith & Holmes (1977) or calculated using the equation:

$$P_{fr}/P_{tot} = 0 \cdot 75/\{1 + (0 \cdot 37/R{:}FR)\} \tag{2}$$

which is a least squares fit for a rectangular hyperbola applied to the curve in Fig. 2 of Smith & Holmes (1977).

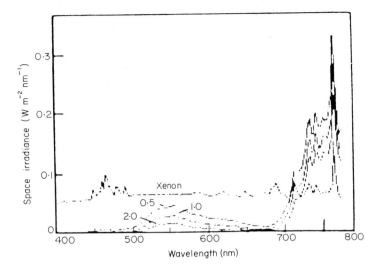

Fɪɢ. 7. Spectral quantum flux at various depths within a *Crassula falcata* leaf. The fibre-optic probe was driven to specific depths within the leaf and the spectrum scanned. The spectral quantum flux is shown for 0·5, 1·0, and 2·0 mm below the irradiated surface. The xenon arc irradiation source is depicted as reference. (From Vogelmann & Björn 1984.)

The measurements on *C. falcata* leaves confirm the predictions (Holmes & Fukshansky 1979) that a gradient in R:FR exists across a green leaf and that the R:FR ratio (and therefore Pfr/Ptot) within the leaf is lower than that in the external, incident radiation. A notable feature is that the thick leaves (*c.* 4 mm) of *C. falcata* produce a substantially greater decrease in R:FR ratio than that calculated for the thin leaves (0·25 mm) of *P. vulgaris* (for details of those calculations, see Holmes & Fukshansky 1979).

Extending the approach used above, phytochrome photoequilibrium within green leaves can be calculated for various ratios of R:FR radiation. In the absence of information on the localization of phytochrome within green leaves, we present calculations for phytochrome situated midway between the upper and lower epidermis of a green *P. vulgaris* leaf, and for phytochrome situated 0·5 mm and 2·0 mm below the irradiated epidermis of a *C. falcata* leaf. It can be seen from Fig. 8 that internal selective screening of radiation by green tissue provides an extension in the range of incident R:FR ratios which would bring about a significant change in phytochrome photoequilibrium. The large shifts in photoequilibrium which have been measured in etiolated tissue (Holmes 1975; Holmes & McCartney 1976; Smith & Holmes 1977) occur within the relatively narrow range of R:FR ratios of between approximately 0·05 and 1·5:1 (Fig. 8). In the centre of a green *P. vulgaris* leaf, the range of R:FR ratios which will produce equivalent shifts in photoequilibrium is extended up to approximately 5:1. In the thick leaves of *C. falcata*, a R:FR ratio of 15:1, or more, is required to produce the same Pfr/Ptot value as that produced by a R:FR ratio of 1·5:1 in etiolated material. Although it

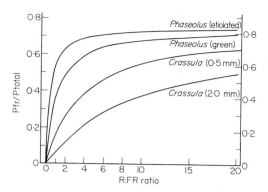

FIG. 8. The relation between the R:FR ratio of the incident radiation and the phytochrome photoequilibrium (Pfr/Ptot) established in dark-grown (etiolated) *Phaseolus vulgaris* tissue (from Holmes 1975 and Smith & Holmes 1977), or in the middle of a green *P. vulgaris* primary leaf (from Holmes & Fukshansky 1979), or at two distances (0·5 mm and 2·0 mm) from the irradiated surface of an approximately 4 mm thick *Crassula falcata* leaf (calculated from the data of Vogelmann & Björn 1984).

must be borne in mind that variables such as location of the photoreceptor, direction of propagation of the incident radiation, and differences between air/tissue and water/tissue interfaces will influence absolute values, the inference from Fig. 8 is that screening by green plant tissue increases the range of spectral sensitivity to R:FR ratios which are typical of many aquatic habitats. Clearly, phytochrome would be capable of detecting the increased R:FR ratio associated with increased water depth, and also localized decreases in R:FR ratio caused by suspended phytoplankton and by macrophytes competing for nutrients and PAR.

Photoperiodism under water

Photoperiodic responses in algae and aquatic angiosperms (see Dring & Lüning (1983) and Chambers & Spence (1984), respectively, for comprehensive lists of species) have received increased attention in the last two decades. Although it is well established that phytochrome is involved in time measurement by plants (Parker *et al.* 1946), the method by which it senses day length is uncertain. Conflicting arguments in favour of changes in photon flux at twilight and changing R:FR ratios at twilight have been proposed (for reviews, see Smith 1982; Vince-Prue 1983). Experimental evidence for both arguments is rare. Experiments have been done supporting the concept that natural changes in the quantity of radiation regulate or influence time measurement in terrestrial plants (Takimoto & Ikeda 1960, 1961; Clements 1968; Salisbury 1981), as have experiments supporting the concept that changes in R:FR ratio influence time measurement (e.g. Borthwick, Hendricks & Parker 1952; Deitzer, Hayes & Jabben 1979).

Phytochrome is involved in the induction of winter bud formation in *Potamogeton crispus* L. (Chambers 1982). Diurnal variations in both the quantity

(PAR) and quality (R:FR ratio) of radiation exist in the natural habitat of *P. crispus* (Chambers & Spence 1984). In view of the larger diurnal changes in R:FR below than above the water (Chambers & Spence 1984), and the observation that green leaves are sensitive to R:FR ratio changes typical of aquatic habitats (Fig. 8), the postulate of Chambers & Spence that the diurnal variation in R:FR is a possible mechanism by which aquatic plants might detect photoperiod deserves experimental attention.

AMPHIBIOUS TEMPERATURE ENVIRONMENT

Distribution of plants across the earth's surface is strongly influenced by temperature. These climatic zones are primarily determined by latitude and altitude and both of these factors have a major influence on the temperature regime of terrestrial, intertidal, and aquatic habitats (Table 5). The influence of latitude derives mainly from the maximum solar elevation achieved, and therefore from the input of direct solar radiation. The influence of altitude derives mainly from the cooling of air as it expands at higher altitudes.

TABLE 5. Comparative influence of environmental factors on temperature in terrestrial, intertidal, and aquatic habitats

	Effect on temperature		
Environmental factor	Terrestrial	Intertidal	Aquatic
Altitude	High	High	High
Solar elevation (latitude)	High	High	High
Solar elevation (diurnal)	High	High	Low
Solar elevation (annual)	High	High	Low
Wind speed	High	High	Low
Cloud cover	High	High	Low
Shading (self/canopy)	High	High	Low

The major temperature difference between aerial and aquatic environments is the magnitude of temperature fluctuation. In the absence of the buffering capacity of water, plants experience large short-term (minutes, hours) and large long-term (days, months) fluctuations in temperature compared to their aquatic counterparts (Table 5). The rapid fluctuations of the aerial temperature environment can cause difficulties in the assessment of long-term experiments on the influence of temperature on plant growth and development. Nevertheless, we do have a fundamental knowledge, if not an understanding, of the influence of temperature on general metabolism, photomorphogenesis, photoperiodism, and dormancy.

Ignoring the complex lunar-solar interactions which influence the temperature environment in intertidal zones (see chapter by Dring), two environmental factors which offer potential information to aquatic plants are the relatively damped diurnal and annual oscillations in temperature (Table 5). Whereas a terrestrial plant has to contend with major random influences of wind speed, cloud cover, and in some instances shading by itself or competitors, the aquatic plant is essentially unaffected by these perturbations. It is therefore not surprising that the highly predictable annual changes in the temperature environment play a major role in the few aquatic angiosperms which have been studied. The early studies of Chrysler (1938) demonstrated the induction of apical turions in *Brasenia schreberi* by low temperatures. Interaction between temperature and photoperiod is seen in *Myriophyllum verticillatus* which forms dormant stem turions in short days, but only if low temperatures are also applied (Weber & Nooden 1976); similar factors are considered to be responsible for stem turion induction in *Potamogeton obtusifolius* (Spence 1981). Interaction of temperature and photoperiod has also been reported for stem turion formation in *P. crispus*, although in this case, a combination of high temperature and long photoperiod are inductive. Clearly, therefore, annual variations in temperature are capable of synchronizing or 'fine-tuning' photoperiodic induction of dormancy. In addition, interactions between temperature and light have been observed in non-dormant stages of growth. Although an increase in temperature alone induces heterophylly in some aquatic macrophytes (Combes 1947; Bostrack & Millington 1962; Johnson 1967), increased temperatures act only as a prerequisite for the photoinduction of aerial-type leaves in *Hippuris vulgaris* (Bodkin, Spence & Weeks 1980). The influence of temperature on germination and growth in aquatic angiosperms is considered in detail in the chapters by Frankland *et al.* and Spence *et al.*, respectively. Finally, it is noteworthy that even with the relatively small periodic variations in temperature in marine habitats, temperature and day length interact in the control of photoperiodism in marine macroalgae (for review, see Dring & Lüning 1983).

REFERENCES

Beggs, C. J., Holmes, M. G., Jabben, M. & Schäfer, E. (1980). Action spectra for the inhibition of hypocotyl growth by continuous irradiation in light and dark-grown *Sinapis alba* L. *Plant Physiology*, **66**, 615–618.

Bodkin, P. C. (1979). *Control of growth in* Hippuris vulgaris L. Ph.D. thesis, University of St Andrews.

Bodkin, P.C., Spence, D. H. N. & Weeks, D. C. (1980). Photoreversible control of heterophylly in *Hippuris vulgaris* L. *New Phytologist*, **84**, 533–542.

Borthwick,H. A., Hendricks, S. B. & Parker, M.W. (1952). The reaction controlling floral initiation. *Proceedings of the National Academy of Sciences*, U.S.A., **38**, 929–934.

Bostrack, J. M. & Millington, W. F. (1962). On the determination of leaf form in an aquatic heterophyllous species of *Ranunculus*. *Bulletin of the Torrey Botanical Club*, **89**, 1–20.

Butler, W. L., Hendricks, S. B. & Siegelman, H. W. (1960). *In vivo* and *in vitro* properties of phytochrome. *Plant Physiology*, **35**, Supplement 32.

Chambers, P. A. (1982). *Light, temperature and the induction of dormancy in* Potamogeton crispus *and* Potamogeton obtusifolius. Ph.D. thesis, University of St Andrews.

Chambers, P. A. & Spence, D. H. N. (1984). Diurnal changes in the ratio of red to far-red light in relation to aquatic plant photoperiodism. *Journal of Ecology*, **72**, 495–503.

Chrysler, M. A. (1938). The structure and relationships of the Potamogetonaceae and allied families. *Botanical Gazette*, **44**, 161–196.

Clements, H. F. (1968). Lengthening versus shortening dark periods and blossoming in sugar cane as affected by temperature. *Plant Physiology*, **43**, 57–60.

Combes, R. (1947). Le mécanisme de l'action du mileu aquatique sur les végétaux. Rôle due facteur température. *Revue générale de Botanique*, **54**, 249–256.

Curcio, J. A. & Petty, C. C. (1951). The near-infrared absorption spectrum of liquid water. *Journal of the Optical Society of America*, **41**, 302–304.

Deitzer, G. F., Hayes, R. & Jabben, M. (1979). Kinetics and time dependence of the effect of far-red light on the photoperiodic induction of flowering in Wintex barley. *Plant Physiology*, **64**, 1015–1021.

Dring M. J. & Lüning, K. (1983). Photomorphogenesis in marine macroalgae. *Encyclopedia of Plant Physiology* (Ed. by W. Shropshire Jr. & H. Mohr), vol. 16B, pp. 545–568. Springer-Verlag, Berlin.

Everett, M. (1974). Dose-response curves for radish seedling phototropism. *Plant Physiology*, **54**, 222–225.

Federer, C. A. & Tanner, C. B. (1966). Spectral distribution of light in the forest. *Ecology*, **47**, 555–560.

Gabrielsen, E. K. (1940). Einfluss der Lichtfactoren auf die Kohlensäureassimilation der Laubblätter. *Dansk Botanische Arkiv*, **10**, 1–177.

Goldberg, B. & Klein, W. H. (1971). Comparison of normal incident solar energy measurements at Washington, D.C. *Solar Energy*, **13**, 311–321.

Goldberg, B. & Klein, W. H. (1977). Variations in the spectral distribution of daylight at various geographical locations on the earth's surface. *Solar Energy*, **19**, 3–13.

Hartmann, E. & Schmidt, K. (1980). Effects of UV and blue light on the bipotential changes in etiolated hypocotyl hooks of dwarf beans. *The Blue Light Syndrome* (Ed. by H. Senger), pp. 221–237. Springer-Verlag, Berlin.

Holmes, M. G. (1975). *Studies on the ecological significance of phytochrome.* Ph.D. Thesis, University of Nottingham.

Holmes, M. G. & Fukshansky, L. (1979). Phytochrome photoequilibria in green leaves under polychromatic radiation: a theoretical approach. *Plant, Cell and Environment*, **2**, 59–65.

Holmes, M. G. & McCartney, H. A (1976). Spectral energy distribution in the natural environment and its implications for phytochrome function. In *Light and Plant Development* (Ed. by H. Smith), pp. 467–476. Butterworth, London.

Holmes, M. G., Sager, J. C. & Klein, W. H. (1986). Sensitivity to far-red radiation in stomata of *Phaseolus vulgaris* L. *Planta*, **168**.

Holmes, M. G. & Smith, H. (1975). The function of phytochrome in plants growing in the natural environment. *Nature*, **254**, 512–514.

Holmes, M. G. & Smith, H. (1977a). The function of phytochrome in the natural environment. I. Characterisation of daylight for studies in photomorphogenesis and photoperiodism. *Photochemistry and Photobiology*, 533–538.

Holmes, M. G. & Smith, H. (1977b). The function of phytochrome in the natural environment. II. The influence of vegetation canopies on the spectral energy distribution of natural daylight . *Photochemistry and Photobiology*, **25**, 539–545.

Holmes, M. G. & Wagner, E. (1980). 1A re-evaluation of phytochrome involvement in time measurement in plants. *Journal of Theoretical Biology*, **83**, 255–265.

Hutchinson, G. E. (1957). A treatise on limnology. In *Geography, Physics and Chemistry.* vol. I, Wiley, New York.

Inada, K. (1978). Photosynthetic enhancement spectra in higher plants. *Plant and Cell Physiology*, **19**, 1007–1017.

Jerlov, N. G. (1968). *Optical Oceanography.* Elsevier, Amsterdam.

Jerlov, N. G. (1976). *Marine Optics.* pp. 1–231. Elsevier, Amsterdam.

Johnson, M. P. (1967). Temperature-dependent leaf morphogenesis in *Ranunculus flabellaris. Nature*, **214**, 1354–1356.

Kain, J. M. (1966). The role of light in the ecology of *Laminaria hyperborea. Light as an Ecological Factor* (Ed. by R. Bainbridge, G. C. Evans & O. Rackham), pp. 319–334. Blackwell Scientific Publications, Oxford.

Kasha, M. (1948). Transmission filters for the ultraviolet. *Journal of the Optical Society of America*, **38**, 929–934.

Kendrick, R. E. & Spruit, C. J. P. (1972). Light maintains high levels of phytochrome intermediates. *Nature*, **237**, 281–282.

Kendrick, R. E. & Spruit, C. J. P. (1976). Intermediates in the photoconversion of phytochrome. *Light and Plant Development*. (Ed. by H. Smith,), pp. 31–43. Butterworths, London.

Klein, W. H., Withrow, R. B., Elstad, V. & Price, L. (1957). Photocontrol of growth and pigment synthesis in the bean seedling as related to irradiance and wavelength. *American Journal of Botany*, **44**, 15–19.

Lüning K. & Dring, M. J. (1979). Continuous underwater light measurements near Helgoland (North Sea) and its significance for characteristic light limits in the sublittoral region. *Helgoland Wissenschaftliche Meersuntersuchung*, **32**, 403–424.

McCree, K. J. (1972a). The action spectrum, absorptance and quantum yield of photosynthesis in crop plants. *Agricultural Meteorology*, **9**, 191–216.

McCree, K. J. (1972b). Significance of enhancement for calculations based on the action spectrum for photosynthesis. *Plant Physiology*, **49**, 704–706.

Morgan, D. C. & Smith, H. (1978). The relationship between phytochrome photoequilibrium and development in light grown *Chenopodium album*, L. *Planta*, **142**, 187–193.

Nichiporovich, A. A. (1960). Measurement of visible radiation in plant physiology and ecology, agrometeorology and plant production. *Fiziologiya Rast.*, **7**, 744–748.

Parker, M. W., Hendricks, S. B., Borthwick, H. A. & Scully, N. J. (1946). Action spectrum for the photoperiodic control of floral initiation of short-day plants. *Botanical Gazette (Chicago)*, **108**, 1–26.

Pfund, A. H. (1939). Transparent and opaque screens for the near infra-red. *Journal of the Optical Society of America*, **29**, 56–58.

Pratt, L. H. & Briggs, W. R. (1966). Photochemical and nonphotochemical reactions of phytochrome *in vivo*. Plant Physiology, **41**, 467–474.

Rozenberg, G. V. (1966). *Twilight — a Study in Atmospheric Optics*. Plenum Press, New York.

Salisbury, F. B. (1981). Twilight effect: Initiating dark measurement in photoperiodism of *Xanthium*. *Plant Physiology*, **67**, 1230–1238.

Seliger, H. H. & Loftus, M. E. (1974). Growth and dissipation of phytoplankton in Chesapeake Bay.II.A statistical analysis of phytoplankton standing crops in the Rhode and West Rivers and an adjacent section of the Chesapeake Bay. *Chesapeake Science*, **15**, 185–204.

Senger, H. (Ed.) (1980). *The Blue Light Syndrome*. Springer-Verlag, Berlin.

Seyfried, M. & Fukshansky, L. (1983). Light gradients in plant tissue. *Applied Optics*, **22**, 1402–1408.

Shropshire, W., Jr. (1973). Photoinduced parental control of seed germination and the spectral quality of solar radiation. *Solar Energy*, 15, 99–105.

Smith, H. (1980). The role of phytochrome in the natural environment. Proceedings of the International Congress of Photobiology, Strasbourg, France.

Smith, H. (1982). Light quality, photoperception, and plant strategy. *Annual Review of Plant Physiology*, **33**, 481–518.

Smith, H. & Holmes, M. G. (1977). The function of phytochrome in the natural environment. III. Measurement and calculation of phytochrome photequilibrium. *Photochemistry and Photobiology*, **25**, 547–550.

Spence, D. H. N. (1975). Light and plant response in fresh water. *Light as an Ecological Factor* (Ed. by G. C. Evans, R. Bainbridge & O. Rackham), pp. 93–113. Blackwell Scientific Publications, Oxford.

Spence, D. H. N. (1981). Light quality and plant response underwater. *Plants and the Daylight Spectrum* (Ed. by H. Smith) pp. 245–276. Academic Press, London.

Takimoto, A. & Ikeda, K. (1960). Studies on the light controlling flower initiation of *Pharbitis nil*.VI. Effect of natural twilight. *Botanical Magazine, Tokyo*, **73**, 175–181.

Takimoto, A. & Ikeda, K. (1961). Effect of twilight on photiperiodic induction in some short-day plants. *Plant and Cell Physiology*, **2**, 213–229.

Tyler, M. J. & Smith, R. C. (1967). Spectroradiometric characteristics of natural light under water. *Journal of the Optical Society of America.* **57**, 595–601.

Vierstra, R. D. & Quail, P. (1983). Purification of 124KD phytochrome. *Biochemistry*, **22**, 2498–2505.

Vince-Prue, D. (1983). The perception of light-dark transitions. *Philosophical Transactions of the Royal Society of London*, **B303**, 523–536.

Vogelmann, T. C. & Björn, L.O. (1984). Measurements of light gradients and spectral regime in plant tissue with a fiber optic probe. *Physiologia Plantarum*, **60**, 361–368.

Weber, J. A. & Nooden, L. D. (1976) Environmental and hormonal control of turion germination in *Myriophyllum verticillatum*. *Plant and Cell Physiology*, **17**, 721–731.

Weinberg, S. (1976). Submarine daylight and ecology. *Marine Biology*, **37**, 291–304.

Withrow, R. B. (1959). A kinetic analysis of photoperiodism. *Photoperiodism and Related Phenomena in Plants and Animals* (Ed. by Withrow, R. B.), pp. 439–471. American Association for the Advancement of Science, Washington D. C.

Withrow, R. B. & Price, L. (1953). Filters for the isolation of narrow regions in the visible and near-visible spectrum. *Plant Physiology*, **28**, 334–338.

Zimmerman, B. K. & Briggs, W. R. (1963). Phototropic dosage-response curves for oat coleoptiles. *Plant Physiology*, **38**, 248–253.

Light climate in intertidal and subtidal zones in relation to photosynthesis and growth of benthic algae: a theoretical model

M.J. DRING

Department of Botany, Queen's University, Belfast BT7 1NN, Northern Ireland

SUMMARY

1 The depth of water overlying different heights in the intertidal and upper subtidal was calculated from a sine wave for frequent intervals through the tidal cycle, and the proportion of surface irradiance reaching these sites was then estimated from the transmission values for a given Jerlov water type.

2 As tidal range increases, critical underwater light levels (e.g. 1% and 0·05% of surface irradiance) occur at progressively shallower depths, and the 1% depth may occur above LAT (i.e. in the intertidal zone) on the shores of estuaries with large tidal ranges (>8 m) and very turbid waters.

3 Predictions of 1% and 0·05% depths for a series of sites in the Bristol Channel gave good agreement with observed lower limits for kelp forest and for foliose algae, respectively.

4 Further developments of the model allowed for diurnal variation in surface irradiance, the progressive change in the time of low and high water, and the variation in tidal range through the spring-neap cycle.

5 At sites in the upper Bristol Channel, the predicted mean irradiance over 15-day periods approached the compensation point for fucoid photosynthesis at 1–2 m above ELWS, so that the observed absence of fucoid cover in the lower intertidal of these shores may be due to light limitation.

INTRODUCTION

The light climate at different depths below open water surfaces has been studied in some detail by optical oceanographers, and different aspects (e.g. total irradiance, spectral distribution, angular distribution, etc.) have been described by Jerlov (1976) and Kirk (1983). Marine waters can be classified into a series of optical water types which are characterized by a set of attenuation coefficients for different wavelengths through the visible spectrum (Jerlov 1976). These have been used to predict the light available for photosynthesis by benthic algae at fixed depths (e.g. Dring 1981), and have contributed to a reassessment of the theory of complementary chromatic adaptation. One oversimplification involved in this approach is that benthic algae on tidal shores do not grow in a fixed depth of water, even though they are at a fixed position on the rock surface. The changes in water level which occur through

23

each tidal cycle result in a continual change in the light climate for intertidal plants and for plants in the upper part of the subtidal zone. This situation is complicated further by the fact that the diurnal cycle of irradiance at the surface is out of phase with the tidal cycle, so that a plant may be covered by a very different depth of water at noon on one day from that which it experiences at the same time of day one week later. This paper describes a model which has been developed to examine the implications of these changes for the light climates in which benthic algae live, and makes a first attempt at assessing their significance for the growth of such plants.

MODEL DEVELOPMENT AND RESULTS

The primary aspect of the light climate considered in the model has been the total visible (350–700 nm) irradiance. This has been calculated for any required water depth using the data for the spectral transmittance of light through sea water of different optical water types (Jerlov 1976). The spectral distribution of light at the surface was assumed to be flat (i.e. equal irradiance at all wavelengths from 350 to 700 nm), and wavelengths <350 nm and >700 nm were ignored. The irradiance at the required water depth was calculated for each wavelength individually and then summed to give the total irradiance at that depth.

TABLE 1. Spectral transmittance (% m^{-1}) of two theoretical water types extrapolated from data of Jerlov (1976)

Water type	Wavelength (nm)							
	350	400	450	500	550	600	650	700
Coastal 11	0·7	3·6	9·5	20·4	44·0	48·4	41·4	26·4
Coastal 13	0·2	1·3	3·8	10·2	36·1	42·1	36·0	20·6

The model has been developed progressively to take account of
(a) the change of water depth through a single tidal cycle;
(b) the change of tidal range in any one site through the spring-neap cycle;
(c) the change of surface irradiance through the diel cycle, and the non-coincidence of tidal and diel cycles.

Change of water depth through tidal cycle

The depth of water overlying a given height on the shore was calculated for every 0·1 h through a tidal cycle, using a sine wave to model the change in water level. The accuracy of the sine wave approximation was checked by also using water heights read from the submersion curve for a particular site (Avonmouth; Fig. 1). The surface irradiance was assumed to be constant at 100 units, and the total visible irradiance was calculated from the water depth for each time interval. The average of all values of visible irradiance over the complete tidal cycle was then calculated

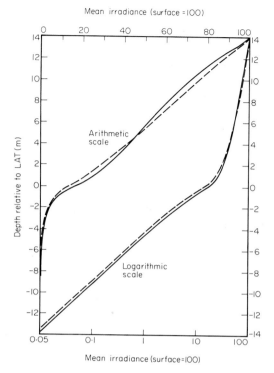

FIG. 1. Mean incident irradiance over a complete tidal cycle at different heights on a shore, as a percentage of mean irradiance at the surface. Water depth was calculated for every 0·1 h through a tidal cycle (i.e. 12·42 h) from a sine wave (solid lines) or from the published emersion curve for spring tides at Avonmouth (broken lines) and irradiance was calculated assuming 'coastal 7' water type. Tidal range: 14 m.

to estimate the mean irradiance at a given height on the shore, expressed as a percentage of the mean irradiance received at the surface over the same time period. The variation of this mean irradiance with height, on a shore which experiences a tidal range of 14 m (cf. Avonmouth, extreme spring range: 14·6 m) and a water type of coastal 7, is shown in Fig 1. The results obtained using the sine wave approximation show little difference from those obtained with the actual submersion curve for Avonmouth tides. Below low water mark (i.e. 0 m), there is an exponential decrease in irradiance with increasing water depth, which is similar to the relationship normally observed in sites of constant water depth, but above 0 m (i.e. in the intertidal zone) the relationship between irradiance and height is more or less linear. The mean irradiance at any height in the intertidal zone can, therefore, be roughly estimated by drawing a line from the mean irradiance at the low water mark to a value of 100 at high water. The actual value predicted for this mean irradiance at low water mark, however, is strongly dependent on the magnitude of the tidal range and on the water type (Fig. 2). Mean irradiance at low

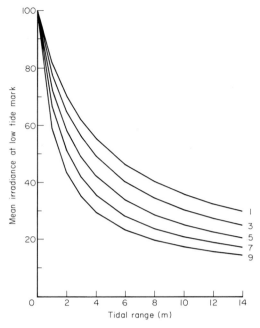

FIG. 2. Effects of tidal range and water type on the mean irradiance at low water mark. Irradiance was calculated for every 0·1 h through a tidal cycle (assuming that water depth varied according to a sine wave) and the mean irradiance is expressed as a percentage of the mean irradiance at the surface. The family of curves labelled 1–9 refer to water types 'coastal 1' to 'coastal 9'.

water mark decreases as the tidal range increases and as the optical quality of the water declines.

The maximum depths at which large brown algae can grow in a wide range of water types have been found to correspond to the depths at which the total irradiance is reduced to 1% of its surface value (the '1% depth'), whereas smaller multicellular algae (e.g. crustose corallines and other foliose or filamentous red algae) can be found down to the '0·05% depth' (Lüning & Dring 1979). One effect of the decrease in mean irradiance at the low water mark which is predicted with increasing tidal range (Fig. 2) is that these critical optical depths will occur at shallower and shallower depths as the tidal range increases. The depth on tidal shores at which the mean irradiance is reduced to 1% of the surface value has been calculated for different tidal ranges and different water types, and the results are shown in Fig. 3. In type 7 coastal water, the 1% depth occurs at about 7·5 m when the tidal variations are small, but is reduced to less than 5 m when the tidal range exceeds 10 m, and similar reductions are predicted for all water types. The implication of this result is that large brown algae (especially laminarians) will not penetrate so deeply in a given water type on coasts with large tidal ranges. This prediction cannot be tested, however, without considering the effects of another factor—the variations in tidal range which occur at any one site through the spring-neap cycle.

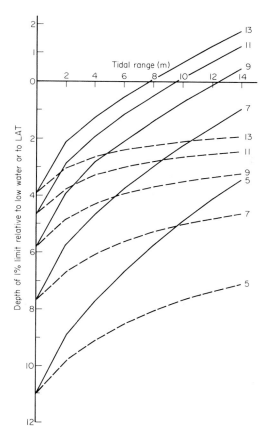

FIG. 3. Effects of tidal range and water type on the depths at which the mean irradiance over a tidal cycle (calculated as in Fig. 2) is reduced to 1% of the surface irradiance. Broken lines: depths expressed relative to the low water mark of each tidal range. Solid lines: depths expressed relative to LAT, assuming that the tidal ranges are the mean ranges for a particular site. Curves labelled 5–13 refer to water types 'coastal 5' to 'coastal 13'.

Change of tidal range through spring-neap cycle

The tidal range on a particular shore is not constant, but varies between high values during spring tides and low values during neap tides, and the average light climate over a long time period will be determined by the mean tidal range. It is this range, therefore, which should be used to estimate the 1% light depth and to predict the depth to which laminarians may penetrate. But, if mean tidal range is used, the broken lines in Fig. 3 will predict the lower limit of the laminarian zone as a depth below the low water mark of mean tides, whereas the conventional reference point for recording heights in the intertidal and subtidal zones is the low water mark of extreme spring tides (ELWS) or the level of the lowest astronomical tide (LAT). The results in Fig 3 must, therefore, be corrected for the difference between LAT and mean low water mark. There is no simple relationship between mean tides and extreme spring tides in any one site, but a linear regression of the height of mean low water above LAT against the mean tidal range was computed for all standard ports in the British Isles ($y = 0.2311x + 0.4642$; Admiralty Tide Tables 1984), and was used to estimate the correction to be applied to the results in Fig. 3. These corrections bring the 1% depth to within 1 m of LAT for a 14 m tidal range in type 7 waters and above LAT (i.e. into the intertidal zone) in type 9 waters (Fig. 3, solid lines). Since laminarians are intolerant of desiccation, but cannot grow in sites where they receive less than 1% of the surface light, this result suggests that these plants will be absent from shores where high tidal ranges are combined with turbid waters.

Large tidal ranges often occur in narrow bays or estuaries, and are often associated with exceptionally turbid waters because of the large volumes of water which must be moved in each tidal cycle. The most turbid waters covered by Jerlov's classification (coastal type 9) are probably clearer than typical inshore waters from estuarine sites (e.g. Colijn 1982). For this reason, Jerlov's classification has been arbitrarily extrapolated to give two 'new' (and entirely theoretical) water types—coastal types 11 and 13. The transmittance data derived for these water types are shown in Table 1. If light penetration through coastal or estuarine waters is as poor as these data suggest, then laminarians should be absent from shores with a mean tidal range of 9.5 m for type 11 and 7.9 m for type 13 (Fig. 3) Mean tidal ranges of this magnitude occur in the upper reaches of the Bristol Channel, and laminarians are absent from sites in this area with tidal ranges greater than 6.9 m (east of Hurlstone Point, Somerset; Crothers 1976; Hiscock 1981). At the same sites, the lowest foliose algae are found less than 1 m below LAT, whereas the 0.05% depth predicted by the model for type 13 water in these sites was about 3 m below LAT. These results suggest, therefore, that the waters in the upper Bristol Channel, which are notorious for their high silt load, are more turbid even than the theoretical type 13 used here.

Change in surface irradiance through the diel cycle, and the non-coincidence of tidal and diel cycles

The final version of the model took account of the changes of irradiance occurring at the surface in the course of a day, instead of assuming that surface irradiance was constant. The surface irradiance at any time was calculated from a sine wave (Kirk 1983):

$$I_t = I_m \sin(\pi . t/N),$$

where I_t = surface irradiance at time t; I_m = surface irradiance at solar noon (i.e. maximal value); t = time from sunrise; N = daylength. I_m was set to constant value of 100, and the mean irradiances calculated by the model were, therefore, expressed as percentages of the maximal noon irradiance at the surface. The variations in tidal range through the spring-neap cycle were allowed for by calculating the mean irradiance over a 15-day period, which extended from the middle of one spring tide series to the middle of the next, and correcting the tidal range each day. For this purpose, three tidal ranges were read in at the start of each run—the maximal range of the first spring series, the minimal range of the neap series, and the maximal range of the second spring series—and the tidal range for each day was assumed to change linearly, decreasing from the first spring to the neap range over the first 8 days, and increasing from the neap to the second spring range over the remaining 7 days. The tidal variations in water level were again calculated from a sine wave, but an actual time for the first low tide in the run was specified at the start, so that the changing relationship between the tidal cycle and the diurnal irradiance cycle could be followed.

Calculations were completed for every 15-min time interval through the 15-day period. The irradiance at the surface was calculated first and, if this was greater than zero (i.e. if t was between sunrise and sunset), the water level was calculated from the tidal sine wave. From this the depth of water overlying a series of different heights on the shore was calculated, and the total irradiance reaching each shore position through these water depths was estimated. The irradiances for each 15-min interval throughout the 15-day period were summed for each position on the shore and then averaged to give the mean irradiance over the complete spring-neap cycle.

The effects of daylength and of water type on the predictions from the model were examined first. Starting data (i.e. tidal ranges for two adjacent spring tide series and the intervening neap series, time of low water for the first spring tide, and daylength) were obtained from tide tables for a specific site (Avonmouth) at four times during the year, and the calculations were repeated for water types 5, 9 and 13 (Fig. 4). Mean irradiance varied almost linearly with height on the shore, as in Fig. 1, but the slope decreased near the top of the shore because these heights are not covered by the neap tides and so are exposed to full surface irradiance throughout the daylight periods at these times. As daylength increased, there was an increase in the light available at the surface from about 22% of the midday maximum in January to 44% in late June, and this difference was reflected in the mean irradiances predicted for positions near the top of the shore. Near the bottom of the intertidal zone, however, there was a smaller proportional increase in mean irradiance between January and June, and this effect was more marked in the poorer water types. The mean irradiance at 1 m above LAT for types 9 and 13 was predicted to be lower in June than in April, in spite of a 3-h increase in daylength. This is because there is a higher probability of high tides occurring during daylight when the daylength is longer, so that a greater proportion of the light available at the surface is attenuated before it reaches the lower intertidal zone. Thus, the lower intertidal is rather unusual among plant habitats in that it does not receive more light in the summer than in spring and autumn.

If the model is used to predict the heights on the shore where specific levels of light are available for plant growth, it is clear that the water type has a smaller influence on the results than in versions of the model which did not allow for diurnal changes in surface irradiance. In general, the substantial decrease in water quality from type 5 to type 13 causes an increase of little more than 1 m in the

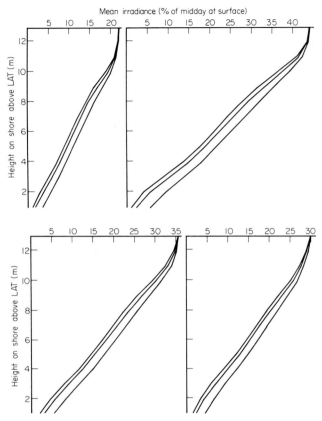

FIG. 4. Mean irradiance (as a percentage of midday irradiance at the surface) over a complete spring-neap cycle at different heights on a shore during four seasons. The tidal ranges, daylengths and initial times of low water (January, 02.15; April, 02.55; June, 02.04; October, 02.37) were determined for Avonmouth, and irradiance was calculated for every 0·25 h of the 15-day period. The three lines in each diagram refer to different water types: coastal 5 (lower line); coastal 9 (middle line), coastal 13 (upper line). Top left: January; daylength 8·24 h; ranges 11·1, 5·9 and 12·2 m. Bottom left: April; daylength 13·55 h; ranges 11·8, 5·8 and 13·7 m. Top right: June; daylength 16·62 h; ranges 11·4, 6·3 and 11·0 m. Bottom right: October; daylength 11·50 h; ranges 14·3, 4·7 and 11·8 m.

height at which a given irradiance is recorded (Fig. 4). This contrasts with differences of 7–8 m in the 1% depths predicted for these two water types in Fig. 3. The influence of water type on the intertidal predictions is slightly greater in longer daylengths, because sites are more likely to be covered by water during daylight hours, but the choice of water type (or its accurate measurement in a particular study site) is less critical for the results from this model than for the predictions of 1% and 0·05% depths discussed in the previous section.

Another characteristic of shores in different geographical locations, which is thought to have a significant influence on the intertidal environment, is the time at which the low waters of spring tides occur. Since tides get approximately 50 min later each day, low waters during one spring tide series must occur about 12 h later than those of the previous series 14 days earlier and, therefore, all of the more extreme low tides in any one site must occur at approximately the same time of day. The exact timing of these tides will markedly affect the light levels on the lower shore, as well as the degree of thermal and desiccation stress to which the organisms are subjected. The influence of this factor on the predictions from the model was examined by repeating the calculations for a 15-day period with all parameters constant except the time of the first low tide. Since the model assumed this tide to be a spring tide with a high tidal range, the time specified for this tide was also the approximate time of all low water springs (LWS).

The largest differences between the light levels predicted when LWS occurred at midnight and midday (0 + 12 h) or in the morning and evening (6 + 18 h) were

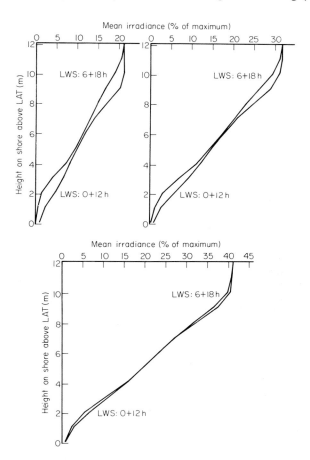

FIG. 5. Effects of the timing of low waters of spring tides (LWS) on mean irradiances in the intertidal zone under three different daylengths. Calculations as Fig. 4, assuming tidal ranges of 9·7, 4·6 and 11·9 m, and 'coastal 13' water type throughout. Top left: daylength 8 h. Top right: daylength 12 h. Bottom: daylength 16 h.

obtained for short daylengths and for positions in the upper and lower regions of the intertidal zone (Fig. 5). Mean irradiances were smaller on the lower shore for morning and evening LWS times because this part of the shore was uncovered only during darkness in an 8-h day. The reductions in irradiance on the upper shore with the same LWS times can be accounted for by the highest tides (i.e. the high waters of spring tides) occurring at about midday and thus causing greatest attenuance at times when irradiance at the surface was highest. If LWS occur at 0 and 12 h, the upper shore will not be submerged in daylight during the winter, and so will usually receive full surface irradiance. In the middle of the shore, these factors appear to be almost in balance, since the predicted mean irradiances are virtually independent of the timing of LWS. Similarly, as daylength was increased, the observed effects of LWS timing decreased and were negligible for 16-h days (Fig. 5). This is because low waters at 06.00 and 18.00 h both occur during daylight, and the lower shore can be well illuminated regardless of the timing of LWS. Therefore, the effects of LWS timing on the light climate for intertidal plants may be significant for many delicate algal species which grow during the winter and early spring, but it is unlikely to be important for the growth of summer-dominant or perennial species.

MODEL VALIDATION AND DISCUSSION

Validation of the predictions for 1% depths made by the model is difficult in the absence of detailed information on the attenuation of light in a series of sites for which the depth limits of kelps and foliose algae are known. However, plants themselves may offer a better indication of the long-term light climate than any instantaneous measurements, and it is possible to use the observed limits of kelp growth as a rough guide to the optical water type. Laminarians are absent from sites in the Bristol Channel east of Hurlstone Point (Crothers 1976; Hiscock 1981) and, as discussed above (p. 28), this suggests that water types in the Bristol Channel are more turbid than the theoretical type 13 used for the calculations presented in Fig. 3. Light measurements in another European estuary indicate that this conclusion is not unreasonable. The 1% depth at the mouth of the Ems-Dollard estuary (with a mean tidal range of only 2–2·5 m) was less than 3 m (Colijn 1982), whereas type 13 water would have a 1% depth of almost 4 m (Fig. 3). Thirty to forty kilometres into the Ems-Dollard estuary, the silt load has increased so much that the 1% depth was reduced to 1 m or less below the water surface (Colijn 1982). Further extrapolations from Jerlov's (1976) data seemed, therefore, to be justifiable, and a number of theoretical water types were constructed to cover the full range of light attenuation observed in the Ems-Dollard estuary.

The water types at different sites in the Bristol Channel were then estimated as the types which, when substituted into the model, gave the closest agreement between the predicted 1% depth (at the appropriate mean tidal range) and the observed depth limit of kelps (Hiscock 1981). These water types were next used to predict the 0·05% depth in the same sites, and these predictions were compared with the observed depth limits for foliose algae (Hiscock 1981). Good agreement was obtained using the data from sixteen separate sites, ranging from the fairly clear waters off Cornwall and Pembrokeshire to the extremely turbid waters of the upper Bristol Channel (e.g. at Nash Point, Glamorgan or Watchet, Somerset). If water types were estimated from the kelp limit without taking account of the tidal variations in water depth, the predicted 0·05% depths were always 2–4 m shallower than the observed limits for foliose algae. Therefore, the model appeared to provide adequate predictions of the depths at which critical light levels occur in tidal sites.

The validation of the final version of the model was again hampered by the lack of detailed light measurements in inshore sites with either poor water quality or high tidal ranges, although this may be less important for this version of the model because the predictions appear to be less sensitive to water type than in simpler versions (see p. 30). Although detailed *in situ* light measurements are clearly required to test many aspects of the model's predictions, a preliminary validation has been attempted using the observation that the lower limit of *Fucus serratus* occurs at 1–2 m above MLWS at three sites in the upper Bristol Channel, north of Weston-super-Mare (Little & Smith 1980). This species normally extends into the subtidal zone (i.e. below LAT) and is finally out-competed by laminarians, but these sites are well beyond the limit of penetration of kelps into the Bristol Channel. A number of other explanations for the unusually high lower limit of *F. serratus* in these sites (e.g. wave action, temperature, salinity, silt load, limpet grazing) were examined and discarded by Little & Smith (1980), but one possibility which remains is that the mean irradiance in these sites is below the compensation point for the photosynthesis and growth of the species.

Photosynthetic compensation was observed in *F. serratus* thalli at irradiances of 1·5–3 Wm^{-2} (or 7–15 $\mu mol\ m^{-2}\ s^{-1}$; King & Schramm 1976; Lüning & Dring 1985), and this is, therefore, the minimum mean irradiance at which positive net photosynthesis (and hence growth) will be possible. Mean irradiances at different heights on the shore at Avonmouth are shown in Fig. 4, expressed as a percentage of the midday irradiance at the surface. The actual surface irradiance at midday will vary with weather and season, but mean values for PAR in this region, based on meteorological records from Aldergrove (54°N, 6°W), are about 50 W m^{-2} in January, 200 W m^{-2} in April, 250 W m^{-2} in June and 150 W m^{-2} in October. Substituting these values into Fig. 4 predicts that *F. serratus* would experience mean irradiances below compensation at sites on the shore lower than about 2 m above LAT in January, 0·5 m in April and 1 m in both June and October.

These values are likely to be underestimates of the exact heights on the shore at which compensation occurs because the plants will frequently be exposed to irradiances which are above the light-saturation level for photosynthesis in *Fucus* (about 100 W m^{-2}), and because emersion will lead to partial desiccation of intertidal plants and a temporary reduction in photosynthesis. These effects have not been (and probably could not be) exactly quantified but, by raising the heights at which compensation might occur, they reinforce the basic prediction of the model—that the lowermost 1–2 m of the intertidal zone at Avonmouth receives too little light for the sustained growth of *Fucus serratus*. This prediction is supported by the observed absence of this species (and, indeed, of all other fucoid species) from this region of the shore.

The model developed in this paper has thus produced realistic predictions about two quite different aspects of the light climate on tidal shores, and it could, therefore, prove to be a useful tool in unravelling some of the seemingly intractable complexities of light in the intertidal environment.

REFERENCES

Colijn, F (1982). Light absorption in the waters of the Ems-Dollard estuary and its consequences for the growth of phytoplankton and microphytobenthos. *Netherlands Journal of Sea Research*, **15**, 196–216.

Crothers, J. H. (1976). On the distribution of some common animals and plants along the rocky shores of West Somerset. *Field Studies*, **4**, 369–389.

Dring, M. J. (1981). Chromatic adaptation of photosynthesis in benthic marine algae: an examination of its ecological significance using a theoretical model. *Limnology and Oceanography*, **26**, 271–284.

Hiscock, K. (1981). *South-West Britain Sublittoral Survey. Final Report*. Nature Conservancy Council, Huntingdon/Oil Pollution Research Unit, Orielton Field Centre, Pembroke.

Jerlov, N. G. (1976). *Marine Optics*. Elsevier, Amsterdam.

King, R. J. & Schramm, W. (1976). Photosynthetic rates of benthic marine algae in relation to light intensity and seasonal variation. *Marine Biology*, **37**, 215–222.

Kirk, J. T. O. (1983). *Light and Photosynthesis in Aquatic Ecosystems*. Cambridge University Press.

Little, C. & Smith, L. P. (1980). Vertical zonation on rocky shores in the Severn Estuary. *Estuarine and Coastal Marine Science*, **11**, 651–669.

Lüning, K. & Dring, M. J. (1979). Continuous underwater light measurement near Helgoland (North Sea) and its significance for characteristic light limits in the sublittoral region. *Helgoländer wissenschaftlichen Meeresuntersuchungen*, **32**, 403–424.

Lüning, K. & Dring, M. J. (1985). Action spectra and spectral quantum yield of photosynthesis in marine macroalgae with thin and thick thalli. *Marine Biology*, **87**, 119–129.

Salinity and seaweed vegetation

G. RUSSELL

Department of Botany, The University, Liverpool L69 3BX

SUMMARY

1 Vegetative tissues of certain intertidal brown algae alter in area and in volume in response to changes in external salinity. Differences in thallus water capacity are somewhat related to zone height and hence to reported species differences in the proportions of cell wall and intercellular gels.

2 The water content of *Fucus* receptacles increases as they reach maturity, mature receptacles always having a greater water content than vegetative thalli. Succulence of *Fucus* receptacles differs with species and with their respective zone heights.

3 The matric content of the thallus does not seem to affect greatly its rate of water uptake or loss but it may serve to buffer the vegetative and reproductive cells against the effects of sudden changes in external salinity.

4 Changes in external salinity may kill algal cells, the amount of cell death varying with species and according to their zone heights. Rock pools are relatively stable saline microhabitats which permit the survival of species at higher levels than are tolerable on open rock. Cell mortality in hyposaline waters may vary with tissue age, distal cells which are active in cell division and growth being most susceptible to damage.

5 Intraspecific variation in salinity tolerance occurs in a number of seaweed species. Biosystematic studies of *Enteromorpha intestinalis* and *Pilayella littoralis* have provided evidence of gene flow between populations. In the latter species, variation is ecoclinal rather than ecotypic in character and is related to an environmental (estuarine) gradient.

6 By combining geographical isolation and the presence of extremely low seawater salinity with a precisely known post-glacial history, the Baltic Sea is of outstanding importance for students of evolutionary aspects of algal ecology. Baltic seaweeds are tolerant to low salinities and are susceptible to damage by high salinities, resulting in a tolerance curve with a lower tolerance optimum than that of algae on N. Altlantic shores. Baltic algae also differ morphologically from N. Atlantic forms but not in a way that can yet be linked directly with salinity tolerance.

7 It is concluded that salinity may operate as an agent of natural selection on marine algae, and that its effect is most pronounced in extremes of salinity. However, ecological deductions need to take into account the limitations of experimental method as well as the evolutionary constraints arising from other environmental factors.

INTRODUCTION

Classification of marine and brackish waters has been attempted by a number of authors, some of whose proposals are summarized or are given in full by Hedgpeth (1957), Segestrale (1959), Anon. (1959) and den Hartog (1968). However, the utility of such classifications is limited by their inability to accommodate environments with fluctuating salinities. The need to distinguish stable and unstable regimes has been given due emphasis by several authors but with evident disagreement on terminology, brackish water being defined both as stable (Hedgpeth 1957) and unstable (den Hartog 1967, 1968, 1970).

In practice, the stable saline environment probably does not exist at all outside the laboratory. Seasonal climatic changes are reflected in the surface salinities of most natural seawaters, although the amplitude of these is normally very small (Riley & Chester 1971). In estuaries and on emergent marine shores, seasonal effects are complicated by other temporal changes (mainly lunar and semidiurnal) arising from tidal patterns. The magnitude of saline fluctuations in such environments may be very great (den Hartog 1967). Superimposed upon these cyclical changes are irregular oscillations caused by random climatic and hydrological events. Marine and brackish waters are consequently almost infinitely variable in the amplitude and frequency of their saline changes, and it is unlikely that they will ever prove completely amenable to classification, as Zenkevitch (1959) has also concluded.

Recent evidence suggests that plants have three primary responses to changes in external salinity. They alter their cell volume and/or turgor, they adjust the cellular content and concentrations of inorganic ions, and they undergo changes in concentration of organic solutes (see reviews by Hellebust 1976; Kauss 1978; Zimmermann 1978; Epstein 1980). Much progress has been made towards identification and quantification of osmotica for members of all major algal groups (Raven 1975; Wyn Jones *et al.* 1977; Kirst & Bisson 1979; Brown & Hellebust 1980; Dickson, Wyn Jones & Davenport 1980, 1982; Raven, Smith & Smith 1980; Reed 1980, 1983a, b; Reed, Collins & Russell 1980a, b, 1981; Reed & Stewart 1983; Chudek *et al.* 1984; Reed *et al.* 1984). The various ways in which a walled marine alga may respond to external salinity changes are summarized in Fig. 1.

Such responses may cause algal thalli to undergo marked changes in their dimensions, a phenomenon first described in detail by Ogata & Takada (1955). Changes in thallus dimensions are liable to involve the cell walls and their associated gels and mucilages (Gessner & Schramm 1971). Rather less attention has been given to these than to intracellular activities in recent years, in spite of the fact that they are important determinants of thallus anatomy, texture and elasticity, although various possible functions have been discussed by Baardseth (1969), Percival (1979), Boney (1981) and Ritchie & Larkum (1982a,b).

The aim of this paper is to examine some of the immediate effects of salinity change upon seaweed thalli, and to consider how this factor may operate in the longer term as an agent of natural selection between and within species. Salinity

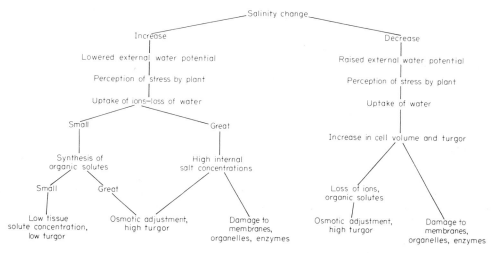

FIG. 1. Short-term responses to salinity changes by seaweeds (after Epstein 1980).

values are expressed in parts-per-thousand (‰) in order to be consistent with work by other authors. However, this convention may soon be replaced by the new Practical Salinity Scale (PSS), details of which are given by Anon. (1981).

WATER RELATIONS OF INTERTIDAL MACROPHYTES

The large perennial brown algae of N. Atlantic rocky shores are commonly found in clearly demarcated zones (Lewis 1964). This characteristic distribution pattern has stimulated a great deal of ecophysiological research, much of which has been devoted to the study of desiccation and its effects (Zanefeld 1937; Kristensen 1965; Schramm 1968; Ried 1969; Berard-Therriault & Cardinal 1973; Dorgelo 1976; Schonbeck & Norton 1978, 1979; Jones & Norton 1979, 1981; Dromgoole 1980; Dring & Brown 1982). Desiccation is normally regarded as water loss from emergent plants under atmospheric conditions although this process is one which in nature is liable to begin as a response to hypersalinity. In environments such as isolated rock pools dehydration may indeed be due entirely to the effect of increased salinity. Conversely, seaweeds on Atlantic rocky shores are very likely to experience reduced salinity as a consequence of rainfall and land run-off (den Hartog 1968), although the ecological importance of this factor has been given rather less experimental attention (Munda & Kremer 1977). Consequently, it may be concluded that intertidal macrophytes experience a continuum of change in external water potential of which desiccation is simply a part; ecological experiments ought therefore to take this range of salinity conditions into greater account.

Recent studies of desiccation rates in intertidal algae by Kristensen (1965), Schonbeck & Norton (1979) and Dromgoole (1980) have confirmed that water loss from the thallus is largely determined by its surface area:volume ratio. These authors' observations have run counter to earlier reports that the desiccation rate of a seaweed is a function of its zone height, with most rapid water loss being experienced by species from the lowest shore levels (Zanefeld 1937). Zanefeld's account remains of interest, however, because it includes data showing that upper-shore fucoids have thicker cell walls than those at successively lower levels, an observation that has since been confirmed by Kristensen (1965). This observation has also been validated by chemical analyses which reveal large concentrations of fucans (Percival 1979) in upper-shore brown algae. Kristensen (1965) could not establish any correlation between cell wall thickness and desiccation rate and Dromgoole (1980) has expressed doubt that this factor could influence water loss in any way. However, if brown algal gels and mucilages do have an ecological role, then it may have less to do with water loss than water uptake, for these are hydrophilic compounds (Percival 1979) which are capable of taking up atmospheric water (Lestang & Quillet 1981).

The following observations have been made on five seaweed species which form a common zonal series on British coasts. (Species authorities here and throughout are as given by Parke & Dixon (1976). In descending sequence they are: *Pelvetia canaliculata*, *Fucus spiralis*, *F. vesiculosus*, *F. serratus* and *Laminaria digitata*. All experiments were carried out on freshly-collected plants which had been hydrated in seawater (34‰) for several hours at the experimental temperature (10°C) before use. Experiments were carried out on apical pieces, about 1·5 cm in length, cut from the vegetative thalli of fucoids. In the case of *Laminaria digitata*, the material used consisted of discs 1 cm in diameter punched from mid-laminar thallus. Tissue portions were washed in clean seawater, and allowed to stand in seawater at 10°C before incubating in waters of different salinity. Incubating thalli received incandescent light of approximately 10 W m^{-2} (see Russell (1985a) for method).

Figure 2 shows area changes in *Fucus spiralis* and *F. serratus* measured at intervals over a 4 h period. Rapid change took place during the first 15 min and after 2 h a fairly stable state was reached by both species. Even after 6 days there was no obvious sign of area recovery in any treatment although some evidence of thallus growth became evident in the *F. spiralis* control. If cellular adjustments as outlined in Fig. 1 had occurred in these tissues, they were plainly insufficient to bring about any restoration of initial thallus area. Of the two species under observation, *F. spiralis* evidently possessed the greater water capacity.

Figure 3 gives percentage changes in tissue mass of all five species incubated for 2 h in a similar salinity range. When *F. spiralis* and *F. serratus* are compared, it is once again evident that the former has the greater capacity. It may also be seen that while percentage decrease in mass and area in the hypersaline treatments are similar in magnitude, hyposaline waters effect a disproportionately greater increase in mass. The superior water capacities of *Pelvetia* and *Fucus spiralis* are therefore largely due to their greater ability to take up freshwater than either *F. vesiculosus* or *F. serratus* which, in turn, take up more than *Laminaria digitata*. The

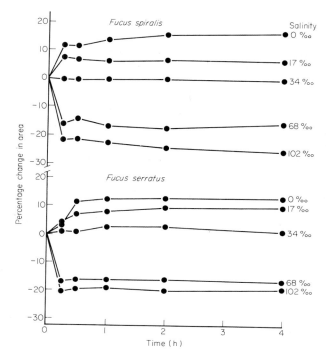

FIG. 2. Percentage changes in area of excised thallus apices of *Fucus spiralis* and *F. serratus* during exposure to water of different salinities. Material was treated for 4 h at 10°C and each graph point represents a mean of five replicate apices. Data obtained by Miss A. Fox.

zonal correlation with water capacity is not very impressive, however, the capacity of *Pelvetia* being lower than that of *Fucus spiralis*. The smaller water capacity of *Pelvetia* than of *Fucus spiralis* is a notable anomaly, but one which is in accordance with data obtained by Schonbeck & Norton (1979). It is possible, as these authors suggest, that *Pelvetia* possess cell wall polysaccharides that are different in character from those of *Fucus* species.

Water uptake in hyposaline media will plainly alter the area: volume ratio, but how this would affect subsequent water loss, if external water potential were to be lowered, is uncertain. The statement that thallus area: volume determines the rate of water loss by desiccation (Dromgoole 1980) seems to have been intended to refer chiefly to species-specific differences in thallus form. Within species, however, it is water content which will determine area: volume.

Desiccation studies are usually conducted on vegetative thalli, although authors do not always specify the reproductive state of their experimental plants. In practice, vegetative and reproductive tissues differ considerably in water content (Pringsheim 1923; Moss 1948, 1950). Figure 4 shows water content in relation to oven-dry weight of vegetative tissue and of ripe receptacles in the three *Fucus* species referred to above. Vegetative *F. serratus* proved to have less dry matter

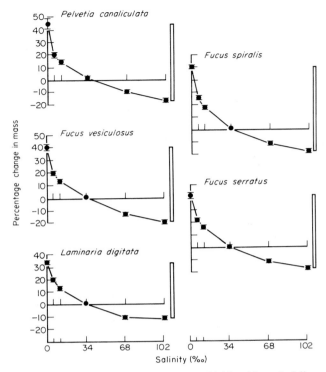

FIG. 3. Percentage changes in mass of excised apices of intertidal fucoids, and of discs of laminar tissue from *Laminaira digitata*, in response to 2 h salinity treatments at 10°C. Each point represents a mean of twenty replicate tissue portions with 95% confidence limits. The heights of the vertical bar serves to illustrate the mean water capacities of tissues under these conditions. Data obtained by Mr D. Thomas.

than *F. spiralis* or *F. vesiculosus*, as has been reported by Kristensen (1965). Receptacles of *F. serratus*, on the other hand, had much less water than those of *F. vesiculosus*, which had, in turn, less than those of *F. spiralis*. Receptacles of *F. serratus* were therefore the least differential in terms of water content from the tissue of the vegetative thallus. Receptacle succulence is obtained as part of a process of thallus transformation from a vegetative to a reproductive function, and its development in Baltic *F. vesiculosus* has recently been recorded by Russell (1985b).

It seems probable that most of the hyposaline water taken up by the vegetative thalli was located within the matrix of cell walls plus intercellular gels, as experiments with formalin-killed thalli showed persistent uptake. There is little evidence to suggest that *F. spiralis* takes up water faster than *F. serratus*, in spite of its greater matric content (Fig. 2) and the pattern of hypersaline water loss by all five species is rather uniform (Fig. 3) as other workers have found. Any changes in external water potential ought to be transmitted to protoplasts in both gelatinous and non-gelatinous thalli alike, and evoke responses along lines indicated in Fig. 1.

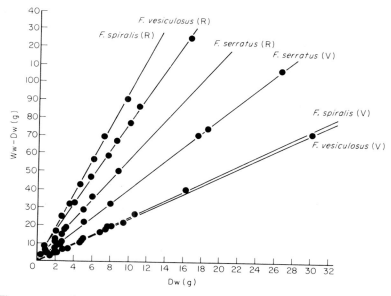

FIG. 4. Tissue water and oven-dry matter correlations in vegetative thalli and ripe receptacles in *Fucus spiralis*, *F. vesiculosus* and *F. serratus*. All correlations are very significant ($P < 0.001$). R = Reproductive tissues (receptacles), V = Vegetative tissues.

However Dickson, Wyn Jones & Davenport (1982) have shown that sinusoidal changes in external salinity resulted in better control of ion fluxes in *Ulva lactuca* and was associated with a smaller amplitude in organic solute (dimethylsulphoniopropionate) level, than when abrupt salinity changes were imposed. The possibility that these gels may serve to buffer the cells against sudden energy demands of a fluctuating saline environment needs looking at in relation to the cytoplasmic responses. The conclusion by Munda & Kremer (1977) that mannitol plays no role in *Fucus* osmoregulation was drawn from observations of mannitol decrease in thalli treated with hyposaline waters, a response that might be expected under such circumstances. More recent work by Reed (1980) and Chudek *et al.* (1984) have confirmed that mannitol and its derivatives may indeed function as osmotica, increasing in concentration with external salinity (Reed 1980).

The protective role of brown algal gels may also apply to receptacles by providing a relatively stable internal environment for the developing gametangia. Most speculation on receptacle function, however, has been concerned with mechanisms of gametic discharge. It has been argued that both dehydration and hydration may cause gametes to be expelled (Fritsch 1945), and temporary immersion of receptacles in freshwater has been recommended as a technique for inducing egg release (Callow, Coughlin & Evans 1978). It remains uncertain whether gametic discharge is wholly dependent on these effects, however; the fact that permanently submerged fucoids can reproduce successfully makes the relationship between external water potential and the mechanics of sexual reproduction one worth further study. Provision of some environmental protection for developing gametangia by means of water located in receptacle gel is as likely to be important in the long-term survival of species. Differences between *Fucus* species in the respective water content of mature receptacles manifest a zonal pattern (Fig. 4), but receptacles of *Pelvetia* are not obviously more succulent than those of *Fucus spiralis*. This anomaly may be explicable in terms of polysaccharide chemistry (Schonbeck & Norton 1979) but may also be due to a more efficient reproductive defence mechanism. Eggs of *Pelvetia* are shed and fertilized while contained within the persistent oogonial wall (mesochiton) (Moss 1974). Mesochiton has been held to protect the zygotes against excessive desiccation (Moss 1974), although it must also serve to contain rapidly expanding protoplasts in the event of a sudden decrease in external salinity.

The need for investigations of water relations to bear in mind the full range of environmental salinities has been stated. It is also important to consider the susceptibilities of the different developmental states of plants, and differences in the way mature plants aggregate and overlap in nature to create favourable microhabitats (Schonbeck & Norton 1979).

SALT TOLERANCE AND SPECIES ZONATION

The desiccation tolerances of the algae referred to in Fig. 3 are now known to increase significantly according to the plants' respective shore heights. This was demonstrated by Dring & Brown (1982) on the basis of oxygen emission from rehydrated thalli following various periods of desiccation, an observation which confirmed an earlier investigation by Ried (1969) using rather similar methods. Photosynthetic activity, expressed as O_2 output or CO_2 fixation, has been used as a measure of salinity tolerance by numerous workers (see Wilkinson 1980; Russell 1985b). Two other techniques have received wide support: measurement of plant growth in controlled culture conditions, and microscopic screening of treated thalli for visible evidence of cell damage or death. This last approach was much favoured by Biebl (1939, 1952, 1967) who did much to advance our knowledge of marine algal ecophysiology. In particular, Biebl (1952) demonstrated that salinity tolerances of upper-shore algae are much broader than those from lower levels, with sublittoral species having the narrowest tolerance limits of all (see reviews by Zanefeld 1969; Gessner & Schramm 1971).

The following observations were made on six species from different habitats using cell mortality as the measure of salt tolerance, the method being a variant of that of Reed (1983b) and Reed & Barron (1983). Small portions of thallus, sufficient for a microscope preparation, were incubated at 10°C in light of 10 W m^{-2} and an 8:$\overline{16}$ h photoperiod, for 48 h. The thallus portions were placed singly in compartments of disposable sterile 'Repliboxes', each compartment also containing 4·5 cm^3 of water. Five replicates of each salinity treatment were employed in random block designs. After incubation, the thalli were stained for 5–10 min in a 0·5% w/v aqueous solution of Evans Blue stain, then transferred to a

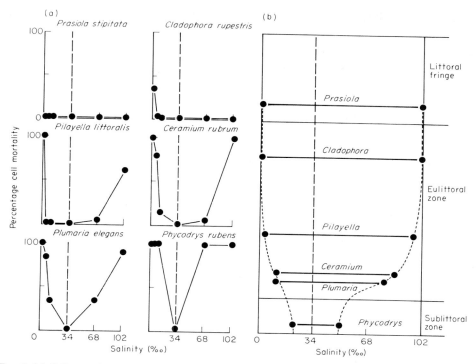

FIG. 5. (a) Cell mortality in six species of marine algae from different habitats in response to different salinities for 48 h at 10°C. Each point represents a mean of five measurements. *Pilayella littoralis* and *Ceramium rubrum* were obtained from rock pools. *Phycodrys rubens* was a permanently submerged sublittoral zone alga, and the remaining species were collected from open rock. (b) LD 50 data from Fig. 5a indicating zonal pattern of salt tolerance in these algae.

microscope slide for examination, and 100 cells scored for the presence or absence of staining. Staining indicates cell death (Reed 1983b; Reed & Barron 1983).

The species investigated were *Prasiola stipitata*, collected from the upper shore on open rock (littoral fringe *sensu* Lewis 1964). Two species were obtained from mid-eulittoral zone rock pools, *Pilayella littoralis* and *Ceramium rubrum*. Two other eulittoral zone species were collected, but from open rock: *Cladophora rupestris* from the upper part of the zone and *Plumaria elegans* from near its lower boundary. *Phycodrys rubens*, the sixth species, was obtained from the sublittoral zone.

The cell mortality curves of these species (Fig. 5a) are consistent with the evidence that salinity tolerance is related to zone height. However, the responses of the mid-eulittoral pool species proved to be more similar to that of low-eulittoral *Plumaria* than to high-eulittoral *Cladophora*, suggesting that the pools were located at lower levels than was actually the case. Large rock pools which are flushed by every tide are likely to be more stable saline environments than adjacent open rock. Consequently selection pressures associated with differences in zone height will be subject to some modification under the influence of local microhabitat factors.

Figure 5b is a salt-tolerance diagram based upon LF50 data culled from Fig 5a; it broadly agrees with the results presented by Biebl (1952).

In the course of these experiments it became evident that algae with apical growth (*Cladophora*, *Ceramium*, *Phycodrys* and *Plumaria*) showed a non-random distribution of cells killed by low salinity, the apices being most susceptible to damage. Had cell mortalities been calculated only from the oldest parts of thalli, quite different graphs could have been obtained. Young distal cells may have thinner walls and protoplasts of lower osmotic potential than their older counterparts; they may also differ in vacuolar content. Such age-linked differences in cell damage should be amenable to investigation by electron microscopy (Young, Collins & Russell 1985). In any event, these rather localized patterns of cell mortality emphasize the important ecological role of thallus bases as a means of plant persistence (Dixon 1973). Age-linked differences in response to desiccation by algal thalli have been reported by Ried (1969), while Kremer (1975) has described a number of physiologically important differences between apical and proximal tissues of *Fucus serratus*.

ECOTYPES AND ECOCLINES

Variation in salinity tolerance below the level of species is known to occur in a number of marine algae. Much of this evidence has demonstrated the existence of low-salinity tolerance in populations subject to brackish conditions (West 1972; Russell & Bolton 1975; Bolton 1979; Reed & Russell 1979; Yarish, Edwards & Casey 1979; Yarish & Edwards 1982). However, Francke & ten Cate (1980) and Francke & Rhebergen (1982) have also reported the existence of saline ecotypes in the freshwater genus *Stigeoclonium*.

Reed & Russell (1979) have measured salinity tolerances of four populations of *Enteromorpha intestinalis* using, as a measure, the ability of excised thallus portions to regenerate rhizoids in a range of saline media. Their results (Fig. 6) show that a marine eulittoral zone population had a narrower tolerance range than another from the littoral fringe (as would be expected from Biebl's work). The tolerance range of an estuarine population was displaced towards the lower salinities, again, as might be expected. The most interesting data came from the *Enteromorpha* of a group of freshwater ponds located close to the littoral fringe populations. These plants had a pattern of tolerance that was indistinguishable from the latter's, suggesting that they had been recruited from that population. The close similarity in salt tolerances of the field samples and of their progeny (Fig. 6) also makes it probable that the particular tolerance characteristics of these populations was heritable.

Further evidence for gene flow between populations has been obtained by Bolton (1979) who measured the tolerance of *Pilayella littoralis* by means of the growth characteristics of clones isolated from a variety of sites. Prolonged conditioning of all these cultures to standard seawater medium before salinity

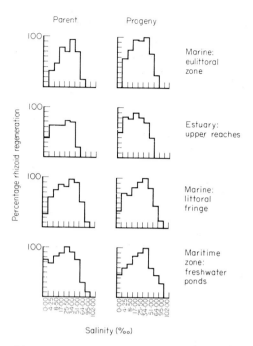

Fig. 6. Salinity tolerances of four populations of *Enteromorpha intestinalis* from different saline environments; tolerances were measured as percentage of excised tissue segments showing rhizoid regeneration in culture media of different salinity. Redrawn from Reed & Russell (1979).

treatment reduced the likelihood of growth differences being due to local acclimation effects. Bolton's results (Fig. 7) show that the most euryhaline isolates were located in the upper and middle reaches of an estuary. In the lower reaches and on the retaining walls of the middle estuary the greatest heterogeneity of salt tolerances was obtained, suggesting downstream recruitment from the upper-estuary population and upstream recruitment from the marine sites. The tolerance polygons are not readily resolved into two classes, however, and Bolton's conclusion that the variation detected is ecoclinal in character rather than ecotypic seems entirely defensible. This work suggests that the well-documented horizontal changes in the species composition of estuaries (see Wilkinson 1980) may be accompanied by interesting genotypic changes within species, in response to increasing selection pressure.

Reed & Barron (1983) have shown that estuarine *Pilayella* has smaller cell volumes than its marine counterpart. Smaller cells have similarly been reported in estuarine *Polysiphonia lanosa* by Reed (1983b). In both cases, the differences in cell dimensions were accompanied by reduced ability to concentrate organic osmotica (mannitol and DMSP respectively) in response to hypersaline stress, as above. Dromgoole (1980) has made the point that cells of drought-tolerant higher plants are often small, and it is possible that algae have responded to saline stress in a

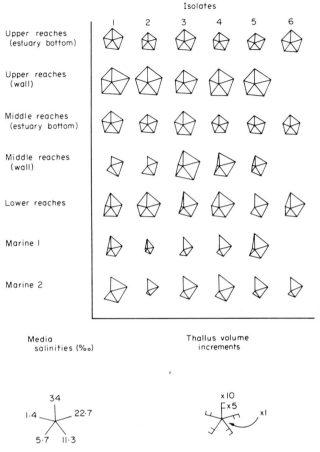

FIG. 7. Salinity tolerance of estuarine and marine populations of *Pilayella littoralis*; tolerances were measured as growth (compacted cell volume) by clonal isolates in culture media of different salinities. Redrawn from Bolton (1979).

similar way. However, cells may be small because loss of turgor has prevented growth or because smallness has evolved as an advantageous characteristic which enables them to cope with saline fluctuations. The mechanical advantage of a small cell as a means of containing increased turgor has been discussed by Haines (1950).

Variation in salt tolerance below the species level is likely to be widespread in algae, and may in some cases be associated with morphological differences either at the cell level (Reed 1983b) or below (Young, Collins & Russell 1985). However, its incidence among natural populations of seaweeds will continue to be poorly documented as long as experimental studies remain based upon single or very few isolates. The need to extend experimental studies, especially those involving algal cultures, to the level of the population is particularly important.

THE BALTIC SEA

The Baltic Sea provides a unique opportunity for investigating evolutionary aspects of salt tolerance in algae. Its post-glacial history, which is known with great precision, involved a number of major saline changes. The present marine algal population seems likely to have been recruited during the Littorina Sea episode (*c.* 7500 B.P.) but it has been exposed to greatly reduced salinity since *c.* 3000 B.P. The Baltic is also very isolated geographically with only a narrow exit to the North Sea through the channel between Denmark and Sweden. Thus isolation, environmental stress and a time scale are combined in a single locality.

Recent observations on Baltic seaweeds by Russell (1985b) indicate that considerable ecotypic (or ecoclinal) evolution has occurred. Figure 8, which gives a salt-tolerance curve calculated from the responses of twelve species, shows that the Baltic flora has diverged considerably from species on British shores (see Fig. 5) in combining tolerance to low salinities with susceptibility to damage from high salinities. The widths of the respective confidence limits show that most variation in salt tolerance was obtained in 34‰ and 51‰. Such values nevertheless greatly exceed the seawater salinities which most of the plants normally experience. However, the universal incidence of low-salinity tolerance in Baltic marine algae may be contrasted with the number of ways in which they differ in thallus organization from their British counterparts. Some species are greatly reduced in thallus size but not in that of their constituent cells, others have smaller cells but not smaller thalli, and others still show size reduction in certain cell types but not in all (Russell

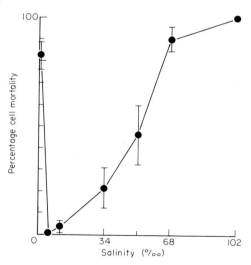

FIG. 8. Provisional salinity-tolerance curve for the algae of the Baltic Sea, obtained from cell mortalities in waters of different salinity by twelve species (cf. Fig. 5). Each point represents a mean of 55–60 measurements with 95% confidence limits. Diagram based on data used by Russell (1985b).

1985a,b). Thus salinity tolerance could not be directly related to any particular morphological or anatomical characteristic.

Some of the morphological peculiarities of these Baltic forms may prove to be readily modifiable by experimental manipulation; and others, though stable in character, may be unrelated except in a pleiotropic sense to the salt tolerance of the plant. Morphological variation in seaweeds, as in other kinds of plant, may arise at random so it should be possible for Baltic forms to occur elsewhere within the natural ranges of these species. It will be interesting to see how frequently they do so, and under which particular environmental conditions. The value of the Baltic flora in comparative ecophysiological investigations is immense, and ought not to be neglected.

CONCLUSIONS

Changes in external salinity bring about a number of plastic responses in algal cells (Table 1, Fig. 2). Emphasis has often been laid on the importance of distinguishing these from others which are due to genetic (e.g. ecotypic) differences. Recently, Reed (1984) has proposed a revision of osmoterminology along these lines. Clarity of terminology is welcome, but only if the underlying biology is classifiable in this way. In practice it may be difficult to distinguish plastic and genetic effects, and indeed the two may be inseparable (Russell 1985b).

A great deal of evidence now exists to support the suggestion put forward by Dahl (1959) that environmental salinity operates as an agent of natural selection. The evidence becomes more clear as the selection pressure becomes more directional. Directional pressure may exist in highly fluctuating salinities such as are obtained in upper zones of rocky shores and in tidal reaches of estuaries, but it may also be present in relatively stable but extreme conditions, as in the inner Baltic Sea.

A note of caution should be struck in making ecological inferences from experiments such as those described. Virtually all the evidence discussed has been obtained from walled tissue, and the results may be quite inapplicable to naked or extremely thin-walled reproductive cells (Burrows 1963; Khfaji & Norton 1979). The ecological errors in ignoring the vulnerability of reproductive cells could be very great. Consequently, there is need for more information on membrane fine structure in relation to osmotic regulation in algae (Wiencke & Läuchli 1983; Knoth & Wiencke 1984).

There is additional scope for error in being too single-minded about the importance of salinity. It is but one component of natural selection among many, and other factors may be expected to introduce an element of evolutionary constraint. The interactive effects of total salinity with those of temperature and specific ions serve to underline that point (Yarish, Edwards & Casey 1980; Yarish & Edwards 1982).

ACKNOWLEDGMENTS

I am grateful to Miss A. Fox and Mr D. Thomas for permission to use their data in Figs 2 and 3 respectively.

REFERENCES

Anon. (1959). The Venice system for the classification of marine waters according to salinity. *Archivio di Oceanografia e Limnologia*, **11**, (suppl.), 243–245.

Anon. (1981). Tenth report of the joint panel on oceanographic tables and standards. *UNESCO Technical Papers in Marine Science*, Vol 36, UNESCO, Paris, 25pp.

Baardseth, E. (1969). Some aspects of the native intercellular substance in Fucaceae. *Proceedings of the International Seaweed Symposium*, **6**, 53–60.

Berard-Therriault, L. & Cardinal A. (1973). Importance de certains facteurs ecologiques sur la resistance a la dessiccation des Fucacees (Phaeophyceae). *Phycologia*, **12**, 41–52.

Biebl, R. (1939). Protoplasmatische Okologie der Meeresalgen. *Bericht der Deutschen botanischen Gesellschaft*, **57**, 78–90.

Biebl, R. (1952). Ecological and non-environmental constitutional resistance of the protoplasm of marine algae. *Journal of the Marine Biological Association of the United Kingdom*, **31**, 307–315.

Biebl, R. (1967). Protoplasmatische Okologie. *Naturwissenschaften Rundschau*, **20**, 248–252.

Bolton, J. J. (1979). Estuarine adaptation in populations of *Pilayella littoralis* (L.) Kjellm. (Phaeophyceae, Ectocarpales). *Estuarine and Coastal Marine Science*, **9**, 273–280.

Boney, A. D. (1981). Mucilage: the ubiquitous algal attribute. *British Phycological Journal*, **16**, 115–132.

Brown, L. M. & Hellebust, J. A. (1980). The contribution of organic solutes to osmotic balance in some green and eustigmatophyte algae. *Journal of Phycology*, **16**, 265–270.

Burrows, E. M. (1963). Ecological experiments with species of *Fucus*. *Proceedings of the International Seaweed Symposium*, **4**, 166–170.

Callow, M. E., Coughlin, S. J. & Evans, L. V. (1978). The role of Golgi bodies in polysaccharide sulphation in *Fucus* zygotes. *Journal of Cell Science*, **32**, 337–356.

Chudek, J. A., Foster, R., Davison, I. R. & Reed, R. H. (1984). Altritol in the brown alga *Himanthalia elongata*. *Phytochemietry*, **23**, 1081–1082.

Dahl, E. (1959). Intertidal ecology in the terms of poikilohalinity. *Archivio di Oceanografia e Limnologia*, **11** (suppl.), 227–236.

Dickson, D. M. J. Wyn Jones, R. G. & Davenport, J. (1980). Steady state osmotic adaptation in *Ulva lactuca*. *Planta*, **150**, 158–165.

Dickson, D. M. J., Wyn Jones, R. G. & Davenport, J. (1982). Osmotic adaptation in *Ulva lactuca* under fluctuating salinity regimes. *Planta*, **155**, 409–415.

Dixon, P. S. (1973). *Biology of the Rhodophyta*. Oliver & Boyd, Edinburgh.

Dorgelo, J. (1976). Intertidal fucoid zonation and desiccation. *Hydrobiological Bulletin*, **10**, 115–122.

Dring, M. J. & Brown, F. A. (1982). Photosynthesis of intertidal brown algae during and after periods of emersion: a renewed search for physiological causes of zonation. *Marine Ecology Progress Series*, **8**, 301–308.

Dromgoole, F. I. (1980). Desiccation resistance of intertidal and subtidal algae. *Botanica Marina*, **23**, 149–159.

Epstein, E. (1980). Responses of plants to saline environments. *Genetic Engineering of Osmoregulation* (Ed. by D. W. Rains, R. C. Valentine & A. Hollaender), pp. 7–21. Plenum Press, New York.

Francke, J. A. & ten Cate, H. J. (1980). Ecotypic differentiation in response to nutritional factors in the algal genus Stigeoclonium Kutz. (Chlorophyceae). *British Phycological Journal*, **15**, 343–355.

Francke, J. A. & Rhebergen, L. J. (1982). Euryhaline ecotypes in some species of *Stigeoclonium*. *British Phycological Journal*, **17**, 135–145.

Fritsch, F. E. (1945). *The Structure and Reproduction of the Algae. Vol. 2* Cambridge University Press, Cambridge.

Gessner, F. & Schramm, W. (1971). Salinity: plants. *Marine Ecology.* (Ed. by O. Kinne), *Vol. 1, Part 2* pp. 705–820. Wiley–Interscience,London.

Haines, F. M. (1950). The relation between cell dimensions, osmotic pressure and turgor pressure. *Annals ofBotany*, **14**, 385–394.

den Hartog, C. (1967). Brackish water as an environment for algae. *Blumea*, **15**, 31–43.

den Hartog, C. (1968). The littoral environment of rocky shores as a border between the sea and the land and between the sea and fresh water. *Blumea*, **16**, 374–393.

den Hartog, C. (1970). Some aspects of brackish-water biology. *Commentationes biological Societas Scientiarum Fennica*, **31**, 1–13.

Hedgpeth, J. W. (1957). Classification of marine environments. *Memoirs of the Geological Society of America*, **67**, 17–28.

Hellebust, J. A. (1976). Osmoregulation. *Annual Review of Plant Physiology*, **27**, 485–505.

Jones, H. G. & Norton, T. A. (1979). Internal factors controlling the rate of evaporation from fronds of some intertidal algae. *New Phytologist*, **83**, 771–781.

Jones, H. G. & Norton, T. A. (1981). The role of internal factors in controlling evaporation from intertidal algae. *Plants and their Atmospheric Environment* (Ed. by J. Grace, E. D. Ford & P. G. Jarvis). Symposia of the British Ecological Society. pp. 231–235. Blackwell Scientific Publications, Oxford.

Kauss, H. (1978). Osmotic regulation in algae. *Progress in Phytochemistry*, **5**, 1–27.

Khfaji, A. K. & Norton, T. A. (1979). The effects of salinity on the distribution of *Fucus ceranoides. Estuarine and Coastal Marine Science*, **8**, 433–439.

Kirst, G. O. & Bisson, M. A. (1979). Regulation of turgor presure in marine algae: Ions and low-molecular-weight organic compounds. *Australian Journal of Plant Physiology*, **6**, 539–556.

Knoth, A. & Wiencke, C. (1984). Dynamic changes of protoplasmic volume and of fine structure during osmotic adaptation in the intertidal red alga *Porphyra umbil calis. Plant, Cell and Environment*, **7**, 113–119.

Kremer, B. P. (1975). Physiologisch-chemische Charakteristik verschiedener Thallus bereiche von *Fucus serratus. Helgolander wissenschaftliche Meeresuntersuchangen*, **27**, 115–127.

Kristensen, I. (1965). Surf influence on the thallus of fucoids and the rate of desiccation. *Sarsia*, **34**, 69–82.

Lestang, G. de & Quillet, M. (1981). The sulfated fucoidan of phaeophyceae as a factor of biological adaptation. *Proceedings of the International Seaweed Symposium*, **8**, 200–204.

Lewis, J. R. (1964). *The Ecology of Rocky Shores.* English Universities Press, London, 323pp.

Moss, B. L. (1948). Studies in the genus *Fucus* I. On the structure and chemical composition of *Fucus vesiculosus* from three Scottish localities. *Annals of Botany*, **12**, 267–279.

Moss, B. L. (1950). Studies in the genus *Fucus* II. The anatomical structure and chemical composition of receptacles of *Fucus vesiculosus* from three contrasting habitats. *Annals of Botany*, **14**, 395–419.

Moss, B. (1974). Attachment and germination of the zygotes of *Pelvetia canaliculata* (L.) Dcne. et Thur. (Phaeophyceae, Fucales). *Phycologia*, **13**, 317–322.

Munda, I. M. & Kremer, B. P. (1977). Chemical composition and physiological properties of fucoids under conditions of reduced salinity. *Marine Biology*, **42**, 9–15.

Ogata, E. & Takada, H. (1955). Elongation and shrinkage in thallus of *Porphyra tenera* and *Ulva pertusa* caused by osmotic changes. *Journal of the Institute of Polytechnics, Osaka City University, Ser. D*, **6**, 29–41.

Parke, M. & Dixon, P. S. (1976). Check-list of British marine algae—third revision. *Journal of the Marine Biological Association of the United Kingdom*, **56**, 527–594.

Percival, E. (1979). The polysaccharides of green, red and brown seaweeds: their basic structure, biosynthesis and function. *British Phycological Journal*, **14**, 103–117.

Pringsheim, E. G. (1923). Uber die Transpiration bei *Fucus. Jahrbucher fur wissenschaftliche Botanik*, **62**, 244–257.

Raven, J. (1975). Algal cells. *Ion Transport in Plant Cells and Tissues* (Ed. by D. A. Baker & J. L. Hall), pp. 125–160. Elsevier N. Holland, Amsterdam.

Raven, J. A., Smith, F. A. & Smith, S. E. (1980). Ions and osmoregulation. *Genetic Engineering of Osmoregulation* (Ed. by D. W. Rain, R. C. Valentine & A. Hollaender), pp. 101–118. Plenum Press, New York.

Reed, R. T. (1980). The influence of salinity upon cellular mannitol concentration of the euryhaline marine alga *Pilayella littoralis* (L.) Kjellm. (Phaeophyta, Ectocarpales): preliminary observations. *Botanica Marina*, 23, 603–605.

Reed, R. H. (1983a). Measurement and osmotic significance of β-dimethyl-sulphoniopropionate in marine macroalgae. *Marine Biology Letters*, 4, 173–181.

Reed, R. H. (1983b). The osmotic responses of *Polysiphonia lanosa* (L.). Tandy from marine and estuarine sites: evidence for incomplete recovery of turgor. *Journal of Experimental Marine Biology and Ecology*, 68, 169–193.

Reed, R. H. (1984). Use and abuse of osmo-terminology. *Plant Cell and Environment*, 7, 165–170.

Reed, R. H. & Barron, J. A. (1983). Physiological adaptation to salinity change in *Pilayella littoralis* from marine and estuarine sites. *Botanica Marina*, 26, 409–416.

Reed, R. H., Collins, J. C. C. & Russell, G. (1980a). The effects of salinity upon cellular volume of the marine red alga *Porphyra purpurea* (Roth.) C. Ag. *Journal of Experimental Botany*, 31, 1521–1537.

Reed, R. H., Collins, J. C. C. & Russell, G. (1980b). The effects of salinity upon galactosyl-glycerol content and concentration of the marine red alga *Porphyra purpurea* (Roth.) C. Ag. *Journal of Experimental Botany*, 31, 1539–1554.

Reed, R. H., Collins, J. C. C. & Russell, G. (1981). The effects of salinity upon ion content and ion transport of the marine red alga *Porphyra purpurea* (Roth.) C. Ag. *Journal of Experimental Botany*, 32, 347–367.

Reed, R. H., Richardson, D. L., Warr, S. C. & Stewart, W. D. P. (1984). Carbohydrate accumulation and osmotic stress in Cyanobacteria. *Journal of General Microbiology*, 103, 1–4.

Reed, R. H. & Russell, G. (1979). Adaptation to salinity stress in populations of *Enteromorpha intestinalis* (L.) Link. *Estuarine and Coastal Marine Science*, 8, 251–258.

Reed, R. H. & Stewart, W. D. P. (1983). Physiological responses of *Rivularia atra* to salinity: osmotic adjustment in hyposaline media. *New Phytologist*, 95, 595–603.

Ried, A. (1969). Physiologische Aspekte der Vertikalzonierung von Algen des marinen Litorals. *Bericht der Deutschen Botanischen Gesellschaft*, 82, 127–141.

Riley, J. P. & Chester, R. (1971). *Introduction to Marine Chemistry*. Academic Press, London, 465pp.

Ritchie, R. J. & Larkum, A. W. D. (1982a). Cation exchange properties of the cell walls of *Enteromorpha intestinalis* (L.) Link (Ulvales, Chlorophyta). *Journal of Experimental Botany*, 33, 125–139.

Ritchie, R. J. & Larkum, A. W. D. (1982b). Ion exchanges fluxes of the cell walls of *Enteromorpha intestinalis* (L.) Link (Ulvales, Chlorophyta). *Journal of Experimental Botany*, 33, 140–153.

Russell, G. (1985a). Some anatomical and physiological differences in *Chorda filum* from coastal waters of Finland and Great Britain. *Journal of the Marine Biological Association of the United Kingdom*, 65, 343–349.

Russell, G. (1985b). Recent evolutionary changes in the algae of the Baltic Sea. *British Phycological Journal*, 20, 87–104.

Russell, G. & Bolton, J. J. (1975). Euryhaline ecotypes of *Ectocarpus siliculosus* (Dillw.) Lyngb. *Estuarine and Coastal Marine Science*, 3, 91–94.

Schonbeck, M. & Norton, T. A. (1978). Factors controlling the upper limits of fucoid algae on the shore. *Journal of Experimental Marine Biology and Ecology*, 31, 303–313.

Schonbeck, M. & Norton, T. A. (1979). An investigation of drought avoidance in intertidal fucoid algae. *Botanica Marina*, 22, 133–144.

Schramm, W. (1968). Okologiech-physiologische Untersuchungen zur Austrocknungs—und Temperaturresistenz an *Fucus vesiculosus* L. der Westlichen Ostee. *International Revue der Gesamten Hydrobiologie under Hydrographie*, 53 469–510.

Segestrale, S. G. (1959). Brackish water classification, a historical survey. *Archivio di Oceanografia e Limnologia*, 11 (suppl.), 7–33.

West, J. A. (1972). Environmental regulation of reproduction in *Rhodochorton purpureum*. *Contributions to the Systematics of Benthic Marine Algae of the North Pacific* (Ed. by I. A. Abbott & M. Kurogi), pp. 212–230. Japanese Society of Phycology. Kobe, Japan.

Wiencke, C. & Läuchli, A. (1983). Tonoplast fine structure and osmotic regulation in *Porphyra umbilicalis*. *Planta*, 159, 342–346.

Wilkinson, M. (1980). Estuarine benthic algae and their environment: a review. *The Shore Environment* (Ed. by J. H. Price, D. E. G. Irvine & W. F. Farnham), Systematics Association Special Volume, 17a, pp. 425–486. Academic Press, London.

Wyn Jones, R. G., Store, P., Leigh, R. A., Ahmad, N. & Pollard, A. (1977). A hypothesis on cytoplasmic osmoregulation. *Regulation of Cell Membrane Activities in Plants* (Ed. by Marre & C. Cifferi). pp. 121–136. Elsevier, Amsterdam.

Yarish, C. & Edwards, P. (1982). A field and cultural investigation of the horizontal and seasonal distribution of estuarine red algae of New Jersey. *Phycologia*, **21**, 112–124.

Yarish, C., Edwards, P. & Casey, S. (1979). A culture study of salinity responses in ecotypes of two estuarine red algae. *Journal of Phycology*, **15**, 341–346.

Yarish, C., Edwards, P. & Casey, S. (1980). The effects of salinity, calcium and potassium variations on the growth of two estuarine red algae. *Journal of Experimental Marine Biology and Ecology*, **47**, 235–249.

Young, A. J., Collins, J. C. & Russell, G. (1985). Ultrastructural characterisation of taxa in the genus *Enteromorpha*. *The Systematics of the Green Algae* (Ed. by D. E. G. Irvine & D. W. John). Systematics Association Special Volume, 23, pp. 343–351. Academic Press, London.

Zanefeld, J.S. (1937). The littoral zonation of some Fucaceae in relation to desiccation. *Journal of Ecology*, **25**, 431–468.

Zanefeld, J. S. (1969). Factors controlling the delimitation of littoral benthic marine algal zonation. *American Zoologist*, **9**, 367–391.

Zenkevitch, L. (1959). The classification of brackish-water basins, as exemplified by the seas of the U.S.S.R. *Archivio di Oceanografia e Limnologia*, 11 (suppl.), 53–62.

Zimmermann, U. (1978). Physics of turgor and osmoregulation. *Annual Review of Plant Physiology*, **29**, 121–148.

Ethylene and growth control in amphibious plants

I. RIDGE

Biology Department, The Open University, Milton Keynes, MK7 6AA, U.K.

SUMMARY

1 Many amphibious plants show increased elongation of petioles or internodes during submergence and this paper considers the nature of this response and the evidence that ethylene gas may initiate and control it.

2 Species can be divided into two broad groups according to the size and speed of their elongation response when leaves are submerged: a large, rapid growth response is characteristic of species that occur mainly as floating-leaved aquatics; marginal species and those growing in marshes or seasonally flooded grassland typically show a smaller, delayed response. It is argued that different responses reflect the balance between opposing selective forces for different species.

3 Ethylene treatment in air reproduces wholly or partially both rapid and delayed submergence responses. For rapid responses, ethylene levels increase during submergence and ethylene appears to be a key regulator: other factors that increase the rate of elongation usually do so only in the presence of ethylene. The only known exceptions are certain non-photosynthetic seedling and sprout tissues where anoxia, and hence inhibition of ethylene synthesis, are likely during submergence; here, growth promotion can occur in the absence of ethylene and is influenced by the levels of oxygen and CO_2.

4 Since rapid growth responses to submergence have evolved that are independent of ethylene, why is ethylene so commonly involved? For photosynthetic organs, control systems based on ethylene may have been selected because accumulation, movement and action of the gas have appropriate kinetics and it is also a reliable regulator in the underwater environment.

INTRODUCTION

The amphibious habit has evolved independently many times in the tracheophytes, and amphibious plants are consequently a diverse and largely artificial group. For the purposes of this review they are defined in terms of: (1) a capacity for shoot growth, at least for a limited period, through water and air; and (2) the necessity, for longer term survival and reproduction in water, that some photosynthetic organs re-gain contact with the aerial environment. This broad definition includes species that occur mainly as floating-leaved plants but may grow on mud and species (such as *Ranunculus repens* or *Plantago major*) that are chiefly terrestrial but able to grow during periods of submergence. Excluded, however, are the most specialized hydrophytes and terrestrial species which do not grow if submerged.

Since the functional requirements for growth in air and water are radically different, a universal characteristic of amphibious plants is their developmental plasticity in response to a switch between aerial and aquatic environments. Submergence of shoots is usually the critical factor that initiates changes and here I assess the role of an endogenous regulator, ethylene gas, in controlling these changes, particularly increased elongation growth of stems and petioles. Wide-ranging reviews of the role of ethylene in submergence responses are also provided by Osborne (1984) and Jackson (1985).

CHANGES THAT OCCUR IN SUBMERGED SHOOTS

The dramatic changes in anatomy, morphology and growth pattern that occur when shoots of amphibious plants are completely submerged have attracted the attention of physiologists for over a hundred years. The book by Arber (1920) gives an excellent review of the early studies and that by Sculthorpe (1967) provides a modern perspective. Table 1 lists the main types of changes that have been observed, although not all may occur in a particular species, and many appear superficially and are often described as 'adaptive'.

Changes induced by environmental conditions do not necessarily serve any useful purpose, however, and the adaptive significance (if any) of (1) and (4) in Table 1 is not clear. It will be argued later that some changes induced by submergence severely reduce fitness in the aerial environment and their overall effect on relative fitness depends on factors such as frequency and duration of flooding. The significance of these changes certainly varies between species, and broad generalizations about their 'value' are not possible. However, there is clear evidence that shoot submergence imposes a severe stress on amphibious plants, partly because of the reduced aeration of roots and shoots and partly because of impaired photosynthesis. In 2-week old seedlings of *Ranunculus sceleratus*, for example, dry matter production was reduced by 83% during 1 week of submergence and there was usually 100% mortality of seedlings in the field during

TABLE 1. Changes that may occur when shoots of amphibious plants are submerged

1 Hypertrophy (mainly in stems of woody species)
2 Increased formation of adventitious roots
3 Increased formation of aerenchyma tissue
4 Reduced lignification
5 Changes in leaf shape
6 Anatomical changes: formation of epidermal chloroplasts
                         loss of palisade mesophyll
                         increase in cortex width
                         reduction in cuticle thickness
                         reduction in hairiness
7 Increased elongation of internodes, petioles or flower stems
8 Altered orientation of leaves and stems (usually from prostrate to vertical)

prolonged periods of flooding (Smith 1986). Thus changes that result in improved aeration and photosynthesis in water are likely to be of adaptive significance and there is some evidence to support this view.

The formation of adventitious roots and aerenchyma have been shown to promote survival under flooded conditions for some species (Gill 1970; Armstrong 1979; Gomes & Kozlowski 1980). Similarly, restoration of leaves to the water surface, whether by increased elongation or by flotation of aerenchyma (as in the water hyacinth, *Eichhornia crassipes* (Pieterse, Aris & Butter 1976)), also appear to improve survival. Williams & Barber (1961) have argued that the functional significance of aerenchyma in submerged conditions may vary for different organs, but the function usually ascribed to it—providing a system for gas storage and improved movement of gases between roots and shoots—may often be incidental to the need for structures that are mechanically strong and yet have a low metabolic requirement. Arber (1920, p. 194) goes even further and suggests that any useful purpose served by secondary aerenchyma in water plants is fortuitous; she regards it simply as a change directly induced by environmental conditions. So caution must be exercised before assigning any function, useful or otherwise, to submergence-induced change, particularly if the long-term effects on survival and reproduction in a fluctuating environment have not been investigated.

Ethylene has been implicated in all the changes listed in Table 1, in so far as exposure of plants in air to an ethylene-enriched atmosphere reproduces the effects of submergence. However, this does not constitute proof that ethylene plays any role either in the initiation or control of submergence-induced changes, and only for two responses, aerenchyma formation and enhanced elongation, has the role of ethylene been examined critically. The first of these has been reviewed by Jackson (1985) and only the second is considered here in detail. Elongation responses to submergence represent an exceptionally clear example of the environmental control of plant growth and there is a reasonable expectation that the interactions between ethylene and other factors that influence growth can be elucidated.

THE NATURE OF ELONGATION RESPONSES

An increase in the length of internodes, petioles or peduncles is one of the most commonly observed effects of submergence in amphibious plants. The response is often described as 'depth accommodation' because, provided the water is not too deep, growth continues until leaves or buds reach the water surface. The definition of 'not too deep' varies with species: it may be 5 m or more in some floating-leaved plants but only 20 cm for flood-tolerant but mainly terrestrial species. Growth may continue after leaves have reached the surface but usually at a much lower rate (Ridge & Amarasinghe 1984).

The extent of depth accommodation varies not only between species, however, but also at different stages of the life cycle and may be small or even absent in young seedlings. Thus we have found very little increase in petiole elongation for

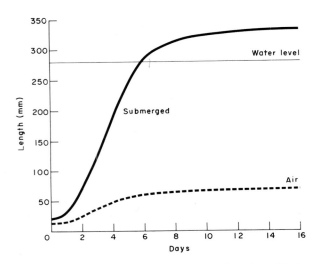

FIG. 1. Petiole growth in water and in air for leaves of *Nymphoides peltata*. Plants were grown in 5 cm pots in the greenhouse either in water (28 cm deep) or in air. For each condition, growth of the youngest leaf was monitored for six plants and average initial and final petiole lengths were 1·9 ± 0·4 and 7·0 ± 0·2 (air), 3·6 ± 0·4 and 33·8 ± 1·1 (water).

young submerged seedlings of the fringed water lily, *Nymphoides peltata*, which usually grows as a floating-leaved plant. Since seedling mortality is high if leaves do not reach air, this species is unlikely to recruit from seed unless germination occurs on damp mud or in very shallow water. This may be generally true for amphibious plants, many of which have germination requirements (light and diurnal temperature variation) that permit germination only on damp mud (Thompson & Grime 1983). With a few notable exceptions, which include rice and species of *Nymphaea* and *Nuphar* that arguably fall outside the definition of amphibious plants (Arber 1920; Funke & Bartels 1937), increased elongation under water is more pronounced for well-established plants than for seedlings, and this may be ecologically significant.

Figure 1 illustrates two key features of a typical submergence response: there is an increase in final length, attained by increasing the rate of elongation and not by extending the duration of growth. Species vary greatly, however, not only in the height to which submerged organs grow but also in the time taken to respond. The latter source of variation is only partly explained by differences in growth rate. Two other factors, changing responsiveness of petioles or internodes at different stages of ontogeny and the rate at which new leaves or internodes are initiated, also affect the outcome.

Figure 2 shows the relationship between the time taken to reach the surface (or attain maximum length) and total height relative to air controls for submerged petioles of several species. The species fall into two main groups: those showing a relatively large, rapid response (group A) and those showing a relatively small, delayed response (Group B). Species 6 (*Plantago major*), where the proportionate

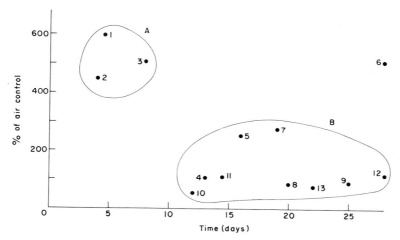

FIG. 2. The minimum time required for petioles to attain their maximum length (group B) or the surface of the water (group A) when growing submerged. Numbers refer to different species which are listed in Table 3. Whole plants were grown in pots in the greenhouse during summer months and submerged in tapwater. Results are expressed as the percentage increase in length above petioles of the same age from control plants grown in air.

increase in petiole length is large but delayed, may be regarded as a special case because of its habit. In air it grows as a flat rosette with short petioles (usually 1–3 cm long), but new leaves developing in submerged plants assume an upright position with petioles up to 13–16 cm long and smaller laminae; the ratio of lamina to petiole length is 0·44 for these leaves compared with 2·2 for leaves of the same age in non-submerged controls.

Two species in Group A, *Regnellidium diphyllum* and *Nymphoides peltata*, grow commonly as floating-leaved plants and both show a remarkably similar response to submergence, although the former is a tropical South American fern (Marsiliaceae) and the latter a temperate member of the Menyanthaceae. The pattern of response for *Nymphoides* was described by Ridge & Amarasinghe (1984) and that for *Regnellidium* is shown in Table 2. All petioles, even those of fully mature fronds, show rapid elongation when laminae are submerged. However, growth capacity is limited in older leaves, and in deep water they stop growing before reaching the water surface; their elongation is due mainly or entirely to an increase in cell length. Leaves that develop during submergence or at an early stage of development (5b in Table 2) grow up to the surface, and increases in both cell length and cell number contribute to these very large elongation responses (Ridge & Amarasinghe 1984).

Other species which on the basis of published data can be assigned to group A are listed in Table 3. For young (detached) petioles of *Ranunculus sceleratus* and *Hydrocharis* and for internodes of *Callitriche platycarpa*, rapid growth during submergence involved only an increase in cell expansion (Musgrave, Jackson & Ling 1972; Cookson & Osborne 1978). Both cell size and cell number increased in submerged internodes of floating rice, however, although internode number was

TABLE 2. Petiole elongation when rooted rhizomes of *Regnellidium diphyllum* were submerged to 40 cm or maintained in air for 5 days

Treatment	Petiole length (cm)							
	1	2	3	4	5		6	7
	Oldest frond						New fronds	
Air								
Initial length	10·4	9·1	6·2	4·5	2·0		—	—
Final length	11·1	10·4	9·6	8·0	7·1		7·2	6·7
Submerged								
Initial length	9·7	9·9	7·0	4·4	2·5	1·4	—	—
Final length	20·4	23·7	23·0	24·9	30·1	47·4	48·7	50·8

Six rhizomes, each with five fronds, were used per treatment and the growth of existing and new fronds (developing from apical and lateral buds) was monitored. Results are the mean lengths of petioles ($n = 6$–12, except for frond 5 in the submerged treatment, which was subdivided into two size classes, $n = 3$). Standard errors for each measurement did not exceed 10% of the mean.

unaffected (Métraux & Kende 1984). Thus for these predominantly aquatic species there is a variable contribution of cell expansion and cell division to the submergence response but, in general, enhanced cell expansion seems to be the more important factor.

The species in group B (Figure 2 and Table 3) grow typically as emergents in shallow water, in marshes, or in areas subject to brief shallow flooding. They differ from group A species in that leaves developing in air rapidly lose the capacity to respond to submergence (or applied ethylene). Only the youngest leaves show a response and maximum lengths are attained only when new leaves develop during submergence (Ridge 1985). This lack of response in existing leaves and the sometimes lengthy period before new leaves develop explains the delay in the submergence response of these species. The nature of the response has been examined in detail for only one species, *Ranunculus repens*, and here it was found that an increase in cell number could account for 80% of the observed increase in petiole length (Ridge 1985). Increased cell division appears, therefore, to be more important for group B than for group A species.

It must be pointed out, however, that some amphibious plants show no depth accommodation at all. Mitchell (1976) described amphibious species of *Polygonum* whose internode growth was unaffected by submergence to 20 cm and we could find no effect of submergence or ethylene treatment on internode growth of *Phalaris arundinacea* and *Alopecurus geniculatus*. When submerged, rooted cuttings of these species simply continued to grow at the same rate as in air. Yet another strategy is found in *Mentha aquatica*, a species in which we found no increase in growth rate when erect shoots were submerged or exposed to ethylene in the laboratory. In the field, however, erect shoots do show depth accommodation, by an increase in internode length in water up to 15 cm deep and by an increase in internode number (without further length increase) in deeper water (I. Ridge, unpublished).

Table 3. Some amphibious or aquatic plants that show elongation responses to submergence

Species	Organ studied	Reference
Group A: Large, rapid response		
* Callitriche platycarpa Kock	internodes	Musgrave et al. (1972)
Callitriche stagnalis Scop.	internodes	McComb (1965)
* Hydrocharis morsus-ranae L.	petiole	Cookson & Osborne (1978)
Nymphaea alba L.	petiole	Funke & Bartels (1937)
*(1) Nymphoides peltata (S. G. Gmelin)	petiole	Funke & Bartels (1937)
O. Kuntze (=Limnanthemum nymphoides)		Ridge & Amarasinghe (1984)
* Oryza sativa L.	coleoptile	Ku et al. (1979)
	mesocotyl	Suge (1971)
* O. sativa floating variety	internodes	Métraux & Kende (1983)
* Potamogeton distinctus A. Benn	winter bud	Suge & Kusanagi (1975)
*(2) Ranunculus sceleratus L.	petiole	Musgrave & Walters (1973)
		Cookson & Osborne (1978)
	internodes	Samarakoon & Horton (1981)
*(3) Regnellidium diphyllum Lindm.	petiole	Musgrave & Walters (1974)
* Sagittaria pygmaea Miq.	1st internodes	Suge & Kusanagi (1975)
Sagittaria sagittifolia L.	petiole	Funke & Bartels (1937)
Victoria amazonica (Poepp.)	petiole	Funke & Bartels (1937)
J. C. Smedley (=Victoria regia)		
Group B: Medium or small, delayed response		
(12)Alisma plantago-aquatica L.	petiole	Funke & Bartels (1937)
*(13)Apium nodiflorum (L.) Lag.	petiole	
* Berula erecta (Hudson) Coville	petiole	
*(5) Caltha palustris L.	petiole	
* Epilobium hirsutum L.	young internodes	
*(9) Geum rivale L.	petiole	
Menyanthes trifoliata L.	petiole	Funke & Bartels (1937)
*(8) Oenanthe crocata L.	petiole	
*(7) Oenanthe fistulosa L.	petiole	
*(6) Plantagago major ssp. major L.	petiole	
Polygonum amphibium L.	internodes	Mitchell (1976)
*(11)Ranunculus flammula	petiole	
ssp. flammula L.	internodes	
*(10)Ranunculus lingua L.	petiole	
*(4) Ranunculus repens L.	petiole	Ridge (1985)
* Nasturtium officinale R. Br.	internodes	Cookson (1976)

*Species in which ethylene has been implicated in the response.
Numbers relate to species in Fig. 2.
Where no reference is given, the data are from I. Ridge and I. Amarasinghe (unpublished).

Clearly, a variety of growth strategies have evolved in amphibious plants and unless an increase in the rate of elongation occurs, there is no reason to implicate ethylene as a controlling factor. It may be significant that applied ethylene does not inhibit growth in any amphibious species tested, whereas it is often a powerful growth inhibitor in purely terrestrial species. For species marked with an asterisk in Table 3, growth rate does increase during submergence and ethylene has been implicated. However, in order to assess the role of ethylene more critically, the questions that need to be considered for submerged shoots are:

(i) Is there an increase in ethylene level (or a change in sensitivity) sufficient to account for the observed growth response and with appropriate kinetics?

(ii) By what mechanism does ethylene enhance growth in amphibious plants, bearing in mind that cell expansion and cell division are involved to different extents in different species?

(iii) What other factors beside ethylene influence growth and how do they interact with ethylene?

These questions are discussed next and the possible significance of variation in growth responses and of control by ethylene are then considered.

ETHYLENE LEVELS

Aerial or floating shoots of amphibious plants contain only low concentrations of ethylene. If the gas is indeed a regulator of growth during submergence, then it must be shown that before or coincident with the increase in elongation rate there is a rise in ethylene to growth-promoting levels and/or an increase in sensitivity to endogneous ethylene. This question has been studied only in the rapidly responding species of group A (Table 3).

Musgrave, Jackson & Ling (1972) were the first to demonstrate that ethylene levels increase during submergence. They collected bubbles of oxygen-rich gas given off when apical rosettes of *Callitriche platycarpa* were submerged in the light, and showed that the bubbles contained growth-promoting levels of ethylene. This was later shown, using the same technique, for leaves of *Ranunculus sceleratus* and fronds of *Regnellidium diphyllum* (Musgrave & Walters 1973, 1974) and for leaves of *Hydrocharis morsus-ranae* and apical rosettes of *Nasturtium officinale* (Cookson 1976). Musgrave, Jackson & Ling (1972) suggested that because ethylene diffuses some 10 000 times more slowly through water than through air, entrapment of the gas within submerged tissues could explain the observed rise in levels.

This delightfully simple hypothesis has been widely accepted but it is underpinned by several assumptions that merit scrutiny. The bubbles of photosynthetic gas in which ethylene was detected derive mainly from leaf blades and, for all the species mentioned above, laminae are the main site of ethylene production. The rate of ethylene release from apical petiole segments of *Regnellidium* and *Nymphoides*, for example, was less than 1% of that from laminae (Malone 1983; I. Ridge, unpublished). Thus if the entrapment hypothesis holds, the kinetics of ethylene accumulation and movement, by diffusion or mass flow (Dacey 1981), should match the kinetics of the submergence response.

Using an auxanometer, Funke & Bartels (1937) showed that the growth rate of the petiole in *Nymphoides peltata* increased 100–110 min after leaf submergence. For fronds of *Regnellidium* there was a longer lag (up to 200 min) between submergence and the increase in extension growth (M. Malone, unpublished) and ethylene appears to accumulate and move sufficiently rapidly to account for these growth kinetics. Thus high concentrations of ethylene (5 Pa, 50 μl l^{-1}) diffuse via laminae (in air) and promote elongation in the petiole (in water) within 10–15 min (Musgrave & Walters 1974). And Malone (1983) showed that isolated petiole segments of *Regnellidium* respond to saturating levels of ethylene within 20 min and also that ethylene accumulates rapidly in submerged fronds. He used gentle vacuum extraction to remove gases from submerged fronds (leaves plus 1 cm of petiole) and found a linear increase in ethylene concentration to 0·1 Pa (1 μl l^{-1}) within 40 min. Since petioles of *Regnellidium* respond to 0·3 μl l^{-1} ethylene and growth saturates at 10 μl l^{-1}, the kinetics of this system adequately support the entrapment hypothesis.

For young leaves of *Regnellidium* and *Nymphoides*, submergence does not appear to influence ethylene synthesis and after 1 or 2 days submergence, the rate of ethylene release (in air) was the same as for controls kept continuously in air (I. Ridge, unpublished). However, ethylene released from mature leaves of these species declined markedly after 1 day of submergence and so did the growth rate. This rapid decline in growth rate did not occur if mature leaves were exposed to exogenous ethylene, so it appears that the pattern of petiole growth when older leaves are submerged is influenced by the rate of ethylene synthesis in leaf blades.

There is a further implication of the entrapment hypothesis. If, without any increase in the rate of synthesis, ethylene accumulates sufficiently rapidly in submerged tissues to explain the observed change in growth rate, then ethylene not only influences growth but also acts as a detection system which signals to the shoot that submergence has occurred. For at least one species that responds rapidly to submergence, however, entrapment alone does not explain the rise in ethylene in elongating organs. The rate of ethylene production increased fourfold in submerged internodes of floating rice, and both entrapment and increased synthesis contributes to the rise in ethylene concentration within internodal lacunae (Raskin & Kende 1984a).

For floating rice, therefore, it is important to know what triggers the increase in ethylene production and what change acts as an indicator of submergence. There is no evidence that ethylene production is autocatalytic in this system (as it is in some plant tissues) but Raskin & Kende (1984a) have shown that the oxygen level declines rapidly within submerged internodes and that hypoxia actually stimulates ethylene synthesis. This is surprising in view of the strict requirement for oxygen of the biosynthetic pathway from methionine to ethylene, which operates in the internodes of floating rice (Métraux & Kende 1983). However, stimulation by hypoxia has also been observed in roots of maize and barley (Jackson 1982; Jackson *et al.* 1984). Thus the detection of submergence and the rise in ethylene levels may depend on the combined effects of gas entrapment and hypoxia.

A preliminary report from Rose-John & Kende (1984) suggests, however, that this explanation may be too simple. They found that when excised internodes of floating rice were submerged, the increase in growth rate during the first 6 h was not changed by supplying either a substrate for ethylene synthesis (ACC, 1-aminocyclopropane-1-carboxylic acid) or an inhibitor of synthesis (AOA, aminooxyacetic acid). Both these substances affected the growth rate after 6 h. These authors suggested that the first effect of submergence in this tissue was a sensitization to ethylene and an induction of ethylene synthesis. This means that ethylene *per se* would not be involved in the 'sensing' of submergence and it raises the possibility that in some amphibious plants, sensitization to ethylene rather than a rise in ethylene level could be the critical factor initiating elongation responses. How this rapid sensitization occurs is not yet known.

To summarize: in amphibious species that show a rapid elongation response to submergence, all those studied to date show an increase in internal ethylene to growth-promoting levels. In some species the increase may occur solely through gas entrapment and in others, notably floating rice, through a combination of entrapment and increased ethylene production. For *Regnellidium*, where entrapment alone operates, the kinetics of ethylene accumulation and movement match the kinetics of growth. Here, therefore, ethylene appears to provide both the submergence signal initiating rapid growth, and the means for sustaining this growth over at least the first few hours of submergence. The situation is more complex for submerged internodes of floating rice, where hypoxia stimulates ethylene synthesis and where the first stage leading to rapid elongation may be sensitization to ethylene.

There have been no measurements of ethylene levels in the species that show a delayed response to submergence (Group B, Fig. 2), and this is an obvious gap in our knowledge. Given the remarkable similarity between the effects of ethylene and those of submergence on group B species (Ridge 1985), it seems probable that ethylene is involved in submergence responses. Furthermore, the delayed response is not due to a delay before internal ethylene levels rise, because it occurs also during ethylene treatment and must, therefore, reflect differences in sensitivity to ethylene. The subject clearly requires further work.

THE MECHANISM OF ETHYLENE ACTION

Two kinds of submergence response were identified earlier: those where elongation depends on an increase in cell length and those where both cell length and cell number increase. Detailed studies of ethylene action have been carried out only for the former and the mechanism(s) by which ethylene increases the rate of cell expansion are discussed first. Since this work has been recently reviewed (Osborne 1984; Jackson 1985), it is considered here only briefly.

Cell expansion

In one of the first studies of ethylene action, when Musgrave, Jackson & Ling (1972) studied internode elongation in apical rosettes of *Callitriche platycarpa*, several key points were established that have not been seriously challenged since. These authors demonstrated that ethylene alone does not promote cell expansion in amphibious plants: at least one other endogenous growth regulator must be present. For submerged internodes of *Callitriche* growing over a 3-day period, this regulator is a gibberellin, GA_3, and Samarakoon & Horton (1981) suggested that an ethylene–gibberellin interaction may occur in flowering internodes of *Ranunculus sceleratus*. It has also been reported for internodes of floating rice (Raskin & Kende 1984b), although cell division contributes to elongation in this system. Ethylene–gibberellin interactions may, therefore, be a general feature of internode elongation in amphibious plants.

For petioles, however, the initial, rapid growth response to submergence requires the presence of auxin, at least for *Regnellidium diphyllum*, *Ranunculus sceleratus* and *Nymphoides peltata* (Walters & Osborne 1979; Horton & Samarakoon 1982; Malone & Ridge 1983). In every case, if the putative source of this 'potentiating hormone' (the lamina) is removed, or if inhibitors of its action or transport are supplied, ethylene fails to promote cell elongation.

This neat division into petioles dependent on auxin–ethylene interactions and stem internodes dependent on gibberellin–ethylene interactions may, however, be an oversimplification. Two of the internode systems (*Callitriche* and *R. sceleratus*) also respond to applied auxin (Horton & Samarakoon 1982; Osborne 1984) and two of the petiole systems (*Ranunculus* and *Nymphoides*) show a long-term response to GA_3 which, for *Nymphoides*, is markedly enhanced by ethylene (I. Ridge, unpublished). Using a linear transducer to monitor growth in *Nymphoides* segments, Malone (1983) found that elongation rate did not increase for at least 10 h in the presence of ethylene and GA_3 although it increased within 20 min for ethylene and IAA. It may be, therefore, that the *initiation* of rapid growth requires the presence of auxin but longer term maintenance of a high growth rate requires the presence of a gibberellin. The essential point is that another regulator must be present before ethylene influences cell expansion.

In all plants studied, this second regulator promotes cell elongation when supplied to isolated organs or segments but this does not occur through an increase in ethylene synthesis (Musgrave, Jackson & Ling 1972; Cookson & Osborne 1978; Malone 1983). Nor does ethylene appear to affect the level of auxin or gibberellin: Walters & Osborne (1979) measured auxin levels directly in fronds of *Regnellidium* and found no change after ethylene pretreatment and other workers reached this conclusion because it was not possible to saturate the growth response by supplying exogenous auxin or GA_3 at higher concentrations (Malone & Ridge 1983; Musgrave, Jackson & Ling 1972). Rather, each regulator has an independent mode of action and one way of describing the interactions is to say that ethylene sensitizes the tissue to auxin or gibberellin.

There have been some attempts to identify the point at which auxin and ethylene exert their individual effects. If the rate of cell expansion increases, then either or both of two parameters must increase (reviewed by Taiz 1984): cell wall extensibility or the driving force for expansion—a function of turgor pressure (and hence internal osmotic potential) and of minimum wall yield stress. In *Regnellidium* and *Nymphoides* both auxin and ethylene increase wall extensibility (Cookson & Osborne 1979; I. Ridge & D. J. Osborne unpublished) and, for the latter species, neither affects the size of the driving force (M. Malone & I. Ridge, unpublished). Furthermore, wall extensibility in *Nymphoides* petioles is regulated by an 'acid growth' mechanism and the rate of cell expansion is proportional to the degree of wall acidification: ethylene causes a decrease in wall pH provided that auxin is present, and thus the auxin-dependent stage in ethylene action must lie at or before the level of the proton pump(s) that acts to acidify cell walls (Malone & Ridge 1983). *Regnellidium* is something of an enigma because, uniquely among plants showing a rapid growth response to auxin, cell expansion appears to be unaffected by wall pH (M. Malone & I. Ridge unpublished).

Very little is known about the interaction between ethylene and gibberellins in stem internodes. Musgrave, Jackson & Ling (1972) suggested that ethylene might affect wall extensibility and gibberellic acid the osmotic potential, and hence the driving force for expansion. The net result of such an interaction would be that each regulator increased the sensitivity of the tissue to the other. This remains an interesting possibility but, if true, it raises the question as to whether the effect of ethylene here is also auxin dependent, as in petioles. It is tempting to speculate that an auxin–ethylene interaction is fundamental to all rapid elongation responses in amphibious plants and may be the only (or major) control mechanism operating in the first few hours of growth during submergence. After this time, and provided that the rate of ethylene synthesis were sustained, the driving force for cell expansion could become critical and a range of factors, including gibberellins, that influence the concentration of osmotically active solutes in cells would influence growth rate. However, this hypothesis is not consistent with the finding that ethylene does not stimulate internode growth in the presence of inhibitors of gibberellin synthesis (Musgrave, Jackson & Ling 1972; Raskin & Kende 1984b). Until more is known about the interaction between gibberellins and auxin and whether gibberellins can independently increase wall extensibility, the mechanism of ethylene–gibberellin interactions in amphibious plants remains obscure.

Cell division

When elongation responses to submergence, whether rapid (Group A) or delayed (Group B), involve an increase in cell number, ethylene treatment in moist air reproduces this effect (Ridge & Amarasinghe 1984; Ridge 1985). The mechanism of ethylene action and whether it is dependent on the presence of other growth regulators or on other factors are unknown. It is possible, but to this author seems unlikely, that ethylene exerts a specific effect on mitosis quite separate from its effect on cell expansion. A more attractive hypothesis is that the two are linked. In young petioles of *R. repens* and *Nymphoides*, cell division occurs over the apical 2–4 cm and is not confined to a highly localized meristematic zone and, for *R. repens*, ethylene or submergence increase the rate of petiole growth early in development without any increase in cell size relative to air controls (Ridge 1985). In this situation, therefore, the rate of cell division relates directly to the rate of cell expansion (see Silk 1984). Thus if the rate of cell expansion increases and if cells divide after attaining a critical size, then the rate of cell division must also increase. A link between the rate of cell expansion and the duration of the cell cycle is, therefore, possible during the early stages of organ development. On this basis, ethylene action on cell division would not be specific: any factor that increased the rate of cell expansion during the period of the development when cells were capable of dividing would, perforce, increase cell number.

OTHER FACTORS INFLUENCING ELONGATION RESPONSES

Plant growth follows a particular pattern as a result of innumerable interactions between internal and external factors; one reason why the growth responses to submergence are of particular interest to physiologists is that internal changes can be related to a definite environmental change—the switch from air to water. Rapid elongation responses have so far been linked with one internal change, the rise in ethylene levels. But other internal changes occur, some of which have been shown to affect shoot elongation, and the aquatic medium can also have direct effects on cell expansion. The significance of these factors in relation to submergence responses is considered next.

In addition to ethylene, five factors have so far been identified which promote extension in submerged tissues; their action is sometimes independent but, more commonly, they act only in the presence of ethylene. These factors are: increased concentration of CO_2, decreased concentration of O_2 (between complete anoxia and atmospheric levels), light, tension arising from the buoyancy of submerged organs, and the availability of water to expanding cells. Species and organs vary greatly in their response to these factors and it will be argued later that this could be of ecological relevance.

CO_2 and O_2

These factors have usually been studied together and are therefore considered together here. The concentrations of CO_2 and O_2 in submerged organs will depend on the concentrations in surrounding water but, more significantly, on the rates of respiration and photosynthesis. If photosynthesis occurs, and especially if there is gas storage within lacunae, then this can be expected to buffer roots and stems against increase in CO_2 and decreases O_2 level. For organs where there is little or no photosynthesis, or in extremely turbid water, changes in the content of these gases would be more pronounced and complete anoxia is possible. It is against this background that the significance of O_2 and CO_2 levels should be considered in relation to rapid growth. In systems where they have an effect, an increase in CO_2 and a decrease in O_2 content promote elongation.

When petioles of *Regnellidium* were exposed to 5% CO_2 in the presence or absence of ethylene, growth rate was unaffected. In the light, moreover, oxygen accumulated in submerged fronds, as shown by a rise buoyancy and the evolution of gas bubbles (Malone 1983), and growth rate was the same in light and darkness. Similarly, neither anoxia nor CO_2 level influenced internode elongation in rosettes of *Callitriche stagnalis* (McComb 1965). However, Funke & Bartels (1937) reported that petiole elongation was reduced in submerged leaves of *Nymphoides* when air was bubbled through the water, suggesting that hypoxia may occur in still water and may promote elongation. On the basis of this limited evidence, the most that can be said at present is that CO_2 level and hypoxia have rarely been shown to influence rapid petiole growth when mature leaves are submerged and photosynthesis occurs. No generalization is possible for internode elongation because the situation in floating rice is quite different to that in *Callitriche*.

In submerged internodes of floating rice, Raskin & Kende (1984a) showed that oxygen content decreased and CO_2 content increased coincident with the rise in ethylene and both these changes promoted elongation. As described earlier, hypoxia stimulates ethylene synthesis and it is by this mechanism that it influences rapid elongation, rather than by a direct effect on growth. Carbon dioxide had no direct effect on growth either, but in some way enhanced the action of ethylene. In this system, therefore, ethylene is the key regulator of rapid growth during submergence and changes in other gases appear only to increase its effectiveness.

The situation is different again for non-photosynthetic organs. In the coleoptile and mesocotyl of rice (reviewed by Jackson (1985)) the key findings are that ethylene, hypoxia and CO_2 all stimulate elongation and all may act independently: coleoptiles elongate even under completely anoxic conditions. Neither hypoxia nor CO_2 affected ethylene synthesis but there was a weak synergistic interaction between ethylene and CO_2. In young sprouts of two paddy field weeds, *Sagittaria pygmaea* and *Potamogeton distinctus*, developing from perennating organs, both ethylene and CO_2 independently promoted elongation (Suge & Kusanagi 1975). However, there was a strong synergistic interaction between the two, and CO_2 appeared to act mainly by increasing the sensitivity of the tissue to ethylene. These authors did not investigate the effects of hypoxia.

The mechanisms by which CO_2 and hypoxia independently promote extension growth in these non-photosynthetic systems are not known although, in terrestrial plants, CO_2 appears to act mainly by acidifying cell walls and increasing their extensibility.

Light

On present evidence, it seems unlikely that changes in the quality or quantity of light initiate rapid growth responses after submergence. The submergence responses of non-photosynthetic organs described earlier were all observed in complete darkness and, for rice coleoptiles, neither red nor white light reduced markedly the growth promoting effects of ethylene (Suge, Katsura & Inada 1971; Raskin & Kende 1983). Similarly, fronds of *Regnellidium diphyllum* showed the same degree of petiole elongation over 24 h whether submerged in light or darkness (Malone 1983). In other species, however, the submergence response of

TABLE 4. Petiole elongation in detached leaves of *Nymphoides peltata* kept submerged in darkness or in continuous light (60 W m^{-2}) at 25 °C

Time (h)		Increase in length of marked 2 cm zones (cm)					% increase in length of apical 10 cm
		1 apical	2	3	4	5 basal	
24	Light	2·74 ±0·20	3·03 ±0·20	2·75 ±0·23	2·26 ±0·33	1·60 ±0·36	124·6 ±10·6
	Dark	1·39 ±0·23	1·59 ±0·25	1·49 ±0·26	1·34 ±0·25	0·99 ±0·22	67·3 ±10·7
48	Light	6·55 ±0·61	7·15 ±0·26	6·35 ±0·21	5·78 ±0·35	4·38 →0·40	300·8 ±8·8
	Dark	2·70 ±0·44	2·88 ±0·47	2·33 ±0·43	1·87 ±0·35	1·42 ±0·33	111·7 ±17·6

Young, floating leaves were removed from plants growing in tanks of water (30 cm deep) in a heated greenhouse. Petioles, trimmed to 12 cm and with 5×2 cm zones marked from the apex, were held submerged (by attaching a weight) in tanks containing 40 cm of distilled water in a controlled environment chamber. Results are mean values ($n = 8$) ± SE.

photosynthetic organs is clearly influenced by light over 24 h or more. Table 4 shows that petiole elongation in *Nymphoides peltata* is reduced by about half during a 24 h incubation in darkness and, compared with controls in continuous light, the inhibiting effect of darkness becomes progressively greater with time. Raskin & Kende (1984a) found a similar inhibitory effect of darkness on the submergence response of floating rice internodes over 3 days and showed also that the increase in endogenous ethylene levels was reduced by 90%, probably because of anoxia in the absence of photosynthetic O_2. However, internodes in the dark were also less sensitive to exogenous ethylene and this may reflect the importance of assimilates for osmoregulation and cell division.

If the leaf sheath enclosing each internode is removed, then internode growth is strongly inhibited by light (Raskin & Kende 1984b) but this effect is overcome completely in the intact system by the opacity of the sheath and the assimilates it supplies. Thus there is no evidence that elongation in submerged organs is influenced by a direct, photomorphogenetic effect and any effects of reduced light intensity in the underwater environment are more likely to result from reductions in the rate of photosynthesis. Such effects may well be significant for the delayed elongation responses characteristic of group B species, but this has not been investigated in detail.

Buoyant tension and the availability of water

The buoyancy of leaves, especially when there is extensive aerenchyma, means that the submerged stems and petioles of rooted plants may be subject to appreciable buoyant tension. This does not influence the submergence response for species in

TABLE 5. Petiole elongation for detached fronds of *Regnellidium diphyllum* after 3 days' treatment (data of Mugrave & Walters 1974)

Treatment	Growth increment* (cm ± S.E.)
In air[†]	1·5 ± 0·8
Submerged	14·9 ± 0·8
Submerged + covers[‡]	5·5 ± 0·7
Ethylene 100 μl l^{-1}	6·8 ± 0·5
Ethylene + 3 g weight	16·0 ± 1·0
In air + weight	0·9 ± 0·5

*Initial petiole length was 5·5 cm.

[†]All non-submerged fronds were suspended with leaflets in air and petioles entirely surrounded by water.

[‡]A perforated Perspex box was inverted over submerged fronds to relieve tension in the petiole due to buoyancy.

group B nor for species such as *Ranunculus sceleratus* and floating rice in group A. But for two group A species, *Regnellidium diphyllum* and *Nymphoides peltata*, buoyant tension may enhance substantially the growth of submerged petioles. Musgrave & Walters (1974) demonstrated the effect first in *Regnellidium* and their results (Table 5) indicate that accumulated ethylene accounts for only about 40% of the observed growth response during 3 days submergence; buoyant tension is apparently responsible for the remaining 60% but is effective only in the presence of ethylene.

Ridge & Amarasinghe (1984) found a similar interaction between ethylene and buoyant tension in petioles of *Nymphoides peltata* and observed two additional features of this system. First, the petioles of young leaves responded to applied tension (a 3 g weight attached to the base) even in the absence of ethylene (i.e.

TABLE 6. The effect of buoyant or applied tension on the elongation of young petioles of *Nymphoides peltata* growing with laminae in ethylene-free air, air + ethylene (10 μl l^{-1}) or submerged to 40 cm

Time (days)	Zone (cm)	Air		Ethylene		Submerged	
		−T	+T	−T	+T	−T	+T
2	2	8 ± 2	89 ± 9	174 ± 18	161 ± 21	79 ± 5	144 ± 22
	10	6 ± 1	67 ± 7	81 ± 5	197 ± 11	25 ± 1	95 ± 15
8	2	46 ± 8	1089 ± 82	670 ± 18	1020 ± 22	217 ± 21	2101 ± 33
	10	24 ± 5	367 ± 14	194 ± 13	407 ± 20	57 ± 7	407 ± 15

Leaves with unexpanded laminae were removed from plants growing in 30 cm of water and the petioles marked 2 cm and 10 cm from apex. Submerged leaves were weighted down or, in order to relieve buoyant tension, held under a perforated Perspex sheet. Non-submerged leaves were suspended with laminae in air and petioles surrounded by water and with or without a 3 g weight attached. Results (percentage increase) are mean values ± S.E., n = 8, for the apical 2 cm and the entire 10 cm marked zones after 2 and 8 days with (+T) or without (−T) tension being exerted on the petiole.

when laminae were in ethylene-free air). Further experiments using even younger leaves, with unexpanded laminae, showed an even larger effect of tension alone (Table 6): provided these young petioles were under tension and surrounded by water, they elongated just as much over 8 days irrespective of the presence of ethylene. Natural buoyancy will be much less than the 3 g weight applied in these experiments, but water currents could exert a force sufficient to induce marked and unnecessary petiole elongation: it is notable that *Nymphoides peltata* grows naturally only in still or very slow-moving waters and the sensitivity to tension of young petioles may be one factor that has influenced this habitat restriction. When laminae have expanded, a buoyant pull equivalent to the force exerted by an attached weight of 0·6 g (Malone 1983) or 1·0 g (Musgrave & Walters 1974) may be exerted in *Regnellidium* fronds and this would indeed be sufficient to affect the size of submergence responses.

The second feature was noted by Ridge & Amarasinghe (1984) in older petioles, where there was an approximately inverse relationship between sensitivity to ethylene and sensitivity to tension along the length of the petiole. Thus in the presence of ethylene, the apical 2 cm zone elongated almost threefold over 48 h irrespective of tension; but 6–10 cm down the petiole, elongation was 10% without applied tension and 187% when under tension. Tension therefore affects primarily the older cells (as judged by their position along the petiole) and the effect is on cell length and not cell number (Ridge 1985). It was suggested that tension acts by supplementing the driving force for cell expansion ($\psi_o - \psi_s - Y$), where ψ_o is the water potential of the source (in this case the external medium), ψ_s the solute potential of cells (which determines cell turgor in this system), and Y the minimum yield stress of cell walls (equivalent to the minimum turgor necessary to extend the walls). Water conductance of the tissue is assumed to be high for submerged organs and, therefore, unlikely to limit cell expansion. The implication is that in submerged petioles where ethylene, or some other agent, increases wall extensibility, but where this 'potential' for expansion is not realized because ψ_s (and hence turgor) is too small or Y is too high, tension will promote elongation. This appears to be the situation in older, more distal cells. Early observations on *Nymphoides* (Funke & Bartels 1937) and recent work on *Regnellidium* (D. J. Osborne, unpublished) show that the concentration of osmotic solutes declines markedly during petiole elongation in water, so a decrease in Y is the most likely mechanism by which applied tension could counter this effect in older cells. If this were so, it would mean that rapid elongation through water could be sustained along much of the petiole without the necessity of maintaining high turgor by osmoregulation in all rapidly expanding cells. This could well result in a significant conservation of resources.

All these experiments on the effect of tension were carried out with petioles completely surrounded by water and only the laminae were in air. The petioles of *Nymphoides* have a poorly developed vascular system, comprising three strands in the inner cortex, and in other species (e.g. *Regnellidium*) there is only a single, central vascular strand. In air, the hydraulic resistance along this route, especially for cells in the epidermis and outer cortex, would be considerably greater than for organs with more conventional anatomy and possibly high enough to restrict rapid cell expansion. We therefore considered the possibility that unless petioles were surrounded by water, the availability of water transported through the vascular system might be inadequate to support the high rates of cell expansion that may be induced in the presence of ethylene.

To test this hypothesis, *Nymphoides* plants raised in air were placed in a humid atmosphere (100% r.h.) containing 1 Pa (10 μl l^{-1}) ethylene. The growth of young petioles was monitored after flooding half the plants up to the laminae of these leaves, which floated as the petioles elongated. Compared with the aerial petioles, floating petioles did indeed grow slightly longer: cell lengths were 60–70% greater in the apical third, the effect decreasing progressively down the petiole, but cell number was 30% higher in the aerial petioles, which partly compensated for the shorter cell lengths. Floating petioles were not under tension and the effects on cell length can be attributed only to 'proximity to water'. Further comparisons with petioles of the same age that had been submerged, indicated that buoyant tension influenced cell elongation mainly in the lower two-thirds and proximity to water in the upper third of the petiole. These two effects explain why *Nymphoides* petioles in ethylene-enriched air do not elongate as much as submerged petioles. It appears that the potential for rapid elongation provided by ethylene cannot be realized over a period of days unless petioles are under tension and surrounded by water.

ORIGIN AND VARIATION OF SUBMERGENCE RESPONSES

When plants growing in a particular habitat possess distinctive features of morphology or physiology which appear to increase relative fitness in that habitat, there are two possibilities: did successful colonization of the habitat depend on possession of the 'features' or did the features evolve in response to selective pressures in the habitat? It does not matter whether the features involved are genetically fixed or arise as plastic responses to the environment, since canalization may fix a plastic response and plasticity itself has a genetic basis (Bradshaw 1965). However, this provides a suitable context in which to consider elongation and perhaps other responses of amphibious plants to submergence and, in particular, why responses and sensitivity to ethylene are so variable.

Although terrestrial plants usually respond to ethylene by growth inhibition (Abeles 1973), Osborne (e.g. 1982) has emphasized that cells which respond differently to ethylene and auxin are differentiated at specific sites, such as abcission zones. Whilst emphasizing that stem or petiole cells which elongate in response to both ethylene and auxin (described as Type 3 cells) are commonly present in amphibious plants, she points out that such cells occur also in limited regions of land plants—for example the adaxial tissue at the base of petioles. Poovaiah & Leopold (1973) also reported that *Poa pratensis* cv 'Sydsport' responded to high levels of ethylene by marked stimulation of internode growth and inhibition of leaf elongation and, although this species is widespread and extremely variable, it is not characteristic of seasonally flooded or damp habitats. Thus Type 3 cells may occur in plants from wet or dry habitats and their high frequency in amphibious plants indicates that there may have been selection for a pre-existing trait.

From Fig. 2 it is clear that possession of Type 3 cells does not lead to a uniform growth response to submergence. Structural limitations may explain some of this variation but they cannot explain all of it. For example, the delayed response in group B species arises partly from loss of sensitivity to ethylene late in organ development, even when cells are still able to expand, and this loss of sensitivity is not consistent with the view that enhanced elongation is always 'advantageous'. If it were, sensitivity to ethylene would be retained and all submerged organs would elongate rapidly to their maximum extent. This is exactly what happens in group A species, where this kind of response appears to increase relative fitness, and its absence in group B species is, therefore, more likely to result from opposing selective forces. Thus group B species should not be regarded as less 'well adapted' than group A species: rather, the likely nature of opposing selective forces should be considered. What are the possible disadvantages of large, rapid elongation responses and why, even within group A species, do older leaves respond much less than younger leaves?

Increased elongation of petioles, especially when it involves an increase in cell number, has a resource cost and this must be weighed against the benefit accruing from improved photosynthesis when leaves are restored to the water surface. For *Nymphoides*, *Regnellidium* and *R. sceleratus*, (group A), leaf turnover is quite high with lifespans typically in the range of 10 to 20 days. Using a cost–benefit approach, the cost of restoring to the surface older leaves nearing the end of their life could easily exceed the benefit, especially if leaves were deeply submerged. It was shown earlier (Table 2) that older leaves of group A species show a rapid but limited response to submergence, due almost entirely to increased cell elongation. Only young leaves attain maximum lengths and are able to elongate through deep columns of water; presumably the photosynthetic benefit here outweighs the resource cost.

The opposing selective forces that may have operated for species with delayed responses to submergence are likely to be different and more complex. Lifespans of 6 to 12 weeks were regularly recorded for leaves of *Ranunculus repens* and *Caltha palustris* growing in the greenhouse, for example. However, the increased petiole length, more upright position, increased formation of aerenchyma, decreased lignification and (apparently) thinner cuticles which characterize leaves that developed during periods of submergence seem to be responsible for their rapid mortality in an aerial environment. Thus if water levels decrease, these modified leaves often desiccate rapidly; they also tend to collapse, are brittle and easily damaged by trampling or wind and, in the field, elongated leaves of *Plantago major* and *Caltha palustris* have been observed to be grazed, apparently selectively, by geese and cattle respectively (I. Ridge, personal observation). Not only do elongation and associated changes carry a resource cost, they also appear to increase the risk of leaf mortality if flood waters subside and, therefore, represent a high-risk strategy in habitats where flooding is often brief and erratic. Amphibious species characteristic of such habitats generally have a delayed elongation response which develops only during prolonged periods of submergence and this would appear to minimize the risks outlined above. The only exception is *Ranunculus sceleratus*, which is a highly opportunistic annual in which the possibility of short term gains may be supposed to outweigh longer term risks.

According to this argument, therefore, the timing and size of elongation responses and the loss of sensitivity to ethylene in older leaves, reflect the balance between opposing selective forces. On the one hand, there is a resource cost and the long-term risk of leaf mortality in air; and, on the other hand, there are functional advantages when leaves are restored to the water surface. It is relevant to point out that species showing high sensitivity to ethylene and large, rapid responses to submergence usually grow in water (*Nymphoides, Callitriche*) or in areas subject to regular and prolonged flooding (*Regnellidium*, floating rice). When water levels increase, these species show a high capacity for rapid depth accommodation; but only during periods of exceptional drought do they grow entirely in air. The balance of selective forces for these species must be quite different to that for group B species, which commonly grow entirely in air.

WHY ETHYLENE?

Ethylene is, so far, the only common factor linking all elongation responses (granted that the evidence is circumstantial for some species), and it may be involved in several other changes that are induced by submergence of shoots. However, ethylene is clearly not the sole regulator of submergence-induced growth: its action depends on the presence of auxin and/or gibberellin, and a range of other factors influence the size of response. These factors have been identified only for rapid growth responses and they vary between species and for different stages of development. But apart from rice coleoptiles and the sprouts of two paddy field weeds, rapid elongation responses have not been shown to occur in the absence of ethylene, which raises the question of why ethylene involvement is so widespread. Two quite different and equally speculative answers to this question are discussed below.

The answer may depend, firstly, on physiological considerations. If the rate of cell division depends on the rate of cell expansion, as argued earlier for submerged shoots, then initiation of rapid cell expansion will initiate all types of rapid growth response. In higher plants, rapid cell expansion is commonly initiated by an increase in wall extensibility and this is known to be the basis of ethylene action in submerged petioles of two species. Thus if ethylene were the only factor whose level changed rapidly on submergence (or for which there was a rapid change in sensitivity) and which also increased wall extensibility, this would explain its common role. Other factors either increase ethylene synthesis (e.g. hypoxia in floating rice internodes), enhance or sensitize ethylene action (as CO_2 may do), or operate at a later stage when osmotic or cell water relationships become important for sustaining rapid growth (e.g. buoyant tension and an aquatic environment). The available evidence is consistent with this view but often derives from investigation of only one or two species, and the main reason for presenting such a speculative idea is to focus attention on areas of uncertainty where further work is needed. Three questions that need answering for amphibious plants are: (i) Does the rate of cell expansion determine the rate of cell division in young tissues? (ii) Is an increase in wall extensibility a universal requirement for the initiation of rapid growth responses to submergence? (iii) Does ethylene act always by increasing wall extensibility in submerged shoots?

A second kind of answer makes no physiological assumptions but rests solely on considerations of kinetics and reliability. Initiation of a rapid growth response to submergence requires systems of detection and control with appropriate rapid kinetics. For at least one system (*Regnellidium diphyllum*) the kinetics of ethylene accumulation match the kinetics of growth initiation and, in several systems, the kinetics of ethylene action are clearly fast enough to account for rapid growth responses. There appear to be no detailed kinetic studies of other factors, such as CO_2 or hypoxia, that influence submergence responses so that, apart from the exceptions discussed below, ethylene is the only regulator known to show appropriate kinetics. If ethylene does not act as a detection system 'sensing' submergence, and this may be true for internodes of floating rice, then possible alternatives are systems based on pressure transduction or on light quality or quantity. At present there is no evidence supporting either of these alternatives: the rate or response does not vary consistently with depth of submergence and responses can usually be initiated (although not necessarily sustained) in light or darkness. These alternatives clearly merit further investigation however.

On the question of reliability, important requirements of any system controlling rapid growth through a water column are that it continues to operate, usually until leaves reach the surface and, furthermore, does not operate in an aerial environment. Ethylene appears to satisfy both these requirements provided that sufficient oxygen is available in submerged tissues to maintain its synthesis. Accumulation or depletion of CO_2 or O_2 would be less reliable in photosynthetic organs because a wide range of factors influence the relative rates of photosynthesis and respiration; the system would have to be sufficiently sensitive to respond to changes in these gases that occur on submergence but unresponsive to changes occurring during day–night transitions in air. There are similar problems of reliability for systems based on light. The underwater light environment is extremely variable, depending on depth, turbidity, dissolved organic matter and shading from other vegetation; and yet many amphibious plants elongate at the same rate when submerged irrespective of these differences but do not show rapid elongation in air when grown under reduced light intensities. However, all amphibious species that respond to ethylene either grow in relatively eutrophic water, and/or are likely to be submerged by turbid floodwater, or grow in dense stands of vegetation. For species that grow in clear, oligotrophic waters with little surrounding vegetation, light could well provide a reliable system for growth control in submerged organs.

This second attempt to explain the almost universal involvement of ethylene in growth responses to submergence simply suggests, therefore, that ethylene provides the only system known to have both appropriate kinetics (for rapid responses) and good reliability. But there are two aquatic environments where ethylene may not necessarily provide a reliable detection or regulating system: rapidly flowing water and, for non-photosynthetic organs, poorly oxygenated water in which plant tissues become anoxic. Rapidly flowing water might be expected to reduce the accumulation of ethylene in submerged tissues, although there is no evidence that this happens. Osborne (1984), however, reported that peduncles of *Ranunculus fluitans*, which elongate above the water surface in flowing water, do not grow in response to ethylene, and the control system here must be independent of it. Anoxia can occur in organs such as rice coleoptiles and young shoots developing from perennating organs, and ethylene synthesis is usually inhibited in such conditions (e.g. Raskin & Kende 1984). However, anoxia or hypoxia and raised CO_2 levels promote elongation of rice coleoptiles independently of ethylene (Ku *et al*. 1979) and CO_2 may act independently in two sprout systems (Suge & Kusanagi 1975). So in situations where ethylene might not provide a reliable mechanism of growth control, other mechanisms operate.

There is obviously a great deal that is not known about growth control in amphibious plants particularly those, probably the majority, that show delayed and unspectacular changes upon submergence. Plant physiologists have concentrated their attention on species showing large, rapid responses because these are very much easier to study, and their findings have shed interesting new light on how plant growth may be controlled. It is now established beyond reasonable doubt that ethylene is an important growth regulator in amphibious plants, although certainly not the only one. Determining what role, if any, ethylene plays in delayed growth responses, and in the wide range of morphological changes that occur during submergence, is a future challenge for physiologists. Determining, more critically than has been done so far, how these changes in growth and morphology affect fitness in the long term, is a task that physiologists and ecologists could usefully tackle together.

ACKNOWLEDGMENTS

My interst in the physiology of amphibious plants was originally stimulated by Dr Daphne Osborne. I am grateful to her and to Dr Jonathan Silvertown for commenting on the manuscript, to Dr Michael Malone for many helpful suggestions and access to his unpublished data, to Ivan Amarasinghe for expert technical help and to John Taylor for drawing the Figures.

REFERENCES

Abeles, F. B. (1973). *Ethylene in Plant Biology.* Academic Press, New York and London.

Arber, A. (1920). *Water Plants: A Study of Aquatic Angiosperms.* Cambridge University Press, Cambridge.

Armstrong, W. (1979). Aeration in higher plants. *Advances in Botanical Research*, **7**, 225–232.

Bradshaw, A. D. (1965). Evolutionary significance of phenotypic plasticity in plants. *Advances in Genetics*, **13**, 115–155.

Cookson, E. C. (1976). Auxin, ethylene and cell growth of water plants. Ph.D. thesis, University of Cambridge.

Cookson, E. C. & Osborne, D. J. (1978). The stimulation of cell extension by ethylene and auxin in aquatic plants. *Planta*, **144**, 39–47.

Cookson, E. C. & Osborne, D. J. (1979). The effect of ethylene and auxin on cell wall extensibility of the semi-aquatic fern, *Regnellidium diphyllum. Planta*, **146**, 303–307.

Dacey, J. W. (1981). Pressurized ventilation in the yellow waterlily. *Ecology*, **62**, 1137–1147.

Funke, G. L. & Bartels, P. M. (1937). Observations on the growth of water plants. *Biologisch Jaarboek*, **4**, 316–344.

Gill, C. J. (1970). The flooding tolerance of woody species—a review. *Forestry Abstracts*, **31**, 671–688.

Gomes, A. R. Sena & Kozlowski, T. T. (1980). Growth responses and adaptations of *Fraxinus pennsylvanica* seedlings to flooding. *Plant Physiology*, **66**, 267–271.

Horton, R. F. & Samarakoon, A. B. (1982). Petiole growth in the celery-leaved crowfoot (*Ranunculus sceleratus*): effects of auxin-transport inhibitors. *Aquatic Botany*, **13**, 97–104.

Jackson, M. B. (1982). Ethylene as a growth promoting hormone under flooded conditions. *Plant Growth Substances 1982* (Ed. by P. F. Wareing), pp. 291–301. Academic Press, London.

Jackson, M. B. (1985). Ethylene and responses of plants to soil water-logging and submergence. *Annual Review of Plant Physiology*, **36**, 145–174.

Jackson, M. B., Dobson, C. M., Herman, B. & Merryweather, A. (1984). Modification of 3,5-diiodo-4-hydroxybenzoic acid (DHIB) activity and stimulation of ethylene production by small concentrations of oxygen in the root environment. *Plant Growth Regulation*, **2**, 251–262.

Ku, H. S., Suge, H., Rappaport, L. & Pratt, H. K. (1979). Stimulation of rice coleoptile growth by ethylene. *Planta*, **90**, 333–339.

Malone, M. (1983). *The mechanism of depth accommodation in* Nymphoides peltata *and other water plants.* Ph.D. thesis, The Open University, Milton Keynes.

Malone, M. & Ridge, I. (1983). Ethylene-induced growth and proton excretion in the aquatic plant *Nymphoides peltata. Planta*, **157**, 71–73.

McComb, A. J. (1965). The control of elongation in *Callitriche* shoots by environment and gibberellic acid. *Annals of Botany*, **29**, 445–459.

Métraux, J-P. & Kende, H. (1983). The role of ethylene in the growth response of submerged deep water rice. *Plant Physiology*, **72**, 441–446.

Métraux, J-P. & Kende, H. (1984). The cellular basis of the elongation response in submerged deep-water rice. *Planta*, **160**, 73–77.

Mitchell, R. S. (1976). Submergence experiments on nine species of semiaquatic *Polygonum. American Journal of Botany*, **63**, 1158–1165.

Musgrave, A., Jackson, M. B. & Ling, E. (1972). *Callitriche* stem elongation is controlled by ethylene and gibberellin. *Nature, New Biology*, **238**, 93–96.

Musgrave, A. & Walters, J. (1973). Ethylene-stimulated growth and auxin transport in *Ranunculus sceleratus* petioles. *New Phytology*, **72**, 783–789.

Musgrave, A. & Walters, J. (1974). Ethylene and buoyancy control rachis elongation of the semi-aquatic fern *Regnellidium diphyllum. Planta*, **121**, 51–56.

Osborne, D. J. (1982). The ethylene regulation of cell growth in specific target tissues of plants. *Plant Growth Substances 1982* (Ed. by P. F. Wareing), pp. 279–290. Academic Press, London.

Osborne, D. J. (1984). Ethylene and plants of aquatic and semi-aquatic environments: a review, *Plant Growth Regulation*, **2**, 167–185.

Palmer, J. H. (1972). Roles of ethylene and indol-3-yl-acetic acid in petiole epinasty in *Helianthus annuus* and the modifying influence of gibberellic acid. *Journal of Experimental Botany*, **23**, 733–743.

Pieterse, A. H., Aris, J. J. A. M. & Butter, M. E. (1976). Inhibition of float formation in water hyacinth by gibberellic acid. *Nature*, **260**, 423–424.

Poovaiah, B. W. & Leopold, A. C. (1973). Effects of ethephon on growth of grasses. *Crop Science*, **13**, 755–758.

Raskin, I. & Kende, H. (1983). Regulation of growth in rice seedlings. *Journal of Plant Growth Regulation*, **2**, 193–203.

Raskin, I. & Kende, H. (1984a). Regulation of growth in stem sections of deep-water rice. *Planta*, **160**, 66–72.

Raskin, I. & Kende, H. (1984b). Role of gibberellin in the growth response of submerged deep-water rice. *Plant Physiology*, **76**, 947–950.

Ridge, I. (1985). Ethylene and petiole development in amphibious plants. *Ethylene and Plant Development* (Ed. by J. A. Roberts & G. A. Tucker), pp. 229–239. Butterworths, London.

Ridge, I. & Amarasinghe, I. (1984). Ethylene and growth control in the fringed waterlily (*Nymphoides peltata*): stimulation of cell division and interaction with buoyant tension in petioles. *Plant Growth Regulation*, **2**, 235–249.

Rose-John S. & Kende, H. (1984). Short-term growth response of deep-water rice internodes to submergence. *Plant Research '83*. MSU-DOE Plant Research Laboratory, Michigan State University, East Lansing.

Samarakoon, A. B. & Horton, R. F. (1981). Flowering and stem growth in the celery-leaved buttercup, *Ranunculus sceleratus*. *Canadian Journal of Botany*, **59**, 1386–1392.

Sculthorpe, C. D. (1967). *The Biology of Aquatic Vascular Plants*. Edward Arnold, London.

Silk, W. K. (1984). Quantative descriptions of development. *Annual Review of Plant Physiology*, **35**, 479–518.

Smith, S. J. (1986). *Ecological strategies of two* Ranunculus *species in relation to seasonal submergence*. D. Phil. thesis, The Open University, Milton Keynes, U.K.

Suge, H. (1971). Stimulation of oat and rice mesocotyl growth by ethylene. *Plant & Cell Physiology*, **12**, 831–837.

Suge, H., Katsura, N. & Inada, K. (1971). Ethylene-light relationship in the growth of rice coleoptile. *Planta*, **101**, 365–368.

Suge, H. & Kusanagi, T. (1975). Ethylene and carbon dioxide: Regulation of growth in two perennial aquatic plants, arrowhead and pondweed. *Plant & Cell Physiology*, **16**, 65–72.

Taiz, L. (1984). Plant cell expansion: regulation of cell wall mechanical properties. *Annual Review of Plant Physiology*, **35**, 585–657.

Thompson, K. & Grime, J. P. (1983). A comparative study of germination responses to diurnally-fluctuating temperatures. *Journal of Applied Ecology*, **20**, 141–156.

Walters, J. & Osborne, D. J. (1979). Ethylene and auxin-induced cell growth in relation to auxin transport and metabolism and ethylene production in the semi-aquatic plant *Regnellidium diphyllum*. *Planta*, **146**, 309–317.

Williams, W. T. & Barber, D. A. (1961). The functional significance of aerenchyma in plants. *Mechanisms in Biological Competition* (Ed. by F. L. Milthorpe). *Symposia of the Society for Experimental Biology*, **15**, 132–144.

B. Photosynthesis under Water

Aquatic plant photosynthesis: strategies that enhance carbon gain

G. BOWES

Department of Botany, University of Florida, Gainesville, Florida 32611, U.S.A.

SUMMARY

1 Submersed leaf photosynthesis is characterized by low rates, shade responses, and high apparent K_m and saturation values for CO_2. This situation is reversed in aerial leaves of amphibious plants.

2 The low photosynthesis and productivity of submersed aquatic macrophytes (SAM) is largely due to the high aqueous resistance to CO_2 and O_2 diffusion. In dense vegetation, the diffusion resistance can result in low CO_2 and high O_2 levels in the water during the day. Thus the aqueous environment imposes a major limitation on SAM photosynthesis.

3 Unlike terrestrial C_3 and C_4 species, SAM plants show variable CO_2 compensation points, which indicates that they exist in a continuum of photorespiratory (PR) states from high to low, depending on the growth conditions. This also applies to the aerial leaves of amphibious plants.

4 The Calvin or photosynthetic carbon reduction (PCR) and photorespiratory carbon oxidation (PCO) cycles occur in SAM plants. However, supplemental biochemical pathways may modulate these two cycles. Thus SAM plants do not correspond to the terrestrial photosynthetic categories.

¯5 SAM plants are postulated to show four major strategies to reduce the adverse effects of the low CO_2 and high diffusion resistance in water. Two are biochemical adaptations: namely, the C_4 acid and bicarbonate utilization strategies; while two are morphological adaptations: namely, the hydrosoil CO_2 utilization and aerial leaf strategies.

6 The C_4 acid and bicarbonate utilization systems are probably CO_2 concentrating mechanisms that elevate the dissolved inorganic carbon (DIC) at the site of fixation by ribulose bisphosphate carboxylase-oxygenase, thereby increasing its fixation efficiency in the presence of O_2.

7 In the C_4 acid system, during day and/or night, phosphoenolpyruvate carboxylase fixes CO_2 into malate, which is subsequently decarboxylated to produce CO_2 for the PCR cycle during the day. This occurs without the Kranz anatomy found in terrestrial C_4 plants. *Hydrilla verticillata* and *Isoetes howellii* use this strategy.

8 The bicarbonate utilization system provides direct access to HCO_3^- ions as a source of DIC, in addition to free CO_2. This is especially advantageous in high pH waters. It is analogous to the inducible system found in unicellular autotrophs. In some SAM plants, secretion of H^+ ions from the abaxial leaf surface, and carbonic

79

anhydrase activity, are important components. *Myriophyllum spicatum* uses this strategy.

9   The C_4 acid and bicarbonate utilization systems are more effectively induced under adverse growth conditions, producing a low-PR state.

10   In the hydrosoil CO_2 utilization strategy, CO_2 enters the roots from the high levels in the sediment. It is piped via lacunal gas channels to the leaves, where it is retained by a thick cuticle until fixed. This strategy occurs in low-growing isoetid forms such as *Lobelia dortmanna* and *Littorella uniflora*, though the latter also uses a C_4 acid system.

11   Amphibious SAM plants are often heterophyllous. The aerial leaf form allows direct access to gaseous CO_2 in the air, whereas the submersed leaf is better adapted for DIC fixation. They are often poor users of HCO_3^- ions. Their photosynthetic biochemistry is poorly understood.

12   The data indicate that DIC is a major limiting nutrient for SAM photosynthesis and growth in freshwater. A given SAM plant may use more than one strategy to reduce this limitation.

INTRODUCTION

The past two decades have seen substantial progress in understanding photosynthetic mechanisms. Although the Calvin or photosynthetic carbon reduction (PCR) cycle is the fundamental entry point for inorganic carbon into the organic sphere, several important supplements to this pathway are now recognized. Thus, terrestrial plants are currently categorized into four major groups with respect to photosynthetic CO_2 fixation: C_3, C_4, C_3–C_4 intermediates, and Crassulacean acid metabolism or CAM (Kluge & Ting 1978; Edwards & Huber 1981; Monson, Edwards & Ku 1984). The groups differ in biochemical, physiological and anatomical features, and many of the differences are correlated with environmental factors; particularly O_2, CO_2, temperature, water and light. Thus, the habitat 'preference' of plants in these groups is, to some extent, linked to their photosynthetic mode, although the ecological advantages bestowed by a particular pathway are not always obvious (Pearcy & Ehleringer 1984).

A major difference within these photosynthetic groups is the ability to reduce the inhibitory effects on the PCR cycle, and the stimulatory effects on the photorespiratory carbon oxidation (PCO) cycle, of atmospheric O_2. Some species employ phosphoenolpyruvate carboxylase (PEPCase) for the initial CO_2 fixation. This enzyme is an efficient scavenger of inorganic carbon, because it has rapid turnover and is not inhibited by O_2. In these plants, the production and subsequent decarboxylation of C_4 acids (primarily malate and aspartate) act as a CO_2 concentrating system. This system provides an elevated CO_2 level to ribulose bisphosphate carboxylase-oxygenase (RuBPCase), thus enabling it to catalyse CO_2 fixation more effectively in the presence of O_2, and simultaneously decreasing the oxygenase activity of RuBPCase, which otherwise provides the substrate for the

PCO cycle. Thus the C_4 acid pathway supplements the PCR cycle. The reduction in O_2 effects in species using this system has broad consequences with respect to net photosynthetic rates, productivity, water use efficiency, CO_2 requirements, quantum requirements, temperature optima, and even nitrogen use efficiency (Pearcy & Ehleringer 1984).

The only other method known to reduce the effects of O_2 on PCR and PCO cycle activities is found in the aquatic environment. It is an inducible HCO_3^- utilization system that similarly elevates the internal dissolved inorganic carbon (DIC) levels at the site of fixation by RuBPCase into the PCR cycle. It has been documented in some unicellular green algae and cyanobacteria (Badger, Kaplan & Berry 1978; Tsuzuki & Miyachi 1979; Beardall & Raven 1981). In the green algae, carbonic anhydrase, which catalyses the interconversion of the DIC forms, is an essential component of the system (Spalding, Spreitzer & Ogren 1983). Growth of the organisms at low CO_2 (air-levels) induces the system, whereas growth at high CO_2 leads to its repression. Physiologically, the effects on photosynthesis and photorespiration are analogous to those of the C_4 system in terrestrial plants. In addition to overcoming the deleterious O_2 effects, the algal DIC concentrating system probably reduces the CO_2 diffusion limitations of the aquatic environment.

For freshwater macrophytes, the photosynthetic pathways, and their ecological implications, are still inadequately documented. An unusual feature common to all the submersed macrophytes we have examined is the ability to vary the photosynthesis/photorespiration ratio in response to changing environmental conditions. This is indicated by substantial variability in CO_2 compensation points, even for a single plant (Holaday, Salvucci & Bowes 1983). The acronym 'SAM' (submersed aquatic macrophyte) has been coined in regard to their photosynthesis to emphasize the distinctive features not found in the terrestrial groups, or even among aquatic unicellular organisms (Bowes et al. 1978; Bowes, Holaday & Haller 1979).

In the case of aquatic angiosperms, several lines of evidence suggest that they have returned from a terrestrial to an aquatic environment in fairly recent evolutionary times (Sculthorpe 1967), and thus they are not primitive but highly specialized plants. Furthermore, a number of aquatic plants are amphibious, being capable of photosynthesis and growth in both the submersed and emersed condition. In this sense they are facultative SAM. Amphibious plants frequently exhibit heterophylly, with considerable variation in leaf morphology even on the same stem, and this may be related to whether the leaf is submersed or aerial (Sculthorpe 1967). Even less is understood about the photosynthetic modes of amphibious plants than of obligate SAM, and how heterophylly and photosynthetic biochemistry are interrelated is largely unknown. Also, the ecological role of the aerial leaves is at present only speculative. A study of amphibious plants could well provide some interesting clues to the evolutionary history of aquatic organisms, and whether fundamental differences exist between aquatic and aerial photosynthesis.

Despite the biosystematic and habitat diversity of aquatic organisms, the fact that submersed plants or organs are bathed in a dense solvent (water), as opposed to a gas (air), has some common implications with respect to photosynthetic constraints and adaptations.

PHOTOSYNTHETIC CONSTRAINTS AND ADAPTATIONS IN AQUATIC HABITATS

The aquatic habitat is sometimes portrayed as a relatively 'benign' milieu, where the extremes of terrestrial environments are moderated. This is a misleading analysis of the environmental constraints imposed upon freshwater autotrophs, as should become apparent.

Light

Light is usually selectively attenuated in its passage through water (Spence 1981), but just below the water surface the irradiance can be as high as in a terrestrial habitat (Van, Haller & Bowes 1976; Bowes, Holaday & Haller 1979). SAM plants invariably can be categorized as shade plants, as leaf photosynthesis is saturated at an irradiance of less than half full sunlight, even for species that inhabit surface waters (Raven & Glidewell 1975; Van, Haller & Bowes 1976; Bowes *et al.* 1977; Salvucci & Bowes 1982; Maberly 1983). Thus the leaves of SAM plants are adapted to the lower radiant energy fluxes of their environment, not the surface irradiance, and there is no evidence that they can respond sufficiently to utilize full sunlight. However, in some heavily vegetated sites, the dense canopy structure may be such as to show increased light interception and growth up to full sunlight.

The submersed leaves of amphibious plants also show shade characteristics as measured by light saturation and compensation points, but the aerial leaves can be full-sun adapted (Lloyd, Canvin & Bristow 1977; Salvucci & Bowes 1982). The role of heterophylly, if any, in these differences is not known.

Consistent with the shade characteristics are low photosynthetic rates, which even at light and DIC saturation rarely exceed 100 μmol CO_2 or O_2/mg Chl·h, and at naturally occurring DIC levels are closer to 10 to 20 (Raven & Glidewell 1975; Van, Haller & Bowes 1976; Salvucci & Bowes 1981). The aerial leaves of amphibious plants show much higher photosynthetic rates at air-levels of CO_2 than submersed leaves (Lloyd, Canvin & Bristow 1977; Salvucci & Bowes 1982; Spencer & Bowes 1985). The severe limitations imposed by the resistance of water to CO_2 diffusion might well mandate the shade and rate photosynthetic characteristics of submersed leaves (Black, Maberly & Spence 1981).

Temperature

Water provides a substantial buffer from rapid fluctuations in air temperature, but seasonal extremes, ranging from near zero to as much as 40°C, can be observed, especially in densely vegetated surface waters (Van, Haller & Bowes 1976; Bowes, Holaday & Haller 1979). Some SAM plants have been shown to be capable of photosynthesis at below 2°C (Boylen & Sheldon 1976), with optima around 20°C; while other species may have optima at 30 to 35°C (Stanley & Naylor 1972; Van, Haller & Bowes 1976). Indirect temperature effects on CO_2 and O_2 solubility can obscure direct effects on the photosynthesis of submersed leaves, especially at natural DIC levels. For this reason, it is unclear whether submersed and aerial leaves of amphibious plants exhibit different temperature responses for photosynthesis. It also makes difficult any correlation with terrestrial C_3 or C_4 photosynthesis as regards temperature response.

Hydrogen ion concentration

In contrast to aerial leaves, the photosynthetic organs of SAM plants are inevitably exposed to an external H^+ ion concentration, due to their immersion in a polar solvent. Few macrophytes can survive below pH 4, which is one reason why acid rain is so devastating, but many tolerate periods at pH 10 to 11 (Brown et al. 1974; Van, Haller & Bowes 1976; Bowes, Holaday & Haller 1979; Keeley 1983). The pH of freshwater is variable from acid to very alkaline, sometimes on a diel basis when photosynthetic and respiratory activity is substantial (Bowes, Holaday & Haller 1979). In lakes the diel pH changes may be localized to areas of dense vegetation. Some SAM plants regulate the microenvironment of the surrounding aqueous boundary layer by secreting H^+ and OH^- ions from the abaxial and adaxial leaf surfaces, respectively, such that the abaxial values may be below pH 5 (Prins et al. 1982). These large microenvironmental pH effects mean that gross measurements of water pH may not provide an accurate reflection of the pH to which the leaves are actually exposed.

Although all SAM plants must be able to compensate for the external pH, their tolerance varies. For example, the amphibious angiosperms *Limnophila sessiliflora* and *Hygrophila polysperma* grow best at pH 5 to 7, and hardly at all at pH 9. In contrast, although *Hydrilla verticillata* grows well at pH 5 to 7, its growth rate is tenfold higher at pH 9 (Spencer & Bowes 1985). This wide pH tolerance may be a factor in the competitive success of a plant.

The pH indirectly affects photosynthesis by influencing the equilibrium between the various forms of DIC. Thus submersed leaves are exposed to CO_3^{2-} and HCO_3^- ions, in addition to free CO_2, as potential carbon sources. It is generally agreed that CO_3^{2-} ions are not used directly, and free CO_2 is preferred to HCO_3^-, based on uptake kinetics, pH drift experiments, and $\delta^{13}C$ values (Steeman-Nielsen 1947; Raven 1970; Van, Haller & Bowes 1976; Browse, Brown & Dromgoole 1979; Allen & Spence 1981; Osmond et al. 1981). Despite the low affinity of photosynthesis for HCO_3^-, the relatively large concentration of this ion in many waters makes access to it potentially a major difference between submersed and aerial photosynthesis. Although HCO_3^- is undoubtedly a carbon source for photosynthesis, whether submersed leaves can, or cannot, take up this ion directly has been argued over since the turn of this century. Distinctions have been drawn between HCO_3^- 'users' and 'nonusers' (Steeman-Nielsen 1947). In general, aquatic

TABLE 1. A comparison of the apparent bicarbonate utilization abilities of submersed leaves from high-photorespiration and low-photorespiration plants of *Limnophila sessiliflora*, *Hygrophila polysperma*, and *Myriophyllum spicatum*

Species	DIC concentration (mM)		pH	Net photosynthesis (%)	
	Free CO_2	HCO_3^-		High-PR	Low-PR
Limnophilia sessiliflora	0·05	0	5·0	17·5	27·9
	0·05	2·45	8·0	20·7	48·9
Hygrophila polysperma	0·05	0	5·0	20·5	21·9
	0·05	2·45	8·0	43·2	51·8
Myriophyllum spicatum	0·05	0	5·0	24·0	38·0
	0·05	2·45	8·0	71·8	83·8

Net photosynthesis is expressed as a percentage of the maximum light and CO_2 saturated rate, measured as O_2 evolution.

mosses and pteridophytes tend to be ineffective at HCO_3^- use (Steeman-Nielsen 1947; Raven 1970; Bain & Proctor 1980; Spence & Maberly 1985). The submersed forms of several amphibious plants similarly have been reported to be poor HCO_3^- users (Bristow 1969; Maberly & Spence 1983; Spence & Maberly 1985).

Table 1 shows data for the abilities of the submersed forms of two amphibious plants, and a more obligate SAM, *Myriophyllum spicatum*, to use HCO_3^-. For each plant, the net photosynthetic rates (expressed as a percentage of the maximum light- and DIC-saturated rate) are compared at the same subsaturating free CO_2 level (0·05 mM), but in the presence or absence of an additional 2·45 mM HCO_3^-. All the plants were previously incubated under summer- or winter-like growth conditions to induce the low- and high-photorespiratory (PR) states, respectively (Holaday, Salvucci & Bowes 1983). In the more typical high-PR state, *Limnophila sessiliflora* showed no evidence of increased photosynthesis with the additional HCO_3^-, though the low-PR state did exhibit some limited capacity for HCO_3^- use. Both *Hygrophila polysperma* and *Myriophyllum spicatum* showed evidence of HCO_3^- utilization, which was further enhanced in the low-PR state. Of the three species tested, the two amphibious plants showed the least photosynthetic response to added HCO_3^-. From a teleological viewpoint, rather than developing an effective HCO_3^- utilization system, amphibious plants may have opted for an aerial strategy to enhance carbon gain in waters that are low in free CO_2. A more comprehensive investigation of HCO_3^- use in amphibious plants is needed to test this possibility.

The distribution of some SAM plants in hard and soft water has been linked to the ability to utilize HCO_3^- (Hutchinson 1970; Bain & Proctor 1980; Sand-Jensen 1983; Spence & Maberly 1985). Although an attractive concept, it is difficult to prove, and there are experimental anomalies. *Hydrilla verticillata*, for example, uses free CO_2 far more effectively than HCO_3^- (Van, Haller & Bowes 1976), yet its growth is greatest at pH 9 (Spencer & Bowes 1985). The data in Table 1, together with other published work (Bain & Proctor 1980; Salvucci & Bowes 1983; Spence

& Maberly 1985) demonstrate that SAM plants, like unicellular algae, exhibit a range of ability to utilize HCO_3^-, even in a single species, that is dependent upon the growth conditions to which the organism has been subjected. The ongoing debate as to whether CO_2 or HCO_3^- is the form that actually enters the cell, blurs the distinction between users and nonusers even further (Smith & Walker 1980).

Diffusion resistance

In aerial as well as submersed leaves, CO_2 and O_2 have to dissolve in order to reach the cellular organelles. However, for submersed leaves the reduced diffusion rate in water (10^4 times slower) combined with the long pathway result in a massive diffusion resistance. Of the several resistances to CO_2 fixation in submersed leaves, the aqueous boundary layer appears to be by far the greatest (Smith & Walker 1980; Black, Maberly & Spence 1981). Submersed leaves face a common dilemma with the respiratory organs of aquatic animals: how to maximize the surface area/volume ratio for gas diffusion, while minimizing potential mechanical damage from the dense medium. For aerial leaves the problem of desiccation is generally more acute than that of damage from agitation.

The morphology of submersed leaves appears to reflect the problem of diffusion resistance. Many are small, thin and/or finely dissected, presumably to increase the surface area, although this morphology has been postulated to be a response to low light (Sculthorpe 1967), or both (Black, Maberly & Spence 1981). Stomates, if present, are non-functional, and the cuticle is generally thin (with certain important exceptions). Chloroplasts are located in the outer cell layers, including the epidermis. Thicker leaves often contain lacunal gas spaces, and are commonly ribbon-like. These features would provide buoyancy and reduce damage from mechanical agitation. Except for one unconfirmed report, submersed leaves lack the Kranz anatomy associated with terrestrial C_4 plants (DeGroote & Kennedy 1977; Hough & Wetzel 1977; Bowes & Salvucci 1984).

Further strategies to lessen the CO_2 diffusion resistance problem will be discussed later.

Carbon dioxide and oxygen

The concentrations of dissolved free CO_2 and O_2 in equilibrium with air are about 0·01 and 0·24 mM, respectively, and these values approximate those encountered by aerial leaves. However, unlike an aerial environment, the diffusion resistance of water combined with respiratory and photosynthetic activity can result in wide fluctuations from air-equilibrium values. In freshwater, free CO_2 may range from zero to over 14 mg l^{-1} or 0·35 mM (Bowes, Holaday & Haller 1979; Keeley 1983); while O_2 can range from near zero to over 200% air-saturation or about 0·5 mM (Brown et al. 1974; Bowes, Holaday & Haller 1979). Such fluctuations in CO_2 and O_2, in opposite diel rhythms, can occur at one location.

TABLE 2. Apparent $K_m(CO_2)$ values for photosynthesis of high-photorespiration and low-photorespiration *Hydrilla verticillata* plants, measured in water or air (100% relative humidity) at 1% and 21% O_2 (gas phase)

Plant material	Measurement medium	Apparent $K_m(CO_2)(\mu M)$	
		1% O_2	21% O_2
Lake-grown and high-PR	Water	120	200
Chamber-grown and high-PR	Water	175	285
	Air	70	—
Chamber-grown and low-PR	Water	29	48
	Air	18	—

SAM plants have a common requirement for much higher DIC levels to saturate photosynthesis than do terrestrial plants or unicellular algae. For SAM plants, approximately 0·5 mM free CO_2 is required for saturation (Steeman-Nielsen 1947; Van, Haller & Bowes 1976). Thus the apparent K_m (CO_2) values for photosynthesis are very high, and even higher for HCO_3^- (Lloyd, Canvin & Bristow 1977; Browse, Dromgoole & Brown 1979; Maberly & Spence 1983). As shown for the obligate SAM plant *Hydrilla verticillata* in Table 2, under typical growth conditions the values in well-stirred water ranged from 120 to 285 μM, depending upon the O_2 concentration in the measurement medium. Determination in the gas phase reduced these values to 70 μM, which demonstrates the large effect of aqueous diffusion resistance. Table 2 also shows that the plants in the low-PR state were much more effective at scavenging CO_2. Corresponding data for the amphibious plant *Myriophyllum brasiliense* indicate that the submersed leaves exhibit $K_m(CO_2)$ values of 706 and 102 μM when under water and in air, respectively (Salvucci & Bowes 1982). However, the air-value for aerial leaves from the same plant is 18 μM. Similarly large differences have been observed between submersed and aerial leaves of another amphibious plant, *Potamogeton amphifolius* (Lloyd, Canvin & Bristow 1977).

The generally low photosynthetic rates of submersed leaves at natural DIC levels are consistent with the high DIC requirements. In addition to the large aqueous boundary layer effect, the low RuBPCase activity in SAM leaves, together with the enzyme $K_m(CO_2)$ which is twice that of the terrestrial plant enzyme, probably add to the high DIC requirements (Van, Haller & Bowes 1976).

Heterophylly

One of the most conspicuous differences between obligate SAM plants and their facultative counterparts is the heterophyllous condition commonly found in the latter. The aerial leaves of an amphibious plant are often similar morphologically to leaves of terrestrial plants, and thus may differ dramatically from submersed leaves (even on the same stem) in terms of thickness, size, lack of dissection, and possession of stomata. It is not known if any aerial leaves possess Kranz anatomy, or an anatomical compartmentation that could perform a similar photosynthetic function.

Although intuitively heterophylly seems to be a device to maximize photosynthesis in both a submersed and aerial habitat, the environmental factors that control the development of submersed versus aerial leaf morphology differ among species, and are not always directly correlated with whether the leaf is submersed or emersed.

Photoperiod and temperature play important roles. For example, with *Proserpinaca palustris* short photoperiods induce the submersed, and long photoperiods the aerial leaf form. This is modified by temperature; irrespective of photoperiod, at low temperatures the submersed leaf form develops, while higher temperatures induce the aerial leaf (Wallenstein & Albert 1963; Davis 1967; Schmidt & Millington 1968; Kane & Albert 1982). Similar effects are seen in *Ranunculus flabellaris* (Johnson 1967). It could be hypothesized that these responses enable the plant to remain submersed and relatively protected during cold winter conditions, while in the summer they would enable the potentially more productive aerial leaf form to prevail. An alternative hypothesis, that increased leaf dissection at reduced temperatures is due to more limited gas diffusion (Johnson 1967), is less plausible as it ignores the photoperiod effect and the increase in dissolved CO_2 at low temperatures.

Light also plays a role, via the phytochrome system, in the heterophylly of *Hippuris vulgaris* (Bodkin, Spence & Weeks 1980), and *Marsilea vestita* (Gaudet 1963). A high red/far-red ratio results in a submersed leaf, and a low ratio produces the aerial leaf form of *Hippuris vulgaris*, regardless of submergence. Far-red light is much more attenuated in passage through water than red; consequently, as summer temperatures and irradiance become suitable, the low red/far-red ratio near the surface initiates the aerial leaf form, even under water (Bodkin, Spence & Weeks 1980).

Light and temperature may act through plant growth regulators, as these also influence heterophylly. Abscisic acid induces the floating leaf form of *Potamogeton nodosus*, while simultaneous treatment with gibberellin or kinetin overcomes this effect (Anderson 1978). Anderson (1978) likened this response to the endogenous production of abscisic acid by water-stressed terrestrial leaves, and suggested that localized desiccation of a submersed leaf, as it becomes emersed, could trigger abscisic acid production. Earlier work with *Proserpinaca palustris* (McCallum 1902) and *Hippuris vulgaris* (McCully & Dale 1961) indicated that leaf water relations could affect the production of the aerial leaf. However, abscisic acid did not affect the submersed leaf of *Prosperpinaca palustris*, although gibberellin did induce the aerial leaf, even under photoperiod and temperature conditions that would normally maintain the submersed leaf form (Kane & Albert 1982).

Ethylene has been implicated in the rapid stem elongation of a number of aquatic plants, such as *Callitriche platycarpa* (Musgrave, Jackson & Ling 1972), *Sagittaria pygmaea*, and *Potamogeton distinctus* (Suge & Kusanagi 1975), and deep-water rice (Métraux & Kende 1983). A further 'growth regulator' for heterophylly is CO_2. Growth of *Marsilea vestita*, *Ranunculus flabellaris* and *Myriophyllum brasiliense* in high CO_2 concentrations (0·6 to 5%) results in the submersed leaf form, even in air (Bristow & Looi 1968; Bristow 1969). This may be correlated with the higher free CO_2 concentrations that can occur in water, as compared to the level in air (0·03%). This CO_2 effect is especially interesting in light of the fact that the photosynthetic characteristics of unicellular autotrophs differ considerably, depending on whether they are grown under high or low CO_2 levels. We have demonstrated a similar CO_2 effect on the photosynthesis of the submersed angiosperm *Hydrilla verticillata*, although no changes in leaf morphology occur (Holaday, Salvucci & Bowes 1983). It is of further interest that CO_2 synergistically enhances the effects of ethylene. These CO_2 effects, however, may not be universal, in that the pioneering studies of McCallum (1902) failed to show a CO_2 effect on *Proserpinaca palustris*, and it has been ruled out as a factor in the heterophylly of *Hippuris vulgaris* (Bodkin, Spence & Weeks 1980).

Water density

Although the denseness of the medium has a number of negative implications for aquatic photosynthesis, it does have some positive aspects. The support provided by water requires minimal investment by SAM plants in structural components such as lignin (Sculthorpe 1967). Furthermore, the need for transport tissues, and the dependence on roots, is often reduced because the plant is bathed in a nutrient solution. These factors result in low dry/fresh weight ratios; a ratio of 1:15 is not uncommon in submersed angiosperms, as compared with a ratio of 1:5 for a typical terrestrial leaf (Van, Haller & Bowes 1976; Bowes, Holaday & Haller 1979). The low dry-weight partially accounts for the prolificness of SAM plants such as *Hydrilla verticillata* and *Myriophyllum spicatum*, both of which can rapidly cover water with an effective photosynthetic canopy, despite low CO_2 fixation rates. Because of the low dry weight, what appears to be dense vegetation actually has low productivity in comparison with most aerial systems (Westlake 1963; Bowes, Holaday & Haller 1979; Spencer & Bowes 1985).

The productivity of amphibious plants has not been widely studied. The maximum standing crop data we have obtained for *Limnophila sessiliflora* and *Hygrophila polysperma* do not suggest much greater productivity than that of entirely submersed species, such as *Hydrilla verticillata* (Table 3). For *Hygrophila polysperma* growing in 1 m of water, the contribution to the total biomass by the aerial portion was surprisingly low (Table 3). Whether this is a typical feature of amphibious plants has yet to be determined. It might be expected that amphibious plants must trade-off some of the support and nutrient benefits of total submersion, in order to support the aerial portions.

It can be concluded that for macrophyte photosynthesis in freshwater, the aqueous diffusion resistance combined with the naturally occurring DIC levels impose a major limitation. This correlates with the finding that the species composition of an aquatic community can change dramatically in response to

TABLE 3. Comparison of maximum standing crop values for *Hydrilla verticillata* (a submersed plant) with those of *Limnophila sessiliflora* and *Hygrophila polysperma* (amphibious plants) in Florida

Species	Plant part weighed	Standing crop values ($kg \, m^{-2}$) Fresh wt	Dry wt
Hydrilla verticillata	Total	4·30	0·89
Limnophila sessiliflora	Total	4·40	0·47
Hygrophila polysperma	Aerial	0·76	0·10
	Submersed	5·40	0·72
	Total	6·16	0·82

changing DIC levels (Roelofs, Schuurkes & Smits 1984). The DIC limitations on SAM plants and its corollary, the need to conserve carbon, are becoming increasingly apparent as important ecological features of aquatic environments (Van, Haller & Bowes 1976; Bowes *et al.* 1978; Holaday & Bowes 1980; Keeley 1981; Moore 1983; Sand-Jensen 1983). This lends weight to the possibility that the amphibious life-style is one of several strategies to mitigate DIC and diffusion resistance limitations on growth, especially during the summer months. This, and other photosynthetic strategies, are covered in the next section.

PHOTOSYNTHETIC STRATEGIES OF AQUATIC PLANTS

There appear to be several strategies used by aquatic plants to ameliorate the effects of unfavourable CO_2 and O_2 conditions that can occur as a result of the high aqueous diffusion resistance. Four such possibilities are outlined in Table 4; two are dependent predominantly on biochemical, and two on morphological features. A given species can resort to more than one of these strategies. This review will

TABLE 4. Aquatic plant photosynthetic strategies postulated to ameliorate the effects of unfavourably low carbon dioxide and high oxygen conditions in the water

Strategy	Mechanism	Examples
C_4 acid utilization	Biochemical. A CO_2 concentrating system. Fixation night and/or day into malate via PEPCase. Decarboxylation and refixation during day into the PCR cycle.	*Hydrilla* *Isoetes* *Crassula* *Littorella*
Bicarbonate utilization	Biochemical. A DIC concentrating system. Direct access to HCO_3^- ions, as well as free CO_2, and possible 'pumping' to elevate internal DIC levels.	*Myriophyllum* *Hydrilla*
Hydrosoil carbon dioxide utilization	Morphological. CO_2 piped (diffuses) from the high hydrosoil levels to the leaves via lacunae. Retained by the cuticle until fixed into the PCR cycle, sometimes via C_4 acids.	*Lobelia* *Littorella*
Aerial leaf	Morphological. Development of heterophylly gives direct access to gaseous CO_2 in the air, as well as dissolved CO_2 in the water.	*Proserpinaca* *Hygrophila* *Limnophila* *Hippuris*

concentrate on the two biochemical strategies, as they have been more extensively studied (Bowes 1985).

There is considerable evidence that the PCR and PCO cycles potentially operate in SAM plants. In many, RuBPCase is the predominant carboxylase, and it exhibits oxygenase activity (Raven 1970; Salvucci & Bowes 1981, 1982; Beer & Wetzel 1982). Photorespiratory enzymes are also present, and photorespiratory CO_2 release along with O_2 inhibition of photosynthesis have been measured (Frederick, Gruber & Tolbert 1973; Stanley & Naylor 1973; Van, Haller & Bowes 1976; Salvucci & Bowes 1983b). Despite these data, SAM plants do not fit the typical C_3 mould. In some, C_4 acid production is a major feature of CO_2 fixation, and even in plants with little C_4 acid formation the physiological characteristics do not always correspond to those of C_3 plants.

A turning point in understanding this situation was our finding that the CO_2 compensation point for a given SAM plant can vary greatly, even in nature, depending on the prior growth conditions (Bowes, Holaday & Haller 1979; Holaday, Salvucci & Bowes 1983; Maberly & Spence 1983). This plasticity is apparently a characteristic of all freshwater SAM plants, including amphibious species (Salvucci & Bowes 1981, 1982; Spencer & Bowes 1985). High temperatures and long photoperiods, typical of summer conditions, play a role in inducing low CO_2 compensation points, while winter-like conditions produce high values; the plants also have to be submerged (Bowes *et al.* 1978; Holaday, Salvucci & Bowes 1983). These environmental conditions are reminiscent of those that result in heterophylly. For *Hydrilla verticillata* the aqueous CO_2 level during growth is important (Holaday, Salvucci & Bowes 1983), which is analogous to the effect CO_2 has in altering algal photosynthetic characteristics as well as amphibious leaf morphology. We have also found that abscisic acid prevents a low value under conditions that would otherwise induce it. Although the factors that influence the CO_2 compensation point and heterophylly are similar, it is unclear if for an amphibious plant there is any direct correlation.

The significance of the variable CO_2 compensation point is that it represents a change in the photosynthesis/photorespiration ratio. A SAM plant with a high CO_2 compensation point has high photorespiration and O_2 inhibition of photosynthesis, like a C_3 plant. As the CO_2 compensation point decreases, so gas exchange data indicate that net photosynthesis increases, while photorespiration and O_2 inhibition are reduced, often to C_4-like levels (Salvucci & Bowes 1981, 1982). Thus, unlike terrestrial species, SAM plants exist in a continuum of photorespiratory states, from high-PR to low-PR. This provides a partial explanation for the discrepancies between SAM photosynthesis and that of the established C_3 and C_4 categories. However, where the aerial leaf photosynthesis of amphibious plants fits into this pattern is unresolved.

During our study of the variable PR-states of aquatic angiosperms it became apparant that two supplemental biochemical mechanisms were involved: one based on the production and decarboxylation of C_4 acids, the other on the utilization of HCO_3^-. These two systems are exemplified by *Hydrilla verticillata* and

Myriophyllum spicatum, respectively. Both systems probably supply an elevated DIC level to RuBPCase and the PCR cycle, and under adverse CO_2/O_2 conditions their operation becomes maximized, as indicated by the low-PR state.

C_4 acid utilization strategy

We have postulated that the system in *Hydrilla verticillata* acts as a CO_2 concentrating mechanism in a manner analogous to that in terrestrial C_4 plants, but without the characteristic Kranz anatomy (Bowes & Salvucci 1984). Three independent lines of evidence support this hypothesis: gas exchange, enzymatic, and labelling data.

As *Hydrilla verticillata* changes to the low-PR state, C_4 enzymes increase up to tenfold in activity, and PEPCase becomes the predominant carboxylase (Salvucci & Bowes 1981, 1983b). Two decarboxylases found in C_4 plants also are present: NAD and NADP malic enzymes; as is pyruvate Pi dikinase. Immumocytochemical localization studies indicate that RuBPCase is chloroplastic and PEPCase is cytosolic in both leaf cell types of *Hydrilla verticillata* (Bowes & Salvucci 1984). Thus these two enzymes exhibit intracellular segregation between the cytosol and chloroplast, rather than intercellular compartmentation as found in C_4 Kranz anatomy.

Labelling data in the light are consistent with the enzymatic and gas exchange data. In the low-PR state up to 60% of the initial ^{14}C incorporated is in malate and aspartate, and they exhibit rapid turnover kinetics (Salvucci & Bowes 1983b; Bowes & Salvucci 1984). This is reminiscent of C_4 photosynthesis, and shows that malate is a photosynthetic intermediate. There is reduced carbon flow through the PCO cycle, even though potential PCO cycle enzyme activities are undiminished (Salvucci & Bowes 1981, 1983b), which is compatible with a CO_2 concentrating mechanism reducing oxygenase activity.

Thus the scheme that best fits the data for *Hydrilla verticillata*, and possibly other plants such as *Elodea canadensis*, *Egeria densa* and *Lagarosiphon major*, involves CO_2 fixation via cytosolic PEPCase in the light. In the low-PR state, malate passes to the chloroplast for decarboxylation by NADP malic enzyme. The CO_2 released is refixed into the PCR cycle by RuBPCase, and pyruvate is converted to PEP by pyruvate Pi dikinase. This places the CO_2 concentrating mechanism in a single cell, rather than between two cell types as in terrestrial C_4 plants. There are still unanswered problems with this scheme. For example, what prevents a rapid egress of CO_2 out of the chloroplast? Furthermore, although the function of C_4 acids in the low-PR state can be explained, their role in the high-PR state is conjecturable.

Recently, three aquatic genera, *Littorella*, *Isoetes* and *Crassula* have been reported to show CAM photosynthesis (Keeley 1981; Keeley & Bowes 1982; Keeley & Morton 1982; Boston & Adams 1983). The CAM designation is primarily based on diel fluctuations in titratable acidity and malic acid, and a substantial capacity for dark CO_2 fixation. Low-PR *Hydrilla verticillata* also exhibits diel acidity fluctuations and dark CO_2 fixation (Holaday & Bowes 1980). For *Isoetes howellii*, dark fixation does contribute to a CAM-like carbohydrate build-up during the day (Keeley 1983), though details of the biochemistry in the light are sketchy. Aqueous and cuticular diffusion resistances, rather than a CAM-like stomatal closure, probably restrict CO_2 loss during decarboxylation, and thus facilitate refixation. It has been suggested that *Isoetes howellii* fixes CO_2 largely via the PCR cycle during the day, as in CAM; but light fixation via PEPCase, as occurs in *Hydrilla verticillata*, cannot be ruled out (Keeley & Bowes 1982), and malate synthesis does occur in the light. It has yet to be clarified to what degree *Isoetes* photosynthesis involves both CAM-like and C_4-like aspects. Irrespective of the exact C_4 acid pathway, the ecological function is similar to that in *Hydrilla verticillata*.

Bicarbonate utilization strategy

Myriophyllum species show no evidence of photosynthetic C_4 acid metabolism, or any capacity for dark fixation (Stanley & Naylor 1973; Salvucci & Bowes 1981, 1982, 1983a, 1983b). However, submersed *Myriophyllum spicatum* and *brasiliense* do show an ability to utilize HCO_3^-, though not as effectively as free CO_2 (Steeman-Nielsen 1947; Van, Haller & Bowes 1976; Salvucci & Bowes 1982), and for *Myriophyllum spicatum* this ability markedly improves in the low-PR state (Table 1). Our data indicate that *Myriophyllum* species exhibit high-PR through low-PR states that physiologically resemble the high-CO_2 and low-CO_2 grown states, respectively, of unicellular autotrophs (Salvucci & Bowes 1981, 1982, 1983a, 1983b). Thus the CO_2 compensation point, O_2 inhibition, photorespiration, and apparent $K_m(CO_2)$ for photosynthesis of the high-PR state are all reduced, and net photosynthesis is increased, as the plants shift to the low-PR state. Also, ^{14}C labelling patterns in the presence of photorespiratory inhibitors indicate a reduction in carbon flow through the PCO cycle of low-PR *Myriophyllum spicatum* (Salvucci & Bowes 1983b) that is not due to a change in RuBPCase or photorespiratory enzymes (Bowes & Salvucci 1984). These observations are compatible with some form of DIC concentrating mechanism serving to overcome the oxygenase activity of RuBPCase and increase the efficiency with which this enzyme fixes CO_2.

Carbonic anhydrase plays an important, though disputed, role in the HCO_3^- utilization system of unicellular green algae (Spalding, Spreitzer & Ogren 1983). This enzyme also appears to be a component of the *Myriophyllum spicatum* system. A doubling of activity occurs as the plant changes to the low-PR state. Furthermore, the carbonic anhydrase inhibitor, ethoxyzolamide, decreases photosynthesis, and increases the apparent $K_m(CO_2)$, the CO_2 compensation point, O_2 inhibition, and the labelling of photorespiratory intermediates in low-PR *Myriophyllum spicatum* (Salvucci & Bowes 1983a, 1983b), in a manner similar to its effects on low-CO_2 grown algae (Tsuzuki & Miyachi 1979). In short, inhibition of carbonic andydrase restricts the maximum operation of the DIC concentrating system, thus pointing to a prominent role for this enzyme in HCO_3^- use by *Myriophyllum* species.

These results with *Myriophyllum* species indicate the presence of an HCO_3^- utilization system that can increase its operating capacity in response to increasingly stressful environmental conditions. Although the data are suggestive, the use of this system to increase DIC levels in other SAM plants has yet to be proven. The algal and SAM mechanisms differ in that the very low apparent $K_m(CO_2)$ values for photosynthesis of the algae are not found with SAM plants; though a 63% decrease is observed as *Myriophyllum brasiliense* shifts to the low-PR state (Salvucci & Bowes 1982). A further difference is that the low-CO_2 grown state appears to be the naturally occurring form for the algae, whereas in SAM plants the high-PR state appears to be much more common in nature (Bowes, Holaday & Haller 1979; Maberly & Spence 1983).

Hydrosoil carbon dioxide utilization strategy

A method employed by some SAM plants to enhance carbon gain in DIC-poor waters entails morphological adaptations, enabling them to use hydrosoil CO_2. Sediment CO_2 levels can be two orders of magnitude greater than those of the water; thus the ability to tap these CO_2 reserves might be advantageous, especially if the plant is a poor HCO_3^- user (Wium-Anderson 1971; Søndergaard & Sand-Jensen 1979). The morphological adaptations include a low-growing, isoetid plant form, with leaves that are short, stem-like, possess large lacunal gas spaces, a thick cuticle, and lack functional stomata. The plants also exhibit a large root surface area in comparison with that of the leaves. Wium-Anderson (1971) first demonstrated for *Lobelia dortmanna* that the plant derived most of its CO_2 for photosynthesis from the hydrosoil via the roots and lacunal gas channels. The morphology of the plant minimizes the diffusion resistance between the hydrosoil and the chloroplast, while the thick cuticle prevents escape of the CO_2 before it is fixed. This system also enables the plant to act as an 'O$_2$ pump', in that photosynthetic O_2 diffuses down to oxidize the sediment in the root zone, instead of being liberated into the water surrounding the leaves (Wium-Anderson 1971).

Littorella uniflora also uses hydrosoil CO_2 for photosynthesis (Søndergaard & Sand-Jensen 1979; Boston & Adams 1983), and its morphology is very similar to that of *Lobelia dortmanna*. However, in the case of *Littorella uniflora* much of the CO_2 is initially fixed into C_4 acids, especially at night, whereas this is not true of *Lobelia dortmanna* (Boston & Adams 1983). This *Littorella uniflora* uses both the C_4 acid and hydrosoil CO_2 strategies to improve carbon uptake and minimize O_2 inhibition effects on photosynthesis. The major pathway for CO_2 movement to the leaves of these plants is by gaseous diffusion, though HCO_3^- transport cannot be ruled out. For *Littorella uniflora*, it is not known if malate transport from roots to leaves plays any role in the process, or even if roots have any substantial capacity for dark CO_2 fixation into C_4 acids.

A less specialized variation on the hydrosoil CO_2 strategy may involve some plants that grow close to organic sediments, for example with a prostrate habit. Their leaves, rather than roots, could take advantage of the locally high CO_2 concentrations released by decomposition at the water-sediment interface. This possibility has yet to be examined.

Aerial leaf strategy

The final strategy, which has been alluded to throughout this paper, involves the morphological adaptations adopted by amphibious SAM plants. As discussed earlier, various factors more typically associated with an aerial existence, such as water stress or a low red/far-red ratio, trigger the submersed shoot into production of aerial leaves in preparation for photosynthesis in a gaseous medium. A number of factors point to the aerial leaf as being a more effective photosynthetic organ than its submersed counterpart. These include: much higher net photosynthetic rates, lower apparent $K_m(CO_2)$ values and CO_2 saturation requirements for photosynthesis, higher RuBPCase activities, and the potential to use high irradiances (Lloyd, Canvin & Bristow 1977; Salvucci & Bowes 1982; Maberly & Spence 1983; Spencer & Bowes 1985). All of these are directly, or indirectly, the result of the greatly reduced CO_2 diffusion resistance in the gas phase. Thus it may be postulated that the function of the aerial leaf is to improve access to inorganic carbon.

However, the enhanced photosynthetic parameters do not, in themselves, constitute proof that the primary function of the aerial shoot is photosynthetic. The high photosynthetic rates could be offset by a more limited season for the aerial shoot, and/or by a limited amount of photosynthetic machinery being placed above the water surface. It is also likely that the contribution of aerial shoot photosynthesis to total carbon gain differs considerably among different amphibious species, or even from year to year. Field observations of *Hygrophila polysperma* suggest that the aerial portion has a greater CO_2 fixation function than that of *Limnophila sessiliflora* (Spencer & Bowes 1985), although this is not well substantiated by data from standing crops (Table 3). It is just as possible that the aerial shoot by its transpiration stream improves nutrient uptake from the hydrosoil. Alternatively, it could serve primarily as a structural support for aerial flower production, pollination, and seed dispersal, with photosynthesis being a secondary role.

It is interesting to note that the aerial leaves of amphibious plants are not completely analogous to the leaves of true terrestrial species. Despite their emergent nature, they still have higher $K_m(CO_2)$ values for photosynthesis, lower net photosynthetic rates, and lower RuBPCase activities than found for most terrestrial C_3 leaves. We have also shown that they can exhibit variable CO_2 compensation points (Salvucci & Bowes 1981, 1982), which may betray their aquatic heritage. It is not known if photosynthetic C_4 acid metabolism occurs in the aerial leaves of any amphibious plant. It has been reported that after emersion of

Isoetes howellii by the drying of the vernal pools the plant reverts from C_4 acid use to more typical PCR cycle activity (Keeley 1983), though this need not be a ubiquitous characteristic of the emergent state of other amphibious plants.

In summary, more research into factors such as seasonal productivity, carbon allocation and fixation patterns, and nutrient requirements, as well as the basic biochemistry of aerial and submersed leaves, needs to be undertaken before the aerial leaf strategy can be unequivocally accepted as primarily photosynthetic. Regardless of this caveat, it is becoming increasingly recognized that DIC is a major limiting nutrient for SAM plant photosynthesis (Van, Haller & Bowes 1976; Bowes *et al.* 1978; Keeley 1981; Moore 1983; Sand-Jensen 1983). Consequently, in any analysis of macrophyte productivity the availability and role of this resource should be given as careful consideration as that traditionally afforded to other inorganic nutrients, such as nitrogen, phosphorus, and potassium.

ACKNOWLEDGMENTS

I thank the following for their valuable contributions to this work: M. E. Salvucci, W. E. Spencer, A. S. Holaday, J. C. Ascencio, T. K. Van, J. B. Reiskind, W. T. Haller, P. A. Seamon and S. A. Berish. I also appreciate the encouragement given by Professor David H. N. Spence to present this paper at the BES symposium in St Andrews. His death has dealt a sad blow to aquatic plant research, and to his many friends around the globe.

This research was supported by the Science and Education Administration of the United States Department of Agriculture under Grants 82-CRCR-1-1147 from the Competitive Research Grants Office, and by the Bureau of Aquatic Plant Research and Control, Florida Department of Natural Resources.

REFERENCES

Allen, E. D. & Spence, D. H. N. (1981). The differential ability of aquatic plants to utilize the inorganic carbon supply in fresh waters. *New Phytologist*, **87**, 269–283.

Anderson, L. W. T. (1978). Abscisic acid induces formation of floating leaves in the heterophyllous aquatic angiosperm *Potamogeton nodosus*. *Science*, **201**, 1135–1138.

Badger, M. R., Kaplan, A. & Berry, J. A. (1978). A mechanism for concentrating CO_2 in *Chlamydomonas reinhardtii* and *Anabaena variabilis* and its role in photosynthetic CO_2 fixation. *Carnegie Institution of Washington Year Book*, **77**, 251–261.

Bain, J. T. & Proctor, M. C. F. (1980). The requirement of aquatic bryophytes for free CO_2 as an inorganic carbon source: some experimental evidence. *New Phytologist*, **86**, 393–400.

Beardall, J. & Raven, J. A. (1981). Transport of inorganic carbon and the 'CO_2 concentrating mechanism' in *Chlorella emersonii* (Chlorophyceae). *Journal of Phycology*, **17**, 134–141.

Beer, S. & Wetzel, R. G. (1982). Photosynthesis in submersed macrophytes of a temperate lake. *Plant Physiology*, **70**, 488–492.

Black, M. A., Maberly, S. C. & Spence, D. H. N. (1981). Resistances to carbon dioxide fixation in four submerged freshwater macrophytes. *New Phytologist*, **89**, 557–568.

Bodkin, P. C., Spence, D. H. N. & Weeks, D. C. (1980). Photoreversible control of heterophylly in *Hippuris vulgaris* L. *New Phytologist*, **84**, 533–542.

Boston, H. L. & Adams, M. S. (1983). Evidence of Crassulacean Acid Metabolism in two north American isoetids. *Aquatic Botany*, **15**, 381–386.

Bowes, G. (1985). Pathways of CO_2 fixation by aquatic organisms. *Inorganic Carbon Uptake by Aquatic Photosynthetic Organisms*, pp. 187–210. American Society of Plant Physiologists, Rockville, Maryland.

Bowes, G., Holaday, A. S., Van, T. K. & Haller, W. T. (1978). Photosynthetic and photorespiratory carbon metabolism in aquatic plants. *Proceedings of the Fourth International Congress on Photosynthesis*, pp. 289–298. The Biochemical Society, London.

Bowes, G., Holaday, A. S. & Haller, W. T. (1979). Seasonal variation in biomass, tuber density, and photosynthetic metabolism of hydrilla in three Florida lakes. *Journal of Aquatic Plant Management*, **17**, 61–65.

Bowes, G., Van, T. K., Garrard, L. A. & Haller, W. T. (1977). Adaptation to low light levels by hydrilla. *Journal of Aquatic Plant Management*, **15**, 32–35.

Bowes, G. & Salvucci, M. E. (1984). *Hydrilla*: inducible C_4-type photosynthesis without Kranz anatomy. *Proceedings of the Sixth International Congress on Photosynthesis*, Vol. 3, pp. 829–832. Martinus Nijhoff/Dr. W. Junk, The Hague.

Boylen, C. W. & Sheldon, R. B. (1976). Submergent macrophytes: growth under winter ice cover. *Science*, **194**, 841–842.

Bristow, J. M. (1969). The effects of carbon dioxide on the growth and development of amphibious plants. *Canadian Journal of Botany*, **47**, 1803–1807.

Bristow, J. M. & Looi, A. S. (1968). Effects of carbon dioxide on the growth and morphogenesis of *Marsilea. American Journal of Botany*, **55**, 884–889.

Brown, J. M. A., Dromgoole, F. I., Towsey, M. W. & Browse, J. (1974). Photosynthesis and photorespiration in aquatic macrophytes. *Mechanisms of Regulation of Plant Growth*. pp. 243–249. The Royal Society of New Zealand, Wellington.

Browse, J. A., Brown, J. M. A. & Dromgoole, F. I. (1979). Photosynthesis in the aquatic macrophyte *Egeria densa*. II. Effects of inorganic carbon conditions on ^{14}C fixation. *Australian Journal of Plant Physiology*, **6**, 1–9.

Browse, J. A., Dromgoole, F. I. & Brown, J. M. A. (1979). Photosynthesis in the aquatic macrophyte *Egeria densa*. III. Gas exchange studies. *Australian Journal of Plant Physiology*, **6**, 499–512.

Davis, G. J. (1967). *Proserpinaca*: photoperiodic and chemical differentiation of leaf development and flowering. *Plant Physiology*, **42**, 667–668.

DeGroote, D. & Kennedy, R. A. (1977). Photosynthesis in *Elodea canadensis* Michx. Four-carbon acid synthesis. *Plant Physiology*, **59**, 1133–1135.

Edwards, G. E. & Huber, S. C. (1981). The C_4 pathway. *The Biochemistry of Plants. Vol. 8. Photosynthesis*. pp. 237–281. Academic Press, New York.

Frederick, S. E., Gruber, P. J. & Tolbert, N. E. (1973). The occurrence of glycolate dehydrogenase and glycolate oxidase in green plants. An evolutionary survey. *Plant Physiology*, **52**, 318–323.

Gaudet, J. J. (1963). *Marsilea vestita*: conversion of the water form to the land form by darkness and far-red light. *Science*, **140**, 975–976.

Holaday, A. S. & Bowes, G. (1980). C_4 metabolism and dark CO_2 fixation in a submersed aquatic macrophyte (*Hydrilla verticillata*). *Plant Physiology*, **65**, 331–335.

Holaday, A. S., Salvucci, M. E. & Bowes, G. (1983). Variable photosynthesis/photorespiration ratios in *Hydrilla* and other submersed aquatic macrophyte species. *Canadian Journal of Botany*, **61**, 229–236.

Hough, R. A. & Wetzel, R. G. (1977). Photosynthetic pathways of some aquatic plants. *Aquatic Botany*, **3**, 297–313.

Hutchinson, G. E. (1970). The chemical ecology of three species of *Myriophyllum* (Angiospermae, Haloragaceae). *Limnology and Oceanography*, **15**, 1–5.

Johnson, M. P. (1967). Temperature dependent leaf morphogenesis in *Ranunculus flabellaris. Nature*, **214**, 1354–1355.

Kane, M. E. & Albert, L. S. (1982). Environmental and growth regulator effects on heterophylly and growth of *Proserpinaca intermedia* (Haloragaceae). *Aquatic Botany*, **13**, 73–85.

Keeley, J. E. (1981). *Isoetes howellii*: a submerged aquatic CAM plant? *American Journal of Botany*, **68**, 420–424.

Keeley, J. E. (1983). Crassulacean acid metabolism in the seasonally submerged aquatic *Isoetes howellii*. *Oecologia*, **58**, 57–62.

Keeley, J. E. & Bowes, G. (1982). Gas exchange characteristics of the submerged aquatic crassulacean acid metabolism plant, *Isoetes howellii*. *Plant Physiology*, **70**, 1455–1458.

Keeley, J. E. & Morton, B. A. (1982). Distribution of diurnal acid metabolism in submerged aquatic plants outside the genus *Isoetes*. *Photosynthetica*, **16**, 546–553.

Kluge, M. & Ting, I. P. (1978). *Crassulacean Acid Metabolism. Analysis of an Ecological Adaptation*. Springer-Verlag, New York.

Lloyd, N. D. H., Canvin, D. T. & Bristow, J. M. (1977). Photosynthesis and photorespiration in submerged aquatic vascular plants. *Canadian Journal of Botany*, **55**, 3001–3005.

Maberly, S. C. (1983). The interdependence of photon irradiance and free carbon dioxide or bicarbonate concentration on the photosynthetic compensation points of freshwater plants. *New Phytologist*, **93**, 1–12.

Maberly, S. C. & Spence, D. H. N. (1983). Photosynthetic inorganic carbon use by freshwater plants. *Journal of Ecology*, **71**, 705–724.

McCallum, W. B. (1902). On the nature of the stimulus causing the change of form and structure in *Proserpinaca palustris*. *Botanical Gazette*, **34**, 93–108.

McCully, M. E. & Dale, H. M. (1961). Heterophylly in *Hippuris*, a problem in identification. *Canadian Journal of Botany*, **39**, 1099–1116.

Métraux, J. P. & Kende, H. (1983). The role of ethylene in the growth response of submerged deep water rice. *Plant Physiology*, **72**, 441–446.

Monson, R. K., Edwards, G. E. & Ku, M. S. B. (1984). C_3–C_4 intermediate photosynthesis in plants. *BioScience*, **34**, 563–574.

Moore, P. D. (1983). Photosynthetic pathways in aquatic plants. *Nature*, **304**, 310.

Musgrave, A., Jackson, M. B. & Ling, E. (1972). *Callitriche* stem elongation is controlled by ethylene and gibberellin. *Nature New Biology*, **238**, 93–96.

Osmond, C. B., Valaane, N., Haslam, S. M., Uotila, P. & Roksandic, Z. (1981). Comparisons of $\delta^{13}C$ values in leaves of aquatic macrophytes from different habitats in Britain and Finland; some implications for photosynthetic processes in aquatic plants. *Oecologia*, **50**, 117–124.

Pearcy, R. W. & Ehleringer, J. (1984). Comparative ecophysiology of C_3 and C_4 plants. *Plant, Cell and Environment*, **7**, 1–13.

Prins, H. B. A., Snel, J. F. H., Zanstra, P. E. & Helder, R. J. (1982). The mechanism of bicarbonate assimilation by the polar leaves of *Potmogeton* and *Elodea*. CO_2 concentrations at the leaf surface. *Plant, Cell and Environment*, **5**, 207–214.

Raven, J. A. (1970). Exogenous inorganic carbon sources in plant photosynthesis. *Biological Reviews*, **45**, 167–221.

Raven, J. A. & Glidewell, S. M. (1975). Photosynthesis, respiration and growth in the shade alga *Hydrodictyon africanum*. *Photosynthetica*, **9**, 361–371.

Roelofs, J. G. M., Schuurkes, J. A. A. R. & Smits, A. J. M. (1984). Impact of acidification and eutrophication on macrophyte communities in soft waters. II. Experimental studies. *Aquatic Botany*, **18**, 389–411.

Salvucci, M. E. & Bowes, G. (1981). Induction of reduced photorespiratory activity in submersed and amphibious aquatic macrophytes. *Plant Physiology*, **67**, 335–340.

Salvucci, M. E. & Bowes, G. (1982). Photosynthetic and photorespiratory responses of the aerial and submerged leaves of *Myriophyllum brasiliense*. *Aquatic Botany*, **13**, 147–164.

Salvucci, M. E. & Bowes, G. (1983a). Ethoxyzolamide repression of the low photorespiration state in two submersed angiosperms. *Planta*, **158**, 27–34.

Salvucci, M. E. & Bowes, G. (1983b). Two photosynthetic mechanisms mediating the low photorespiratory state in submersed aquatic angiosperms. *Plant Physiology*, **73**, 488–496.

Sand-Jensen, K. (1983). Photosynthetic carbon sources of stream macrophytes. *Journal of Experimental Botany*, **34**, 198–210.

Schmidt, B. L. & Millington, W. F. (1968). Regulation of leaf shape in *Proserpinaca palustris*. *Bulletin of the Torrey Botanical Club*, **95**, 264–286.

Sculthorpe, C. D. (1967). *The Biology of Aquatic Vascular Plants*. Edward Arnold Limited, London.

Smith, F. A. & Walker, N. A. (1980). Photosynthesis by aquatic plants: effects of unstirred layers in relation to assimilation of CO_2 and HCO_3^- and to carbon isotopic discrimination. *New Phytologist*, **86**, 245–259.

Søndergaard, M. & Sand-Jenson, K. (1979). Carbon uptake by leaves and roots of *Littorella uniflora* (L.) Aschers. *Aquatic Botany*, **6**, 1–12.

Spalding, M. H., Spreitzer, R. J. & Ogren, W. L. (1983). Genetic and physiological analysis of the CO_2-concentrating system of *Chlamydomonas reinhardii*. *Planta*, **159**, 261–266.

Spence, D. H. N. (1981). Light quality and plant responses underwater. *Plants and the Daylight Spectrum*. pp. 245–275. Academic Press, London.

Spence, D. H. N. & Maberly, S. C. (1985). Occurrence and ecological importance of HCO_3^- use among aquatic higher plants. *Inorganic Carbon Uptake by Aquatic Photosynthetic Organisms*. pp. 125–143. American Society of Plant Physiologists, Rockville, Maryland.

Spencer, W. E. & Bowes, G. (1985). *Limnophila* and *Hygrophila*: a review and physiological assessment of their weed potential in Florida. *Journal of Aquatic Plant Management*, **23**, 7–16.

Stanley, R. A. & Naylor, A. W. (1972). Photosynthesis in Eurasian watermilfoil (*Myriophyllum spicatum* L.). *Plant Physiology*, **50**, 149–151.

Stanley, R. A. & Naylor, A. W. (1973). Glycolate metabolism in Eurasian watermilfoil (*Myriophyllum spicatum*). *Physiologia Plantarum*, **29**, 60–63.

Steemann-Nielsen, E. (1947). Photosynthesis of aquatic plants with special reference to the carbon-sources. *Dansk Botanisk Arkiv*, **12**, 3–71.

Suge, H. & Kusanagi, T. (1975). Ethylene and carbon dioxide: regulation of growth in two perennial aquatic plants, arrowhead and pondweed. *Plant & Cell Physiology*, **16**, 65–72.

Tsuzuki, M. & Miyachi, S. (1979). Effects of CO_2 concentration during growth and of ethoxyzolamide on CO_2 compensation point in *Chlorella*. *Federation of European Biochemical Societies Letters*, **103**, 221–223.

Van, T. K., Haller, W. T. & Bowes, G. (1976). Comparison of the photosynthetic characteristics of three submersed aquatic plants. *Plant Physiology*, **58**, 761–768.

Wallenstein, A. & Albert, L. S. (1963). Plant morphology: its control in *Proserpinaca* by photoperiod, temperature, and gibberellic acid. *Science*, **140**, 998–1000.

Westlake, D. F. (1963). Comparisons of plant productivity. *Biological Reviews*, **38**, 385–425.

Wium-Anderson, S. (1971). Photosynthetic uptake of free CO_2 by the roots of *Lobelia dortmanna*. *Physiologia Plantarum*, **25**, 245–248.

Environmental control of bicarbonate use among freshwater and marine macrophytes

KAJ SAND-JENSEN

Freshwater Biological Laboratory, University of Copenhagen, Helsingørsgade 51, 3400 Hillerød, Denmark

SUMMARY

1 Morphological and anatomical adaptations to enhance the carbon gain are integrated in the physiology and ecology of aquatic plants and may also play a role in mineral nutrition, light utilization and O_2 transport.

2 Use of HCO_3^- is widespread among fully submerged macrophytes in freshwater and marine habitats and is present within most taxonomic groups.

3 Macrophytes with no apparent affinity for HCO_3^- are few in seawater and more common in freshwater especially softwater habitats. Affinity for HCO_3^- also appears to be higher among marine species.

4 Macrophytes with no apparent affinity for HCO_3^- can use alternative CO_2 sources in the atmosphere or the sediment. When they rely on CO_2 in water they often grow in microhabitats enriched with CO_2 or have very low CO_2 compensation points.

5 Use of HCO_3^- is plastic in *Elodea canadensis* depending on environmental conditions. Use of HCO_3^- is particularly beneficial when light, nutrients and HCO_3^- are high. It is suppressed when CO_2 is high or light is low.

6 High affinity for HCO_3^- and high photosynthetic capacities of *Elodea* are coupled and linearly related to tissue concentrations of chlorophyll.

7 Use of HCO_3^- is a universal carbon concentrating mechanism and probably in part responsible for the low photorespiratory state induced by high light and temperature during summer (Salvucci & Bowes 1983).

CARBON SOURCES AND ADAPTATION

Transport of dissolved inorganic carbon (DIC) to the surface of submerged macrophytes is restricted by the slow rate of diffusion through the stagnant boundary layer surrounding the plant. The supply of DIC is therefore potentially rate limiting for fixation and growth. Photosynthetic rates of submerged plants brought into the laboratory in natural water and exposed to optimum light are often carbon limited. This means that photosynthetic rates are stimulated when DIC concentrations are increased and when HCO_3^- is converted into CO_2 which is used more readily (Sand-Jensen 1983). In the field, the situation is more complex and rate limitation by different environmental variables can change with species, time and the particular microenvironment (Maberly 1985). Sustained carbon limitation of photosynthesis and growth has never been demonstrated in the field.

99

Many characteristics have been suggested as adaptations to enhance the external supply of inorganic carbon and maintain net fixation under severe depletion (Salvucci & Bowes 1983). The plant, however, is an integrated functional unit that can exploit an array of environmental variables, so rather than looking at those characteristics purely as adaptations to enhance the carbon gain, they can be viewed as general characteristics of the physiology and ecology of the plant.

Macroscopic plants grow in the transition between sediment, water and air and have alternative carbon sources to those in water. Macrophytes with floating or emergent leaves have access to the large CO_2 reservoir in the atmosphere where diffusion rates are high. Exposure to air can also increase light utilization, increase the supply of inorganic nutrients by inducing a transpiration stream from the roots to the leaves and increase O_2 transport to the below-ground parts especially when solar heating of leaves generates a mass flow of air in the lacunae (Dacey 1980). Isoetid species which dominate in softwater and nutrient-poor lakes have short leaves and roots with extensive longitudinal air channels (Sand-Jensen & Prahl 1982). They possess an efficient pathway for transport of gases between the sediment and chloroplasts in cells lining the leaf lacunae, by which they can utilize the rich source of CO_2 in the sediment and provide the roots and the sediment with O_2 (Wium-Andersen & Andersen 1972). Oxygen release to the sediment can maintain an efficient oxic mineralization of organic compounds and supply O_2 to mycorrhizal fungi which may assist the plant to exploit the mineral sources (Sand-Jensen & Prahl 1982). So in both examples the particular morphology and

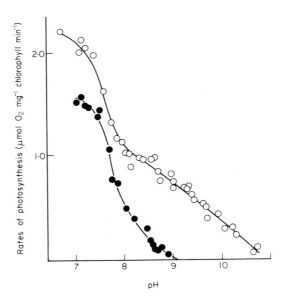

FIG. 1. Photosynthetic rates as a function of pH for *Potamogeton pectinatus* ($\bigcirc$, a HCO_3^- user) and for *Callitriche stagnalis* ($\bullet$, a CO_2 user) at a constant DIC concentration of 5 mM. (Redrawn from Sand-Jensen (1983)).

anatomy can play a role in carbon and mineral nutrition as well as in transport of O_2.

Fully submerged macrophytes without extensive air-filled roots rely on the supply of carbon from the water. They usually have thin leaves, sometimes finely dissected, and a thin cuticle to optimize exposure of the assimilating plant surface to the surrounding water. This morphology is often combined with an ability to use HCO_3^- by active transport to supplement the passive entry of CO_2 (Raven 1970). The concentration of HCO_3^- varies among freshwater habitats from zero in acid softwater localities to several mM HCO_3^- in hardwater localities. In seawater the HCO_3^- concentration is high, at about 2 mM, and more constant (Stumm & Morgan 1970). The concentration of carbon dioxide is only about 15 μM at 15°C in water equilibrated with the atmosphere and is surpassed severalfold by HCO_3^- in most waters. The ability to use HCO_3^- therefore enables the submerged macrophyte to photosynthesize more rapidly in alkaline water at normal pH, to extract more carbon from the initial DIC pool and maintain photosynthetic activity at pH values above 10 (Fig. 1).

TECHNIQUES FOR MEASURING HCO_3^- USE

It is a problem to distinguish between use of CO_2 and HCO_3^- because the two compounds exist together and are interconvertible. Most techniques make use of the fact that the distribution among $CO_2 \leftrightarrows HCO_3^- \leftrightarrows CO_3^{2-}$ is accurately defined by ionic strength, temperature and in particular pH (Rebsdorf 1972). The equilibrium is pushed to the right by increasing pH and to the left by decreasing pH.

An early technique to examine use of HCO_3^- in submerged macrophytes was to measure the dependency of photosynthesis on DIC concentrations at around pH 5 where only CO_2 occurs and at around pH 8 or 9 where HCO_3^- predominates (Steemann-Nielsen 1947). The relationship between photosynthesis and CO_2 concentrations at low pH is then used to subtract the photosynthetic rate, based on the low CO_2 concentrations present at high pH, and the difference is believed to represent use of HCO_3^-. One problem with this technique is that lowering the pH to 5 sometimes inhibits photosynthesis. Another problem, which this technique shares with the following one, is the replenishment of CO_2 from HCO_3^- as CO_2 is assimilated. Generation of CO_2 can be evaluated by calculating the rate either in the bulk medium or the stagnant boundary layer surrounding the plant surfaces given the pH and temperature (Stumm & Morgan 1970). The generation rate often cannot account for the photosynthetic rates, or the relationship to photosynthetic rates is not likely to describe a direct dependency (Beer, Eshel & Waisel 1977).

In the pH drift experiments, changes in pH with time in a closed photosynthetic chamber are used to calculate DIC uptake rates as a function of the composition of inorganic carbon species in the medium (Allen & Spence 1981; Maberley & Spence 1983). It is best to check that estimated and actual changes in DIC agree. Plants that use HCO_3^- will increase pH in an alkaline medium to above 10 where CO_2 is negligible and HCO_3^- is reduced to low concentrations. They will also extract a large proportion of the initial DIC pool before they stop photosynthesizing. The drawbacks are the extensive time needed for the plant to reach its 'final pH' during which the HCO_3^- affinity may change (K. Sand-Jensen & T. V. Madsen, unpublished data), and the build-up of high O_2 concentrations which tends to reduce carbon fixation.

We have examined HCO_3^- use in submerged macrophytes by measuring photosynthetic rates in a closed chamber at selected DIC concentrations and pH values between 6 and 11 set by adding HC1 or NaOH (Sand-Jensen 1983; Sand-Jensen & Gordon 1984, following Beer, Eshel & Waisel 1977). Those experiments were always run at two DIC concentrations. A species with no apparent use of HCO_3^- has a photosynthesis-pH relationship resembling the concentration of CO_2 versus pH. A HCO_3^--user has a plateau of photosynthesis between 8 and 9 and maintains activity at much higher pH values (Figs 1 and 2). This difference can be demonstrated more clearly if we plot photosynthetic rates (*y*-axis) versus CO_2 concentrations (*x*-axis) for pH values below 8·5 (Fig. 3). We

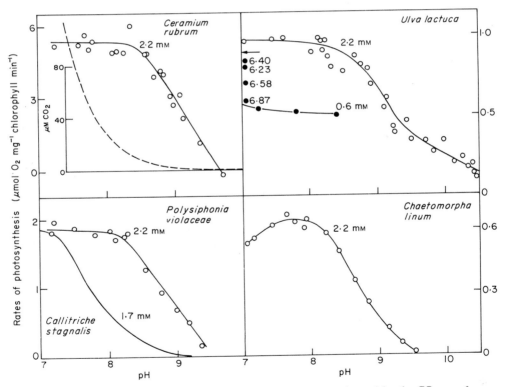

FIG. 2. Photosynthetic rates as a function of pH for four marine species and for the CO_2 user from freshwater, *Callitriche stagnalis*. The CO_2 concentration as a function of pH at 2·2 mM of DIC (– – –, upper left). The DIC concentration is shown on the figure and photosynthetic rates below the arrow should be shifted to the pH indicated. From Sand-Jensen & Gordon (1984).

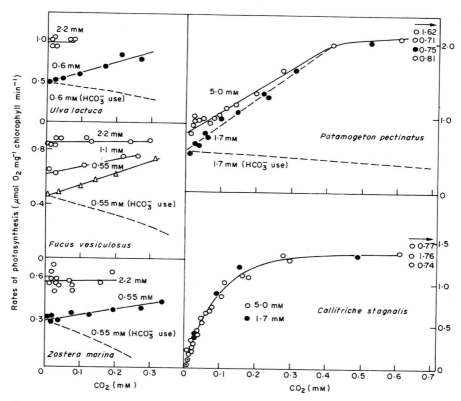

FIG. 3. Photosynthetic rates as a function of CO_2 created by varying pH from 8·5 and downwards in a medium of constant DIC as shown on the figure. The stippled line shows the contribution of HCO_3^- to the photosynthetic rate at 0·55 or 0·60 mM of DIC (left column) and at 1·7 mM of DIC for *P. pectinatus*. Photosynthetic rates below the arrow should be shifted to the CO_2 concentration indicated. From Sand-Jensen & Gordon (1984).

exclude the higher pH values because CO_2 concentrations here are very small and photosynthetic rates decline because of limitation by high pH. For a HCO_3^--user the linear relationship between photosynthetic rates and low CO_2 concentrations will intercept the *y*-axis at positive photosynthetic rates at zero CO_2 (Fig. 3, upper right). Use of HCO_3^- is furthermore shown by the higher photosynthetic rates with increasing HCO_3^- concentrations for the same rate limiting CO_2 concentration. A species which does not use HCO_3^- will have photosynthetic rates that follow a normal CO_2 saturation curve and different levels of HCO_3^- will not affect the photosynthetic rate (Fig. 3, lower right).

APPARENT NON-USERS OF HCO_3^-

A number of submerged macrophytes are apparently not able to use HCO_3^- (reviewed by Spence & Maberly 1985). Instead they use alternative CO_2 sources

TABLE 1. Species of freshwater vascular plants and bryophytes classified as CO_2 users and grouped according to their usual source of CO_2

Examples of different sources of CO_2	C_T/Alk	γ(μM CO_2)
CO_2 use from sediment:		
Isoetes lacustris	1·01	60*
Littorella uniflora	1·04	70–110*
Lobelia dortmanna	0·99	60–95*
CO_2 use from air and/or water:		
Callitriche stagnalis		2·5
Hippuris vulgaris	0·98	4·0
Nuphar lutea	0·97	
Pilularia globulifera	0·98	
Potamogeton natans	0·97	2·1
P. polygonifolius	0·96	5·1
Subularia aquatica	1·00	
CO_2 use from water:		
Drepanocladus aduncus		5·5
Eurhynchium rusciforme	0·96	3·5–3·7
Fontinalis antipyretica	0·91–0·99	2–12*
Juncus bulbosus var. *fluitans*	0·93	5·9
Pellia epiphylla		5·4
Racomitrium aciculare		3·2
Rhynchostegium riparoides		6·0
Sparganium simplex		8·3
Sphagnum cuspidatum		
Utricularia intermedia	0·97	

The ability to extract inorganic carbon from water (C_T/Alk) and the CO_2 compensation point (γ) are shown when available. This grouping should not be taken too rigorously since some species can change between alternative CO_2 sources (e.g. amphibious *Littorella uniflora*) and some have not been examined for their ability to exploit sediment sources. This list is an extract of Tables 3 and 6 in Spence & Maberly (1985) and citations from there with a few additions and alterations based on K. Sand-Jensen (unpublished data)*. A number of species particularly in the last group take advantage of living in a microzone enriched with CO_2 from sediment release.

from air or sediment, grow in habitats where CO_2 is oversaturated or possess very low CO_2 compensation points (Table 1). Isoetid species which have an efficient uptake of CO_2 from the sediment and species with floating leaves all belong to this group (Fig. 4). Amphibious species, including several isoetids, also appear to use CO_2 only, even when submerged. The inability to use HCO_3^- can be due to a high resistance to transport induced by a thick cuticle (e.g. *Lobelia dortmanna*), the lack of porters for active transport of ions and a long and difficult diffusion path from the surrounding water to the chloroplast (e.g. thick leaves and passage through internal lacunae).

Many of the fully submerged macrophytes which use CO_2 from the water grow in habitats that are oversaturated with CO_2 or they have very low CO_2 compensation points (Table 1). The aquatic mosses, for example, are often small and may grow in a microzone enriched with CO_2 from respiratory processes in the

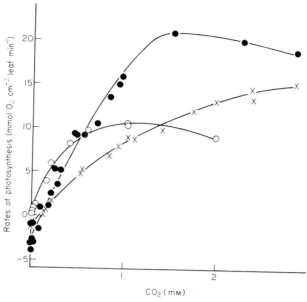

Fig. 4. Photosynthetic rates of three CO_2 users as a function of different CO_2 concentrations induced by varying the pH from 8·5 and downwards in a medium of constant DIC. Entire plants with intact roots were incubated and leaves and roots were exposed to the same CO_2 concentrations. *Littorella uniflora* and *Lobelia dortmanna* use CO_2 from both leaf and root surfaces and have high CO_2 compensation points and high saturating CO_2 concentrations. *Juncus bulbosus* var. *fluitans* has thin filamentous leaves and can exploit CO_2 by the leaves quite efficiently. All species were collected in an oligotrophic softwater lake (pH 4·7–5·5). From Sand-Jensen (unpublished data). (●) *Littorella uniflora*; (X) *Lobelia dortmanna*; (○) *Juncus bulbosus*.

mud below (Maberly 1985). They frequently grow at the lower depth boundary of submerged vegetation where light rather than inorganic carbon is growth-limiting. If these species are plastic in their affinity for inorganic carbon (like *Elodea canadensis*, Sand-Jensen & Gordon 1986) their inability to use HCO_3^- may simply reflect their growth habitat. Until recently all aquatic mosses tested had been unable to use HCO_3^- (Bain & Proctor 1980), but a new examination of *Fontinalis antipyretica* collected from a Spanish hardwater stream did demonstrate an ability to use HCO_3^- (Penuelas 1985).

We have examples from hardwater streams of HCO_3^--users and apparent non-users growing next to each other (Sand-Jensen 1983). The flowing water and the high buffer capacity tend to prevent severe depletion of CO_2 around the submerged plant surfaces. When this is combined with a low CO_2 compensation point and oversaturation with CO_2 in the morning following night-time respiration, this is apparently sufficient for non-users to grow quite rapidly although their photosynthetic rates at light saturation and CO_2 equilibrium with the atmosphere are only 6–10% of the maximum capacity and 4–5 times lower than the HCO_3^-

users (Fig. 1, and Sand-Jensen 1983). Based on photosynthetic performance under light saturation we would predict competitive exclusion of the non-users. Competition probably occurs during summer when dense plant stands develop, but the ability to survive during winter, regenerate after weed cutting and exploit efficiently other environmental variables (e.g. light and nutrients) are probably equally important and make the outcome of competition unpredictable.

USE OF HCO_3^- AMONG FRESHWATER AND MARINE MACROPHYTES

Representatives of HCO_3^--users are found within most taxonomic groups of macrophytes in freshwater (Spence & Maberly 1985). Bicarbonate use is also widespread among algae and angiosperms in seawater and it is difficult to find species here whose photosynthetic characteristics accord entirely with diffusive entry of CO_2 (Sand-Jensen & Gordon 1984).

TABLE 2. Apparent half-saturation constants for CO_2 and HCO_3^- use and ratios of CO_2/HCO_3^- affinity for photosynthesis of marine (M) and freshwater macrophytes (F) under rate limiting concentrations

Macrophyte species and habitat	$K_\frac{1}{2}(CO_2)$ (mM)	$K_\frac{1}{2}(HCO_3^-)$ (mM)	CO_2/HCO_3^- affinity
Ceramium rubrum M		0·80	
Fucus vesiculosus M	0·28	0·54	$2\cdot1^a\ 1\cdot9^b\ 1\cdot6^c$
Ulva lactuca M	0·26	0·60	$2\cdot5^a\ 2\cdot3^b\ 2\cdot4^c$
Zostera marina M	0·27	0·60	$1\cdot7^a\ 2\cdot2^b\ 1\cdot3^c$
Potamogeton pectinatus F*	0·28	(1·25)	$9\cdot8^c$
Potamogeton crispus F*	0·18		
Sparganium simplex F*	0·12		
Callitriche stagnalis F*	0·07		
	0·12		
Elodea canadensis F[†]	0·30	16·0	52
Nitella flexilis F[†]	0·10	8·0	10·1
Potamogeton crispus F[†]	0·20	23·0	66
Egeria densa[‡]	0·08		8–12
Ceratophyllum demersum F[§]	0·17	6·4	38^b
Hydrilla verticillata F[§]	0·17	6·4	38^b
Myriophyllum spicatum F[§]	0·15	4·5	30^b
Myriophyllum spicatum F[‖]	0·18	(0·6)	$5\cdot4^c$
Potamogeton lucens F[¶]	0·2	(0·7)	$6\cdot0^c$

(a) The ratio of photosynthetic rates at 0·55 or 0·60 mM of pure CO_2 (linear extrapolation of the curves in Fig. 3, left column) to rates at the same concentrations of pure HCO_3^- (*y*-axis intercept, Fig. 3). (b) The ratio of HCO_3^- and CO_2 concentrations producing half the maximum rate of photosynthesis. (c) The ratio of initial slopes of photosynthetic rates versus increasing CO_2 and HCO_3^- concentrations. The top eight species (this work and Sand-Jensen, 1983*) were examined by the same technique and intensity of stirring. The bottom nine species are from [†]Allen & Spence (1981), [‡]Browse *et al.* (1979), [§]Van *et al.* (1976), [‖]Steemann-Nielsen (1947) and [¶]Lucas *et al.* (1978). $K_\frac{1}{2}(HCO_3^-)$ values shown in parentheses are not comparable with other values of $K_\frac{1}{2}(HCO_3^-)$ or with $K_\frac{1}{2}(CO_2)$ since $V_{max}(HCO_3^-)$ in these cases is lower than $V_{max}(CO_2)$.

Seawater is well buffered with high HCO_3^- concentrations about 2 mM and the use of alternative CO_2 sources in the sediment or the atmosphere have apparently not been selected for, as they have in freshwater habitats. Intertidal species that are alternately exposed to air and water would be the most likely candidates for non-users. A few intertidal species apparently lack the ability to use HCO_3^- (e.g. *Carpophyllum* species, Dromgoole 1978), but others are very efficient HCO_3^--users (e.g. *Enteromorpha* sp., Sand-Jensen & Gordon 1984; and *Ascophyllum nodosum*, K. Johnston pers. comm.). If HCO_3^- use is under environmental control in marine macroalgae we would suggest it would be the less beneficial at high CO_2 concentrations and low light. Most marine algae grow attached to hard substrata under well-mixed conditions and are not exposed to elevated CO_2 concentrations whereas growth in a light-limited environment, of course, occurs deep down into the photic zone.

A comparison between marine and freshwater macrophytes showed a higher affinity for HCO_3^- among marine species and the same affinity for CO_2 (Table 2). The high affinity for HCO_3^- among marine species is shown by the high photosynthetic rates in pure HCO_3^- when CO_2 concentrations are interpolated to zero (Fig. 3, left column) and by the saturation of photosynthesis at the natural levels of 2 mM HCO_3^- (Fig. 2). The high affinity is consistent with the high and constant availability of HCO_3^- and low availability of CO_2 in seawater. In freshwaters, availability of HCO_3^- and CO_2 varies considerably and alternative CO_2 sources are used depending on habitat. It is not clear whether different transport mechanisms involved in HCO_3^- use operate in freshwater and marine plants.

PLASTICITY OF HCO_3^- USE IN *ELODEA CANADENSIS*

Bicarbonate use represents an investment in transport apparatus and energy which is beneficial when light, nutrients and ratios of HCO_3^- to CO_2 are high and which is wasteful when they are low. Use of HCO_3^- is active and the affinity for CO_2 is always at least twice as high (Table 2). Until recently we thought that the affinity for HCO_3^- was constant within a particular species, but it has become clear that there is a gradation in the affinity for HCO_3^- among submerged species (Maberly & Spence 1983) and, more importantly, that the affinity depends on environmental conditions (Sand-Jensen & Gordon 1986).

We examined the HCO_3^- affinity of *Elodea canadensis* collected from shallow water in seven lakes and streams during summer. The response of *Elodea* varied from a high affinity for HCO_3^- to no measurable affinity at all (Table 3). The affinity for HCO_3^- tended to decrease with reduced natural levels of HCO_3^- in the habitats but CO_2 concentrations in the microenvironment around the plant appeared to be more important. In the same hardwater system the affinity was high in Esrom Stream where CO_2 concentrations were about atmospheric equilibrium, and low at a site in Lake Esrom where *Elodea* grew on organic mud and was enriched with CO_2 (Table 3).

TABLE 3. Affinity for HCO_3^- of *Elodea canadensis* collected from different localities during summer

Locality and its DIC (mM)	Experimental DIC (mM)	Affinity for HCO_3^- $(P_{HCO_3^-}/P_{max})$
Havelse Stream	2·98	0·53
4·38	1·49	0·36
Esrom Stream	3·14	0·64
1·57	1·57	0·34
Slotshave Pond	2·96	0·49
2·98	1·49	0·31
Burre Lake	5·94	0·10
3·50	2·97	0·04
Esrom Lake	2·98	0·05
1·44	1·49	0·03
Almind Lake	2·98	−0·02
0·50	1·49	−0·02
Boest Stream	2·98	−0·02
0·90	1·49	−0·02

The HCO_3^- affinity is expressed as the quotient of photosynthesis in pure HCO_3^- ($P_{HCO_3^-}$) to the photosynthetic capacity under carbon saturation (P_{max}). Data from Sand-Jensen & Gordon (1986).

We could demonstrate more clearly the importance of CO_2 concentrations for HCO_3^- affinity when we grew *Elodea* in an alkaline medium in the laboratory. The affinity for HCO_3^- increased when CO_2 was depleted below atmospheric equilibrium and it decreased gradually when the medium was enriched with CO_2 (Fig. 5). At the highest CO_2 concentration (2·2 mM) there was no affinity for HCO_3^- and even the CO_2 compensation point increased from 4–22 μM CO_2 to 360 μM CO_2. By manipulating the light conditions we could also suppress the HCO_3^- affinity by growing *Elodea* in low light, and nutrient limitation had a similar effect (K. Sand-Jensen, unpublished data). It is therefore evident that affinity for HCO_3^- is very plastic in *Elodea* and that it is a reflection of the enviromental conditions of the plant.

To examine how phenotypically plastic *Elodea* is under field conditions we grew it from the same initial plant material in plastic pots filled with sandy sediment with and without additional fertilizers in shallow water of a number of lakes (Madsen & Sand-Jensen 1986). Photosynthetic capacities were higher in fertilized plants. Photosynthetic rates and HCO_3^- affinites were coupled, and both increased linearly with increasing chlorophyll anconcentrations in the plants (Fig. 6). This supports the idea that HCO_3^- affinity is an integrated part of the physiology of the plant. We explain the linear coupling of photosynthetic rates and HCO_3^- affinities to chlorophyll by the hypothesis that high chlorophyll concentrations may sustain an efficient electron transport which produces ATP and reduced equivalents both for uptake of HCO_3^- and concomitant fixation of inorganic carbon.

Holaday, Salvucci & Bowes (1983) and Salvucci & Bowes (1983) have also demonstrated a high plasticity in CO_2 compensation points, and photosynthetic enzymes of submerged macrophytes depending on environmental conditions. The

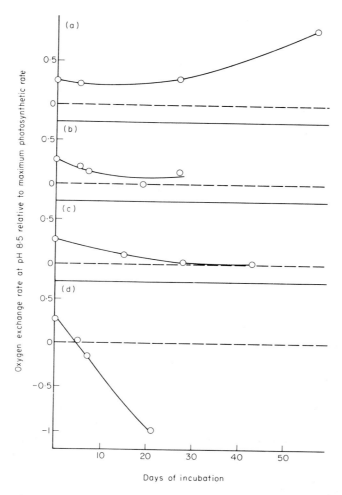

Fig. 5. Changes of HCO_3^- affinity of *Elodea canadensis* with time in days following growth in a hardwater medium with four different concentrations of CO_2 bubbled through during pretreatment: (a) 14 μM; (b) 37 μM; (c) 1000 μM; (d) 2000 μM. The HCO_3^--affinity is calculated as the ratio of photosynthesis at pH 8·5 and 6·0. This ratio gives approximately the same results as the $P_{HCO_3^-}/P_{max}$ ratio defined previously, and it involves the measurements at only two pH values. Redrawn from Sand-Jensen & Gordon (1986).

same conditions that induce a high affinity for HCO_3^- also induce a low CO_2 compensation point and a low photorespiratory state (high light treatment, high temperature—low CO_2 equilibrium concentrations). The low photorespiratory state is dependent on an efficient C-concentrating mechanism probably based on HCO_3^- use and combined with an efficient internal fixation through C_4 intermediates in some species. So these authors are probably examining the same phenomenona as we are, but from a more biochemical viewpoint.

Bicarbonate use

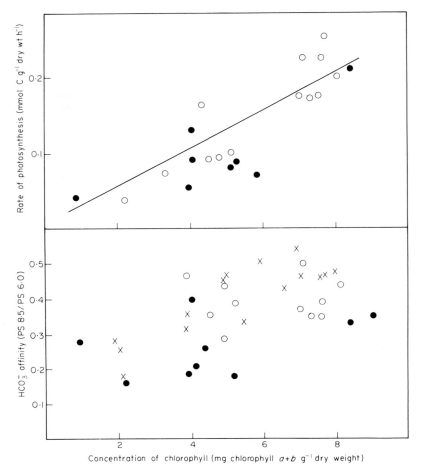

FIG. 6. Photosynthetic rates and HCO_3^- affinity of *Elodea canadensis* as a function of tissue concentrations of chlorophyll. The HCO_3^- affinity is calculated as the ratio of photosynthesis at pH 8·5 and 6·0 as described in the legend to Fig. 5. *Elodea* was grown from the same initial population in different lakes during summer and examined after 3–6 weeks. The symbols represent different experimental series. Redrawn from Madsen & Sand-Jensen (1986).

CONCLUSIONS

The ability to use HCO_3^- is widespread among fully submerged macrophytes in freshwater and marine habitats, and it appears to be present within most taxonomic groups. The affinity for HCO_3^- is usually high in marine species, and seawater is a well-buffered medium of high HCO_3^- and low CO_2 concentrations. The affinity is lower and more variable among freshwater macrophytes, depending on habitat. Those freshwater macrophytes with no apparent affinity for HCO_3^- often grow in habitats low in HCO_3^-, inorganic nutrients or light. They often use alternative CO_2

sources in the atmosphere or the sediment. When they rely entirely on CO_2 in the water, they either grow in microhabitats enriched with CO_2 or have very low CO_2 compensation points.

Recently we discovered that HCO_3^- affinity is not constant within a particular species, but under environmental control. *Elodea canadensis* collected in the field showed a response ranging from high affinity for HCO_3^- to no affinity at all. Use of HCO_3^- is mediated by light, and CO_2 is preferred when possible. Bicarbonate affinity can, therefore, be suppressed, when CO_2 is high and light is limiting. Photosynthetic capacity and HCO_3^- affinity were coupled in plants collected in the field, and they were linearly related to tissue concentrations of chlorophyll. This in turn indicates a direct relationship between HCO_3^- affinity and photosynthetic electron transport rates. Use of HCO_3^- is therefore an integrated part of the eco-physiology of the submerged plant.

REFERENCES

Allen, E. D. & Spence, D. N. (1981). The differential ability of aquatic plants to utilize the inorganic carbon supply in freshwaters. *The New Phytologist*, **87**, 269–283.

Bain, J. T. & Proctor, M. C. F. (1980). The requirement of aquatic bryophytes for free CO_2 as an inorganic carbon source: some experimental evidence. *The New Phytologist*, **86**, 393–400.

Beer, S., Eshel, A. & Waisel, Y. (1977). Carbon metabolism in seagrasses. I. The utilization of exogenous inorganic carbon species in photosynthesis. *Journal of Experimental Botany*, **28**, 1180–1189.

Browse, J. A., Dromgoole, F. I. & Brown, J. M. A. (1979). Photosynthesis in the aquatic macrophyte, *Egeria densa*. III. Gas exchange studies. *Australian Journal of Plant Physiology*, **6**, 499–512.

Dacey, J. W. H. (1980). Internal winds in water lilies: an adaptation for life in anaerobic sediments. *Science*, **210**, 1017–1019.

Dromgoole, F. I. (1978). The effects of pH and inorganic carbon on photosynthesis and dark respiration of *Carpophyllum* (Fucales, Phaeophyceae). *Aquatic Botany*, **4**, 11–22.

Holaday, A. S., Salvucci, M. E. & Bowes, G. (1983). Variable photosynthesis/photorespiration ratios in *Hydrilla* and other submersed aquatic macrophyte species. *Canadian Journal of Botany*, **61**, 229–236.

Lucas, W. J., Tyree, M. T. & Petrov, A. (1978). Characterization of photosynthetic [14]carbon assimilation by *Potamogeton lucens* L. *Journal of Experimental Botany*, **29**, 1409–1421.

Maberly, S. C. (1985). Photosynthesis by *Fontinalis antipyretica*. II. Assessment of environmental factors limiting photosynthesis and production. *The New Phytologist*, **100**, 141–155.

Maberly, S. C. & Spence, D. H. N. (1983). Photosynthetic inorganic carbon use by freshwater plants. *Journal of Ecology*, **71**, 705–724.

Madsen, T. V. & Sand-Jensen, K. (1986). Photosynthetic capacity and bicarbonate affinity of *Elodea canadensis* Michaux grown in lakes of different concentrations of inorganic carbon. Submitted.

Penuelas, J. (1985). HCO_3^- as an exogenous carbon source for aquatic bryophytes *Fontinalis antipyretica* and *Fissidens grandifrons*. *Journal of Experimental Botany*, **36**, 441–448.

Raven, J. A. (1970). Exogenous inorganic carbon sources in plant photosynthesis. *Biological Reviews*, **45**, 167–221.

Rebsdorf, Aa. (1972). *The Carbon Dioxide System of Freshwater. A Set of Tables for Easy Computation of Total Carbon Dioxide and Other Components of the Carbon Dioxide System*. Freshwater Biological Laboratory, Hillerød, Denmark.

Salvucci, M. E. & Bowes, G. (1983). Two photosynthetic mechanisms mediating the low photorespiratory state in submersed aquatic macrophytes. *Plant Physiology*, **73**, 488–496.

Sand-Jensen, K. (1983). Photosynthetic carbon sources of stream macrophytes. *Journal of Experimental Botany*, **34**, 198–210.

Sand-Jensen, K. & Gordon, D. M. (1984). Differential ability of marine and freshwater macrophytes to utilize HCO_3^- and CO_2. *Marine Biology*, **80**, 247–253.

Sand-Jensen, K. & Gordon, D. M. (1986). Variable HCO_3^- affinity of *Elodea canadensis* Michaux in response to different HCO_3^- and CO_2 concentrations during growth. *Oecologia*, in press.

Sand-Jensen, K. & Prahl, C. (1982). Oxygen exchange with the lacunae and across leaves and roots of the submerged vascular macrophyte, *Lobelia dortmanna* L. *The New Phytologist*, **91**, 103–120.

Spence, D. H. N. & Maberly, S. C. (1985). Occurrence and ecological importance of HCO_3^- use amongst aquatic higher plants. In: *Bicarbonate Utilization by Photosynthetic Organisms*. (Ed. by J. Berry & W. J. Lucas), pp. 125–143. American Society of Plant Physiologists.

Steemann-Nielsen, E. (1947). Photosynthesis of aquatic plants with special reference to the carbon sources. *Dansk Botanisk Arkiv*, **12**, 1–71.

Stumm, W. & Morgan, J. J. (1970). *Aquatic Chemistry*. Wiley, New York.

Van, T. K., Haller, W. T. & Bowes, G. (1976). Comparison of the photosynthetic characteristics of three submersed aquatic plants. *Plant Physiology*, **58**, 761–768.

Wium-Andersen, S. & Andersen, J. M. (1972). The influence of vegetation on the redox potential in the sediment of Grane Langsø, a Danish *Lobelia* lake. *Limnology and Oceanography*, **17**, 943–947.

The adaptive radiation of photosynthetic modes in the genus *Isoetes* (Isoetaceae)

J.E. KEELEY

Department of Biology, Occidental College, Los Angeles, California 90041, U.S.A.

SUMMARY

1 Crassulacean Acid Metabolism (CAM) involves the capture of ambient CO_2 at night and the daytime release of that CO_2 within the mesophyll cell. Two very different ecological settings have selected for this ability: particular xeric environments where daytime stomatal closure limits CO_2 availability and particular aquatic habitats where ambient CO_2 levels are limiting to photosynthesis.

2 Amphibious species of *Isoetes* in seasonal pools have a well developed CAM pathway biochemically identical to that found in xeric adapted succulents. Studies of these pools show that free-CO_2 levels range from $0 \cdot 25$–$0 \cdot 70$ mol m^{-3} at night and by noon on most days CO_2 is depleted in these waters. *Isoetes howellii* does not take up bicarbonate and therefore carbon uptake rates in this species are highly correlated with free-CO_2 levels in the water. Total daytime carbon uptake ranges from 40–70 μmol mg^{-1} Chl whereas night-time malic acid accumulation is commonly double that level; carbon uptake at night accounts for up to half of the malic acid accumulation and the remainder apparently is due to refixation of respiratory CO_2. Starch levels drop substantially overnight, but early in the season the diel fluctuation in starch is insufficient to account for all of the substrate for malic acid production. It is suggested that early in the season CAM is dependent upon the large starch stores in the corms.

3 Few of the other species with which amphibious *Isoetes* coexists in densely vegetated seasonal pools possess CAM. It is suggested that since night-time capture of carbon is of selective value as a means of avoiding daytime competition for carbon, this advantage would be greatly reduced if all plants in these pools were competing for the same night-time carbon pool.

4 As the water table drops and these amphibious *Isoetes* are exposed to the atmosphere they loose all capacity for CAM. This is consistent with the hypothesis that CAM was selected for by the carbon limitations of this particular aquatic environment.

5 These characteristics are contrasted with *Isoetes* species from other habitats. Many species in the genus are lacustrine and thus they are permanently submerged. These species also have a well developed CAM pathway. However, the soft water oligotrophic lakes to which these species are restricted are very different environments from the seasonal pool habitats of amphibious species. Carbon levels in the water do not fluctuate dielly and thus overnight levels are no higher than daytime levels. These environments are characterized by being at the extreme end in total inorganic carbon for aquatic habitats. It is hypothesized that due to the low carbon conditions and high diffusive resistance of water, the macrophytes in these environments are carbon limited and that capturing carbon at night through CAM effectively doubles the total time available for carbon uptake. Studies of such species show that CAM accounts for half of the total carbon budget.

6 While lacustrine and amphibious *Isoetes* species are largely restricted to abiotically stressful carbon-limited environments, terrestrial species have radiated into seasonal drought environments. Two very different syndromes are evident: summer deciduous C_3 species of low-elevation temperate latitudes, and evergreen high-elevation tropical species which lack stomata and are hermetically sealed from the air and obtain the bulk of their carbon from the sediment. The former species have no capacity for CAM whereas the latter do, although quantitatively it does not appear to contribute substantially to the carbon budget of such species.

INTRODUCTION

Isoetes is a worldwide genus of more than 150 taxa most of which are either amphibious or lacustrine (Pfeiffer 1922; Tryon & Tryon 1982). A particularly intriguing aspect of these species is the presence of Crassulacean Acid Metabolism (CAM) (Keeley 1981a), a photosynthetic pathway commonly associated with terrestrial xerophytes. This phenomenon was initially discovered in the amphibious species *Isoetes howellii* Engelm, of western North America. Here I will focus on: (i) the evidence for CAM in this amphibious species; (ii) the adaptive significance of CAM in amphibious *Isoetes*; (iii) a survey of the photosynthetic characteristics of other species in the genus, distributed in a range of environments from aquatic to terrestrial habitats.

EVIDENCE FOR CRASSULACEAN ACID METABOLISM

The discovery of CAM in an aquatic species was unexpected and some investigators have been reluctant to accept it, often referring to it as CAM-like or coining new acronyms such as AAM for aquatic acid metabolism (Cockburn 1983). Much of the confusion is semantic and revolves around differing definitions as well as inherent problems in applying typological concepts to natural phenomenona.

Kluge & Ting (1978) suggested that for plants with 'true' CAM metabolism, stomatal opening occurs at night rather than during the day. While this definition is conceptually useful it does raise problems in classifying many species, not only aquatic macrophytes and some epiphytic orchids which lack stomata on photosynthetic tissues, but also many terrestrial species as well. Few terrestrial CAM plants show abrupt and complete stomatal closure throughout the day; normally species which obtain the bulk of their CO_2 during the night are considered 'good' CAM plants. However, across the range of terrestrial species with CAM, there is continuous variation with respect to the proportion of light CO_2 uptake versus dark

CO_2 uptake which contributes to the net carbon gain. Terms such as CAM-idling, CAM-C_3 intermediates, CAM-cycling, C_3-idling etc. have been coined to catagorize this variation.

In light of this inherent variation, Teeri (1982) has suggested that the minimum characteristics necessary to consider a plant CAM are that the 'photosynthetic cells have the ability to fix CO_2 in the dark via PEP carboxylase, forming malic acid which accumulates in the vacuole. During the following light period the malic acid is decarboxylated, and the CO_2 enters the PCR cycle in the same cell'.

For plants with CAM, as defined by Teeri, the primary question is: what does CAM contribute to the carbon budget of the plant?

Isoetes howellii

The conclusion that Crassulacean Acid Metabolism is present in *I. howellii* is based on the following: (i) dark CO_2-fixation occurs in photosynthetic tissues but not in corms; (ii) malic acid accumulates in these tissues overnight; (iii) there is a diel cycle of night-time acidification/daytime deacidification up to 300 μmol H^+ g^{-1} fresh weight; (iv) PEP carboxylase activity is sufficient to account for observed rates of acid accumulation; (v) PEP carboxykinase (ATP dependent) activities are sufficient to account for decarboxylation of malic acid (malate enzyme levels are low); (vi) carbon released from malic acid in light is incorporated in PGA and phosphorylated sugars (Keeley 1981a; Keeley & Bowes 1982; Keeley & Busch 1984; J. E. Keeley unpubl. data).

SIGNIFICANCE OF CAM TO THE AMPHIBIOUS
ISOETES HOWELLII

In terrestrial CAM plants the contribution of dark versus light CO_2 uptake is regulated by changes in stomatal conductance. Although many amphibious species possess stomata they are typically non-functional while submerged (Sculthorpe 1967). *Isoetes howellii* leaves have stomata but on submerged leaves they remain closed and the aperture is lined with a wax. These leaves have a relatively thin cuticle and it is assumed that gas exchange occurs by diffusion across the epidermis. Consequently, stomatal behaviour does not regulate the contribution of dark versus light carbon uptake, rather it is largely a function of the ambient inorganic carbon conditions.

Isoetes howellii *and its habitat*

This species is distributed throughout western North America and is typically found in low elevational seasonal pools which fill during the winter rains and dry down in summer. Observations suggest that *I. howellii* requires an average time of

Photosynthesis in Isoetes

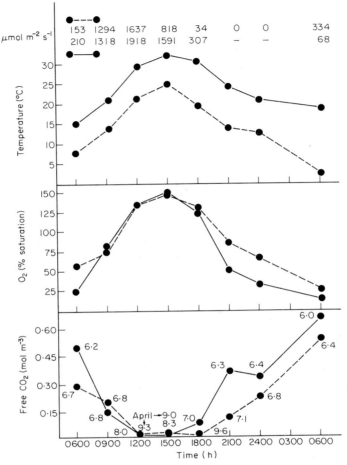

FIG. 1. Diel changes in physical characteristics of a seasonal pool inhabited by *Isoetes howellii* on Mesa de Colorado, Riverside County, California for 13–14 April (dashed line) and 24–25 May (solid line) 1983. pH of the water is indicated adjacent to the line in lower box. Specific conductance was 43 and 101 μmhos cm^{-1} respectively (methods are as described in Keeley (1983) and **Keeley & Busch (1984)**). Further details of this pool (Cl) and others associated with it are described in Lathrop & Thorne (1983).

inundation of approximately 2 months; it is uncommon in pools filled for a shorter time (Zedler 1984) or a longer time. Such pools are relatively shallow (20–40 cm depth) and densely vegetated.

Physical characteristics for one such pool are shown in Fig. 1. Diel patterns typically show highest CO_2 levels early in the morning. As light and temperature increase through the morning the photosynthetic demand for carbon by the total pool flora, coupled with lower solubility, results in a rapid reduction in free-CO_2. Circumstantial evidence that the CO_2 depletion is a result of photosynthetic consumption is the observation of a highly significant negative correlation between

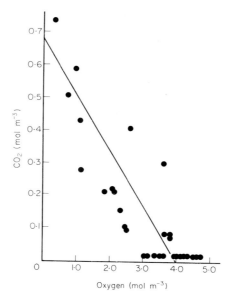

FIG. 2. Relationship between CO_2 and O_2 concentrations in the vernal pool described in Fig. 1, from 06.00 to 18.00 h on several days throughout the spring growing season of 1983. $r = -0.89$; $r^2 = 0.79$; $n = 32$.

CO_2 and O_2 levels in the water throughout the daylight hours (Fig. 2). During much of the day the water is supersaturated with O_2 and depleted of free-CO_2. Overnight conditions are reversed due largely to respiration by the pool flora and invertebrate fauna.

The magnitude of these diel changes is far greater than the seasonal changes, although some patterns are evident. In late spring, peak light levels and temperatures are greater and the rate at which CO_2 is depleted through the day is greater. The inorganic carbon chemistry also changes through the season (Table 1). Early in the spring a much greater proportion of the total carbon is as bicarbonate and carbonate. Later in the season, although the total pool of inorganic carbon remains unchanged, a much greater proportion is as free-CO_2. Since, throughout

TABLE 1. Comparison of carbon conditions in early April and late May of 1983 for seasonal pool shown in Fig. 1

	Hour	pH	Free CO_2 (mol m^{-3})	Alkalinity (mol m^{-3} as CaCO$_3$)
April 13				
	06.00	6·7	0·277	0·316
	12.00	9·3	0·000	0·272
May 24				
	06.00	6·2	0·508	0·184
	12.00	8·0	0·009	0·167

the season, free CO_2 is largely depleted by noon, the total inorganic carbon during the afternoon is much less in late spring than in early spring.

As a consequence of the diel patterns, ambient conditions for photosynthesis are poor throughout a good portion of the day; with very little free-CO_2 in the water, photosynthesis requires uptake of bicarbonate ions. Disadvantages to bicarbonate uptake are that it is an active uptake process and therefore is energetically costly, and that the affinity for bicarbonate ions is lower than for CO_2 (Wetzel & Grace 1983). Additionally the high ambient O_2 levels during much of the day suggest that elevated internal CO_2 levels will be required to overcome photorespiratory competition. Advantages to carbon uptake at night are centered around the fact that CO_2 levels at night are equal to or greater than the total inorganic carbon levels available during much of the day.

Carbon uptake rates for *Isoetes howellii* are closely tied to the changes in ambient CO_2 levels (Fig. 3). Peak rates occur early in the morning (through the C_3 cycle, Keeley 1981a) but are short-lived. Throughout the season of the study reported in Fig. 3 there was a highly significant correlation ($r = 0.91$, $P < 0.01$, $n = 29$) between carbon uptake rates in the light and free-CO_2 levels in the water. Dark CO_2 uptake rates are relatively low in the early evening but increase as free-CO_2 increases in the pool; the correlation between these two parameters is less than for the daytime ($r = 0.58$, $P < 0.01$, $n = 24$). Part of the reason for this may lie in the observation that dark CO_2 uptake rates always declined between midnight and 06.00; high internal acid levels near the end of the dark period may play a role in dampening dark CO_2 uptake as is the case in terrestrial CAM plants (Kluge & Ting 1978). An alternative explanation, however, may be that PEP substrate levels are limiting. This is suggested by the observation that although starch levels undergo a large drop overnight they are typically less than the moles of glucose equivalents required for the malic acid accumulation (Keeley 1983a). It is unknown whether other carbohydrates are involved or whether carbon is transported from the corms.

Total carbon uptake over the two 24-hour periods shown in Fig. 3 ranged from 64 μmol mg^{-1} Chl early in the season, to double that level late in the season (Table 2). In general, dark CO_2 uptake contributed between 1/3 and 1/2 of the total carbon gained, and dark uptake exceeded the contribution due to light uptake during the daylight period from 09.00 to 18.00 h. In short, there is a brief 'window' of time at the beginning of each day when ambient conditions lead to relatively high (C_3) photosynthetic rates. As ambient CO_2 levels become limiting to photosynthesis, malic acid stores are used up, thus maintaining elevated CO_2 levels in the cell. Evidence for environmental control of deacidification is provided by experimental manipulation of the ambient CO_2 levels (Table 2); lower ambient CO_2 levels lead to more rapid deacidification in the leaves.

Crassulacean Acid Metabolism plays a major role in the carbon economy of *Isoetes howellii*. Not only does it account for a large percentage of the gross carbon gain but apparently it is important in recapturing respiratory CO_2. Under CO_2 levels typically found at night in these pools, *I. howellii* shows no net CO_2 evolution

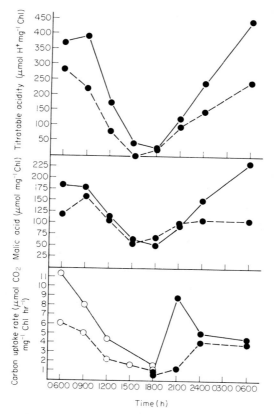

FIG. 3. Diel changes in titratable acidity (to pH 6·4), malic acid and carbon uptake measured by ^{14}C incorporation for leaves of *Isoetes howellii* in the pool described in Fig. 1 for 13–14 April (dashed line) and 24–25 May (solid line) (methods as described in Keeley & Busch 1984).

in the dark (Keeley & Bowes 1982). Additionally, malic acid accumulation is consistently larger than can be accounted for by overnight CO_2 uptake (Table 4). I interpret this to mean that a large portion of the malic acid production is derived from refixation of respiratory CO_2. Experimental manipulations support this interpretation (Table 5); leaves sparged overnight with CO_2-free gas still accumulated substantial levels of malic acid, about half of the level under the 1% CO_2 treatment or observed in the field. It appears that early in the season this

TABLE 2. Total 24 h carbon assimilation for *I. howellii* leaves in southern California seasonal pool estimated from carbon uptake curves in Fig. 3

	Total 24 h carbon uptake (μmol mg^{-1} Chl)	% due to dark uptake (18.00–06.00 h)	% due to light uptake	
			(06.00–09.00 h)	(09.00–18.00 h)
13–14 April	64·1	39	26	35
24–25 May	126·5	47	26	27

TABLE 3. Daytime deacidification rates in *I. howellii* leaves sparged from 06.00 to 09.00 h with either 1% CO_2 or CO_2-free gas at two light intensities; high light = constant 1000 μmol m^{-2}s^{-1} and low light = a stepwise increase of 100 (06.00–07.00 h), 300 (07.00–08.00 h), and 500 (08.00–09.00 h) μmol m^{-2}s^{-1} (treatments were compared with the Student's *t*-test)

| | Malic acid (μmol mg^{-1} Chl) | |
	High light $\bar{X} \pm$ SD (n)	Low light $\bar{X} \pm$ SD (n)
06·00	147 ± 9 (3)	147 ± 33 (33)
09·00		
Sparged with 1% CO_2 (21% O_2)	112 ± 7 (3)	90 ± 11 (3)
Sparged with CO_2-free gas (21% O_2)	75 ± 13 (3)	59 ± 9 (3)
P	<0·01	<0·01

TABLE 4. Seasonal comparison between carbon uptake, malic acid fluctuation and starch fluctuation (μmol mg^{-1} Chl) in *Isoetes howellii* leaves (calculated from data presented in Fig. 3 except starch levels are from Keeley (1983) and are from a different year)

	Carbon uptake in the light	Carbon uptake in the dark	Diurnal malic acid fluctuation	Diurnal fluctuation (as glucose equivalents)
April	39	25	112	32
May	68	59	122	144

TABLE 5. Overnight changes in titratable acidity (to pH 6·4) and malic acid in *I. howellii* leaves sparged from 18.00 to 06.00 h with either CO_2-free gas or 1% CO_2 gas (treatments were compared with the Student's *t*-test)

Hour	Titratable acidity (μmol H$^+$ mg^{-1} Chl) $\bar{X} \pm$ SD (*n*)	Malic acid (μmol mg^{-1} Chl) $\bar{X} \pm$ SD (*n*)
18.00	9 ± 3 (6)	45 ± 6 (6)
06.00		
Sparged overnight with 1% CO_2 gas (21% O_2)	334 ± 67 (6)	161 ± 34 (6)
Sparged overnight with CO_2-free gas (21% O_2)	162 ± 30 (6)	334 ± 70 (6)
P	<0·01	<0·01

overnight malic acid production is fed by starch reserves in the corm as the diel fluctuation in starch is insufficient to account for the level of acid accumulated (Table 4).

Photosynthetic characteristics of associated species

Throughout its range *Isoetes howellii* coexists with a variety of other species. One such community is shown in Table 6. Although vegetation zonation in seasonal pools has not been studied in great detail there are some clear differences between the centre and the periphery. Both sections had 70% total coverage but the number of species and composition varied; there were fifteen species in the centre portion whereas towards the periphery, due to the occurrence of a number of grassland elements, there were twenty-three species.

TABLE 6. Representative cover values for centre and periphery of southern California seasonal pool on Mesa de Colorado (Riverside County) and overnight changes in titratable acidity and malic acid

Species (Family)	Relative coverage		Titratable acidity (μmol H$^+$ g^{-1} FW)		Malic acid (μmol g^{-1} FW)		(n)
	Periphery	Centre	PM $\bar{X} \pm$ SD	AM $\bar{X} \pm$ SD	PM $\bar{X} \pm$ SD	AM $\bar{X} \pm$ SD	
Callitriche ssp.* (Callitrichicaeae)	2·5	12·8	0 ± 0	2 ± 1	2 ± 1	2 ± 1	(2)
Crassula aquatica (Crassulaceae)	12·4	0·0	9 ± 10	112 ± 9	17 ± 5	53 ± 9	(4)
Downingia cuspidata (Campanulaceae)	4·9	5·7	1 ± 1	0 ± 0	8 ± 4	4 ± 1	(2)
Elatine ssp.† (Elatinaceae)	6·2	11·4	1 ± 1	2 ± 1	4 ± 1	10 ± 2	(2)
Eleocharis acicularis (Cyperaceae)	8·6	5·7	3 ± 2	9 ± 4	5 ± 3	6 ± 1	(6)
E. macrostachya	8·6	11·4	(data not available)				
Isoetes ssp.‡ (Isoetaceae)	2·5	21·4	26 ± 3	275 ± 18	34 ± 2	143 ± 10	(3)
Lilaea scilloides (Lilaeaceae)	2·5	2·9	1 ± 1	1 ± 1	1 ± 1	2 ± 1	(2)
Pilularia americana (Marsileaceae)	9·9	2·9	0 ± 0	0 ± 0	6 ± 3	9 ± 4	(2)
Plagiobothrys undulatus (Boraginaceae)	9·9	15·7	0 ± 0	0 ± 0	8 ± 3	10 ± 2	(4)
Ranunculus aquatilis (Ranunculaceae)	0·0	5·7	0 ± 0	3 ± 4	4 ± 1	10 ± 3	(2)
Other species	32·0	4·4					

*Cover data includes *C. longipedunculata* and *C. marginata*; only the former was assayed.
†Cover data includes *E. californica* and *E. chilensis*; only the latter was assayed.
‡Cover data includes *I. howellii* and *I. orcuttii*; data presented is for *I. howellii*, although the two species are indistinguishable in magnitude of acid metabolism (Keeley 1982).
Data for coverage from Kopecko & Lathrop (1975); data for acidity and malic acid from Keeley & Morton (1982); Keeley (1983), unpublished data; nomenclature is according to Munz (1974).

In our survey of plants for Crassulacean Acid Metabolism we have found two other species which show a marked ability for overnight acid accumulation: *Isoetes orcuttii* Engelm. and *Crassula aquatica*. The robust *Isoetes howellii* and diminutive *I. orcuttii* coexist in many pools throughout California and are indistinguishable in CAM characteristics; one study reported 69 and 70 μmol g^{-1} fresh weight, respectively, in overnight malic acid accumulation (Keeley 1981b). Typically *Isoetes* species occupy the deepest (central) parts of pools (Zedler 1984), as is the case for the pool shown in Table 6. In this portion of the pool, *Isoetes* species are the only ones which appear to obtain a major portion of their carbon from CAM. One explanation for the lack of CAM in other species may lie in phylogenetic constraints since CAM is not known in any of the families of species in the centre of the pool (other than *Isoetes*). I support an alternative explanation that community coexistence of a dense cover of all CAM plants would not be likely in seasonal pools, the reason being that the significance of CAM in *I. howellii* is that it avoids daytime competition for carbon by obtaining much of its carbon at night when the rest of the flora is evolving VO$_2$. The selective advantage for doing this would be greatly reduced if all plants were competing for the same night-time carbon pool. On the surface this sounds like 'group selection' but it need not be. One can imagine that as more species in the pool evolved CAM the selective coefficients for CAM would decrease, making the evolution of it less likely.

Towards the periphery of the pool *Isoetes* spp. are less abundant but the diminutive annual *Crassula aquatica* is the dominant CAM species. Evidence of CAM in this member of the Crassulaceae family is perhaps not suprising. If the aquatic habit was the derived condition in the family, CAM in *C. aquatica* could be a conservative trait which presently is of little adaptive value. Several observations argue against this interpretation. This species does not develop CAM if grown out of water. Also the closely related (nearly indistinguisable) terrestrial *C. erecta*, which grows (only metres away) on elevated ground in nearby rock outcrops, lacks CAM, even if artifically submerged (Keeley & Morton 1982). A few other species in Table 6 showed a very slight overnight increase in malic acid. Similar small diel fluctuations in acidity have been reported for a number of other aquatic macrophytes (Holaday & Bowes 1980; Beer & Wetzel 1981; Keeley & Morton 1982) suggesting the presence of low-level CAM activity; in these cases the contribution of CAM to the carbon economy of the plants in question is probably minor.

The terrestrial part of *Isoetes howellii's* *life cycle*

When CAM was originally reported in *I. howellii*, a number of colleagues suggested that it was undoubtedly of adaptive value as a means of conserving water after the pools have dried up; a notion which is clearly wrong as CAM is lost by leaves as they emerge (Table 7). This apparently happens on a cell-by-cell basis as the tops of leaves may be metabolically quite different from the bases (Table 7). This loss of

TABLE 7. Overnight changes in titratable acidity and malic acid in submerged and emergent leaves of *I. howellii* from seasonal pool on Mesa de Colorado, Riverside County, California (methods as in Keeley (1983), $n = 3$)

Conditions	Titratable acidity (μmol H$^+$ g^{-1} FW)		Malic acid (μmol g^{-1} FW)		Chlorophyll (mg g^{-1} FW)
	18.00 h $\bar{X} \pm$ SD	06.00 h $\bar{X} \pm$ SD	18.00 h $\bar{X} \pm$ SD	06.00 h $\bar{X} \pm$ SD	
Submerged	4 ± 1	141 ± 12	21 ± 4	79 ± 9	0·46
Partially submerged					
Submerged bases	3 ± 2	105 ± 28	9 ± 2	61 ± 28	0·16
Emergent tops	3 ± 2	7 ± 3	32 ± 7	39 ± 1	1·14
Emergent					
In moist soil	6 ± 2	12 ± 1	43 ± 4	44 ± 2	0·87
In dry soil	5 ± 2	10 ± 5	44 ± 4	50 ± 8	0·54

CAM upon emergence appears to be cued by changes in the water potential of the leaves (J. E. Keeley, unpubl. data).

Why does CAM not play a dual role, i.e. as a mechanism to alleviate CO_2 depletion in the water, and to alleviate water stress on land? Possibly it is because of the relatively short time for which it survives on land coupled with the very sharp changes in growing conditions. For approximately 2–3 weeks after the leaves are fully emergent the ground is quite moist but then drying and death of the *Isoetes* leaves occurs rapidly. Under these conditions it would be best to shift entirely to C_3 photosynthesis which is capable of much greater rates of carbon gain; emergent *I. howellii* leaves have substantially higher chlorophyll levels, RUBISCO activities and CO_2 uptake rates, than submerged leaves (Keeley & Busch 1984; J. E. Keeley, unpubl. data). This would far outweigh the advantages of remaining CAM and surviving for slightly longer into the summer drought. In all likelihood the same arguments apply to why annual plants, e.g., *Crassula erecta* described above, are seldom CAM.

ADAPTIVE RADIATION IN THE GENUS *ISOETES*

In addition to amphibious habitats, the genus *Isoetes* has radiated into lacustrine and terrestrial environments. A convenient means of dividing species within each of these different habitats is by their evergreen or deciduous habit (Table 8).

Amphibious Isoetes

Amphibious species are defined as ones which occur in environments which alternate between submerged and emergent conditions. This may occur seasonally as is the case for vernal pools or it may occur dielly as in the case of tidal creeks.

Isoetes species which occur in seasonal pools require a period of several months submergence followed by a period of drying, during which they survive as dormant underground corms or spores. Commonly such species are collected after the

TABLE 8. Habit and habitat characteristics for species of *Isoetes* (including stylites)

Habit	Lacustrine		Amphibious species		Terrestrial species	
	Winter deciduous (>5 taxa)	Evergreen (>20 taxa)	Summer deciduous (>25 taxa)	Evergreen (>3 taxa)	Summer deciduous (>5 taxa)	Evergreen (>4 taxa)
Distribution						
Latitude	Temperate and boreal	All latitudes	Temperate and tropical	Temperate and tropical	Temperate	Tropical
Elevation	>3000 m	Sea-level to 5000 m	Sea-level to 3000 m	Sea-level	<1000 m	>3600 m
Habitat	Soft water oligotrophic lakes	Soft water oligotrophic lakes	Seasonal pools	Freshwater tidal creeks	Borders of rock outcrops	Seasonally wet bogs
Abiotic stress	High due to very low inorganic carbon levels	High due to very low inorganic carbon levels	High due to summer drought	High (?) due to diurnal submergence	High due to summer drought	High due to 'summer' drought and 'winter' flooding
Biotic stress	Low (usually sparsely vegetated)	Low (usually sparsely vegetated)	High (dense vegetation produces CO_2 limitation)	Low (sparsely vegetated)	Low (sparsely vegetated)	High (densely vegetated)
Functional attributes						
Photosynthesis	CAM contribution substantial	CAM contribution substantial	CAM contribution substantial	CAM contribution may be substantial	CAM not detectable	CAM present
Source of ambient carbon uptake	Largely leaves, perhaps 1/3 root contribution	Largely leaves, perhaps 1/3 root contribution	Largely leaves, some root contribution	?	Largely (if not entirely) leaves	Largely (if not entirely) roots
Structural attributes						
Stomata	Absent but inducible under aerial conditions	Absent but inducible under aerial conditions	Present but non-functional when submerged	Present but behaviour unknown	Present (presumably functional)	Absent
Cuticle	Present but thin	Present but thin	Present but thin	Present but thin	Present	Present and thick
Leaf cross-sectional shape	Terete to quadrangular	Terete to quadrangular	Terete to quadrangular	Terete to quadrangular	Quadrangular	Triquetrous
Four lacunal chambers	Present	Present	Present	Present	Present	Present
Root:shoot ratio	≥1	≥1	≤1	?	<1 ?	>>>1

water-table has dropped, when they are more conspicuous, and thus they are commonly described in the literature as 'terrestrial' (cf. Keeley 1984). However, submergence is a critical part of their life cycle in that it provides a form of refuge from non-aquatic competitors; likewise the drying-down of the pool provides a refuge from true-aquatic competitors which dominate permanent bodies of water. Seasonal pools occur wherever shallow basins form over a relatively impervious substrate, in regions of seasonal rainfall. Consequently, amphibious *Isoetes* occur throughout the world: e.g., *I. lithophila* Pfeiffer in seasonal pools on granite outcrops in Texas, U.S.A. or *I. australis* Williams in similar habitats in Western Australia, *I. abyssinica* Chiov. in shallow pools of central Africa, *I. panamensis* Max. & Mort. in lowland Neotropics or *I. hystrix* Bory & Dur. in southern Europe. The first two are known to be CAM species (Keeley 1982, 1983c) and I presume all such species will prove to be similar.

Tidal creek species are submerged at high tide and emergent at low tide. They do not have a dormant season and thus are evergreen. Species such as *I. echinospora* ssp. *maritima* (Underw.) Love in British Columbia or *I. riparia* Engelm. on the eastern seaboard of the U.S.A. occur far enough upstream in tidal creeks to be considered freshwater species and they possess a well developed CAM pathway when submerged (Keeley 1982). The South American tidal species *I. clavata* U. Weber is described as occurring in salt water (Hickey 1985), a report which needs confirmation since such conditions would most likely preclude CAM photosynthesis.

Lacustrine Isoetes

Lacustrine species are those that normally spend their entire lifecycle submerged, and by and large are restricted to softwater oligotrophic lakes. The distribution of these species, as well as the conditions which promote low productivity lakes, follows a pattern in which elevational distribution is inversely related to latitude. Lacustrine *Isoetes* may be at sea-level in boreal latitudes but in the tropics they are normally restricted to >3000 m.

In temperate latitudes the growing conditions for these lacustrine *Isoetes* are normally cut short during the winter. Most species remain evergreen and some, such as *I. macrospora* Duriev., even remain photosynthetically active under winter ice cover (Boylen & Sheldon 1976). The exact number of taxa which are deciduous is unknown but *I. bolanderi* Engelman from high elevations in California has been reported to be deciduous in winter (Keeley, Walker & Mathews 1983), probably due to the very high winter snow pack covering lakes in that region.

Oligotrophic lakes occupied by *Isoetes* species are typically sparsely vegetated compared to habitats occupied by amphibious species. Also the physical conditions are much different than conditions in low elevation seasonal pools (Table 9). There is no large diel fluctuation in free-CO_2 level, and therefore it is initially suprising that lacustrine *Isoetes* species have Crassulacean Acid Metabolism of the same order of magnitude as observed for amphibious species such as *I. howellii* (Keeley,

TABLE 9. Seasonal pattern of pH, free-CO_2 and alkalinity of Siesta Lake in the Sierra Nevada, California (elevation 2440 m) inhabited by *Isoetes bolanderi* (data from Keeley, Walker & Mathews 1983)

	pH		Free CO_2 (mol m^{-3})		Alkalinity (mol m^{-3} as $CaCO_3$)	
	06.00 h	12.00 h	06.00 h	12.00 h	06.00 h	12.00 h
16 June	6·2	5·7	0·116	0·150	0·078	0·040
1 July	6·8	6·6	0·114	0·086	0·156	0·098
27 July	6·6	6·2	0·099	0·052	0·122	0·136
18 August	6·2	6·6	0·100	0·109	0·130	0·136
5 September	6·6	6·6	0·111	0·073	0·100	0·100

Walker & Mathews 1983; Boston & Adams 1983; Richardson *et al*. 1983). For aquatic habitats, however, these oligotrophic lakes are at the extreme end in terms of total inorganic carbon. I suggest that due to low carbon levels and the high diffusive resistance of water, the macrophytes in these environments are carbon limited and that capturing carbon at night through CAM effectively doubles the total time available for carbon uptake. Detailed studies on *I. macrospora* from Wisconsin lakes indicate that CAM contributes 45–50% of the annual carbon gain (Boston & Adams 1985).

Lacustrine *Isoetes* species will frequently form nearly pure stands or coexist with another CAM species such as *Littorella* species (Plantaginaceae) (Keeley & Morton 1982; Boston & Adams 1985) in temperate latitudes or with *Crassula (Tillaea)* species in tropical latitudes (J. E. Keeley, unpubl. data). This is quite different from the pattern observed in seasonal pools where CAM species do not normally coexist in close association with other CAM species. This pattern, however, is consistent with the hypotheses proposed above for CAM in these two environments. In oligotrophic lakes the major advantage to CAM is not that there is a large biogenic source of CO_2 at night but that CAM doubles the time available for uptake. This, coupled with the dependence upon sediment CO_2 uptake through the roots (Richardson *et al*. 1983), suggests that night-time competition for carbon by other CAM species is likely to be less intense than in seasonal pools.

Terrestrial Isoetes

All terrestrial *Isoetes* species possess four lacunal chambers which run the length of the leaf, a feature uncommon in terrestrial plants and one which suggests an aquatic ancestry.

At low elevations in temperate latitudes these are several species of *Isoetes* which are strictly terrestrial. These species are restricted to seasonally moist sites and all are deciduous during the summer drought; examples include *I. nuttallii* A. Br. ex. Engelm. from western North America, *I. butleri* Engelm. from southeastern North America and *I. durieui* Bory from Europe. These species are not CAM and CAM can not be induced even if they are artificially submerged (Keeley 1983b).

At extremely high elevations in tropical latitudes there are several terrestrial *Isoetes* species restricted to seasonally wet bogs. These species include *I. andicola*

Amstutz (formerly *Stylites a.*) of the central Andes of South America, *I. andina* Hook and *I. novo-granadensis* Fuchs of the northern Andes and *I. hopei* Croft from Papua New Guinea. All are characterized by having fibrous, astomatous leaves with a thick cuticle, and most of the leaf length occurs below ground level, only the tips (which often have spines) being green and aerial. These photosynthetic surfaces however are hermetically sealed from the atmosphere and, as shown for stylites, the bulk of the carbon for photosynthesis is derived from the sediment (Keeley, Osmond & Raven 1984). This species has some capacity for CAM but it does not appear to contribute substantially to the carbon economy of the plant.

The explanation for this unusual mode of carbon nutrition is unknown. These species may represent a line of early land plants which failed to evolve stomata and thus gas exchange with the sediment (with its much greater diffusional resistances) being the price for avoiding dessication on periodically dry terrestrial sites. This hypothesis is not supported by cladistic analysis which suggest that the evolution of this syndrome in the above species was polyphyletic (Hickey 1985). A more likely hypothesis is that there are unique features of these tropical seasonal bogs at high elevations which have selected for this strategy. This problem is the major focus of my current research.

ACKNOWLEDGMENTS

I thank G. Busch and T. Montygierd for laboratory assistance and J. Hickey for a critical review of the manuscript and useful discussion of *Isoetes*. This work was supported by J. Teeri and grants DEB-8004614, DEB-8206887 and BSR-8407935 from the National Science Foundation.

REFERENCES

Beer, S. & Wetzel, R. G. (1981). Photosynthetic carbon metabolism in the submerged aquatic angiosperm *Scirpus subterminalis*. *Plant Science Letters*, **21**, 199–207.

Boston, H. L. & Adams, M. S. (1983). Evidence of crassulacean acid metabolism in two North American isoetids. *Aquatic Botany*, **15**, 381–386.

Boston, H. L. & Adams, M. S. (1985). Seasonal diurnal acid rhythms in two aquatic crassulacean acid metabolism plants. *Oceologia*, **65**, 573–579.

Boylen, C. W. & Sheldon, R. B. (1976). Submergent macrophytes: growth under winter ice cover. *Science, (New York)*, **182**, 841–842.

Cockburn, W. (1983). Stomatal mechanism as the basis of the evolution of CAM and C_4 photosynthesis: a review. *Plant, Cell and Environment*, **6**, 275–280.

Hickey, R. J. (1985). *Revisionary studies of neotropical* Isoetes. Ph.D. dissertation, University of Connecticut, Storrs.

Holaday, A. S. & Bowes, G. (1980). C_4 acid metabolism and dark CO_2 fixation in a submersed aquatic macrophyte (*Hydrilla verticillata*). *Plant Physiology*, **65**, 331–335.

Keeley, J. E. (1981a). *Isoetes howellii*: a submerged aquatic CAM plant? *American Journal of Botany*, **68**, 420–424.

Keeley, J. E. (1981b). Diurnal acid metabolism in vernal pool *Isoetes*. *Madrono*, **28**, 167–171.

Keeley, J. E. (1982). Distribution of diurnal acid metabolism in the genus *Isoetes*. *American Journal of Botany*, **69**, 254–257.

Keeley, J. E. (1983a). Crassulacean acid metabolism in the seasonally submerged aquatic *Isoetes howellii. Oecologia*, **58**, 57–62.

Keeley, J. E. (1983b). Lack of diurnal acid metabolism in terrestrial *Isoetes* (Isoetaceae). *Photosynthetica*, **17**, 93–94.

Keeley, J. E. (1983c). Report of diurnal acid metabolism in two aquatic Australian species of *Isoetes. Australian Journal of Ecology*, **8**, 203–204.

Keeley, J. E. (1984). Search theory and convergent spore morphology. *American Naturalist*, **124**, 307–308.

Keeley, J. E. & Bowes, G. (1982). Gas exchange characteristics of the submerged aquatic Crassulacean Acid Metabolism plant, *Isoetes howellii. Plant Physiology*, **70**, 1455–1458.

Keeley, J. E. & Busch, G. (1984). Carbon assimilation characteristics of the aquatic CAM plant *Isoetes howellii. Plant Physiology*, **76**, 525–530.

Keeley, J. E. & Morton, B. A. (1982). Distribution of diurnal acid metabolism in submerged aquatic plants outside the genus *Isoetes. Photosynthetica*, **16**, 546–553.

Keeley, J. E., Osmond, C. B. & Raven, J. A. (1984). *Stylites*, a vascular land plant without stomata absorbs CO_2 via its roots. *Nature*, (*London*), **310**, 694–695.

Keeley, J. E., Walker, C. M. & Mathews, R. P. (1983). Crassulacean acid metabolism in *Isoetes bolanderi* in high elevation oligotrophic lakes. *Oecologia*, **58**, 63–69.

Kluge, M. & Ting, I. P. (1978). *Crassulacean Acid Metabolism.* Springer-Verlag, Berlin.

Kopecko, K. J. & Lathrop, E. W. (1975). Vegetation zonation in a vernal marsh on the Santa Rosa Plateau of Riverside County, California. *Aliso*, **8**, 281–288.

Lathrop, E. W. & Thorne, R. F. (1983). A flora of the vernal pools on the Santa Rosa Plateau, Riverside County, California. *Aliso*, **10**, 449–469.

Munz, P. A. (1974). *A Flora of Southern California.* University of California Press, Berkeley.

Pfeiffer, N. E. (1922). Monograph on the Isoetaceae. *Annals of the Missouri Botanical Garden*, **9**, 79–103.

Richardson, K., Griffiths, H., Reed, M. L., Raven, J. A. & Griffiths, N. M. (1983). Inorganic carbon assimilation in the isoetids, *Isoetes lacustris* L. and *Lobelia dortmanna* L. *Oecologia*, **61**, 115–121.

Sculthorpe, C. D. (1967). *The Biology of Aquatic Vascular Plants.* Edward Arnold, London.

Teeri, J. A. (1982). Carbon isotopes and the evolution of C_4 photosynthesis and crassulacean acid metabolism. *Biochemical Aspects of Evolutionary Biology* (Ed. by M. H. Nitecki), pp. 93–129. University of Chicago Press, Chicago, Illinois.

Tryon, R. M. & Tryon, A. F. (1982). *Ferns and Allied Plants.* Springer Verlag, New York.

Wetzel, R. G. & Grace, J. B. (1983). Aquatic plant communities. *CO_2 and Plants* (Ed. by E. R. Lemon), pp. 233–280. Westview, Boulder, Colorado.

Zedler, P. H. (1984). Micro-distribution of vernal pool plants of Kearny Mesa, San Diego County. *Vernal Pools and Intermittent Streams* (Ed. by S. Jain & P. Moyle). Institute of Ecology Publication No. 28, University of California, Davis.

The application of carbon isotope discrimination techniques

JOHN A. RAVEN AND JEFFREY J. MACFARLANE

Department of Biological Sciences, University of Dundee, Dundee DD1 4HN, Scotland

HOWARD GRIFFITHS

Department of Plant Biology, University of Newcastle, Newcastle upon Tyne NE1 7RU

SUMMARY

The measurement of $\delta^{13}C$ values for terrestrial vascular plants has been of considerable use in understanding the biochemistry, biophysics, ecophysiology and taxonomy of inorganic carbon acquisition by these plants. Aquatic and amphibious plants present additional problems of measurement and interpretation, related to the frequent presence of HCO_3^- as well as CO_2 as 'source' inorganic C; variations in the $\delta^{13}C$ value of dissolved inorganic C; the frequently large, but not readily measured, diffusive limitations on the rate of photosynthesis; and the unknown discrimination between ^{12}C and ^{13}C exhibited by the mechanism(s) of active transport of inorganic C across membranes. Despite these problems, the measurement of carbon isotope ratios has, in conjunction with data from other sources, permitted a number of important conclusions to be drawn. Among these are the extent to which boundary layers limit the rate of photosynthesis by non-HCO_3^--users such as *Lemanea* and *Fontinalis*; the contribution of HCO_3^- to *in situ* photosynthesis by Batrachian Ranunculi; and the carbon nutrition of *Anabaena* symbiotic with *Azolla*. Possible areas where present techniques could permit a contribution of $\delta^{13}C$ measurements to the quantitative ecophysiology of aquatic and amphibious plants include the relative magnitude of emersed and submersed photosynthesis by amphibious and intertidal plants, and the contribution of CAM-like carbon acquisition in certain Isoetids.

INTRODUCTION

This paper addresses the use of natural abundances of carbon isotopes in investigating the mechanism of submersed and emersed photosynthesis by amphibious plants, and in assessing the contribution of emersed and submersed carbon acquisition to the overall carbon gain by the plants.

The two naturally-occurring stable isotopes of carbon are ^{12}C and ^{13}C, with the lighter isotope being about one hundred times more abundant on earth than the heavier isotope. The interconversion of different molecular species of carbon involves two categories of discrimination between the two isotopes, inter-related via rate and equilibrium constants. For reactions at (or close to) equilibrium there is

129

the thermodynamic influence, with different equilibrium constants (and free energy changes) for reactions depending on whether ^{12}C or ^{13}C is involved. For reactions far from equilibrium, kinetic discrimination (differences in rate constants) is dominant. The sum of these two effects on the catena of transport and biochemical reactions of carbon in photosynthesis involves a discrimination against the heavier isotope, and differences in the *extent* of this discrimination permit inferences to be drawn as to the mechanism of carbon fixation in photosynthesis, and the relative importance of the various steps in determining the overall rate of photosynthesis. O'Leary (1981) provides an excellent review of the physico-chemical background to the use of carbon isotope ratios in the study of photosynthesis; a brief résumé is given here.

CARBON ISOTOPE DISCRIMINATION BY COMPONENT REACTIONS OF PHOTOSYNTHETIC CARBON ACQUISITION

In general we deal, in photosynthetic carbon fixation, with kinetic discrimination between the carbon isotopes in transport and chemical reactions (O'Leary 1981). Determination of *absolute* discrimination is troublesome; discrimination is usually expressed as the $\delta^{13}C$ values defined by eqn (1):

$$\delta^{13}C(\text{‰}) = \left[\frac{^{13}C/^{12}C_{(sample)}}{^{13}C/^{12}C_{(standard)}} - 1 \right] \times 100 \tag{1}$$

The standard used is $CaCO_3$ from the Pee Dee Belemnite (PDB) which has a $^{13}C/^{12}C$ of 0·01124 and a $\delta^{13}C$ (by definition: eqn (1)) of 0. Near-zero $\delta^{13}C$ values characterize the inorganic C in seawater, while atmospheric CO_2 has a $\delta^{13}C$ value of about -7‰, i.e. is enriched in ^{12}C relative to seawater. This is largely a result of the different $^{13}C/^{12}C$ ratios between HCO_3^- and CO_2 at equilibrium in solution (-7‰ at 25°C, and increasing at lower temperatures: Mook, Bommerson & Staverman 1974) rather than differences between the $^{13}C/^{12}C$ of gaseous and dissolved CO_2 at equilibrium ($\delta^{13}C$ of $-0·9$‰ with ^{13}C concentrated in dissolved CO_2: O'Leary 1981). Fresh waters may have $\delta^{13}C$ values substantially more negative than those of seawater as a result of the input of respiratory CO_2, derived with little discrimination from 'light' terrestrial organic C, in groundwater (e.g. Lazerte & Szalados 1982; Raven, Beardall & Griffiths 1982).

Table 1 provides, in addition to the data on $\delta^{13}C$ values related to equilibria of inorganic carbon, some $\delta^{13}C$ values for net transport and chemical conversion processes related to photosynthesis. We note that the $\delta^{13}C$ values associated with diffusion in air or in solution, with CO_2 fixation by PEP carboxylase, and (usually) with conversion of plant organic C into CO_2 in respiration, are small as are the sum of the reactions of intermediate biochemistry which convert the C fixed into the carboxyl group of PGA by RUBISCO into the sum of plant organic compounds (Table 1; O'Leary 1981). The only process listed in Table 1 which invariably has a large (greater than 10‰) $\delta^{13}C$ value is the carboxylase activity of RUBISCO, for

TABLE 1. Values of carbon isotope discrimination (δ^{13}C, ‰) for various equilibria, and net transport of inorganic C and net chemical conversions, related to photosynthesis and respiration (after O'Leary 1981; Roeske & O'Leary 1984; Rosenberg & O'Leary 1985)

Equilibrium or process	δ^{13}C (‰)
Gaseous CO_2 relative to an equilibrium solution of CO_2	-0.9
Dissolved CO_2 relative to HCO_3^- in an equilibrium solution at 25°C	-7.0
Net diffusive transport of CO_2 in the gas phase: CO_2 at sink end of pathway relative to that at source end	-4.4
Net diffusive transport of CO_2 or HCO_3^- in aqueous solution, or of CO_2 across bilayer membranes: CO_2 or HCO_3^- at sink end of pathway relative to that at source end	0 (?)
Net mediated transport of HCO_3^- or CO_2 across membranes: CO_2 or HCO_3^- on sink side of membrane relative to that on source side	?
Net fixation of CO_2 by RUBISCO carboxylase activity: C in carboxyl group of PGA which was derived from CO_2 relative to C in an infinite substrate pool of dissolved CO_2 (CO_2 is the 'true' substrate for RUBISCO)	-30
Net fixation of CO_2 by PEP carboxylase: C in carboxyl group of OAA which was derived from CO_2 relative to C in an infinite substrate pool of dissolved CO_2 (HCO_3^- is the 'true' substrate for PEP carboxylase)	$+5.0$
Net CO_2 production by decarboxylation reactions in respiration: C in respiratory CO_2 relative to C in infinite pools of respiratory substrates	-3.3 -9.9
Net CO_2 production by decarboxylation reactions in respiration: C is respiratory CO_2 relative to C in very small and rapidly turning over pools of respiratory substrates.	0

which the value given is based on an analysis of some rather disparate values to be found in the literature (O'Leary 1981). The value of δ^{13}C associated with the mediated transport of CO_2 or HCO_3^- across biological membranes is listed as unknown in Table 1: while it is often assumed to be small (Beardall, Griffiths & Raven 1982) there is little experimental, and less theoretical (Smith & Walker 1980), evidence for this assumption.

CARBON ISOTOPE DISCRIMINATION IN CARBON ACQUISITION BY TERRESTRIAL VASCULAR PLANTS

In addition to providing an invaluable background to our later considerations of emersed photosynthesis by vascular amphibious plants, a brief summary of the contribution of δ^{13}C measurements to studies of the photosynthesis of terrestrial vascular plants is important inasmuch as the phenomena are relatively well understood, and provide a basis for the analysis of δ^{13}C values resulting from submersed photosynthesis.

C_3 photosynthesis by terrestrial vascular plants would result in a $\Delta\delta^{13}$C value of -4‰ if the rate of CO_2 fixations were entirely determined by the rate of CO_2 diffusion in the gas phase, and of -30‰ if the rate were entirely determined by RUBISCO carboxylase activity. These predicted $\Delta\delta^{13}$C values, relative to CO_2 in air at a $\delta^{13}C$ of $-7‰$, would result in plant $\delta^{13}C$ values of $-11\cdot4‰$ and $-37‰$ respectively. Observed $\delta^{13}C$ values for terrestrial C_3 plants are in the range -21 to $-36‰$, with a mean at $-27\cdot8‰$ (Troughton, Card & Hendy 1974), suggesting that RUBISCO activity is usually a greater determinant of the rate of CO_2 fixation than is CO_2 diffusion. This has been quantified in terms of the resistance model of C_3 photosynthesis (cf. Jones 1985) by Farquhar, O'Leary & Berry (1982; cf. Farquhar 1980) who derived eqn (2):

$$\delta_{plant} = \delta_{atm} + a + \frac{(b - a)c_i}{c_a} \tag{2}$$

where δ_{plant} is the $\delta^{13}C$ value of the organic material of a C_3 plant, δ_{atm} is the $\delta^{13}C$ value of atmospheric CO_2, a is the $\delta^{13}C$ value associated with gaseous diffusion of CO_2 (Table 1), b is the $\delta^{13}C$ value associated with RUBISCO activity (Table 1), and c_i and c_a are respectively the steady-state intercellular space and bulk air partial pressures (or concentrations) of CO_2 during photosynthesis (measured in Pa or in mol m^{-3}). Accordingly, plant $\delta^{13}C$ values can be used to estimate the extent of stomatal limitation of C_3 photosynthesis, and also give information about water use efficiency (Farquhar, O'Leary & Berry 1982; Raven, Allen & Griffiths 1984).

For C_4 and CAM terrestrial vascular plants ^{13}C values are much more positive than those for terrestrial C_3 plants, with mean values (Troughton, Card & Hendy 1974) of $-27\cdot8‰$ for C_3, $-13\cdot6‰$ for C_4 and $-17\cdot6‰$ for CAM. These values are, however, significantly more negative than the values predicted (Table 1) for limitation by PEP carboxylase activity ($-7‰ + +5‰$ or $-2‰$) or, in most cases, by gas-phase CO_2 diffusion ($-7‰ + -4\cdot4‰$ or $-11\cdot4‰$). The observed values can be explained by co-limitation of the initial CO_2 fixation by gas-phase diffusion and by PEP carboxylase activity, yielding a carboxyl group, derived from CO_2, in OAA with a $\delta^{13}C$ in the range $-2‰ - -11\cdot4‰$, with the final, more negative, $\delta^{13}C$ values of plant organic C resulting from a partial expression of the large intrinsic discrimination of RUBISCO carboxylase. The expression of discrimination by RUBISCO is a result of the extent of leakage of CO_2 from the high steady-state CO_2 concentration maintained in the bundle sheath of C_4 plants by the biochemical 'CO_2 pump' based on the C_4–C_3 acid cycle (Farquhar 1983), and, in CAM plants, by any leakage via stomata in the deacidification (light) phase as well as, more importantly, any C_3 photosynthesis using exogenous CO_2 which may occur (Griffiths & Smith 1983). We note that the 'explanations' are sufficiently mechanistic to have predictive value (Farquhar 1983; Griffiths & Smith 1983).

In summary, $\delta^{13}C$ measurements on terrestrial vascular plants can shed important light on the mechanism of C_3 and C_4 photosynthesis, and on the extent of CAM and C_3 fixation of CO_2 by facultative CAM plants; they can be used to assign herbarium and, under certain circumstances, even fossil, material to the C_3 or C_4/CAM categories (DeNiro & Hastorf 1985).

CARBON ISOTOPE DISCRIMINATION IN CARBON ACQUISITION BY AQUATIC, AMPHIBIOUS AND INTERTIDAL PLANTS

General considerations

Some of the additional problems associated with the interpretation of $\delta^{13}C$ in aquatic as compared to terrestrial plants have been discussed by Smith & Walker (1980), Osmond *et al.* (1981), Lazerte & Szalados (1982), Beardall, Griffiths & Raven (1982) and Raven, Beardall & Griffiths (1982). A major difference concerns the molecular or ionic species of inorganic carbon, and their $\delta^{13}C$ values, in the two environments. In addition to the occurrence of HCO_3^- in all but the more acid aquatic habitats, there is also (as a result of limited CO_2 exchange across air-water interfaces) the possibility of the CO_2 concentration in solution being less than the air-equilibrium value if phototrophic reactions dominate the inorganic C content of the water, and in excess of the air-equilibrium value if chemo-organotrophic reactions dominate. The chemo-organotrophic (respiratory) CO_2 input via metabolism of terrestrial plant detritus supplies carbon with a more negative $\delta^{13}C$ value than that derived from dissolution of CO_2 from air or by dissociation of HCO_3^- (see Table 1). While it is desirable to determine the $\delta^{13}C$ for source CO_2 (the atmosphere) in work on carbon isotope discrimination by terrestrial plants, especially if cities, glass-houses or dense plant canopies are involved, it is essential that such measurements be made when aquatic systems are being investigated. We note that freshwater systems show much more variability of pH, total dissolved inorganic carbon, and $\delta^{13}C$ values for inorganic carbon, than do marine environments.

A second important difference between terrestrial and aquatic carbon acquisition relates to boundary layer effects. These are at once more quantitatively important for most aquatic macrophytes than for terrestrial plants, and less readily quantified in aquatic systems. This provides an opportunity for $\delta^{13}C$ measurements to provide, via a version of eqn (2) and with measurements of photosynthetic rates under near-natural conditions, an estimate of the unstirred layer thickness (see Raven, Beardall & Griffiths 1982, and below).

A third important difference concerns the nature of the inorganic carbon source used and the mode of transport across the plasmalemma. Terrestrial vascular plants use diffusive entry of CO_2, with the possible exception of 'C_3–C_4 intermediates', some of which have C_4-like gas exchange characteristics with C_3 biochemistry (see Holaday, Lee & Chollet 1984), and which might have an inorganic carbon concentrating mechanism based on active transport of CO_2 or HCO_3^-. Many aquatics can 'use' HCO_3^- in the sense of having a more rapid rate of photosynthesis than can be accommodated by the uncatalysed rate of HCO_3^- to CO_2 conversion in their environment, and many of these have an active influx of HCO_3^- (or CO_2) at the plasmalemma (Raven, Osborne & Johnston 1985). In addition to the unknown $\delta^{13}C$ associated with the active transport reaction, there are also effects on $\delta^{13}C$ resulting from the use of exogenous HCO_3^- (which has a less negative $\delta^{13}C$ value than the CO_2 with which it is in equilibrium: Table 1) and of the extracellular, plant-catalysed conversion of HCO_3^- to CO_2 in some of these plants (see Raven, Osborne & Johnston 1985).

These complications mean that quite a lot must be known about the physiology and biochemistry of photosynthesis of a submerged aquatic plant before full use can be made of measurements of its $\delta^{13}C$ value. The same is true, *a fortiori*, if we consider amphibious and intertidal plants, where part of their organic carbon is derived from submersed, and part from emersed, photosynthesis. In the discussion of individual cases which follows, we shall concentrate on data which we have been involved in obtaining, and on the realised and the potential uses of $\delta^{13}C$ measurements in understanding the mechanisms of submersed and emersed photosynthesis and the contribution which these two processes make to the growth and reproduction of amphibious and intertidal plants.

Lemanea mamillosa: *a C_3 plant which has diffusive entry of CO_2*

Lemanea mamillosa Kützing is a haptophytic freshwater Rhodophyte (Floridiophyceae: Batrachospermales) with a perennial, encrusting *Chantransia* stage which annually (February–June in Dundee) produces 'bristles' up to 0·5 m long (see Sheath 1984). Our data are for plants growing in shallow, fast-flowing (>1 m s^{-1}) portions of the Dichty Burn (Raven & Beardall 1981; Raven, Beardall & Griffiths 1982; Macfarlane & Raven 1985; Raven, Osborne & Johnston 1985) where its claim to be an amphibious plant rests on the emersion of plants when the water level in the Burn falls, although the published data are for plants whose bristles had grown totally submersed before collection. The photosynthetic characteristics of *Lemanea mamillosa* are those of a plant with C_3 biochemistry and diffusive CO_2 entry, with no significant 'HCO_3^- use' or CAM-like metabolism (Raven & Beardall 1981; Raven, Beardall & Griffiths 1982; MacFarlane & Raven 1985; Raven, Osborne & Johnston 1985; J. J. MacFarlane, unpublished observations).

Table 2 gives data on the $\delta^{13}C$ for CO_2 in the Dichty Burn, and for *Lemanea mamillosa* organic C. Application of eqn (2), modified for use in an aquatic system (i.e. with δ_{atm} replaced by the $\delta^{13}C$ value of dissolved CO_2, and with the $\delta^{13}C$ value associated with aqueous rather than aerial CO_2 diffusion, used for the coefficient *a*: see Table 1). The data shows that c_i/c_a (in an aquatic system equal to the CO_2 concentration at the site of RUBISCO activity divided by that in the bulk water) is 0·77 (Raven, Beardall & Griffiths 1982). This suggests that only 0·23 of the limitation on photosynthetic rate is associated with CO_2 diffusion from the bulk phase to the carboxylase, with the remaining 0·77 imposed by biochemical restrictions. Measurements of photosynthetic rate in the laboratory under as nearly natural conditions of medium composition and flow as can be managed, and at light saturation, permit the c_i/c_a values to be used in computing a value for the thickness of the unstirred layer outside the plant plus the intraplant diffusion distance to RUBISCO (Raven, Beardall & Griffiths 1982). This value is 11 μm, and is a

TABLE 2. $\delta^{13}C$ values for submerged macrophytes, and for CO_2 and HCO_3^-, in the Dichty Burn; and for related plants in other habitats

Organism	Source	$\delta^{13}C_{plant}$	$\delta^{13}C_{CO_2}$	$\delta^{13}C_{HCO_3^-}$	Velocity of water movement ($m\ s^{-1}$)	References
Lemanea mamillosa	Dichty Burn, Angus, Scotland.	$-39.9 \pm 0.08\%_0$	$-15.9 \pm 0.5\%_0$	$-5.2 \pm 0.5\%_0$	>1	Raven, Beardall & Griffiths (1982)
Fontinalis antipyretica	Dichty burn, Angus, Scotland.	$-43.5 \pm 0.3\%_0$	$-15.9 \pm 0.5\%_0$	$-5.2 \pm 0.5\%_0$	~1	H. Griffiths (hitherto unpublished)
	River Frome, Dorset, England.	$-43.9\%_0$	$-15.6\%_0$	$-5.1\%_0$	≥0.1	Osmond et al. (1981)
	Springs, Ruovesi, Finland.	$-49.4\%_0$	$-17.8\%_0$	$-7.3\%_0$	>0.1	Osmond et al. (1981)
		$-50.7\%_0$	$-22.3\%_0$	$-11.8\%_0$	>0.1	Osmond et al. (1981)
Ranunculus penicillatus var. *calcareus*	Dichty Burn, Angus, Scotland.	$-22.5 \pm 0.9\%_0$	$-15.9 \pm 0.5\%_0$	$-5.2 \pm 0.5\%_0$	<1	Raven, Beardall & Griffiths (1982)
	River Frome, Dorset, England.	$-29.2\%_0$	$-15.6\%_0$	$-5.1\%_0$	≥0.1	Osmond et al. (1981)
Ranunculus baudotii Godron	Inlets of Baltic Sea, Finland.	$-11.6\%_0$	$-14.8\%_0$	$-5.4\%_0$	~0	Osmond et al. (1981)
Ranunculus trichophyllous Chaix	Lake Memphemagog, Quebec, Canada.	$-17.0\%_0$	$-16.5\%_0$	$-6.5\%_0$	~0 (?)	Lazerte & Szalados (1982)

$\delta^{13}C_{CO_2}$ and $\delta^{13}C_{HCO_3^-}$ computed as in Raven, Beardall & Griffiths (1982) using the reported values for the $\delta^{13}C$ of total dissolved inorganic carbon and pH, and assumed temperatures of 12°C (Osmond et al. 1981) or 12°C (Lazerte & Szalados 1982), and the relationship of Mook, Bommerson & Staverman (1974) for the $\delta^{13}C$ values of CO_2 and of HCO_3^- at equilibrium as a function of temperature. Errors (where quoted) are standard errors of the mean.

minimal value since the plant *in situ* is unlikely to carry out all of its photosynthesis under light-saturated conditions and (J. J. Macfarlane, unpublished) the CO_2 concentration in the Dichty Burn earlier in the bristle-growing season can be substantially higher than the $c_a = 30$ mmol m^{-3} of Raven, Beardall & Griffiths (1982). An *in situ* unstirred layer thickness of 10–20 μm is compatible with the $\delta^{13}C$ results, is hydrodynamically reasonable, and is also in fair agreement with kinetic analysis of the photosynthesis–CO_2 concentration relationship (MacFarlane & Raven 1985). We would argue that $\delta^{13}C$ measurements have materially contributed to our understanding of inorganic carbon acquisition by *Lemanea mamillosa*.

Emersed photosynthesis by *Lemanea* would be expected to produce organic C with a $\delta^{13}C$ value less negative than $-37\permil$, the most negative value which can be attained in the absence of diffusive limitation when atmospheric CO_2 is being fixed by RUBISCO (see Table 1). Since the observed $\delta^{13}C$ value for *Lemanea* which has grown completely submersed is $-38\cdot9\permil$ (Table 2), emersed photosynthesis would make the plant $\delta^{13}C$ value less negative. The contribution of emersed photosynthesis to the growth of haptophytes, with restricted water and nutrient supply when emersed, will be considered again below in the context of the intertidal *Ascophyllum nodosum*. We note that the contribution of emersed photosynthesis to the fitness of *Lemanea* is likely to be less than is the case for a perennial intertidal organism such as *Ascophyllum*: not only is emersion a much less predictable event for *Lemanea* bristles, but it is also much more likely to occur late in the bristle growing season when carpospore production, and any revictualing of the *Chantransia* stage by translocation from the bristles, may already have occurred.

Fontinalis antipyretica: a C_3 *plant which has diffusive entry of* CO_2

Fontinalis antipyretica Hedw. is a haptophytic freshwater Bryophyte (Bryopsida: Isobryales) with a number of infraspecific taxa with preferences for different habitats (Smith 1978). Our specimens were from the Dichty Burn where they occupied similar habitats to those occupied by *Lemanea mamillosa*; and its claim to be a functionally amphibious organism is perhaps stronger than is that of *Lemanea*, since both sexual and vegetative reproduction by *Fontinalis* are favoured by emersion (Glime *et al.* 1979) while survival *in situ* may be the main emersed concern of *Lemanea*. *Fontinalis antipyretica* has photosynthetic characteristics consistent with C_3 biochemistry, a predominance of CO_2 diffusion as the means of inorganic C entry, and no CAM-like metabolism (Allen & Spence 1981; Bain & Proctor 1980; Keeley & Morton 1982; Peñuelas 1985; Raven, Osborne & Johnston 1985).

Table 2 gives data on the $\delta^{13}C$ for CO_2 in the Dichty Burn, and for *Fontinalis antipyretica* organic C. Application of eqn (2), with the modifications for aquatic photosynthesis described above, suggests that only $0\cdot08$ of the limitation on photosynthetic rate is associated with CO_2 diffusion from the bulk phase to the carboxylase, with the remaining $0\cdot92$ imposed by biochemical restrictions.

Measurement in the laboratory of the rate of photosynthesis under near-natural conditions of medium composition and flow rate, and at light saturation, yielded a value of $1\cdot3$ μmol CO_2 m^{-2} s^{-1} when the assumed surface area is that of a solid with the cross-section of an equilateral triangle with 2 mm sides, and $0\cdot42$ μmol CO_2 m^{-2} s^{-1} when the assumed surface area is that of one side (the abaxial side which is closest to the medium) of the leaves. With a bulk medium free CO_2 concentration (c_a) of 30 mmol m^{-3}, and a c_i/c_a of $0\cdot92$ from eqn (2), the computed concentration difference between the bulk medium and the site of RUBISCO activity is $2\cdot4$ mmol CO_2 m^{-3}. With a diffusion coefficient for CO_2 of $1\cdot7 \times 10^{-9}$ m^2 s^{-1}, Fick's law yields diffusion pathlengths of $3\cdot1$ μm (for the smaller assumed plant area) and $9\cdot7$ μm (for the larger assumed plant area). The 'real' plant area exposed to the medium is between the two extremes (the plant is *not* a regular solid, *and* the leaves overlap), so the *minimum* value for the diffusion pathlength is not *less* than $3\cdot1$ μm: the actual pathlength is, for reasons given under the discussion of the *Lemanea* data, larger than this. Thus the CO_2 diffusion pathlengths deduced from $\delta^{13}C$ measurements for *Fontinalis antipyretica* are similar to those for *Lemanea mamillosa* (above) growing under similar conditions. Other data on the $\delta^{13}C$ values for *Fontinalis antipyretica*, in conjunction with computed values for the *Fontinalis antipyretica*, and computed values for the $\delta^{13}C$ of free CO_2 in their environment, give values of ($\delta^{13}C_{plant} - \delta^{13}C_{CO_2}$) which are similar to those found for the Dichty Burn material, and similarly imply a small diffusive limitation of photosynthesis for *Fontinalis antipyretica* in other habitats (Table 2).

Using arguments similar to those used for *Lemanea mamillosa* (above), an even stronger case can be made for emersed photosynthesis by *Fontinalis antipyretica* leading to a less negative value of the organic C produced, since the $\delta^{13}C$ value for photosynthate produced by submersed photosynthesis ($-43\cdot5\permil$: Table 2) is even more negative than is that of *Lemanea mamillosa*, and so is more different from the maximum theoretical $\delta^{13}C$ value of $-37\permil$ computed for C_3 photosynthesis at the expense of CO_2 from air with a $\delta^{13}C$ of $-7\permil$. We have already seen that emersed photosynthesis could contribute to both sexual and vegetative reproduction of *Fontinalis*. With respect to sexual reproduction, we note that the 'immersed' condition (in the so-called perichaetial leaves) of the *Fontinalis* sporophyte (Smith 1978), and the absence of stomata on the capsule (Paton & Pearce 1957) militate against substantial contributions to sporophyte C nutrition by its own photosynthesis, so that any contribution of emersed photosynthesis to spore production would have to be via translocation to the sporophyte of photosynthate from the gametophyte. As with other haptophytes, the supply of water and mineral nutrients poses problems for long-term emersed photosynthesis by *Fontinalis*; work of Jenkins (quoted by Proctor 1984) showed little or no growth by emersed *Fontinalis antipyretica*.

Lobelia dortmanna: *a C_3 Isoetid with root uptake of CO_2*

Lobelia dortmanna L. is a rhizophytic Tracheophyte (Magnoliopsida: Campanales) which generally grows submersed in acid bodies of freshwater, but which produces an emersed inflorescence from water depths of up to 2 m (Sculthorpe 1967), and is thus amphibious with respect to sexual reproduction although vegetatively submersed. Its main photosynthetic distinguishing feature is the uptake of the great majority of its inorganic C via its roots from the relatively high CO_2 concentration in the root environment (by courtesy of chemo-organotrophic action on sedimented organic material, and restricted diffusive loss of CO_2 to the bulk water above). Ninety-nine per cent or more of the net carbon fixed by submersed plants under natural conditions can enter through the roots which are exposed to $0.8–5.0$ mol CO_2 m^{-3}: shoots are exposed to a solution of $0.006–0.43$ mol CO_2 m^{-3} (Wium-Anderson 1971; Sand-Jensen & Søndergaard 1979; Richardson *et al.* 1984). Acquisition of CO_2 via the roots is common in plants of the Isoetid life-form although most (an exception is *Stylites (Isoetes) andicola*) do not have such a large fraction of their CO_2 entering via the roots as does *Lobelia dortmanna* (Keeley, Osmond & Raven 1984; Raven, Osborne & Johnston 1985). Inorganic C probably enters the roots as CO_2, and diffuses in the gas phase up to the chloroplasts in cells surrounding the gas lacunae in the leaves (Sand-Jensen & Prahl 1982). Evidence that the photosynthetic biochemistry of *Lobelia dortmanna* is C_3 (rather than C_4) is based on the high CO_2 compensation concentration found by Wium-Anderson (1971) for cut leaves (extrapolation to zero net photosynthesis of the photosynthesis–CO_2 concentration relationships in his Fig. 2); it is clear that *Lobelia dortmanna* lacks CAM-like metabolism (Boston & Adams 1983; Richardson *et al.* 1984).

Table 3 gives data for the $\delta^{13}C$ values of *Lobelia dortmanna*, and of the dissolved CO_2 supply, in Loch Brandy (Richardson *et al.* 1984). It will be seen that the plant $\delta^{13}C$ is $12.5–14.2‰$ more negative than is the $\delta^{13}C$ of the root medium, so that some $0.51–0.58$ (eqn (2)) of the total limitation of photosynthesis can be attributed to diffusion of CO_2 from root medium to the site of action of RUBISCO. A rather larger fractional limitation by CO_2 diffusion can be deduced from a comparison of the CO_2 fixation–CO_2 concentration relationship for cut leaves (Fig. 1 of Wium-Anderson 1971) with that for the CO_2 supply to the roots of intact plants (Fig. 2 of Wium-Anderson 1971). $\delta^{13}C$ values are helpful in apportioning limitation on submersed photosynthesis by *Lobelia dortmanna*.

The emergent stems of *Lobelia dortmanna* bear the flowers and fruits as well as scale leaves, and are plentifully supplied with apparently functional stomata (Sculthorpe 1967). C_3 photosynthesis by these emersed portions of the plant would be expected to give a larger difference between plant $\delta^{13}C$ and the $\delta^{13}C$ of CO_2 from the air as is the case for the difference between plant $\delta^{13}C$ and the $\delta^{13}C$ of dissolved CO_2 during submersed photosynthesis, since it is unlikely that more than half of the limitation on CO_2 fixation during emersed photosynthesis would (as is the case for submersed photosynthesis) be associated with CO_2 diffusion. This is

TABLE 3. $\delta^{13}C$ values for *Isoetes lacustris* and *Lobelia dortmanna*, and for the CO_2 available for photosynthesis, in Loch Brandy, and for related plants in other habitats

Organism	Source	$\delta^{13}C_{plant}$	$\delta^{13}C_{CO_2}$	Reference
Lobelia dortmanna	Loch Brandy, Angus, Scotland.	$-31.7 \pm 0.4\%_0$ (submerged leaves)	$-17.5\%_0$	Richardson *et al.* (1984)
		$-30.0 \pm 0.6\%_0$ (submerged roots)	$-17.5\%_0$	
	Lake Kuivajärvi, Ruovesi, Finland.	$-29.6\%_0$ (emergent stem)	$-7\%_0$ (assuming use of CO_2 from air)	Osmond *et al.* (1981)
		$-30.2\%_0$ (submerged stem)	?	
		$-33.2\%_0$ (submerged leaves)	?	
Isoetes lacustris	Loch Brandy, Angus, Scotland.	$-23.5 \pm 1.2\%_0$ (submerged leaves)	$-17.5\%_0$	Richardson *et al.* (1984)
		$-23.1 \pm 0.9\%_0$ (submerged leaves)	$-17.5\%_0$	
Isoetes durieui	Terrestrial site, Sicily.	$-26.5\%_0$ (leaves)	$-7\%_0$ (air)	Richardson *et al.* (1984)
		$-26.6\%_0$ (roots)	$-7\%_0$ (air)	
Isoetes howellii	Large Pool, Mesa de Colorado, California, U.S.A.	$-28.7\%_0$ (submerged leaves)	$-21.6\%_0--28.6\%_0$	Keeley & Busch (1984)
		$-28.7\%_0$ (submerged roots)	$-21.6\%_0--28.6\%_0$	
		$-29.8\%_0$ (leaves of emergent plant)	$-7\%_0$ (air)	
		$-29.4\%_0$ (roots of emergent plant)	$-7\%_0$ (air)	

$\delta^{13}C$ values for CO_2 in the *Isoetes howellii* were computed as described in Table 2, using the range of pH and temperature for the pool given by Keeley & Busch (1984). These values only apply to the bulk water in the pool, and may not be applicable to root-derived CO_2. Errors, where quoted, are standard errors of the mean ($n = 4$).

especially the case in view of the aquatic habitat: given adequate xylem conductivity, a high stomatal resistance (and hence less negative plant $\delta^{13}C$ value) resulting from water deficits in the emersed parts of the plant is unlikely. The only data on $\delta^{13}C$ values for the emergent stem of *Lobelia dortmanna* gives values very close to those for submersed parts of the plants (Table 3: data of Osmond *et al.* 1981). No $\delta^{13}C$ values for the sediment CO_2 are given for the Finnish site for which the plant data are reported: however, given a similar value for ($\delta^{13}C$ submersed plant $-\delta^{13}C$ sediment CO_2) to that found by Richardson *et al.* (1984), i.e. $-12\cdot5 - -14.2‰$ we see that this value is substantially smaller than the difference between the $\delta^{13}C$ of the emergent part of the plant and that of CO_2 in air, i.e. $-22\cdot6‰$. This is consistent with C_3 photosynthesis by the emergent stem, using CO_2 from the air but, alas, does not prove it, since the data are equally compatible with growth of the emergent part of the stem using organic C translocated from the submerged parts of the plant and resulting from submersed photosynthesis. Thus the $\delta^{13}C$ data for *Lobelia dortmanna* are consistent with the carbon nutrition of the flowers and developing fruits (which do *not*, unlike those of many submerged angiosperms with emergent flowers, develop under water following bending of the stem which bears them: Sculthorpe 1967) either resulting from fixation of CO_2 from the surrounding air, or by translocation of the products of submersed photosynthesis.

Ranunculus penicillatus var calcareus: *a freshwater* C_3 *plant which actively transports* HCO_3^-

Ranunculus penicillatus (Dumort) Bab. var. *calcereus* (R. W. Butcher) is a rhizophytic Tracheophyte (Magnoliopsida: Ranales) which grows submerged in fast-flowing streams: our material came (like *Lemanea mamillosa* and *Fontinalis antipyretica*) from the Dichty Burn, albeit from a portion with a slower flow velocity (Table 2). It is amphibious in that its flowers are emergent although Batrachian Ranunculi, like many other submerged aquatic plants with emergent flowers, but unlike *Lobelia dortmanna*, have their developing fruits submerged by bending of the peduncle (Sculthorpe 1967). This organism has C_3 biochemistry (based on a relatively high CO_2 compensation point in media in pH 6·5), can 'use' HCO_3^- (based on C_4-like gas exchange characteristics at high external pH values), has negligible CO_2 uptake through its roots, and lacks CAM-like metabolism (Raven, Beardall & Griffiths 1982; Raven, Osborne & Johnston 1985).

   Table 2 shows the $\delta^{13}C$ values of *Ranunculus penicillatus* var *calcareus*, and of CO_2 and HCO_3^-, in the Dichty Burn. If the major C source for its photosynthesis in the Burn is CO_2, then the plant is only $6\cdot6‰$ more negative than the $\delta^{13}C$ of its source CO_2, suggesting (eqn (2)) that photosynthesis is mainly (0·78) limited by CO_2 diffusion, a reasonable state of affairs in view of the density of the *Ranunculus* beds and the consequent likelihood of poor mixing within the canopy. However, since *Ranunculus penicillatus* var *calcareus* can also use HCO_3^-, we must also consider the possibility that HCO_3^- is the major C source at the alkaline pH value of the Dichty Burn (pH 8); in this case the plant has a 17·3‰ more negative $\delta^{13}C$ value than its source carbon (Table 2). This poses some difficulties if we wish to maintain that the HCO_3^- pump has a negligible discrimination between $H^{12}CO_3^-$ and $H^{13}CO_3^-$ (thus delivering HCO_3^- with a $\delta^{13}C$ value equal to that of the 'source' HCO_3^-, i.e. −5·2‰, to the intracellular compartment) *and* that there is negligible leakage from the intracellular pool, a circumstance which means that RUBISCO carboxylase activity has to fix *all* of the CO_2 produced from *all* of the pumped HCO_3^- at a $\delta^{13}C$ value of −5·2‰. We note that a qualitatively similar situation is found for *Ranunculus penicillatus* var *calcareus* from the River Frome (Table 2), although the data for *Ranunculus baudotii* and *Ranunculus trichophyllous* (Table 2) reveal plant $\delta^{13}C$ values more *positive* than those of the CO_2 in the relatively stagnant medium and thus favour a dominant role for HCO_3^- as C source for photosynthesis. More work is clearly needed to resolve the questions as to the C source for *Ranunculus penicillatus* var *calcareus*: $\delta^{13}C$ measurements would have an important role to play. Another role for $\delta^{13}C$ measurements in investigating the ecophysiology of the Batrachian Ranunculi is in deciding on the C source for fruit growth: C_3 photosynthesis with air CO_2 as C source should yield a significantly more negative $\delta^{13}C$ value for plant carbon than is found for the submerged vegetative parts of these plants (Table 2).

Ascophyllum nodosum: *a marine C_3 plant which actively transports HCO_3^-*

Ascophyllum nodosum (L.) Le Jol is a haptophytic Phaeophyte (Phaeophyceae: Fucales) which is an intertidal perennial, producing gametes annually, on relatively sheltered rocky shores of the North Atlantic (Baardseth 1970). Like other fucoids, *Ascophyllum nodosum* is biochemically a C_3 plant (see Kerby & Raven 1985). However, *Ascophyllum* is, from the point of view of gas exchange, C_4-like, with a high CO_2 affinity, negligible O_2 inhibition, and a low CO_2 compensation concentration, both when submerged in seawater and when emersed (Johnston 1984). It can also use HCO_3^- for submerged photosynthesis (Johnston 1984) as can many other intertidal seaweeds (Sand-Jensen & Gordon 1984; Kerby & Raven 1985). It would appear that *Ascophyllum* has a 'CO_2 concentrating mechanism' (Johnston 1984). A link with submerged *Isoetes* spp. (below) is a CAM-like diel change in titratable acidity and of malic acid concentration, together with a dark $^{14}CO_2$ fixation rate which is a larger fraction of the rate of photosynthesis than is the case for green and red seaweeds (Johnston 1984; Kerby & Raven 1985). However, the diel fluctuations in acidity are much smaller than in submerged *Isoetes* spp. and the contribution of CAM-like metabolism to total C fixation is small.

There are a number of reasons for expecting a more negative $\delta^{13}C$ value to be associated with emersed as opposed to submersed photosynthesis in *Ascophyllum*. Submersed photosynthesis must, to a substantial extent, use HCO_3^- (the major C source in seawater): CO_2 diffusion to the surface of the plant, and uncatalysed generation of CO_2 from HCO_3^-, are not adequate to support the observed rate of photosynthesis in terms of CO_2 entry (Johnston 1984). Since HCO_3^- in seawater has a $\delta^{13}C$ of around 0‰, a substantial discrimination in favour of $H^{12}CO_3^-$ by the HCO_3^- pump, and/or a substantial leak of CO_2 from the intracellular inorganic C pool, thus allowing RUBISCO carboxylase to express at least some of its intrinsic discrimination (Table 1), would be needed to generate a very negative $\delta^{13}C_{plant}$. Any exogenous CO_2 that *is* used in submersed photosynthesis will be subject to a substantial boundary layer resistance which will reduce the $\delta^{13}C_{plant}$ (eqn (2)). Finally, the low boundary layer resistance in air (Raven & Richardson 1986) means that the only C source (CO_2) would be much more readily available than would CO_2 in seawater: granted the same discrimination of the inorganic C active transport mechanism as occurs in seawater, the emersed plant will have a more negative $\delta^{13}C$ value for inorganic C supplied to RUBISCO than will the submersed plant.

Table 4 shows values of $\delta^{13}C$ for *Ascophyllum*, and for the inorganic C components of seawater. *Ascophyllum* from the lowermost part of its vertical range has a less negative $\delta^{13}C$ value than plants from the top of the range, in agreement with prediction. However, the $\delta^{13}C$ values do not permit quantification of the roles of emersed and submersed photosynthesis to overall plant C gain: we note that there are a number of reasons (Raven & Richardson 1986) for believing that emersed photosynthesis should be restricted relative to that found for a similar time of submergence. Table 4 also gives data for another intertidal alga, the

TABLE 4. $\delta^{13}C$ values for intertidal and subtidal marine macrophytes from Broughty Ferry, Angus, with values for related organisms from other sites

Organism	Source	$\delta^{13}C_{plant}$	Reference
Ascophyllum nodosum	Broughty Ferry, Angus; intertidal:		H. Griffiths (previously unpublished)
	Bottom of range	−16·4‰	
	Middle of range	−18·4‰	
	Top of range	−19·5‰	
Enteromorpha intestinalis Link	Broughty Ferry, Angus; intertidal:		H. Griffiths (previously unpublished)
	Bottom of range	−20·6‰	
	Top of range	−20·0‰	
Laminaria digitata Lamour	Broughty Ferry, Angus; subtidal	−17·8‰	H. Griffiths (previously unpublished)
Laminaria hyperborea Foslie	Broughty Ferry, Angus; subtidal	−18·0‰	H. Griffiths (previously unpublished)
Laminaria longicruris	Nova Scotia; subtidal	−12‰ −20‰	Stephenson, Tan & Mann (1984)

Seawater $\delta^{13}C_{CO_2} \simeq -7‰$, $\delta^{13}C_{HCO_3} \simeq 0‰$.
Further values for $\delta^{13}C_{plant}$ for marine algae may be found in Table III of Kerby & Raven (1985).

Ulvophycean *Enteromorpha intestinalis*, which is also a HCO_3^- user with a 'CO$_2$ concentrating mechanism' and a good capacity for emersed photosynthesis (provided desiccation does not intervene): Raven (1984), Kerby & Raven (1985). In this case there is no difference between the $\delta^{13}C$ values for plants at the top and the bottom of the intertidal. We note (Table 4) that the magnitude of the $\delta^{13}C$ value of *Ascophyllum nodosum* is similar to those found for the subtidal Phaeophyte (Laminariales) algae *Laminaria digitata, Laminaria hyperborea* and *Laminaria longicruris* and (Table III of Kerby & Raven 1985) to those of many other seaweeds.

Isoetes lacustris: a freshwater plant with substantial CAM-like metabolism and root CO$_2$ uptake

Isoetes lacustris L. is a rhizophytic Tracheophyte (Lycopsida: Isoetales), with the Isoetid life form, which lives submerged in slow-flowing waters (Page 1982); it is not dependent on emergence for sexual reproduction (Sculthorpe 1967). It has a substantial contribution (in excess of 40%) of root uptake of CO$_2$ to its total C acquisition (Richardson *et al.* 1984): we have seen (under *Lobelia dortmanna* above) that this is a common feature of submerged plant with the Isoetid life form. *Isoetes lacustris* also has a substantial contribution from CAM-like metabolism to its overall C acquisition (Keeley 1982; Richardson *et al.* 1984). We assume that the fixation of exogenous CO$_2$ in the light involves the C$_3$ pathway.

Table 3 shows the $\delta^{13}C$ values for *Isoetes lacustris* and for CO$_2$ from its immediate environment, together with the values of *Lobelia dortmanna* from the same loch. The difference between the $\delta^{13}C$ value of *Isoetes lacustris* and that of the surrounding CO$_2$ ($-5.6 - -6.0‰$) is less than that for *Lobelia dortmanna* ($-12.5 - -14.2‰$) from the same habitat with the same life-form and of a similar size, and also acquiring a large fraction of its CO$_2$ through its roots. It is tempting, in view of these similarities, to associate the difference in $\delta^{13}C$ values with the presence of CAM-like metabolism in *Isoetes lacustris* and its absence from *Lobelia dortmanna* (see Table 1 for $\delta^{13}C$ values associated with CO$_2$ fixation by PEP carboxylase and by RUBISCO). We note that comparisons *between* life-forms for plants from a given habitat do *not* permit the use of $\delta^{13}C$ values to distinguish CAM from non-CAM aquatic macrophytes (Sternberg, DeNiro & Keeley 1984).

Table 3 also gives values, derived from data presented by Keeley & Busch (1984), for submerged and emergent leaves of *Isoetes howellii*, an amphibious species. Emerged leaves have functional stomata and carry out C$_3$ photosynthesis using atmospheric CO$_2$, while submerged leaves have a substantial contribution of CAM-like metabolism to their carbon acquisition. The difference between the $\delta^{13}C$ values of the plant and those of the ambient CO$_2$ are substantially greater for the emergent than for the submerged plants ($-22.4 - -22.8‰$ for emergent plants against $-0.1 - -7.1‰$ for submerged plants). While part of this difference may be related to the occurrence of CAM-like processes in submerged, but not in emergent, plants, it is clear that there are also likely to be substantial contributions from the difference in limitation of carbon acquisition by CO_2 diffusion. Similar differences in $\delta^{13}C_{plant}$ and $\delta^{13}C_{CO_2}$ were noted by Richardson *et al.* (1984) who compared the submerged *Isoetes lacustris* and the terrestrial *Isoetes durieui* (Table 3).

A very interesting amphibious member of the Isoetales is the Peruvian endemic *Stylites (Isoetes) andicola* which has astomatous leaves, takes up essentially all of its CO_2 through the roots, and has low-amplitude CAM-like diel changes in titratable acidity and malic acid content, regardless of whether the plants have grown emergent or submerged (Keeley, Osmond & Raven 1984). The $\delta^{13}C$ values for the plant are very similar to those of the peat, and of the CO_2 in the interstitial waters of the peat in which the plant is growing, suggesting a large diffusional limitation (Keeley, Osmond & Raven 1984): there are, apparently, no comparisons of $\delta^{13}C$ values for plants, peat and CO_2 in interstitial waters between emergent and submerged plants.

Azolla caroliniana: an interfacial freshwater plant with C_3 biochemistry and symbiotic, N_2-fixing cyanobacteria

Azolla caroliniana Wild. is a floating aquatic Tracheophyte (Filicopsida: Salviniales) with stomata (albeit of a rather peculiar structure: Busby & Gunning 1984) and, usually, N_2-fixing cyanobacteria (*Anabaena azollae*) in mucilage-filled cavities in the leaves (Sprent 1979). *Azolla caroliniana* has the C_3 pathway of photosynthesis and lacks CAM-like change on titratable acidity (Ray *et al.* 1979; Raven, Osborne & Johnston 1985). The differences between the $\delta^{13}C$ of the plant and that of the CO_2 in air (Table 5) is consistent with C_3 fixation of atmospheric CO_2 for both symbiotic and asymbiotic plants, although it does not rule out the possibility of recycling of organic C via inorganic C in the rooting medium of the type described by another floating tracheophyte, *Eichhornia crassipes* (Ultsch & Anthony 1973; cf. Smith & Epstein 1971).

Of particular interest from the viewpoint of the interaction between the symbionts is the finding (Table 5) that the $\delta^{13}C$ of *Anabaena azollae* expressed from the cavities of *Azolla caroliniana* leaves is similar to that of its exhabitant. Ray *et al.* (1979) have shown that the freshly-expressed cyanobacterium is photosynthetically competent and has the C_3 biochemistry but C_4-like gas exchange characteristics found in cyanobacteria (and many algae) grown at air levels of CO_2 (Raven 1984). Table 5 also presents values for the difference in $\delta^{13}C$ values between the photolithotrophically cultured cyanobacterium (Newton & Herman 1979; Gates *et al.* 1980) and the source CO_2 for high- and low-CO_2-grown cultures. The low-CO_2-grown culture has a small $\Delta\delta^{13}C$, typical of such growth conditions (Beardall, Griffiths & Raven 1982), while the high-CO_2 grown culture, which lacks the 'CO_2 concentrating mechanism', has a larger $\Delta\delta^{13}C$ characteristic of C_3

TABLE 5. $\delta^{13}C$ values for *Azolla caroliniana* and its symbiont, *Anabaena azollae*, and for the CO_2 on which they were grown

Organism	$\delta^{13}C_{plant}$	$\delta^{13}C_{CO_2}$	$\Delta\delta^{13}C_{(plant-CO_2)}$
Azolla caroliniana growing on N_2 with *Anabaena azollae*	$-30\cdot9‰$	$-7‰$	$-23\cdot9‰$
Azolla caroliniana growing on NO_3^- as N source, lacking *Anabaena azollae*	$-32\cdot5‰$	$-7‰$	$-25\cdot5‰$
Anabaena azollae expressed from cavities in leaves of *Azolla caroliniana*	$-27\cdot3‰$	$-7‰$	$-20\cdot3‰$
Anabaena azollae grown asymbiotically and photolithotrophically with high (5%) CO_2	$-56\cdot3‰$	$-31‰$	$-25\cdot3‰$
Anabaena azollae grown asymbiotically and photolithotrophically with air (0·03‰) CO_2	$-17\cdot9‰$	$-7‰$	$-10\cdot9‰$

Previously unpublished results of H. Griffiths on organisms cultured at the University of Dundee.

biochemisty with diffusive CO_2 entry. If the cyanobacterium is indeed functioning photolithotrophically in the *Azolla* then it is clearly not obtaining its CO_2 directly from the atmosphere, since the presence of the 'CO$_2$ concentrating mechanism' would lead to a $\delta^{13}C$ of $-17\cdot9°/_{00}$ or so rather than the $-27\cdot3°/_{00}$ observed (Table 5). A strictly diffusion-limited supply of CO_2 from exhabitant respiration (at $\delta^{13}C$ of some $-30\cdot9‰$) could best account for the observed $\delta^{13}C$ of the inhabitant cyanobacterium: essentially all of the CO_2 reaching the cells would be fixed, allowing no display of the (possible) discrimination of the 'CO$_2$ concentrating mechanism' or the (known) discrimination by RUBISCO carboxylase activity (Table 1). However, it is also known that *Anabaena azollae* is capable of chemo-organotrophic growth, and N_2 fixation, with fructose as energy and carbon source (Newton & Herman 1979), so the possibility that a transfer of organic compounds from exhabitant to inhabitant underlies their similar $\delta^{13}C$ values cannot as yet be excluded. However, it is clear that the measurement of $\delta^{13}C$ values has (in conjunction with other data) ruled out the possibility of direct use of atmospheric CO_2 for photolithotrophy by the inhabitant.

CONCLUSIONS

The utility of $\delta^{13}C$ measurements in investigating photosynthesis in aquatic and amphibious plants is less obvious than their use for such investigations in terrestrial vascular plants. The reasons relate to the variable $\delta^{13}C$ values for source CO_2 (and the additional presence, in many systems, of another inorganic C source, HCO_3^-); the variable, and often large, but not readily measured, ratio of transport limitation to total limitation of the rate of photosynthesis; and the unknown $\delta^{13}C$ value associated with the active transmembrane transport of inorganic carbon species. It is clear that variations in plant anatomy and morphology, and the species of inorganic C used by the plant, can render difficult the differentiation, on the basis of $\delta^{13}C$ (or even $\Delta\delta^{13}C$) values alone, of such mechanisms of C acquisition as C_3 and CAM in aquatic plants. However, the example discussed in this paper show that, provided they are used in conjunction with other sources of information, $\delta^{13}C$ values can be useful in understanding the photosynthetic carbon acquisition processes in aquatic plants, e.g. the significance of boundary layers in limiting the rate of photosynthesis in *Lemanea* and *Fontinalis*, the *in situ* inorganic C source for Batrachian Ranunculi, and the carbon nutrition of the cyanobacterial symbionts of *Azolla*. Much more work needs to be done: it would, for example, be useful to have $\delta^{13}C$ measurements on *Hydrilla verticillata* in its C_3 and C_4 modes (Salvucci & Bowes 1983; cf. the unspecified growth conditions used by Benedict and co-workers to obtain their $\delta^{13}C$ values; Benedict 1978).

ACKNOWLEDGMENTS

This work was supported by Research Grants from SERC and NERC. We wish to thank Ms. Shona McInroy for assistance in the field and in the laboratory; Professor W. D. P. Stewart, FRS, Dr P. Rowell and Dr A. N. Rai for provision of *Azolla caroliniana* cultures with and without *Anabaena azollae*; and Dr Sarah Webster for confirming the identity of *Ranunculus penicillatus* var. *calcareus*.

REFERENCES

Allen, E. D. & Spence, D. H. N. (1981). The differential ability of aquatic plants to utilize the inorganic carbon supply in fresh waters. *New Phytologist,* 87, 269–283.
Baardseth, E. (1970). Synopsis of biological data on knobbed wrack *Ascophyllum nodosum* (L) Le Jol. *FAO Fisheries Synopsis* 38.
Bain, J. T. & Proctor, M. C. F. (1980). The requirements of aquatic bryophytes for free CO_2 as an inorganic carbon source: some experimental evidence. *New Phytologist,* 86, 393–400.
Beardall, J., Griffiths, H. & Raven, J. A. (1982). Carbon isotope discrimination and the CO_2 accumulating mechanism in *Chlorella emersonii. Journal of Experimental Botany,* 33, 729–737.
Benedict, C. R. (1978). Nature of obligate photo-autotrophy. *Annual Review of Plant Physiology,* 29, 67–93.
Boston, H. I. & Adams, M. S. (1983). Evidence of Crassulacean Acid Metabolism in two North American Isoetids. *Aquatic Botany,* 15, 381–386.

Bushby, C. H. & Gunning, P. E. S. (1984). Microtubules and morphogenesis in stomata of the water fern *Azolla:* an unusual mode of guard cell and pore formation. *Protoplasma,* **122,** 108–119.

DeNiro, M. J. & Hastorf, C. A. (1985). Alteration of $^{15}C/^{14}N$ and $^{13}C/^{12}C$ ratios of plant matter during the initial stages of diagenesis: studies utilizing archaeological specimens from Peru. *Geochimica et Cosmochimica Acta,* **49,** 97–115.

Farquhar, G. D. (1980). Carbon isotope discrimination by plants: effects of carbon dioxide concentration and temperature via the ratio of intercellular and atmospheric CO_2 concentrations. In: *Carbon Dioxide and Climate: Australian Research* (Ed. by G. I. Pearman), pp. 105–110. Australian Academy of Sciences, Canberra.

Farquhar, G. D. (1983). On the nature of carbon isotope discrimination in C_4 species. *Australian Journal of Plant Physiology,* **10,** 205–236.

Farquhar, G. D., O'Leary, M. H. & Berry, J. A. (1982). On the relationship between carbon isotope discrimination and the intercellular space carbon dioxide concentration in leaves. *Australian Journal of Plant Physiology,* **9,** 121–137.

Gates, J. E., Fisher, R. W., Goggin, T. W. & Azalar, N. I. (1980). Antigenic differences between *Anabaena azollae* fresh from the *Azolla* leaf and free-living cyanobacteria. *Archives of Microbiology,* **128,** 126–129.

Glime, N. M., Nissila, P. C., Trynoski, S. E. & Fornwall, M. D. (1979). A model for attachment of aquatic mosses. *Journal of Bryology,* **10,** 313–320.

Griffiths, H. & Smith, J. A. C. (1983). Photosynthetic pathways in the Bromeliaceae of Trinidad: relations between life-forms, habitat preference and the occurrence of CAM. *Oecologia,* **60,** 176–184.

Holaday, A. S., Lee, K. W. & Chollet, R. (1984). C_3–C_4 intermediate species in the genus *Flaveria*: leaf anatomy, ultrastructure and the effect of O_2 on the CO_2 compensation concentration. *Planta,* **160,** 25–32.

Johnston, A. M. (1984). *The assimilation of inorganic carbon by* Ascophyllum nodosum *(L.) Le Jol.* pp. 158. Ph.D. thesis, University of Dundee.

Jones, H. G. (1985). Partitioning stomatal and non-stomatal limitations to photosynthesis. *Plant, Cell and Environment,* **8,** 95–104.

Keeley, J. E. (1982). Distribution of diurnal acid metabolism in the genus *Isoetes. American Journal of Botany,* **69,** 254–257.

Keeley, J. E. & Busch, G. (1984). Carbon assimilation characteristics of the aquatic CAM plant, *Isoetes howellii. Plant Physiology,* **76,** 525–530.

Keeley, J. E. & Morton, B. A. (1982). Distribution of diurnal acid metabolism in submerged aquatic plants outside the genus *Isoetes. Photosynthetica,* **16,** 546–553.

Keeley, J. E., Osmond, C. B. & Raven, J. A. (1984). *Stylites,* a vascular land plant without stomata absorbs CO_2 via its roots. *Nature,* **310,** 694–695.

Kerby, N. W. & Raven, J. A. (1985). Transport and fixation of inorganic carbon by marine algae. *Advances in Botanical Research,* **11,** 71–123.

Lazerte, B. D. & Szalados, J. E. (1982). Stable carbon isotope ratio of submerged freshwater macrophytes. *Limnology and Oceanography,* **27,** 413–438.

Macfarlane, J. J. & Raven, J. A. (1985). External and internal CO_2 transport in *Lemanea*: interactions with the kinetics of ribulose bisphosphate carboxylase. *Journal of Experimental Botany,* **36,** 610–622.

Mook, W. G., Bommerson, J. C. & Staverman, W. H. (1974). Carbon isotope fractionation between dissolved bicarbonate and gaseous carbon dioxide. *Earth and Planetary Science Letters,* **22,** 169–176.

Newton, J. W. & Herman, A. I. (1979). Isolation of cyanobacteria from the aquatic ferm *Azolla. Archives of Microbiology,* **120,** 161–165.

O'Leary, M. H. (1981). Carbon isotope fractionation in plants. *Phytochemistry,* **20,** 553–568.

Osmond, C. B., Valaane, N., Haslam, S. M., Uotila, O. & Roksandic, Z. (1981). Comparisons of $\delta^{13}C$ values in leaves of aquatic macrophytes from different habitants in Britain and Finaland: some implications for photosynthetic processes in aquatic plants. *Oecologia,* **50,** 117–124.

Page, C. N. (1982). *The Ferns of Britain and Ireland.* pp. xii + 447. Cambridge University Press.

Paton, J. A. & Pearce, J. V. (1957). The occurrence, structure and function of the stomata of British bryophytes. *Transactions of the British Bryological Society,* **3**, 228–259.

Peñuelas, J. (1985). HCO_3^- as an exogenous carbon for aquatic bryophytes *Fontinalis antipyretica* and *Fissidens grandifrons. Journal of Experimental botany,* **36**, 441–448.

Proctor, M. C. F. (1984). Structure and ecological adaptation. In: *The Experimental Biology of Bryophytes.* (Ed. by A. F. Dyer & J. G. Duckett), pp. 9–37. Academic Press, London.

Raven J. A. (1984). *Energetics and Transport in Aquatic Plants,* pp. xi + 587. A. R. Liss, New York.

Raven, J. A. & Beardall, J. (1981). Carbon dioxide as the exogenous inorganic source for *Batrachospermum* and *Lemanea. British Phycological Journal,* **16**, 165–175.

Raven, J. A. & Richardson, K. (1986). Photosynthesis in marine environments. In: *Photosynthesis in Specific Environment.* (Ed. by N. R. Baker & S. P. Long), pp. 337–398. Elsevier, Amsterdam.

Raven, J. A., Allen, S. & Griffiths, H. (1984). N source, transpiration rate and stomatal aperture in *Ricinus* In: *Membrane Transport in Plants* (Ed. by W. J. Cram, K. Janázcek, R. Rybová & K. Sigler) pp. 161–162. Academia, Praha.

Raven, J. A., Beardall, J. & Griffiths, H. (1982). Inorganic C sources for *Lemanea, Cladophora* and *Ranunculus* in a fast-flowing stream: measurements of gas exchange and of carbon isotope ratio and their ecological implications. *Oecologia,* **53**, 68–78.

Raven, J. A., Osborne, B. A. & Johnston, A. M. (1985). Uptake of CO_2 by aquatic vegetation. *Plant, Cell and Environment,* **8**, 417–425.

Ray, T. B., Mayne, B. C., Toia, R. E., Jr. & Peters, G. A. (1979). *Azolla–Anabaena* relationships. VIII. Photosynthetic characterisation of the association and individual partners. *Plant Physiology,* **64**, 791–795.

Richardson, K. Griffiths, H., Reed, M. I., Raven, J. A. & Griffiths, N. M. (1984). Inorganic carbon assimilation in the Isoetids, *Isoetes lacustris* L. and *Lobelia dortmanna* L. *Oecologia,* **61**, 115–121.

Roeske, C. A. & O'Leary, M. (1984). Carbon isotope effects on the enzyme-catalysed carboxylation of ribulose bisphosphate. *Biochemistry,* **23**, 6272–6284.

Rosenberg, R. M. & O'Leary, M. H. (1985). Aspartate β-decarboxylase from *Alcaligenes faecalis*: carbon-13 kinetic isotope effect and deuterium exchange experiments. *Biochemistry,* **24**, 1598–1602.

Salvucci, M. E. & Bowes, G. (1983). Two photosynthetic mechanisms mediating the low photorespiratory state in submersed aquatic angiosperms. *Plant Physiology,* **73**, 488–496.

Sand-Jensen, K. & Gordon, D. M. (1984). Differential ability of marine and freshwater macrophytes to utilize HCO_3^- and CO_2. *Marine Biology,* **80**, 247–253.

Sand-Jensen, K. & Prahl, C. (1982). Oxygen exchanges with the lacunae and across leaves and roots of the submerged vascular macrophyte, *Lobelia dortmanna* L. *New Phytologist,* **91**, 103–120.

Sand-Jensen, K. & Søndergaard, M. (1979). Distribution and quantitative development of aquatic macrophytes in relation to sediment characteristics in the oligotrophic Lake Kalgaard. *Freshwater Biology,* **9**, 1–12.

Sculthorpe, C. D. (1967). *The Biology of Aquatic Vascular Plants,* pp. xviii + 610. Edward Arnold, London.

Sheath, R. G. (1984). The biology of freshwater red algae. In: *Progress in Phycological Research Vol. 3* (Ed by F. E. Round & D. J. Chapman) pp. 89–157. Biopress, Bristol.

Smith, A. J. E. (1978). *The Moss Flora of Britain and Ireland,* pp. 706. Cambridge University Press.

Smith, B. N. & Epstein, S. (1971). Two categories of $^{13}C/^{12}C$ ratio for higher plants. *Plant Physiology,* **47**, 380–384.

Smith, F. A. & Walker, N. A. (1980). Photosynthesis by aquatic plants: effects of unstirred layers in relation to assimilation of CO_2 and HCO_3^- and to carbon isotopic discrimination. *New Phytologist,* **86**, 245–259.

Sprent, J. I. (1979). *The Biology of Nitrogen-Fixing Organisms.* pp. 196. McGraw-Hill, Maidenhead.

Stephenson, R. L., Tan, F. C. & Mann, K. H. (1984). Stable carbon isotope variability in marine macrophytes and its implications for food web studies. *Marine Biology,* **81**, 223–230.

Sternberg, L., DeNiro, M. J. & Keeley, J. E. (1984). Hydrogen, oxygen and carbon isotope ratios of cellulose from submerged aquatic CAM and non-CAM plants. *Plant Physiology,* **76**, 68–70.

Troughton, J. H., Card, K. A. & Hendy, C. H. (1974). Photosynthetic pathways and carbon isotope discrimination by plants. *Yearbook of the Carnegie Institution of Washington*, **73**, 768–780.

Ultsch, G. R. & Anthony, D. S. (1973). The role of the aquatic exchange of carbon dioxide in the ecology of the water hyacinth (*Eichhornia crassipes*). *Florida Scientist*, **36**, 16–22.

Wium-Anderson, S. (1971). Photosynthetic uptake of free CO_2 by the roots of *Lobelia dortmanna*. *Physiologia Plantarum*, **25**, 245–248.

C. Growth, Development and Dispersal in Aquatic Plants

Photomorphogenic processes in freshwater angiosperms

DAVID H. N. SPENCE, MICHAEL R. BARTLEY* AND RICHARD CHILD

Department of Plant Biology and Ecology, The University, St Andrews, Fife KY16 9TH

SUMMARY

Photomorphogenesis in aquatic macrophytes is interesting because these species can be subject to a much wider range of light climates than strictly terrestrial plants. Detailed measurements of light climates were made in two Scottish lochs which have contrasting optical properties. The light regimes in these lochs are discussed with respect to the implications for photoreception by the plant photomorphogenic photoreceptors, cryptochrome and phytochrome. These results serve as a basis to explain a number of photomorphogenic responses observed in the field or under laboratory conditions and our main conclusions are: (i) Phytochrome photoequilibrium is unlikely to act as a measure of depth under water, because of the high ratios of red light to far-red light encountered at depths greater than about 1 m. Photoequilibrium may, however, be a useful indicator of the proximity of the water surface and seems important in the induction of aerial-type leaves on shoots of the amphibious plant, *Hippuris vulgaris*, and in the control of turion formation in *Potamogeton crispus*. (ii) It is likely that submerged plants perceive depth through fluence rate-dependent processes, such as phytochrome cycling rate and blue light absorption by cryptochrome: stem elongation in *Potamogeton richardsonii* and *Hippuris vulgaris* was sensitive to fluence rate. We propose an hypothesis to explain the control of shoot elongation in *Hippuris vulgaris*. This hypothesis involves an interaction between photosynthesis and photomorphogenesis and we provide preliminary evidence of a role for cryptochrome and phytochrome cycling rate in the photomorphogenic control of shoot elongation.

INTRODUCTION

In its broadest sense, photomorphogenesis has been defined as the control of plant development by light independently of photosynthesis (see Smith 1975). One of the chief functions of photomorphogenesis is to produce morphological and physiological responses which adapt the photosynthesizing plant to its local light climate. A second function is to synchronize the phases of plant development with the changing seasons. To achieve these, a plant must possess photoreceptors

*Present address: I.C.I. Plant Protection Division, Jealott's Hill Research Station, Bracknell, Berkshire, RG12 6EY England. To whom correspondence should be addressed.

capable of perceiving changes in the light environment and of transducing this information into modified cellular processes.

There are two important reasons for studying photomorphogenesis in freshwater angiosperms. The first relates to the far wider range of light climates which these plants occupy compared with strictly terrestrial plants. This range includes the qualitatively distinct underwater shade and terrestrial vegetational shade (Spence 1981; Smith 1982; Holmes & Klein, this volume). The second reason is the high degree of phenotypic plasticity which can be shown by individual plants of many aquatic angiosperm species, especially amphibious species. The ability of individuals of a species to photosynthesize in air and underwater, either simultaneously or at different times, is the hallmark of amphibious plants. Besides being an unusual feature of angiosperms generally, the possession of this ability by amphibious species confers upon them their wide range of phenotypic plasticity. This paper deals with aspects of photomorphogenic responses in several fully submerged and amphibious freshwater angiosperms.

Terrestrial and underwater shade

The light regimes under vegetation canopies and under water are both associated with an overall reduction in photon fluence rate compared with natural daylight and because of this they are shade environments. In terms of light quality, however, the two environments are very different.

In vegetational shade, selective absorption of blue and red wavelengths by the photosynthetic pigments of the canopy results in light enriched in the green and far-red regions of the spectrum compared with unfiltered daylight (Smith 1982). Light quality under water can also be affected by the presence of vegetation canopies, but a much greater influence is exerted by the optical properties of water itself together with any dissolved and suspended substances. Water selectively absorbs the longer wavelengths of light; so that in clear water the spectrum is progressively depleted in far-red relative to blue, green and, to a lesser extent, red as depth increases. As a result, the light climate under water typically has much more red than far-red and it is, therefore, in direct contrast to terrestrial canopy shade. Indeed, the spectrum under water has more red relative to far-red than unfiltered daylight.

Terrestrial plants appear to perceive the amount of canopy shade by measuring the change in light quality as it controls the proportion of the photoreceptor pigment phytochrome present in the far-red absorbing form (Pfr). In saturating light conditions—that is, those that produce a true equilibrium—the proportion of Pfr to total phytochrome is described as the phytochrome photequilibrium ($\varphi = \text{Pfr}/(\text{Pfr} + \text{Pr})$, where Pr is the red-absorbing form of phytochrome). The photoequilibrium (φ) varies between about 0·58 in unfiltered daylight and almost zero in deep canopy shade; and it is particularly sensitive to changes in the degree of shade (Smith 1982). In terrestrial habitats, a good index of the degree of canopy shade is the ratio of red light to far-red light (R:FR); and the relationship between

φ and R:FR ratio shows φ to be especially sensitive to variations in the R:FR ratio in the range found in vegetational shade (R:FR *ca* 1·2 to *ca* 0·05) (Smith & Holmes 1977; Holmes 1981; Smith 1982). Because of the ability of φ to monitor terrestrial shade, it is important to know the range of φ values likely to be encountered under water.

Spectroradiometer scans were made in April 1984 at 1 m intervals down to 5 m in Loch Borralie, Durness (58°38′N, 4°47′W), representing unusually clear, blue-green freshwater, and in nearby Loch na Thull (58°24′N, 5°W) with more highly attenuating, brown water typical of most freshwater in the United Kingdom (Spence 1976, 1981). Values for vertical attenuation coefficients (K_d, l nm^{-1}) per 10 nm wavelength between 400 and 750 nm were derived from these scans. These K_d values were combined with phytochrome photoconversion cross-section data, derived from measurements on dark-grown oat seedlings made by Butler, Hendricks & Siegelman (1964), to calculate φ using the formula (Bartley & Frankland 1982):

$$\varphi = \frac{\Sigma N_\lambda \sigma_1}{\Sigma N_\lambda \sigma_{1\lambda} + \Sigma N_\lambda \sigma_{2\lambda}}$$

where: φ = phytochrome photoequilibrium, Pfr/(Pfr + Pr); N = photon fluence rate (μmol m^{-2}s^{-1}); σ_1 = phytochrome photoconversion cross-section for Pr $\rightarrow$ Pfr; σ_2 = phytochrome photoconversion cross-section for Pfr $\rightarrow$ Pr; λ = wavelength.

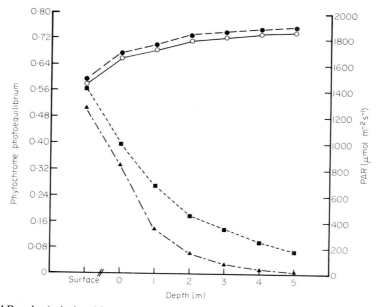

FIG. 1. PAR and φ (calculated from spectroradiometric scans) at different depths in Loch Borralie and Loch na Thull. Values at 0 m are subsurface values calculated from extrapolation of semi-logarithmic plots of fluence rate at each wavelength against depth. (○) φ, Loch Borralie, (●) φ, Loch na Thull, (■) PAR, Loch Borralie, (▲) PAR, Loch na Thull.

The relationship between φ and water depth in Loch Borralie and Loch na Thull is illustrated in Fig. 1. With increasing depth, very large falls in the photon fluence rate of photosynthetically active radiation (PAR) were accompanied by little change in φ, which in turn approached its asymptotic maximum over most of the depth range occupied by aquatic plants in both lochs. Any small, localized decreases in R:FR ratio caused by vegetational self-shading are likely to be swamped by the much greater increases in the R:FR ratio down the water column. Also, even if self-shading were to decrease the R:FR ratio at the sites of photoperception ten times, this will have little significant effect on φ at depth (Bartley & Spence 1986; cf. Holmes & Klein, this volume).

Qualitative growth responses under water

In shallow water there is evidence that phytochrome may be important in mediating at least three qualitative growth responses: the change from submerged-type to aerial-type leaves in the heterophyllous species, *Hippuris vulgaris*; aspects of turion formation in *Potamogeton crispus*; and production of stem tubers in *Hydrilla verticillata*.

Hippuris vulgaris is capable of growing entirely in air, entirely submerged rooted at water depths as great 6·5 m, or amphibiously with emergent portions of the stem, showing the characteristics of aerial shoots, above the water surface. The plant produces whorls of short, aerial-type leaves in air and whorls of long, thin, strap-shaped leaves under water. These leaf-types lack intermediate forms and are normally characteristic of aerial or submerged stems, because the transition from submerged- to aerial-type usually occurs when a shoot emerges from water into air. In two clear limestone lochs, Loch Balnagowan, Isle of Lismore, and Loch Borralie, however, this transition occurs under water about 1 m below the surface, on shoots which have grown up from sediments in less than 2 m of water—that is, the shoots are up to 1 m tall when the change occurs. It is apparent from this observation that the induction of aerial-type leaves is not merely a consequence of a shoot reaching the air/water interface. In controlled conditions, aerial-type leaves have been induced on submerged shoots of *Hippuris vulgaris* in long, high fluence rate (1000 μmol m^{-2} s^{-1}) photoperiods with a R:FR ratio of 1 (Bodkin, Spence & Weeks 1980). The effect was reversed by end-of-day light treatments having a high R:FR ratio, and this modulation of leaf-type by the R:FR ratio implies control by phytochrome.

The formation of turion by *Potamogeton crispus* provides an example of interactions between temperature, photoperiod, photon fluence rate and R:FR ratio (Chambers, Spence & Weeks 1985). In this species, long days (12 h+) and high temperatures (16°C+) are prerequisites for turion formation in plants previously grown in short days. A single long day or a single night-break with red light will suffice to initiate turion formation provided the temperature is adequate. This indicates that phytochrome may be the photoreceptor pigment which mediates this response. In low fluence rates, however, few turions are formed, but the

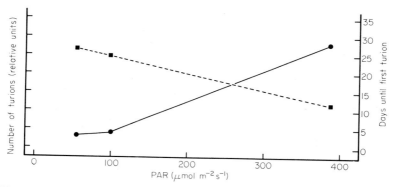

Fig. 2. Effect of PAR photon fluence rate on turion production in *Potamogeton crispus* at 30°C (18 h photoperiod). (●) Relative number of turions. (■) Number of days before first turion is visible. (After Chambers, Spence & Weeks 1985.)

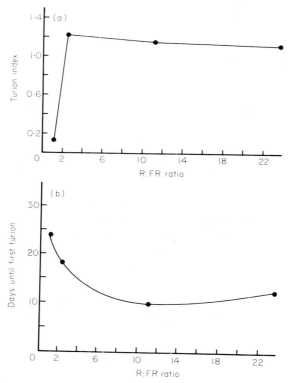

Fig. 3. Effect of R:FR ratio on turion production in *Potamogeton crispus* at 30°C (18 h photoperiod, 400 μmol m^{-2}s^{-1} PAR).

(a) Turion index $= \sum\limits_{1}^{n} \dfrac{\text{no. of turions in a pot}}{\text{no. of shoots in that pot}}$

(b) Minimum response time: no. of days until first turion appeared in each treatment.

number formed per plant is promoted by increasing fluence rate *per se*, whatever the total daily light dose (Fig. 2). In addition, with the other conditions optimal, R:FR ratios below 2·3 decreases turion formation (Fig. 3a), and the response time for turion initiation on isolated plants without self-shading is inversely related to the R:FR ratio for values less than 8 (Fig. 3b). Thus under suitable conditions, the fluence rate and R:FR ratio of light exert qualitative and quantitative effects by controlling the occurrence, magnitude and rate of turion production. These experimental findings in controlled environments have allowed fairly accurate predictions to be made about the field response of *Potamogeton crispus* in one part of its range. In plants growing in various lochs in Fife, turions were only observed from June until August, which is during the maximum temperature and photoperiod of the year. Furthermore, production of turions was confined to the upper 1·5 m of the water column, where the fluence rate was high.

Whereas *Potamogeton crispus* may be regarded as a long-day plant, for turion production, *Hydrilla verticillata* produces stem tubers in response to short days (Klaine & Ward 1984). Under experimental conditions, tuber initiation is prevented in short days by a pulse of red light given during darkness. The effects of this red light night-break are reversed by a following far-red light pulse. This red/far-red reversibility satisfies the major criterion for phytochrome involvement in photoperiodic responses.

These examples involve qualitative growth responses to the spectral quality of light and implicate phytochrome acting via photoequilibrium close to its maximum value 0·80. This involvement of φ as a depth sensor must be confined to shallow water (Fig. 1).

Quantitative growth responses under water

A classic photomorphogenic response to terrestrial canopy shade is the graded increase in stem elongation shown by responsive species, particularly arable weeds, to increases in the degree of vegetational shade (Morgan & Smith 1976, 1978, 1979, 1981a; Child, Morgan & Smith 1981). In this response, the rate and amount of stem elongation negatively correlate with φ. It is unlikely, however, that φ can mediate similar responses to underwater shade, because of the apparent insensitivity of φ to the changes in light quality found below about 1 m depth (Fig. 1). A similar conclusion, that φ probably has limited value as an indicator of depth under water, was reached by Morgan & Smith (1981b) from an analysis of the spectroradiometer scans of the very clear Crater Lake, and the phytoplankton-dense San Vincente Lake, both in the U.S.A., made by Tyler & Smith (1970). This being the case, it follows from our current knowledge of plant photoreceptor systems, that underwater shade is probably perceived through the decrease in fluence rate with depth. Fluence rate dependency of internode growth has previously been reported in the fully submerged species, *Potamogeton richardsonii* (Spence & Dale 1978; Spence 1981). In summer, internode length is logarithmically related to depth in high-density populations. Moreover, in young plants, grown without self-shading in large outdoor tanks, internode lengths were greater in 12% daylight than in shoots grown in full daylight, because of quicker elongation of the same number of internodes.

In a more recent study, whole specimens of *Hippuris vulgaris* were collected by scuba diver at several depths each month from April to September inclusive in Loch Borralie. *Hippuris vulgaris* is a rhizomatous perennial, with each rhizome growing horizontally and, in Loch Borralie, producing five or six shoots per growing season. The shoots overwinter either as small photosynthetic shoots or as small white shoots and apical rhizome buds in the lake sediment. Shoots emerge from the sediment in early autumn and in spring, and this growth habit means that particular care was required for the computation of mean total height, node number and mean internode length (MIL, height/number of internodes) for equivalent shoots from different rhizomes. Loch Borralie is a well-mixed lake (Spence, Barclay & Allan 1984), and water temperature is uniform with depth at any one time; so it is assumed that the differences in shoot growth with depth were due to changes in the light environment. During the period of collection, attenuation coefficients for photosynthetically active radiation (PAR) and for blue, red and far-red wavelengths were calculated from measurements of fluence rate made at several depths. These attenuation coefficients showed very little monthly variation, which indicates the absence of seasonal algal blooms. Plants collected in April showed differences in the height of green shoots (Fig. 4a). At 2·5 m and 4·5 m rooting depth, the shoots were relatively short compared with plants at 3·5 m, but node number per shoot was similar at each depth (Fig. 4b). It seems, therefore, that the differences in height in spring could be attributed to variations in MIL (Fig. 4c). In June, however, shoot height was inversely related to depth. At this time the relationship between MIL and depth was similar to that in April, but node number was inversely related to depth: shoots rooted at 2·5 m had about twice as many nodes as plants at 3·5 m. By August, the relationship between shoot height and depth resembled that found in spring—that is, with the tallest plants at 3·5 m—but MIL was about the same at each depth and node number was greatest at 3·5 m. Thus, shoot height at this time appears to be determined by the production of internodes rather than differences in internode elongation. This contrasts with the pattern in spring when shoot height was dependent upon internode elongation.

Our hypothesis to explain these variations in shoot growth involves an interaction between photosynthesis and photomorphogenesis. We propose that a graded photomorphogenic response to increased shading with depth tends to increase shoot height through the stimulation of internode elongation; but that the expression of this response is modified by the photosynthetic status of the plants through the production of nodes. In other words, shoot growth may be limited by the capacity of the plant to photosynthesize in deep water where fluence rates are low. As a result, the peak for shoot height and MIL at 3·5 m during spring may be a consequence of photomorphogenic inhibition of stem elongation at depths less than 3·5 m, whereas an insufficiency of stored carbohydrate in the overwintering rhizome may limit the growth of shoots in depths greater than 3·5 m. As the season progresses and the water temperature rises (in Loch Borralie the water

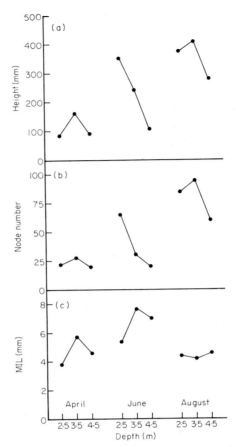

Fig. 4. Variations in *Hippuris vulgaris* of (a) shoot height, (b) node number and (c) mean internode length (MIL) with depth in Loch Borralie. Whole plants were collected monthly by scuba diver from rooting depths of 2·5, 3·5 and 4·5 m. Values presented are for shoot no. 2 on the rhizome, which is the second shoot initiated in the growing season.

temperature rises from 5°C in April to 15°C by the start of June), respiration will make greater demands on carbohydrate resources and the limitation of photosynthesis by low light levels in deep water could override the accompanying reduction in photomorphogenic inhibition. Photosynthetic limitation appears to affect node number in June, when shoot height is inversely related to depth, even though plants rooted at 3·5 m still have the longest internodes. By August, the photomorphogenic internode elongation response is not evident, but node production at 3·5 m has increased dramatically, such that plants at this depth have the longest shoots.

To test our hypothesis we have started to examine the possible mechanisms of photoperception for the photomorphogenic control of elongation in submerged shoots of *Hippuris vulgaris*. Green plants possess two photomorphogenic photoreceptor systems: phytochrome, and the blue light-absorbing pigments collectively referred to as cryptochrome. Cryptochrome is only capable of measuring the fluence rate of blue light, but phytochrome, in addition to being able to monitor light quality as already described, is also capable, at least in theory, of measuring fluence rate via cycling rate. Phytochrome cycling rate is the rate of the Pr ↔ Pfr photointerconversion reactions which are equal and opposite at photoequilibrium. It was originally evoked as a mechanism of photoperception in order to explain fluence rate-dependent responses in emergent seedlings (Smith 1975; Johnson & Tasker 1979) and there is now strong correlative evidence for its involvement in the control of seed germination in some species (Bartley & Frankland 1982). There is as yet, however, no evidence for any cycling rate-controlled phenomena in mature, green plants.

Smith (1982) has suggested that at low fluence rates, for example, in very deep or turbid water, the dark reactions of phytochrome (phytochrome destruction and Pfr reversion to Pr) may become significant in relation to the light reactions. This would lead to the actual Pfr/P ratio (where P is total phytochrome) being significantly lower than that at true photoequilibrium, which is calculated ignoring any thermal dark reactions. Thus depth, or PAR, may be monitored through this Pfr/P ratio. The rate constants for Pfr destruction and Pfr reversion to Pr have been measured in dark-grown *Hippuris vulgaris* apical rhizome segments, and it is possible to determine the theoretical Pfr/P ratio in *Hippuris vulgaris* at different depths from these rate constants and from the rate constants for the light reactions calculated from spectral scans. At 25°C, which is higher than the maximum temperature reached in Loch Borralie at depths at which *Hippuris vulgaris* occurs, the rate constant for Pfr destruction (k_d) was measured as $8.7 \times 10^{-5} \, \text{s}^{-1}$. The rate constant for thermal reversion of Pfr to Pr (k_r) was measured as $43 \times 10^{-5} \, \text{s}^{-1}$. Using the equation $Pfr/P = k_1/(k_1 + k_2 + k_d + k_r)$ (Fukshansky & Schafer 1983) where k_1 is the rate constant for phototransformation of Pr to Pfr and k_2 is the rate constant for phototransformation of Pfr to Pr (both of which are fluence rate and wavelength dependent), expected values for Pfr/P can be calculated for *Hippuris vulgaris* shoots growing at depth in both Loch Borralie and Loch na Thull. At 5 m in Loch Borralie, Pfr/P is virtually identical to φ, the true photoequilibrium. At 5 m in the turbid water of Loch na Thull, Pfr/P is 0·74 compared with a value for φ of 0·76. Although these calculations are based on a high surface fluence rate (see Fig. 1), any significant prolonged reduction in Pfr/P below photoequilibrium values are unlikely within the depth range at which *Hippuris* occurs.

Figure 5 shows that the amount of blue light and the calculated phytochrome cycling rate decline in parallel with PAR down the water column in both Loch Borralie and Loch na Thull. To determine the involvement of a specific blue light photoreceptor we have used a method based on that of Thomas & Dickinson (1979). This method involves adding supplementary blue light to a background of orange light from low pressure sodium discharge lamps. The orange light provides PAR and maintains a high φ of about 0·75, which is similar to φ values likely to be generated below 1 m. Also, the quantum efficiency of phytochrome

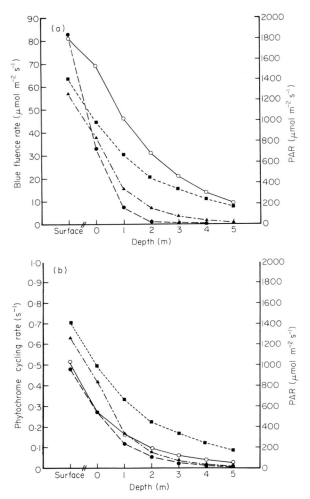

FIG. 5. PAR and blue light (a) and calculated phytochrome cycling rate (b) at different depths in Loch Borralie and Loch na Thull. Blue light is defined as waveband 440–460 nm. (a) (○) Blue, Loch Borralie, (●) Blue, Loch na Thull; (■) PAR, Loch Borralie; (▲) PAR, Loch na Thull. (b) (○) cycling rate, Loch Borralie; (●) cycling rate, Loch na Thull; (■) PAR, Loch Borralie; (▲) PAR, Loch na Thull.

photoconversion by orange light (λ_{max} 585 nm) is much greater than that of blue light, so the supplementary blue light of low fluence rate should not affect φ or phytochrome cycling rate significantly, and any specific blue light effects can be attributed to photocontrol through cryptochrome. The preliminary results of this work show that the elongation rate of shoots grown in different fluence rates of orange light alone was inversely related to fluence rate (Fig. 6) and that height was further reduced by the addition of blue light (Fig. 7, Table 1). Node production was not affected significantly by the light treatments, which suggests that it was mainly

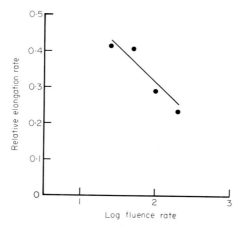

FIG. 6. The relationship between shoot relative elongation rate and fluence rate of orange light in *Hippuris vulgaris*.

$$\text{Relative elongation rate} = \frac{\text{ln final height (mm)} - \text{ln initial height (mm)}}{\text{ln initial height (mm)}}$$

The plants were grown for 12 days in orange light ($\lambda_{max} = 585$ nm) from low pressure sodium discharge tubes.

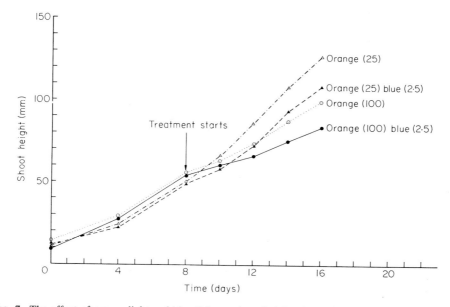

FIG. 7. The effect of orange light and blue light on shoot height of *Hippuris vulgaris*. The plants were grown for 8 days under orange light (50 μmol^{-2} s^{-1}) alone and then for a further 8 days in the treatments indicated. Blue light was from Thorn 20W 660 mm Northlight fluorescent tubes wrapped in one layer of Primary blue Np. 20 Cinemoid.

TABLE 1. The effect of orange and blue light on relative elongation rate and node number on day 16 of the experiment described in Fig. 7

Light treatment	Relative elongation rate ($\bar{x} \pm$ S.E.)	Node number ($\bar{x} \pm$ S.E.)
Orange (25 μmol m^{-2}s^{-1})	0·256 ± 0·018	27·3 ± 1·6
Orange (25 μmol m^{-2}s^{-1}) + blue (2·5 μmol m^{-2}s^{-1})	0·217 ± 0·018	25·3 ± 1·7
Orange (100 μmol m^{-2}s^{-1})	0·156 ± 0·017	28·3 ± 1·5
Orange (100 μmol m^{-2}s^{-1}) + blue (2·5 μmol m^{-2}s^{-1})	0·121 ± 0·016	24·3 ± 1·5

Values are mean ± S.E. for eight replicates.

internode elongation that was affected by light treatments. The data are consistent with cryptochrome as a photoreceptor for the inhibition of internode elongation by blue light. The photoreceptor for the fluence rate-dependent inhibition of elongation by orange light alone, however, does pose more of a problem. It seems counter-intuitive that this form of inhibition could be brought about through photosynthesis when shoot elongation was inversely related to PAR. Because the fluence rates used for the experiment were more than sufficient to saturate φ, then phytochrome cycling rate is an attractive candidate for the role of the orange light photoreceptor. Further experiments are necessary to test for the involvement of cycling rate, but the possibility exists that *Hippuris vulgaris* possesses three photomorphogenic photoreceptor systems which govern shoot growth in adult plants: they are cryptochrome, phytochrome acting via photoequilibrium and phytochrome acting via cycling rate.

Cryptochrome and phytochrome cycling would be capable of detecting shade under water and in air, but could not distinguish neutral shade, such as cloud cover, from vegetational shade. In contrast, photoequilibrium would be of limited value as an indicator of shade under water and, therefore, a poor measure of depth; it would be most effective in detecting canopy shade in submerged shoots close to the water surface and in emergent aerial shoots. In Table 2 we summarize how these different

TABLE 2. A scheme for the photomorphogenic control of shoot elongation in *Hippuris vulgaris*

	Terrestrial		Underwater	
	Open	Shade	Subsurface (first few cm normally)	Deep water
R:FR	≈1	<1	≈1	13 → ∞
PAR	high	low	high	low
φ	>0·55	<0·55	≈0·67–0·72	>0·72
Cycling rate	high	low	high	low
MIL	very short	long	short	very long
Control mechanism	cryptochrome, phytochrome cycling	phytochrome photoequilibrium, cryptochrome, phytochrome cycling	phytochrome photoequilibrium, cryptochrome, phytochrome cycling	cryptochrome, phytochrome cycling

photoreceptor systems may function in the terrestrial and underwater environments. Emergent shoots of *Hippuris vulgaris* offer an interesting example of the inter-relationships between photoreceptor systems in photomorphogenesis, in adapting amphibious plants to their environment.

ACKNOWLEDGMENTS

We are grateful to Dr David Weeks for much helpful advice and discussion, and to Mr Harry Hodge and Miss Alison Lovegrove for excellent technical assistance. We would like to thank Mr and Mrs Donald Munro for their kind hospitality and assistance during fieldwork at Durness. Financial support was from NERC.

REFERENCES

Bartley, M. R. & Frankland, B. (1982). Analysis of dual role of phytochrome in the photoinhibition of seed germination. *Nature*, **300**, 750–752.

Bartley, M. R. & Spence, D. H. N. (1986). The light environments of two contrasting Scottish lochs: implications for photomorphogenesis under water. In press.

Bodkin, P. C., Spence, D. H. N. & Weeks, D. C. (1980). Photoreversible control of heterophylly in *Hippuris vulgaris* L. *New Phytologist*, **84**, 533–542.

Butler, W. L., Hendricks, S. B. & Siegelman, H. W. (1964). Action spectra of phytochrome *in vitro*. *Photochemistry and Photobiology*, **3**, 521–528.

Chambers, P. A., Spence, D. H. N. & Weeks, D. C. (1985). Photocontrol of turion formation by *Potamogeton crispus* L. in the laboratory and natural water. *New Phytologist*, **99**, 183–194.

Child, R., Morgan, D. C. & Smith, H. (1981). Morphogenesis in simulated shadelight quality. *Plants and the Daylight Spectrum* (Ed. by H. Smith), pp. 409–420. Academic Press, London.

Fukshansky, L. & Schafer, E. (1983). Models in photomorphogenesis. *Photomorphogenesis, Encyclopedia of Plant Physiology New Series Vol. 16A* (Ed. by W. Shropshire, Jr. & H. Mohr), pp. 69–75. Springer-Verlag, Berlin.

Holmes, M. G. (1981). Spectral distribution of radiation within plant canopies. *Plants and the Daylight Spectrum* (Ed. by H. Smith), pp. 147–158. Academic Press, London.

Johnson, C. B. & Tasker, R. (1979). A scheme to account quantitatively for the action of phytochrome in etiolated and light-grown plants. *Plant, Cell and Environment*, **2**, 259–265.

Klaine, S. J. & Ward, C. H. (1984). Environmental and chemical control of vegetative dormant bud production in *Hydrilla verticillata*. *Annals of Botany*, **53**, 503–514.

Morgan, D. C. & Smith, H. (1976). Linear relationship between phytochrome photoequilibrium and growth in plants under simulated natural radiation. *Nature*, **262**, 210–212.

Morgan, D. C. & Smith, H. (1978). The relationship between phytochrome photoequilibrium and development in light-grown *Chenopodium album* L. *Planta*, **142**, 187–193.

Morgan, D. C. & Smith, H. (1979). A systematic relationship between phytochrome-controlled development and species habitat, for plants growing in simulated natural radiation. *Planta*, **145**, 253–258.

Morgan, D. C. & Smith, H. (1981a). Control of development in *Chenopodium album* L. by shadelight: the effect of light quantity (total fluence rate) and light quality (red:far-red ratio). *New Phytologist*, **88**, 239–248.

Morgan, D. C. & Smith, H. (1981b). Non-photosynthetic responses to light quality. *Physiological Plant Ecology 1, Encyclopedia of Plant Physiology New Series Vol. 12A* (Ed. by O. L. Lange, P. S. Nobel, C. B. Osmond & H. Ziegler), pp. 109–134. Springer-Verlag, Berlin.

Smith, H. (1975). *Phytochrome and Photomorphogenesis*. McGraw-Hill, London.

Smith, H. (1982). Light quality, photoperception, and plant strategy. *Annual Review of Plant Physiology*, **33**, 481–518.

Smith, H. & Holmes, M. G. (1977). The function of phytochrome in the natural environment: III measurement and calculation of phytochrome equilibria. *Photochemistry and Photobiology*, **25**, 547–550.

Spence, D. H. N. (1976). Light and plant response in fresh water. *Light as an Ecological Factor II* (Ed. by G. C. Evans, R. Bainbridge & O. Rackham), pp. 93–133. Blackwell Scientific Publications, Oxford.

Spence, D. H. N. (1981). Light quality and plant responses underwater. *Plants and the Daylight Spectrum* (Ed. by H. Smith), pp. 245–275. Academic Press, London.

Spence, D. H. N., Barclay, A. M. & Allen, E. D. (1984). Limnology and macrophyte vegetation of a deep, clear limestone lake, Loch Borralie. *Transactions of the Botanical Society of Edinburgh*, **44**, 187–204.

Spence, D. H. N. & Dale, H. M. (1978). Variations in the shallow water form of *Potamogeton richardsonii* induced by some environmental factors. *Freshwater Biology*, **8**, 251–268.

Thomas, B. & Dickinson, H. G. (1979). Evidence for two photoreceptors controlling growth in de-etiolated seedlings. *Planta*, **146**, 545–550.

Tyler, J. E. & Smith, R. C. (1970). *Measurement of Spectral Irradiance Underwater*. Gordon and Breach, New York.

Germination under water

B. FRANKLAND

School of Biological Sciences, Queen Mary College, London E1 4NS

M. R. BARTLEY

I.C.I. plc, Jealott's Hill Research Station, Bracknell, Berkshire, RG12 6EY

D. H. N. SPENCE

Department of Plant Biology and Ecology, The University, St Andrews, Fife, KY16 9TH

SUMMARY

Factors controlling induction of germination of dormant seeds of aquatic plants are reviewed with emphasis on environmental signals which may be important under water. Seed germination in general is affected by hydration, aeration, temperature and light. Light has a dual effect, both stimulating and inhibiting germination, and this can be quantitatively interpreted in terms of the properties of the photoreceptor pigment phytochrome, given the quantity and spectral quality of the light. On theoretical grounds it is predicted that light transmitted through green leaves underwater will tend to promote rather than inhibit germination in contrast to the strong inhibition that occurs under leaf shade in an aerial environment. Other significant differences in the aquatic environment are reduced oxygen concentration and reduced temperature fluctuations. Studies on the effects of light and temperature on the onset of growth in turions and other propagules are briefly reviewed.

GENERAL ASPECTS OF CONTROL OF SEED GERMINATION BY ENVIRONMENTAL FACTORS

Four important factors are involved in the environmental control of seed germination: hydration, aeration, temperature, light (Bewley & Black 1982). A seed which fails to germinate under environmental conditions which would be suitable for seedling growth is said to be dormant. When germination is impossible because the seed is too dry, temperature too extreme or oxygen absent the seed is said to be 'quiescent'. Dormant seeds may be induced to germinate by some specific environmental stimulus or by a period under specific environmental conditions. Seeds maintained under conditions unsuitable for germination may sometimes enter into secondary dormancy. Dormancy and the requirement for specific conditions for germination are mechanisms for ensuring dispersal of the seed in space and/or time.

167

In many plants mature seeds have a low (8%) water content and contact with a moist substrate is an essential prerequisite for germination. Hydration can be a factor even in the germination of seeds of aquatic plants. Some seeds are unable to take up water because of impermeable coats. Other seeds, which are resistant to desiccation, may persist in the quiescent state and so be able to recolonize lakes or ponds which are subject to intermittent drying-out. However, seeds of many true aquatic plants rapidly lose viability in the absence of water.

Light effects on germination are varied and complex and this is the factor which on first impression is the most difficult to comprehend. If effects of white light are considered the following situations arise: (i) short irradiation stimulates, (ii) long irradiation stimulates, (iii) long irradiation inhibits (Frankland & Taylorson 1983). Although stimulation will be most obvious in seeds with a low proportion germinating in the dark, and inhibition in those with a high germination rate in the dark, it is often possible to demonstrate all these effects in one species. The separation of seeds into 'positively photoblastic' and 'negatively photoblastic' is to be avoided (Kendrick 1976). Photoperiodic effects on seed germination have been reported with 'short day' and 'long day' seeds being distinguished. There is no evidence that these effects involve true photoperiodism and they may simply be aspects of the dual effect of long irradiations. The stimulating effect of a long irradiation can usually be replaced by repeated or intermittent short irradiations or pulses; this is not so for the inhibitory effects of long irradiations. Emphasis is usually placed on the final germination percentage attained although delays in germination (increase in time to 50% of final germination) may be observed.

The chromoprotein phytochrome is now well established as the key photoreceptor pigment in seeds. The involvement of photosynthetic pigments, chlorophylls and carotenoids, can be fairly safely discounted in the case of seeds but must be considered in relation to regulation of dormancy in propagules containing photosynthetic tissue. Another pigment, a blue-absorbing flavoprotein, may be involved in seed germination but the evidence for this is slender and this photoreceptor system is in general much less well characterized than phytochrome. Phytochrome exists in two forms: Pr with an absorption peak in the red (R) region of the spectrum and a far-red (FR) absorbing Pfr form. Pfr is the active form and is necessary for a key step in the sequence of events leading to the elongation of the embryonic root and hence germination.

$$
\begin{array}{c}
\text{Pr} \\
\text{R} \; \Big\Updownarrow \; \text{FR} \quad \text{)D} \\
\text{Pfr} \\
\longrightarrow \;\; + \;\; \longrightarrow \; \longrightarrow \; \longrightarrow \quad \text{germination} \\
\text{X} \\
\nearrow
\end{array}
$$

There is at least one case where a short irradiation with white or red light delays germination and hence Pfr must be inhibiting germination; in most seeds Pfr promotes germination. Pfr is relatively unstable and can show thermal reversion back to Pr in the dark. This is the reason why many seeds require long or repeated irradiations for induction of germination; the amount of Pfr action required is not met by a single pulse. A light requirement for germination enables seeds to germinate at or near the soil surface and ensures that small seeds do not germinate at such depth that food reserves are used up in the growth of the seedling to the surface; it also provides a mechanism for maintaining a bank of dormant seeds in the soil. Only small amounts of light are needed to convert Pr to Pfr. The chief reason that seeds fail to germinate under a layer of soil is that the rate of photochemical conversion of Pr to Pfr is less than the rate of thermal reversion to Pfr to Pr.

Sufficient is now known about the photochemical properties of phytochrome for any light environment of known fluence rate (quantity) and spectral quality to be fully described in terms of two phytochrome parameters. The proportion of phytochrome in the active Pfr form at photoequilibrium is given the symbol φ and is equal to $k_1/(k_1 + k_2)$ where k_1 and k_2 are the first order rate constants for the photochemical interconversions of Pr and Pfr. Each rate constant is the product of the photon fluence rate and the 'photoconversion cross-section' ($k_1 = N\sigma_1$, $k_2 = N\sigma_2$).

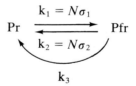

When the quantity of light is low, account must be taken of the rate of thermal reversion of Pfr to Pr (rate constant k_3). This Pft/P total ratio, referred to here as φ', is equal to $k_1/(k_1 + k_2 + k_3)$. (Strictly speaking the spectral transmission characteristics of the seed coat should also be taken into account. For instance, blue light does not readily penetrate into seeds with black or brown coats.) The rate of interconversion of Pr and Pfr under light is often referred to as the 'cycling rate'. The relative cycling rate, H (or C) = $(1-\varphi) k_1 = \varphi k_2$. An alternative parameter is θ, the rate constant for approach to photoequilibrium which is equal to $k_1 + k_2$ (Fukshansky & Schafer 1983). Both H and θ are dependent on fluence rate whereas φ is independent (φ' is fluence rate dependent at low levels of light). Light of a particular spectral quality will be characterized by a particular φ value. This value is high over the spectral range 400 to 700 nm (photosynthetically active radiation) but low beyond 700 nm (far-red radiation). This provides a mechanism by which seeds perceive and hence avoid germination under leaf shade where the proportion of light suitable for photosynthesis, PAR/(PAR + FR), is low (Frankland 1981).

With long periods of exposure to light germination may be inhibited. This inhibitory effect is dependent on fluence rate as well as dependent on a long irradiation time. The effect can be observed with light maintaining a relatively high level of Pfr. Germination can be blocked at a late stage and after the completion of Pfr action. The dual action of light may be represented as photocontrol at two steps in the sequence of events leading to germination, one promoted by Pfr, the other inhibited by a 'high irradiance reaction', symbolized as H. The H effect can be shown to be quantitatively correlated with the rate of phytochrome cycling (Bartley & Frankland 1982). Light inhibition acts to prevent germination on the soil surface, particularly under conditions of high irradiance when the seedling would be likely to experience water stress.

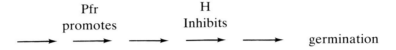

Under prolonged irradiation there is an antagonistic interaction between Pfr and phytochrome cycling. This has been studied in *Sinapis arvensis, Plantago major* and *Amaranthus caudatus* using crossed gradients of red and far-red light to vary light quality, and neutral filters to vary light quantity (Bartley & Frankland 1982). There is an inverse linear relationship between probit of germination and logarithm of cycling rate H and this inhibition is reduced with increase in φ. The interaction can be described by the empirical expression $\varphi\text{-}x\log H$. This expression suggests that photoinhibition will be greatest at about 710 nm which provides the most effective combination of low Pfr with high cycling rate; experimental studies on wavelength dependency confirm that this is so. Photoinhibition can be most conveniently studied in seeds with a high dark germination; an example is cucumber (*Cucumis sativus*) which has sufficient Pfr for germination 'trapped' in the mature, dry seeds. Studies on seeds of this species (G. Angelakis & B. Frankland, unpubl. data) confirm the interaction between φ and H established in previous work. In addition, it was found that blue light has an inhibitory effect greater than could be predicted from its φ value and this could be taken as evidence for the involvement of a pigment other than phytochrome.

Seeds may exhibit dormancy in the sense that they require a temperature range for germination which is narrower than that for seedling growth. Seeds of some species germinate best at fairly low temperature (10–15°C). This is particularly characteristic of species from dry (Mediterranean) areas where the low temperature requirement may be an environmental cue for the coming of the rainy season. Seeds of other species may germinate best at fairly high temperature (30–35°C). There are various types of interaction between temperature and light. Some seeds do not germinate in the dark at any temperature. Others show germination over a wide temperature range in the presence of a stimulating light treatment or a narrow temperature range with an inhibiting light treatment. In some cases the effects of light and temperature are roughly additive.

There are many cases where fluctuating temperatures stimulate germination. Daily alternations of temperature are of particular ecological significance. Again there are interactions with light, some seeds being stimulated by fluctuating temperature in both dark and light, whereas others require both light and fluctuating temperatures. The latter situation is particularly characteristic of marsh plants (e.g. *Bidens tripartita*, *Lycopus europaeus*, *Lythrum salicaria*). In temperate regions seedling establishment is restricted to late spring when the water table recedes, causing bare soil to be exposed and the surface layers to become aerobic. It has been suggested (Thompson, Grime & Mason 1977) that a fluctuating temperature requirement could delay germination until such time as 'the declining water content causes the seed to be no longer insulated against the increasing daily radiation load'. A response to alternating temperatures in some marsh plants may be an adaptation ensuring germination in relatively shallow water where daily temperature fluctuations are much greater than in deeper water. Experimental studies have shown several cases where a single temperature shift is sufficient to stimulate germination. Several seeds are stimulated by sowing at a low or high temperature and then after one or two days transferring to a moderate, germination temperature.

There are many seeds, particularly those with a deep dormancy which cannot be overcome by light or fluctuating temperature, which require to be maintained under moist conditions at a low temperature (0–5°C) for a relatively long period of time which may range from a few days to several months. Germination does not take place until seeds are transferred to a higher temperature following the period of low temperature induction. This chilling requirement ensures that seeds germinate after and not before the cold season.

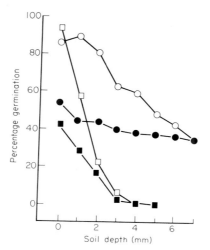

FIG. 1. Germination of seeds of *Sinapis arvensis* at various depths in well-drained, aerated (circles) or very wet, waterlogged (squares) soil. Germination in pots both in dark (closed symbols) or under white fluorescent light (open symbols) at 25°C. Unpublished data of W. K. Poo & B. Frankland.

Seeds of most mesophytes do not germinate under anaerobic conditions. If seeds of a light-requiring species are sown at different depths in a well-aerated soil there is a progressive decline in percentage germination which is correlated with the progressive attenuation of light passing into the soil. If seeds are sown in waterlogged soil the percentage germinating falls much more rapidly with increase in depth (Fig. 1). This effect is probably due to lack of oxygen.

GERMINATION OF SEEDS UNDER WATER

Seeds of marsh plants may germinate above the water table on bare mud or soil. However, seeds of rooted hydrophytes, including emergent, floating and submerged species, must have the capacity to germinate under water. This means that in contrast to seeds of mesophytes they must be able to tolerate low concentrations of oxygen (300 to 400 μM in standing water) or even germinate in the virtual absence of oxygen if they are on or in anaerobic mud. The processes involved in determining whether a seed floats or sinks are very important in dispersal and survival. Seeds may be dispersed by floating but do not germinate until after they have sunk. Very little is known of the factors which prevent a floating seed from germinating. These could involve features of the seed or fruit coat (Spence *et al.* 1971) or embryo dormancy, or germination could be blocked by high light irradiance or high oxygen concentration.

Seeds of *Typha latifolia* germinate to some extent on a moist surface exposed to air although they germinate much better when covered by water or if exposed to a reduced oxygen concentration (Sifton 1959) (Fig. 2). Seeds could germinate both under water or on exposed wet mud although seedlings are not often found in nature. Seeds of this species have an absolute light requirement for germination

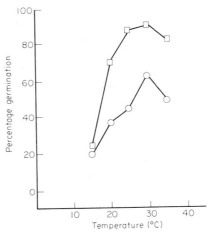

FIG. 2. Germination of seeds of *Typha latifolia* at various temperatures. Seeds exposed to light and either on (O) a moist surface or (□) fully immersed in water. Data derived from Sifton (1959).

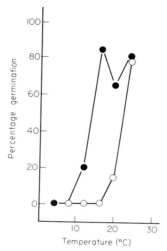

FIG. 3. Germination of seeds of *Najas marina* in (●) dark or (○) light (16 h per day) at various temperatures. Seeds maintained under water on sediment rich in organic material and hence with oxygen concentration of zero (redox potential −300 to −440 mV). There was no germination on sand or clay substrates where the redox potentials were just above or just below zero. Data derived from Van Vierssen (1982).

with the effect of light being greater at higher temperatures. It has been reported that freshly harvested seed of *Typha latifolia* have an absolute requirement for alternating temperature before germination occurs. Large daily fluctuations do not occur under water (at a depth of 2 m daily fluctuations do not exceed 0·5°C whereas at the surface the maximum day temperature may be 10°C higher than the night temperature) but in this species a 1°C shift in temperature can result in a 50% germination response (Thompson, Grime & Mason 1977), and this temperature differential certainly occurs in shallow water. Germination of seeds of *Najas marina* is also favoured by higher temperatures but here long periods of light each day inhibit germination (Fig. 3). Seeds of this species do not germinate unless oxygen is totally absent; this has been demonstrated experimentally by placing seeds under water in organic sediment with a redox potential of about −400 mV (Van Vierssen 1982). Simpson (1966) found that lowering oxygen concentration in the water in which wild rice (*Zizania aquatica*) seeds were after-ripening accelerated the rate of germination. This effect was thought to be associated with changes in the metabolism of the seed embryo, as seeds after-ripened at low oxygen had a higher optimum temperature for germination compared to those seeds after-ripened in well-aerated water. This species also responds to a period of chilling in aerated water, such that the rate and extent of germination is greater compared to seed maintained at warm temperatures before returning to favourable germination conditions. Fluctuating temperatures also increase the rate of germination of this species, but air-drying reduces the viability of the seed. All these factors together would favour germination in spring in the cool temperate climates of North America where this species occurs, at the bottom of lake margins. There are other cases where prolonged chilling of the wet seeds is required to break dormancy. In *Eleocharis coloradoensis* (Yeo & Dow 1978), for instance, the requirement is 24 months at 2°C or only 3 months if the seed coat is 'scarified'.

There has been little detailed work on the features of photocontrol of germination in aquatic plants. It is presumed that, as with terrestrial species, phytochrome is the key photoreceptor pigment. In *Potamogeton schweinfurthii* and *P. richardi* (Spence *et al*. 1971), for instance, seeds have an absolute requirement for light, with germination being favoured by red light but inhibited by far-red light. Phytochrome control of germination, as evidenced by red/far-red reversibility, has been recently demonstrated for *Nuphar lutea, Typha latifolia* and *T. angustifolia* (M. J. Vagges & D. H. W. Spence, unpubl. data) (Table 1). In *Typha angustata* (Gopal & Sharma 1983) it has been shown that yellow and red light promote germination whereas blue light strongly inhibits germination. This is suggestive of a flavoprotein system operating antagonistically to phytochrome although the evidence is not conclusive.

Spectroradiometric studies of the light environment at various depths in freshwater show that red light is transmitted better than far-red light (Spence 1981; Spence *et al*., this volume; Bartley & Spence, 1986). There are differences in spectral quality between clear water and water with a high content of organic matter. In the latter case very little blue light penetrates into deep water. From data on light under leaf shade and in soil it is possible to predict in a general way the nature of the light environment under water both below vegetation and within the sediment (Table 2). It appears that under water φ would have a high value both with and without shading from green leaves; in other words, phytochrome is not a good detector of leaf shade under water. Predicted germination under water is high partly because of high Pfr and partly because of reduced light quantity and hence reduced photoinhibition. Light-requiring seeds buried more than 10 mm in the sediment would fail to germinate for the same reasons that seeds do not germinate in soil. All this is conjecture based on the assumption that light is the only controlling factor and there is a great need for experimental studies on seeds of a range of hydrophytes.

TABLE 1. Red/far-red reversibility in photocontrol of seed germination in *Nuphar lutea*

Irradiation treatment	Percentage germination
dark	28
R	70
FR	22
R-FR	30
R-FR-R	74
R-FR-R-FR	29
FR-R-FR-R	68

Seeds are stored in aerated water at 5°C, then placed in Petri dishes in distilled water 1 cm deep. They were exposed to short irradiations with red (R) and/or far-red (FR) light before returning to dark at 25°C. The percentage germination was recorded after 6 days. Unpublished data of M. J. Vagges & D. H. W. Spence.

TABLE 2. Predicted effects of light transmitted through water, green leaves and soil on phytochrome parameters and seed germination

	Q (400–800) (μmol m^{-2} s^{-1})	B (400–500 nm)	G (500–600 nm)	R (600–700 nm)	F (700–800 nm)	φ'	log$H$	%G
(a) sunlight	1700	0·23	0·26	0·26	0·25	0·59	−0·32	33
(b) 1 m clear water	707	0·30	0·39	0·26	0·05	0·69	−0·80	51
(c) 1 m brown water	373	0·11	0·40	0·41	0·08	0·71	−0·95	55
(d) leaf shade (LAI = 4)	60	0·04	0·15	0·11	0·70	0·31	−1·87	20
(e) 5 mm soil	0·002	0·01	0·05	0·17	0·76	0·02	−6·38	19
(f) clear water + leaf shade	13	0·10	0·43	0·21	0·26	0·58	−2·47	57
(g) brown water + leaf shade	8	0·03	0·36	0·27	0·34	0·57	−2·61	57
(h) clear water + soil	0·0004	0·03	0·18	0·42	0·37	0·01	−6·71	19
(i) brown water + soil	0·0003	0·01	0·13	0·45	0·41	0·01	−6·82	19

Light quantity and quality given for some representative situations in the natural environment. Q: quantity of light (400–800 mm) in μmol m^{-2} s^{-1}. B,G,R,F: proportion of light in the spectral regions 400–500, 500–600, 600–700, 700–800 nm respectively. Values for a, b and c recalculated from Bartley *et al.* (1975b); values for d and e recalculated from Poo (1980) and Frankland (1981); values for f, g, h and i are predictions obtained by using appropriate combinations of values a to e.

φ' and H are respectively the proportion of phytochrome as Pfr (assuming slow dark reversion of Pfr to Pr with a half-life of 5 h) and the phytochrome cycling rate (in s^{-1}) as calculated from the light quantity and quality data. %G is the predicted percentage of the seeds of a hypothetical species (with light-requiring seeds and a dark germination of 18%) as determined by computer simulation.

GERMINATION OF TURIONS AND OTHER PROPAGULES

Propagation from seeds is not a common method of spread in most aquatic plants; freedom from water shortage has removed the necessity for such a drought-avoidance mechanism, and vegetative propagation is the norm. This may involve random mechanical breakage of pieces of the plant or the abscission of morphologically distinct shoot or root propagules, such as stem or root tubers or the specialized condensed shoots known as turions. Some propagules act both as dispersal units and as perennating organs (hibernaculae), others act only as perennating organs and still others only as dispersal units. Apart from the latter case turions are formed as dormant organs which require, like seeds, some special environmental cue to bring about resumption of growth or 'germination' (Bartley & Spence 1985).

In general turions of temperate zone species would be expected to be formed in late summer or autumn in response to short days or decreasing temperatures. Dormancy is broken by winter chilling (Czopek 1962; Weber & Nooden 1976; Newton *et al.* 1978) but also light may be required. Once dormancy is broken resumption of growth requires warmer temperatures, light for photosynthesis and, in some cases, availability of nutrients such as nitrate (Sibasaki & Oda 1979; Winston & Gorham 1979). In contrast turions of *Potamogeton crispus* are formed in early summer in response to long photoperiods (Chambers, Spence & Weeks 1984) and germinate in autumn.

There has been little detailed study of the photocontrol of germination of turions although phytochrome has been implicated in the case of *Spirodela polyrrhiza* (Czopek 1962) on the basis of the red/far-red reversibility of daily pulses of light. A light requirement for the germination of turions has also been demonstrated in a number of other species, such as *Potamogeton crispus* (Kadono 1982). Temperature greatly modifies this response; at 25°C all turions germinated in the light and none in the dark, whilst at 12°C, although there was some dark germination, the response to light was much reduced. Germination of *Hydrilla verticillata* tubers has been shown to occur more readily in the light than in the dark (Miller, Garrard & Haller 1976) but there was no attempt to demonstrate the nature of the photoreceptor involved in this process.

The examples given here demonstrate the broad physiological similarity between the control of the release from dormancy and germination in aquatic vegetative propagules and seeds. Similar environmental cues have developed, such as light, chilling and shifts in temperature, which may act individually or collectively to regulate germination. That these vegetative and sexual propagules adopt similar dormancy and germination strategies is perfectly reasonable, since these adaptations generally ensure survival of the propagules through unfavourable environmental conditions followed by establishment of new individuals during favourable growing conditions.

REFERENCES

Bartley, M. R. & Frankland B. (1982). Analysis of the dual role of phytochrome in the photoinhibition of seed germination. *Nature,* **300**, 750–752.

Bartley, M. R. & Spence, D. H. N. (1985). Dormancy and propagation in helophytes and hydrophytes. In press.

Bartley, M. R. & Spence, D. H. N. (1986). The light environments of two contrasting Scottish lochs: implications for photomorphogenesis under water. In press.

Bewley, J. D. & Black, M. (1982). *Physiology and Biochemistry of Seeds. Vol. 2. Viability, Dormancy and Environmental Control.* Springer-Verlag, Berlin.

Chambers, P. A., Spence, D. H. N. & Weeks, D. C. (1984). Photocontrol of turion formation in *Potamogeton crispus* L. in the laboratory and natural water. *New Phytologist,* **99**, 183–194.

Czopek, M. (1962). The oligodynamic action of light on germination turions of *Spirodela polyrrhiza* (L.) Schleiden. *Acta Societatis Botanicorum Poloniae,* **31**, 705–722.

Frankland, B. (1981). Germination in shade. In *Plants and the Daylight Spectrum* (Ed. by H. Smith), pp. 187–204. Academic Press, London.

Frankland B. & Taylorson, R. (1983). Light control of seed germination. In *Encyclopedia of Plant Physiology New Series Vol 16A* (Ed. by W. Shropshire Jr. & H. Mohr). pp. 428–456. Springer-Verlag, Berlin.

Fukshanksy, L. & Schafer, E. (1983). Models in photomorphogenesis. In *Encyclopedia of Plant Physiology New Series Vol 16A,* (Ed. by W. Shropshire Jr & H. Mohr) pp. 69–95. Springer-Verlag, Berlin.

Gopal, B. & Sharma, K. P. (1983). Light regulated seed germination in *Typha angustata* Bory et Chamb. *Aquatic Botany,* **16**, 377–384.

Kadono, Y. (1982). Germination of the turion of *Potamogeton crispus* L. *Physiological Ecology, Japan,* **19**, 1–5.

Kendrick, R. E. (1976). Photocontrol of seed germination. *Science Progress,* **63**, 347–367.

Miller, J. L., Garrard, L. A. & Haller, W. T. (1976). Some characteristics of *Hydrilla* tubers taken from Lake Ocklawaha during drawdown. *Journal of Aquatic Plant Management,* **14**, 29–31.

Newton, R. J., Shelton, D. R., Disharon, S. & Duffey, J. E. (1978). Turion formation and germination in *Spirodela polyrrhiza. American Journal of Botany,* **65**, 421–428.

Poo, W. K. (1980). *Phytochrome control of seed germination in relation to the natural light environment.* Ph.D. thesis, University of London.

Sibasaki, T. & Oda, Y. (1979). Heterogeneity of dormancy in the turions of *Spirodela polyrrhiza. Plant and Cell Physiology,* **20**, 563–571.

Sifton, H. B. (1959). The germination of light-sensitive seeds of *Typha latifolia* L. *Canadian Journal of Botany,* **37**, 719–739.

Simpson, G. M., (1966). A study of germination in the seed of wild rice *(Zizania aquatica). Canadian Journal of Botany,* **44**, 1–9.

Spence, D. H. N. (1981). Light quality and plant responses under water. In *Plants and the Daylight Spectrum* (Ed. by H. Smith), pp. 245–276. Academic Press, London.

Spence, D. H. N., Bartley, M. R. & Child, R. (1985). Photomorphogenic processes in freshwater angiosperms, this volume.

Spence, D. H. N., Milburn, T. R., Ndawula-Senyimba, M. & Roberts, E. (1971). Fruit biology and germination of two tropical *Potamogeton* species. *New Phytologist,* **70**, 197–212.

Thompson, K., Grime, J. P. & Mason G. (1977). Seed germination in response to diurnal fluctuations of temperature. *Nature,* **267**, 147–149.

Van Vierssen, W. (1982). Some notes on the germination of seeds of *Najas marina* L. *Aquatic Botany,* **12**, 201–203.

Weber, J. A. & Nooden, L. D. (1976). Environmental and hormonal control of turion formation in *Myriophyllum verticillatum. Plant and Cell Physiology* **17**, 721–731.

Winston, R. D. & Gorham, P. R. (1979). Turions and dormancy states in *Utricularia vulgaris. Canadian Journal of Botany,* **57**, 2740–2749.

Yeo, R. R. & Dow, E. J. (1978). Germination of seeds of dwarf spikerush *Eleocharis coloradoensis. Weed Science,* **26**, 425–431.

Dispersion in aquatic and amphibious vascular plants

CHRISTOPHER D. K. COOK

Institut für Systematische Botanik, Universität Zürich, Zollikerstrasse 107, Zürich, CH 8008, Switzerland

SUMMARY

Amphibious vascular plants are ecologically specialized and usually confined to a relatively small range of water depths. Suitable habitats for establishment are usually found near to the coast of water bodies or in regularly inundated regions. These habitats usually represent a very small part of the total area of land and water in a particular region. As might be expected, many diaspore dispersal mechanisms are specialized and unlike those in patristically related terrestrial plants. Of the three principal agents of dispersal—wind, water and animals (including man)—wind plays a very small role.

WIND

Wind dispersal is common among land plants but very rare among amphibious plants and, hardly surprising, non-existent among non-amphibious aquatics although some such as *Hydrilla verticillata* (L. fil.) Royle and some species of *Potamogeton* have wind-dispersed pollen. The Asteraceae (Compositae) often show a complex development of the pappus for wind dispersal; all the amphibious members (*Cadiscus*, *Cotula* aquatic ssp., *Eclipta*, *Enhydra*, *Erigeron* aquatic ssp., *Hydropectis*, *Jaegeria*, *Megalodonta*, *Sclerolepis* and *Shinnersia*) are not wind dispersed and have a pappus which is either reduced, or specialized for animal dispersal.

It is usually considered that dispersal by wind is a disadvantage because the diaspores are more likely to be blown to terrestrial than to aquatic sites. In spite of this, some amphibious plants such as *Arundo*, *Gynerium*, *Phragmites*, *Saccharum* (Poaceae), *Eriophorum* (Cyperaceae) and *Typha* (Typhaceae) are dispersed by wind. *Typha* shows some startling adaptations that ensure that the seed is liberated only in water. These are described in detail by Krattinger (1978). The fruits are liberated when the distal swellings of the sterile female flowers (Fig. 1) shrink allowing the inflorescence to dry out. As the inflorescence dries the hairs on the individual flowers spread and 'burst' the inflorescence releasing the diaspores. Morphologically the diaspores are stalked capsules with persistent styles and numerous long, simple hairs (see Fig. 1). Depending upon the species of *Typha* the terminal velocity of the seed carrying diaspores lies between 8 and 30 cm s^{-1} with the means lying between 12 and 17 cm s^{-1}. Seed-carrying diaspores often get

179

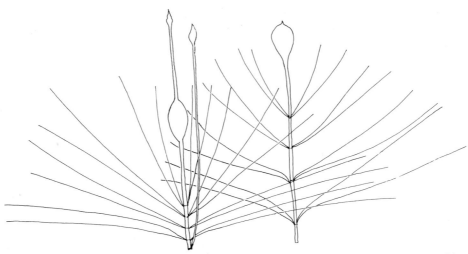

Fig. 1. A diagrammatic representation of female flowers of *Typha*: left side, a fertile flower with bract, hairs on the pedicel, carpel, and style with stigma; right side, a sterile flower with hairs and a distally swollen tip. (The number of hairs is very much reduced in this illustration.)

entangled in the much lighter sterile flowers and sometimes in unpollinated fertile flowers; these masses have slower terminal velocities than individual fertile flowers.

As the hairs spread only when dry there is a guarantee that the diaspores will be liberated in dry weather. If the diaspore lands on land the hairs remain spread and it will soon be blown away. When the diaspore lands on water it rests on the tips of the somewhat hydrophobic hairs and the stigma. In this state it may be blown for a considerable distance across the water. After a few minutes (about 15 min on smooth water) the hairs fold back and lie along the floral axis. As soon as the pericarp comes into contact with the water it springs open and releases the solitary seed which then sinks, chalazal end first, at a speed of $0.1–1.0$ cm s^{-1}

WATER

The diaspores of most amphibious plants are liberated into water. In some plants the dispersal is so tied to water that the seeds die if dried: for example, *Aponogeton*, *Orontium*, *Peltandra*, *Vallisneria*, *Victoria*, *Zizania*. In some species the diaspores sink at once while in others they float. The mechanisms ensuring buoyancy have been studied in detail; the most important works are cited by Sculthorpe (1967). (Among the plants listed by Sculthorpe *Baldellia ranunculoides* (L.) Parlatore is a 'floater' and not a 'sinker'.) However, sooner or later the floating diaspores will sink but their subsequent fate has very rarely been studied.

Just because the diaspores float on the surface of the water it does not mean that they are exclusively dispersed by movements of water. Wind can play an important secondary role and blow floating material over large distances. It is probably the most important factor carrying diaspores over large lakes. However, it is unlikely that many diaspores of amphibious plants have travelled over oceans by this means; no freshwater amphibious plants have developed diaspores like the coconut. Also I know of no aquatics having floating diaspores with upright sails or wings, in spite of the fact that the male flowers of *Lagarosiphon* have well-developed and efficient sails (Cook, 1982).

It seems reasonable to put floating diaspores into two categories: globose and disc-like. The globose ones usually lie partly above the water and are presumably more likely to be transported by wind than flattened or disc-shaped ones which lie flat on the surface. Disc-like diaspores are found in phylogenetically unrelated groups and there is considerable transference of function. The disc-like diaspores are seeds in *Nymphoides* and *Veronica*, one-seeded pod segments in *Aschynomene*, several-seeded pods in *Neptunia*, caryopses with flattened lemmas in *Neostapfia*, schizogenous nutlets in *Callitriche*, schizogenous merocarps in *Centella* and some other umbellifers, drupes in *Alisma* and *Sagittaria* and achenes (cypselas) in *Eclipta*. This convergence in evolution can be taken as an indication of adaptive advantage which, in turn, should indicate a particular function. It can be seen that these disc-like diaspores are slightly hydrophobic and readily stick to hands, boots and clothing. Why not also to birds and larger animals? I suggest that these flattened diaspores are often dispersed by animals. The seeds of *Nymphoides peltata* (S. G. Gmelin) O. Kuntze, described in detail by Velde & Heijden (1981), bear around the edge processes with microscopically small barbs which may help them to become attached to fur or feathers. It is also interesting to note that the amphibious species of *Veronica* have disc-like seeds while the terrestrial relatives have globose or cup-like seeds. The composite *Eclipta* has dimorphic achenes (cypselas): one form is flat and corky and floats while the other is rounded and smooth and sinks. A similar situation is found in the seeds of some species of *Ludwigia*, perhaps also in fruits of *Lilaea*.

ANIMALS

Animals, particularly water birds, play a large role in the short- to middle-range dispersal of many aquatics. Some diaspores are clearly adapted for endozoic transport. Fassett (1960) listed the use of aquatic plants by birds and mammals and cited, up to date in 1939, ninety-seven references, and a detailed evaluation of food of ducks was published by Martin & Uhler (1939). For some plants the germination is improved after the seeds have passed through an animal. Fish also eat plant diaspores but the effectiveness of fish as means of dispersal (for plants) is unknown. Unfortunately, there is still very little information on how far diaspores may be endozoically transported.

Spines, hooks and barbs are common and no doubt play an important role in the exozoic transport of diaspores. As in land plants the attachment organs of aquatics show considerable transference of function and may arise from the seed in *Blyxa* and *Nymphoides*, the pappus of achenes (cypselas) in *Megaladonta* (*Bidens*), the capsule in *Reussia*, the follicle in *Damasonium*, the nut (calybium) in *Trapa*, the nutlet in *Echinodorus* and many other genera, the caryopsis in *Echinocloa* and some other grasses, nutlet and perianth in *Eleocharis* and *Websteria*, and the inflorescence in *Navarretia*. All these and many others are illustrated in Cook *et al.* (1974). There is little direct evidence of the distance these diaspores are transported. The distribution pattern of plants such as *Aldrovanda vesiculosa* L., *Lemna minor* L., *Ranunculus rionii* Lagger and *Utricularia australis* R. Brown, which are found along bird migration routes, is evidence of the range of bird transportation particularly because these species are repeatedly found in climatically unsuitable regions where they may persist for some years but are unable to complete their generative cycles.

SUNKEN DIASPORES

The diaspores of many amphibious plants sink as soon as they come into contact with water but, sooner or later, even buoyant diaspores will sink. The mobility of sunken diaspores is unknown. Some will, perhaps, be endozoically transported by bottom feeding animals, others may be transported in ice as Fauth (1903) suggests in an excellent paper on the biology of aquatic diaspores. In others, the bottom substrate may partially dry and the diaspores may be transported in mud. I have found no quantitative work on the 'seed banks' under water; however, L. Bates & D. Webb (pers. comm. 1983) have recovered enormous numbers of viable diaspores from mud in dammed parts of the Tennessee River.

The seeds of most amphibious plants germinate under water. As soon as the young seedlings start to assimilate and develop aerenchyma they tend to become lighter than water and start to float. There seem to be two alternative strategies: either the seedlings float or they remain submerged by anchors or weights before they are rooted and established.

Among the species with floating seedlings some, such as *Typha* (see Krattinger 1978) or *Nymphoides* (see Fauth 1903), do not show any obvious adaptations to floating while others such as *Hydrocharis* (Cook & Lüönd 1982) and *Limnobium* (Cook & Urmi-König 1983b), are highly specialized. The first indication of germination is that the seed, lacking its exotesta, floats shedding its endotesta. The cotyledon is green, relatively massive and resembles *Wolffia*. The radical emerges first followed by the first foliage leaf and plumule; the radicle remains relatively short and may loose its rootcap. The very young seedling has a lemnid habit with the upper surface emerging above the air–water interface. These lemna-like seedlings are also found in *Pistia*, *Ceratopteris* and *Salvinia*. Schneller (1976) has pointed out that the standard 'text book' illustration of the *Salvinia* megaprothallus is drawn upside down!

Many species show specialized organs to anchor seedlings. The radicle or cotyledon stalk (hypocotyl) often swells distally (Wurzelknoten) and develops numerous hairs which anchor the seedling before the roots and leaves develop (see Fig. 2 a–c). These prop hairs usually disintegrate as soon as the roots develop. The prop hairs on swollen 'Wurzelknoten' are found in many patristically unrelated

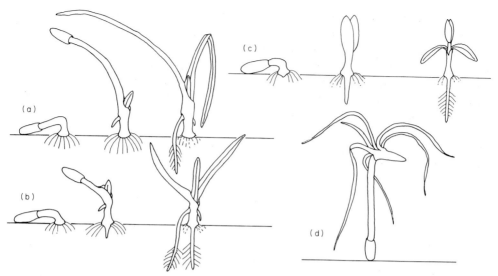

FIG. 2. Diagrammatic representations of seedlings: (a) *Limnocharis* (Limnocharitaceae); (b) *Vallisneria* (Hydrocharitaceae); (c) *Elatine* (Elatinaceae); (d) *Stratiotes* (Hydrocharitaceae). Not drawn to scale.

groups in both monocots and dicots. Other species such as, for example, *Hippuris*, *Myriophyllum*, *Sparganium* and *Trapa* have very heavy endocarps which hold the seedlings down until they are rooted. In *Littorella* and some species of *Aponogeton* the seeds germinate while attached to the mother plant.

Some seedlings have numerous leaves at germination; *Stratiotes* (described by Cook & Urmi-König 1983a) has one cotyledon and up to ten leaves in the embryo; at germination these leaves spread (Fig. 2d) and may help to anchor the seedlings. A similar kind of seedling is found in *Ceratophyllum* and perhaps functionally also in *Utricularia*. In *Amphibolis*, a totally submerged marine plant, the anchoring organs are outgrowths of the pericarp.

SPECIALIZED SEEDS

Many amphibious plants have specialized seeds but the function of the specialization is often unknown. For example, many have mucilaginous excretions, particularly in the Podostemonaceae and in some cases the excretions are highly developed, for instance, evaginating mucilaginous hairs develop in *Rotala* and some other amphibious Lythraceae (Wisselingh 1920). In *Stratiotes* (Cook & Urmi-König 1983a), *Egeria* (Cook & Urmi-König 1984a), *Ottelia* (Cook & Urmi-König 1984b) and some species of *Elodea* (Cook & Urmi-König 1985) the exotesta has numerous simple hairs which are flaccid and somewhat mucilaginous but the whole exotesta is shed shortly after the seeds are liberated from the fruit. In *Hydrocharis* and *Limnobium* (Cook & Urmi-König 1983b) the hairs are very complex; they are formed by the union of two or three adjacent epidermal cells

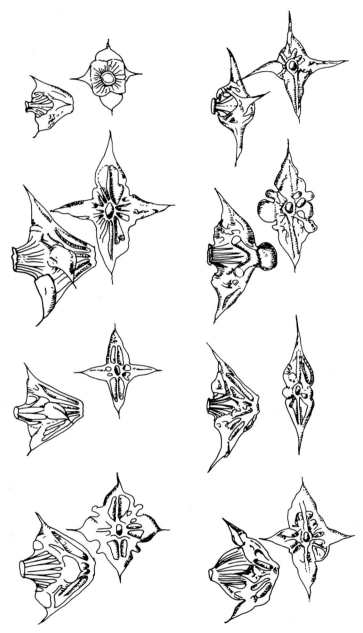

Fɪɢ. 3. Variation in fruit shape in *Trapa* in USSR. Taken from Komarov, Flora URSS.

with complex patterns of thickening arising from the regions of union. These hairs are also shed soon after the seeds are freed from the fruit. *Halophila* (Birch 1981) also has a complex exotesta which is shed before germination. As pointed out by Sculthorpe (1967) mucilage is sticky when dry, or almost so, but non-sticky when wet. All these seeds develop mucilage when wet and the mucilage is usually shed before germination!

POLYMORPHIC DIASPORES

Another feature of aquatic diaspores is that they are often highly polymorphic. However, the polymorphism is found not only in patristically unrelated groups but also in analogous organs. Considering both these factors, it is highly likely that the polymorphism is of some selective advantage. Superficially similar spined diaspores are found in *Trapa* (Fig. 3), *Ceratophyllum* and *Blyxa*. In *Ceratophyllum* (Fig. 4a) the diaspore is a nutlet and the spines develop directly from the pericarp, in *Trapa* the diaspore is a nut developing from an inferior ovary (calybium) and the hypanthium contributes to the development of the spines, while in *Blyxa* (Fig. 4b) the diaspore is a seed and the spines develop from the testa. Often there are no other morphological characteristics associated with these diaspore variations but this has not prevented their taxonomic recognition; there are about thirty described species of *Trapa* and about thirty 'fruit' species of *Ceratophyllum*. In the *Sparganium*

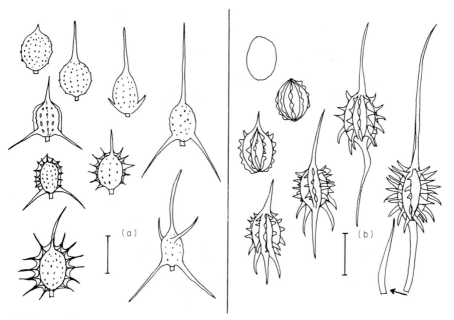

FIG. 4. Polymorphism in diaspores: (a) nutlets of *Ceratophyllum*, taken from Komarov, Flora URSS (Scale-bar = 5 mm); (b) seeds of *Blyxa aubertii* L. C. M. Richard, taken from Cook & Lüönd (1982) (Scale-bar = 1 mm).

erectum L. group there is also considerable polymorphism in fruit form apparently not correlated with vegetative form. The selective advantage of this polymorphism is difficult to interpret in terms of dispersal function.

RANGE EXTENSIONS

Aquatic and amphibious plants are notorious for extending their ranges and often settling where they are unwelcome. Cook (1985) has reviewed the range extensions of 172 aquatic vascular plants and discusses the distribution of an additional twenty-six widespread or disjunct species which are unlikely to be native through-out their range today but it is not known or it is disputed where they are native and alien. Most of these invading plants have well-defined native ranges. The endemism of aquatics is probably more pronounced than many earlier reviewers have believed; Cook (1983) has documented the distribution of sixty-one taxa of aquatics endemic to Europe and the Mediterranean.

The detailed spread of aquatic plants has been studied in very few cases. The best known examples are those of plants introduced to North America in historical times.

Stuckey & Phillips (1970) have documented the spread of *Lycopus europaeus* L. in North America. It is not a plant of horticultural or economic value as its English names suggest: water-horehound, bitter bugle-weed, gipsy-wort. It seems to have reached North America in the ballast of ships as its first records are around seaports: 1860 Chesapeake Bay, 1867 Delaware Bay, 1877 New York and 1880 Boston. Following the initial introduction at different stations along the coast the spread of *L. europaeus* has been slow inland and it is established locally in Virginia, Pennsylvania, New Jersey, New York and Massachusetts (Fig. 5). In 1903 it was recorded in Lake Ontario on Toronto Island where it has spread more rapidly. Within 65 years it has reached westwards to the west coast of Lake Erie and eastwards along the St Lawrence River to Quebec.

The fruits are schizocarpic nutlets and are reported to float for 12 to 15 months. Stuckey & Phillips (1970) present a convincing argument that following the original introduction in several east coast seaports it has spread slowly along waterways and has become established inland by 'natural' means. The spread through the Great Lakes and along the St Lawrence River has been more rapid because there are more interconnecting waterways and they have considerable boat traffic.

The pattern of spread of *Lycopus* along the St Lawrence River is being repeated by *Hydrocharis morsus-ranae* L. According to Catling & Dore (1982) it was planted in Ottawa in 1932 and since then has spread westwards to Rondeau Park on the north shore of Lake Erie, northwards along the Ottawa River and eastwards to Quebec. *Hydrocharis morsus-ranae* does not set seed in North America and its spread is almost certainly through specialized hibernacula often called turions or winter buds (Cook & Lüönd 1982); they are specialized shoot apices which sink in autumn and rise to the surface and germinate in spring or early summer. The floating hibernacula may be transported by water movements but they are some-

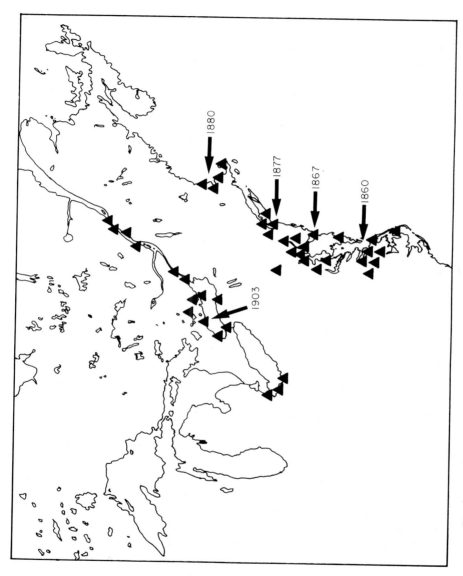

FIG. 5. The spread of *Lycopus europaeus* L. in North America, taken from Stuckey & Phillips (1970).

what hydrophobic and tend to stick to objects emerging from water; transportation by water birds or animals or by boats is also likely.

Butomus umbellatus L. is another species showing a similar pattern of spread around the Great Lakes and the St Lawrence River. Details are given by Stuckey (1968) who correctly notes that the seeds of *Butomus* sink after release from the follicles. However, *Butomus* is self-incompatible and I have found no evidence of its setting seed in North America. It is spread by pea-sized bulbils probably due to the agency of water movements or by aquatic animals; Stuckey (1968) suggests that muskrats may play a role. The spread of *Nymphoides peltata* (Stuckey 1973–74) and *Lythrum salicaria* L. (Stuckey 1980) are also well-documented in North America but the pattern is much more complex as both species are attractive garden plants and thus commonly cultivated. Nevertheless, the spread along waterways is clearly quicker than the spread between land-locked bodies of water. In *L. salicaria*, hardly an aquatic or amphibious plant but rather a plant of wet lands, seeds are the agency of dispersal and it is likely that railways have also played a role in its spread in North America.

The spread of aquatics does not necessarily depend on specialized diaspores. A good example is *Egeria densa* Planchon, described in detail by Cook & Urmi-König (1984a). In spite of diligent searching in its native range ripe fruits and seeds have never been found in nature; they are known only from plants artificially pollinated in Zürich botanic gardens in 1982. Outside its native range only males are known. It has no specially differentiated dispersal organs such as hibernacula, bulbils, tubers or specialized buds. Also the stem is morphologically complex and only the so-called 'double nodes' are capable of developing roots and new shoots. In spite of this negative catalogue of means for dispersal it has spread from a relatively small native range in temperate South America to all continents in less than 90 years. At the beginning of this century it was looked upon as an excellent 'oxygenator', and oxygenator plants were considered essential for raising fish. As fish were widely used to control mosquito larvae it was, therefore, often planted as part of malaria eradication programmes. It is ironic that recent work has shown that it often depresses oxygen levels which, in turn, may increase the number of mosquito larvae! *Egeria* is also perhaps the most universally available aquarium plant. Its spread is probably directly the result of man's activity. The following plants, like *E. densa*, seem to have recently extended their range without seed and without specialized vegetative organs: *Acorus calamus* L. (partly), *Cotula coronopifolia* L. (partly), *Elodea callitrichoides* (L.C.M. Richard) Caspary, *Elodea nuttallii* (Planchon) St John, *Lagarosiphon major* (Ridley) Moss, *Salvinia molesta* Mitchell.

CONCLUSIONS

The spread of many aquatic plants has been largely dependent on vegetatively developed diaspores. These are often complex in structure and sometimes show dormancy and resistance to drying. They have, unfortunately, rarely been studied in detail; exceptions are *Hydrilla*, for which see Basiouny, Haller & Garrard

(1978a, 1978b) and Pieterse (1981), and *Hydrocharis* published work summarized by Cook & Lüönd (1982). However, it is of little importance to the dispersion of the plants if the diaspores are the result of a sexual process or not. A frequent consequence is that new territory is colonized by one sex or by one self-incompatible mating type.

Considering the higher plants as a whole, the study of sporophytic mobility has been rather neglected when compared to studies made on the gametophytic mobility. Considering the aquatic and amphibious plants as a group our ignorance is even more crass. However, it is clear that there are specialized structures which differ from those in terrestrial relatives and there are processes that, at least, increase the chance of the seedling becoming established in the 'correct' ecological niche. Experiments with labelled diaspores or, perhaps, with models have yet to be done.

REFERENCES

Basiouny, F. M., Haller, W. T. & Garrard, L. A. (1978a). The influence of growth regulators on sprouting of *Hydrilla* tubers and turions. *Journal of Experimental Botany*, 29(110), 663–699.

Basiouny, F. M., Haller, W. T. & Garrard, L. A. (1978b). Survival of Hydrilla (*Hydrilla verticillata*) plants and propagules after removal from the aquatic habitat. *Weed Science*, 26(5), 502–504.

Birch, W. R. (1981). Morphology of germinating seeds of the seagrass *Halophila spinulosa* (R. Brown) Ascherson (Hydrocharitaceae). *Aquatic Botany*, 11, 79–90.

Catling, P. M. & Dore, W. G. (1982). Status and identification of *Hydrocharis morsus-ranae* and *Limnobium spongia* (Hydrocharitaceae) in northeastern North America. *Rhodora*, 84, 523–545.

Cook, C. D. K. (1982). Pollination mechanisms in the Hydrocharitaceae. In *Studies on Aquatic Vascular Plants*. (Ed. by J. J. Symoens, S. S. Hooper & P. Compère). pp. 1–15. Royal Botanical Society of Belgium, Brussels.

Cook, C. D. K. (1983). Aquatic plants endemic to Europe and the Mediterranean. *Botanische Jahrbücher*, 103(4), 539–582.

Cook, C. D. K. (1985). Range extensions of aquatic vascular plant species. *Journal of Aquatic Plant Management*, 23(1), 1–6.

Cook, C. D. K., Gut, B. J., Rix, E. M., Schneller, J. & Seitz, M. (1974). *Water Plants of the World*. pp. 1–561. Dr W. Junk, The Hague.

Cook, C. D. K. & Lüönd, R. (1982). A revision of the genus *Hydrocharis* (Hydrocharitaceae). *Aquatic Botany*, 14, 177–204.

Cook, C. D. K. & Urmi-König, K. (1983a). A revision of the genus *Stratiotes* (Hydrocharitaceae). *Aquatic Botany*, 16, 213–249.

Cook, C. D. K. & Urmi-König, K. (1983b). A revision of the genus *Limnobium* including *Hydromystria* (Hydrocharitaceae). *Aquatic Botany*, 17, 1–27.

Cook, C. D. K. & Urmi-König, K. (1984a). A revision of the genus *Egeria* (Hydrocharitaceae). *Aquatic Botany*, 19, 73–96.

Cook, C. D. K. & Urmi-König, K. (1984b). A revision of the genus *Ottelia* (Hydrocharitaceae). 2. The species of Eurasia, Australia and America. *Aquatic Botany*, 20, 131–177.

Cook, C. D. K. & Urmi-König, K. (1985). A revision of the genus *Elodea* (Hydrocharitaceae). *Aquatic Botany*, 21(2), 111–156.

Fassett, N. C. (1960). *A Manual of Aquatic Plants* with Revision Appendix by Ogden, E. C. pp. 1–405. The University of Wisconsin Press, Madison.

Fauth, A. (1903). Beiträge zur Anatomie und Biologie der Früchte und Samen einiger einheimischer Wasser- und Sumpfpflanzen. *Beihefte zum Botanischen Centralblatt*, 14, 327–373.

Krattinger, K. (1978). *Biosystematische Untersuchungen innerhalb der Gattung Typha*. Inaugural-Dissertation der Universität Zürich, pp. 1–240.

Martin, A. C. & Uhler, F. M. (1939). Food of game ducks in the United States and Canada. *Technical Bulletin of the United States Department of Agriculture*, **634**, 1–157.

Pieterse, A. J. (1981). *Hydrilla verticillata*—a review. *Abstracts on Tropical Agriculture, Koninklijk Instituut voor de Tropen (Amsterdam)*, **7**, 9–34.

Schneller, J. (1976). The position of the megaprothallus of *Salvinia natans*. *Fern Gazette*, **11**(4), 217–219.

Sculthorpe, C. D. (1967). *The Biology of Aquatic Vascular Plants*, pp. 1–610. Edward Arnold Ltd., London.

Stuckey, R. L. (1968). Distributional history of *Butomus umbellatus* (flowering-rush) in western Lake Erie and Lake St Clair Region. *The Michigan Botanist*, **7**, 134–142.

Stuckey, R. L. (1973–1974). The introduction of *Nymphoides peltatum* (Menyanthaceae) in North America, *Bartonia*, **42**, 14–23.

Stuckey, R. L. (1980). Distributional history of *Lythrum salicaria* (purple loose strife) in North America. *Bartonia*, **47**, 3–20.

Stuckey, R. L. & Phillips, W. L. (1970). Distributional history of *Lycopus europaeus* (European Water-horehound) in North America. *Rhodora*, **72**(791), 351–369.

Velde, G. van der & Heijden, L. A. van der (1981). The floral biology and seed production of *Nymphoides peltata* (Gmel.) O. Kuntze (Menyanthaceae). *Aquatic Botany*, **10**(3), 261–294.

Wisselingh, C. van (1920). Untersuchungen über Osmose. *Flora*, **113**, 357–420.

PART II
AMPHIBIOUS PLANTS AND FLOODING TOLERANCE

A. Oxygen Stress in
Seeds and Seedlings

Germination physiology of rice and rice weeds: metabolic adaptations to anoxia

ROBERT A. KENNEDY, MARY E. RUMPHO, THEODORE C. FOX

Department of Horticulture, The Ohio State University, 2001 Fyffe Court, Columbus, Ohio 43210, U.S.A.

SUMMARY

Metabolic responses to flooding were investigated in rice and several rice weeds (*Echinochloa* spp.). The effects of temperature and oxygen concentration on germination were determined. Interestingly, the shortest time for 50% emergence of both radicle and coleoptile occurred at 10% O_2 for all species examined. Although we found no correlation between flood tolerance and ethanol production, alcohol dehydrogenase (ADH) activity, or number of ADH isozymes, we did observe induction of an additional ADH isozyme in seeds germinated in N_2 that did correlate with increased ADH activity. Furthermore, flood-tolerant species were able to synthesize adenylates under anoxic conditions, but flood-intolerant species were unable to do so. Finally, studies with metabolic inhibitors showed that cyanide inhibited germination in *Echinochloa* spp. but not in rice. Cyanide plus SHAM inhibited all species tested and CO, a specific metabolic inhibitor of cytochrome oxidase, only prevented germination of flood-intolerant species.

INTRODUCTION

A wide range of flood tolerance can be found within the rice ecosystem. This is particularly true in the *Echinochloa* complex, a large and diverse group of grasses which are the worst weeds in rice world-wide, whether in upland (non-flooded) or lowland (flooded) culture. For several years we have been studying germination physiology of several *Echinochloa* species, with a special interest in comparisons between species which are restricted to flooded rice fields and species which only grow in dry sites around the fields. In this context, we have been most interested in *E. crus-galli* var. *oryzicola* (corresponding to *E. phyllopogon* by Michael 1973) and *E. crus-galli* var. *crus-galli*. In rice fields of California, *E. crus-galli* var. *oryzicola* only occurs as a weed in the fields *per se*, while *E. crus-galli* var. *crus-galli* only grows on the dry levees surrounding the paddies. Other *Echinochloa* taxa are found in still different sites and some of these species are included in our studies as well. A general discussion of the weed flora of California rice fields, including *Echinochloa* spp., was published by Barrett & Seamon (1980).

Because of the obvious differences in habitat 'preference' by these *Echinochloa* species, and since they can be found growing within a few metres of one another in and around the rice fields, the *Echinochloa* genus presents a unique opportunity for comparative studies on metabolic adaptations to flood tolerance. Understanding metabolic differences between these closely related plants could be valuable, not only in understanding the role of oxygen in plant metabolism, but also in controlling weeds in aquatic environments, especially in rice fields. Here we report on experiments aimed at obtaining that information.

MATERIALS AND METHODS

Seeds of *Echinochloa crus-galli* var. *crus-galli* (hereafter, crus-galli), *E. crus-galli* var. *oryzicola* (hereafter, oryzicola), *E. crus-pavonis*, *E. muricata* and *Oryza sativa* were grown as previously described (Kennedy, Rumpho & VanderZee 1983) except that we used an anaerobic chamber (Forma Scientific) in these experiments.

Inhibitor experiments were conducted as reported by Kennedy *et al.* (1983) and ethanol determinations and enzyme assays were performed according to Rumpho & Kennedy (1983). Adenylates were quantified as reported by Rumpho *et al.* (1984). Polyacrylamide gel electrophoresis was carried out according to Davis (1964), as modified by Rumpho (1982).

RESULTS

We were interested in comparing the temperature and oxygen requirements for germination of two of the worst weeds, oryzicola and crus-galli, to that of rice. As seen in Fig. 1, all three plants germinated well under aerobic conditions. Germination proceeded faster in warmer temperatures. However, while oryzicola and rice had similar germination characteristics over the time and temperature ranges studied, crus-galli took longer to germinate than the two aquatic plants, especially at colder temperatures. Even larger differences were seen between these plants under anaerobic conditions: the optimum germination environment for crus-galli shifted toward higher temperatures. Oryzicola, on the other hand, tolerated colder temperatures better; its germination percentage and rate were inhibited at the highest temperatures when compared to rice.

The time required for 50% of the seeds to germinate (emergence of shoot and root) under various oxygen concentrations of oryzicola, crus-galli, and *E. muricata* are given in Fig. 2. Within these three *Echinochloa* taxa, the lowest T_{50} for emergence of both radicle and coleoptile occurred at 10% O_2. Otherwise, the T_{50} for emergence of the coleoptile was constant regardless of oxygen concentration and radicle emergence was more sensitive to oxygen than the coleoptile. Among the three plants surveyed, *E. muricata* required the lowest oxygen levels for emergence of both root and shoot while oryzicola was intermediate. Crus-galli required the longest time to germinate at any O_2 level.

We also compared ethanol production and alcohol dehydrogenase (ADH)

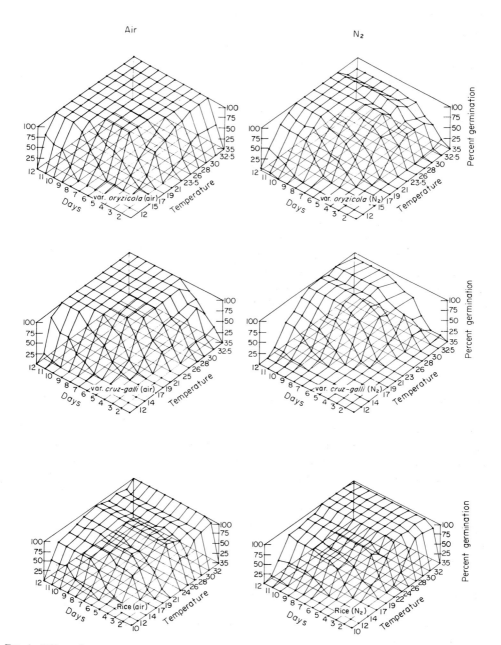

FIG. 1. Effect of temperature on percent germination and time required to germinate in *E. crus-galli* var. *oryzicola*, *E. crus-galli* var. *crus-galli*, and rice germinated under aerobic and anaerobic conditions.

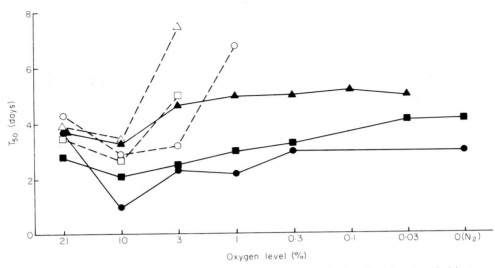

FIG. 2. Time required for emergence of radicle (open symbols) and coleoptile (closed symbols) at various oxygen concentrations in (△, ▲) *E. crus-galli* var. *crus-galli*; (□, ■) *E. crus-galli* var. *oryzicola*; and (○, ●) *E. muricata*.

activities of *Echinochloa* to that of rice (Table 1). Among the plants tested, a large range of ethanol accumulation was found within the seed after 7 days anaerobic germination. These values varied from 12·3 μmol ethanol g^{-1} dry weight in *E. crus-pavonis* to 89·2 μmol g^{-1} dry weight in *E. muricata*. Even greater amounts of ethanol were recovered in the imbibition solution, but the relative amount produced by each plant was similar to that found in the seeds. That is, *E. crus-pavonis* accumulated the least amount of external ethanol, and *E. muricata* the greatest amount. Ethanol production for rice was similar to that of oryzicola except when calculated on a per seed basis. Then, because of the larger size of the rice seed, rice produced 12 to 50 times more ethanol.

We were also interested in determining if there was any relationship between

TABLE 1. Ethanol produced by several *Echinochloa* species and rice during 7 days germination under anaerobic conditions

Plant species	Ethanol concentration (μmol g^{-1} dry weight)	
	Seeds	Imbibition solution
Echinochloa crus-galli var. *oryzicola*	43·5 ± 4·9 (0·22)	603·3 ± 40·5 (3·02)
Echinochloa crus-galli var. *crus-galli*	38·5 ± 4·1 (0·07)	565·6 ± 41·2 (1·04)
Echinochloa muricata	89·2 ± 8·1 (0·15)	779·0 ± 14·8 (1·57)
Echinochloa crus-pavonis	12·3 ± 0·8 (0·02)	171·7 ± 5·7 (0·25)
Rice	43·0 ± 4·0 (1·08)	517·4 ± 13·6 (12·44)

Values represent the mean of three separate experiments ± S.E. Figures in parenthesis indicate μmol ethanol per seed.

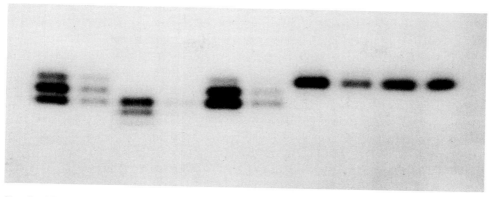

FIG. 3. ADH isozymes in air- and nitrogen-grown seeds of several *Echinochloa* species. Seedlings were 7 days old when analysed. From left to right: *E. crus-pavonis*, air and nitrogen; *E. muricata* ('early'), air and nitrogen; *E. crus-galli* var. *crus-galli*, air and nitrogen; *E. crus-galli* var. *oryzicola* ('early'), air and nitrogen; and *E. crus-galli* var. *oryzicola* ('late'), air and nitrogen.

ADH isozymes and tolerance to flooding. Figure 3 shows ADH isozymes in air- and nitrogen-grown seedlings of several *Echinochloa* taxa. The number of isozymes varies greatly, from one in *E. muricata* to five in *E. utilis*. More importantly, the number of isozymes bears no correlation with tolerance to anoxia; the species with the largest number of isozymes (*E. utilis*, unpublished data) is intolerant of anoxia. Both *E. muricata* and *E. crus-pavonis*, on the other hand, had only one ADH isozyme band. *Echinochloa muricata* is perhaps the most tolerant of all species tested to anoxia but *E. crus-pavonis*, like *E. utilis*, does not germinate under anoxia.

Although the total number of isozymes may not correlate with tolerance to flooding in *Echinochloa*, induction of new isozymes might. As shown in Table 2,

TABLE 2. Alcohol dehydrogenase activity in several *Echinochloa* species after 7 days germination in air or nitrogen

Plant species	ADH activity (μmol g^{-1} fresh weight min^{-1})	
	Air	Nitrogen
Echinochloa crus-galli var. *oryzicola*, 'late'	$4 \cdot 10 \pm 0 \cdot 20$ ($0 \cdot 09$)	$20 \cdot 1 \ \pm 0 \cdot 85$ ($0 \cdot 19$)
Echinochloa oryzicola var. *oryzicola*, 'early'	$2 \cdot 25 \pm 0 \cdot 12$ ($0 \cdot 06$)	$17 \cdot 24 \pm 0 \cdot 34$ ($0 \cdot 18$)
Echinochloa crus-galli var. *crus-galli*	$3 \cdot 59 \pm 0 \cdot 30$ ($0 \cdot 04$)	$22 \cdot 64 \pm 1 \cdot 46$ ($0 \cdot 08$)
Echinochloa muricata	$4 \cdot 90 \pm 0 \cdot 16$ ($0 \cdot 09$)	$19 \cdot 98 \pm 0 \cdot 17$ ($0 \cdot 18$)
Echinochloa crus-pavonis	$2 \cdot 66 \pm 0 \cdot 13$ ($0 \cdot 02$)	$12 \cdot 19 \pm 05N09$ ($0 \cdot 32$)

Values represent the mean of three separate experiments $\pm$ S.E. Figures in parenthesis indicate enzyme activity per seed. All seeds were germinated in the light.

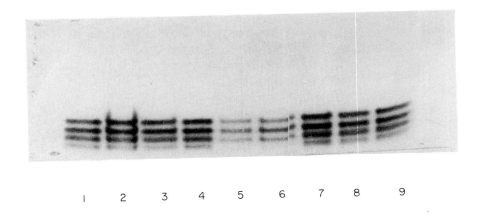

FIG. 4. Polyacrylamide gel electrophoretic patterns of ADH from crude extracts of oryzicola seeds. Lanes 5 and 6 are fractions from dry, unimbibed seeds, and lanes 4 and 7, 3 and 8, 2 and 9, and 1 are from seeds germinated under N_2 for 14, 15, 18 and 19 days respectively. The amounts of protein applied to lanes 1 to 9, in order, were 37·8, 33·0, 34·5, 39·0, 38·1, 63·5, 52·0, 46·0 and 44·0 μg protein.

ADH activity increased markedly during anoxia in oryzicola. Possibly more important, however, an additional isozyme band that was absent or much reduced in dry- or air-grown seedlings was observed in anaerobically-grown seedlings (Fig. 4).

Adenylate content for several *Echinochloa* species is given in Fig. 5. While there was little difference in energy charge among the *Echinochloa* species tested, the plants did separate into two distinct groups based on their total adenylates. Crus-galli and *E. crus-pavonis* showed little ability to synthesize adenine nucleotides during anoxia, but *E. muricata*, oryzicola, and rice had linear increases of adenylates with time under anoxia. Most interestingly, the two former species do not germinate under flooded conditions, while the latter three plants are very tolerant of flooding.

Lastly, we examined the response of rice and several rice weeds to metabolic inhibitors. As shown in Fig. 6, in air, germination of all of the weed species was inhibited by cyanide. Rice germination was largely unaffected by the presence of CN^-. In an N_2 environment, cyanide or cyanide plus SHAM inhibited germination of all of the grass species. Various combinations of CO, CN^- and SHAM caused different effects on the various weeds (Fig. 6). It is interesting that CO inhibited germination of only two species of *Echinochloa*, both flood-intolerant, with little or

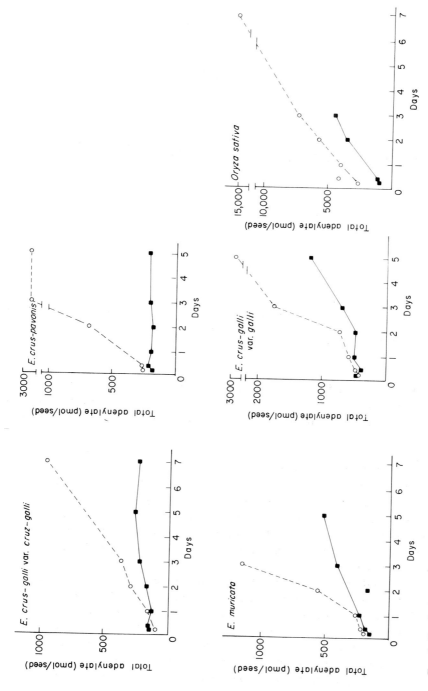

FIG. 5. Total adenylate content of *E. crus-galli* var. *crus-galli*, *E. crus-pavonis*, *E. muricata*, *E. crus-galli* var. *oryzicola*, and *Oryza sativa* seedlings germinated under aerobic and anaerobic conditions for 7 days.

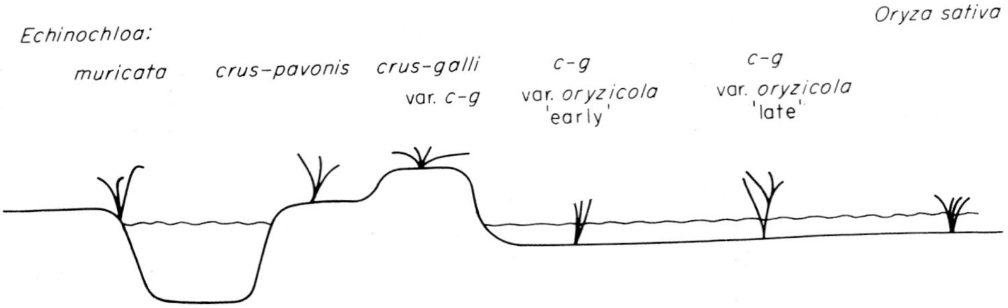

	Echinochloa: muricata	crus-pavonis	crus-galli var. c-g	c-g var. oryzicola 'early'	c-g var. oryzicola 'late'	Oryza sativa
Control-O_2	+++	+++	+++	+++	+++	+++
Control-N_2	+++	0	+++	+++	+++	+++
KCN-O_2	0	0	0	0	0	+++
KCN-N_2	0	0	0	0	0	0
KCN/SHAM-O_2	0	0	0	0	0	0
CO	+++	0	0	++	+++	+++
CO/KCN	0	0	0	0	0	+++
CO/SHAM	++	0	+	+	++	+++
SHAM-O_2	+++	+++	+++	+++	+++	+++
SHAM-N_2	+++	0	0	++	+++	+++

FIG. 6. Effect of metabolic inhibitors on germination of *E. muricata*, *E. crus-pavonis*, *E. crus-galli* var. *crus-galli*, *E. crus-galli* var. *oryzicola* 'early', *E. crus-galli* var. *oryzicola* 'late', and *Oryza sativa* in aerobic and anaerobic environments. The drawing depicts the 'preferred' natural habitat of each species. Symbols indicate ability to germinate: +++, good; ++, fair; +, poor; 0, none.

no effect on flood-tolerant species. Also, CO plus cyanide gave identical results to CN⁻ in air. Again, all of the weed seeds were inhibited, with no affect on rice.

DISCUSSION

The ability of rice to tolerate and grow under flooded conditions has received considerable attention. Yet, the basic question, 'what metabolic adaptations permit this phenomenon?', remains unanswered. We have been studying the germination physiology of several species of *Echinochloa* which are more tolerant and grow more vigorously under anaerobic conditions than rice. As a group, the *Echinochloa* complex represents the worst rice weeds in the world. Within this genus, however, there are considerable differences in the ability of individual species to tolerate anoxia, some being totally intolerant to flooding or anoxia. Thus, the *Echinochloa* genus provides a unique experimental system to study metabolic adaptations to flooding among related species. Furthermore, comparative studies on the germination physiology of rice to that of its worst weeds are easily done. Through these studies we hope to ascertain metabolic differences between flood-tolerant and flood-intolerant species and, eventually, to suggest control practices that minimize the detrimental effects of *Echinochloa* spp. on rice production.

Initially, we investigated the effect of environmental parameters on germination of rice and *Echinochloa* seeds. In response to temperature, germination occurred as predicted: germination proceeded faster at higher temperatures except at the highest temperatures investigated. Rice and oryzicola responded in similar fashion except at cooler temperatures where oryzicola germinated better; crus-galli lagged behind them. Exposure to anaerobiosis delayed germination in all three cases.

Surprisingly, the lowest T_{50} for emergence of both radicle and coleoptile occurred, not at ambient levels, but at 10% oxygen. This suggests that hypoxic conditions may stimulate germination in the three *Echinochloa* species studied. Below 10% oxygen, the T_{50} for coleoptile emergence did not change with oxygen concentration.

Despite an abundance of literature, no clear relationship exists between ADH activity (Table 2) and ethanol production (Table 1). For example, the species (*E. crus-galli*) that had the highest ADH activities under nitrogen only accumulated intermediate levels of ethanol. In addition, ADH levels in the least tolerant species (*E. crus-pavonis*) were similar to other *Echinochloa* taxa tested, but it produced far less ethanol.

Among the several *Echinochloa* taxa, there were differences in the number of isozymes present, but the number did not correlate with flood tolerance. However, the increased activity of ADH under anaerobic conditions did correlate with the appearance of an additional ADH isozyme. Even in air, ADH activity increased significantly if ethanol was included in the inhibition medium (Rumpho 1982). This increased activity again correlated with the appearance of an additional ADH isozyme, indicating that the number of ADH isozymes may be less important to flood tolerance than induction of new forms of the enzyme.

While the exact relationship between ADH activity and flood tolerance is unknown, little doubt remains about the overall importance of fermentative metabolism, and substrate level phosphorylation derived from it, to anaerobic metabolism. The net effectiveness of the overall metabolism in an organism or its response to stress can best be seen in its energy charge or ability to synthesize adenylates (Pradet & Raymond 1983). Although the energy charge for all species was similar, the plants separated into two groups based on total adenylates. It is interesting to note that *E. crus-galli* and *E. crus-pavonis*, which do not germinate under flooded conditions, had little ability to synthesize adenylates under anoxia while *E. muricata*, oryzicola and rice, which are very flood-tolerant, had linear increases in adenylates with time under anoxia.

Earlier, we showed (Kennedy, Rumpho & VanderZee 1983) that in the presence of cyanide, rice could shift to the alternate electron pathway present in most plants, whereas oryzicola appeared unable to do so. In the present experiments, we extended the use of this inhibitor and others to several

Echinochloa species. Cyanide inhibited germination in *Echinochloa* spp., but not in rice. Thus, rice and *Echinochloa*, although similar in many aspects of their anaerobic metabolism, possibly differ significantly in their response to metabolic inhibitors; rice can shift to the alternate electron pathway, but *Echinochloa* spp. cannot. Furthermore, earlier results suggest that *Echinochloa* may utilize an alternative electron acceptor to O_2, possibly NO_3^- (Kennedy *et al.* 1980; Kennedy, Rumpho & VanderZee 1983). It is interesting to note that CO inhibited germination of only those two species of *Echinochloa* which are flood intolerant, with little or no effect on those which are tolerant to flooding. Also, CO plus cyanide and SHAM in nitrogen gave identical results to CN^- in air. These results indicate that, at least in flood-tolerant species of *Echinochloa*, cyanide inhibits germination at sites other than cytochrome oxidase.

CONCLUSIONS

Our studies thus far indicate important differences in the metabolic responses of flood-tolerant and flood-intolerant *Echinochloa* spp. Although there was no clear relationship between flood tolerance and ethanol production, ADH activity, or number of ADH isozymes, induction of an additional ADH isozyme was observed in seeds germinated anaerobically. The ability to synthesize adenylates under anoxic conditions also correlated with flood tolerance; flood tolerant species were able to synthesize adenylates under anoxia whereas flood-intolerant species could not. Flood-tolerant species were able to germinate in the presence of CO, a metabolic inhibitor of cytochrome oxidase, but flood-intolerant species could not. Finally, we observed that although rice and *Echinochloa* possess many similar adaptations to flooding, metabolic differences do exist. Most notably, cyanide inhibits germination of *Echinochloa* whereas germination of rice proceeds unimpeded. The extent to which these and other adaptations are important to flood tolerance in rice and *Echinochloa* spp. is being further investigated.

ACKNOWLEDGMENT

This work was supported by NSF grant PCM 83–04978 awarded to R.A.K.

REFERENCES

Barrett, S. C. H. & Seamon, D. E. (1980). *The weed flora of Californian rice fields. Aquatic Botany*, **9**, 351–376.

Davis, B. J. (1964). Disc electrophoresis. II. Method and application of human serum proteins. *Annals of the New York Academy of Science*, **121**, 404–427.

Kennedy, R. A., Barrett, S. C. H., VanderZee, D. & Rumpho, M. E. (1980). Germination and seedling growth under anaerobic conditions in *Echinochloa crus-galli* (barnyard grass). *Plant, Cell and Environment*, **3**, 243–248.

Kennedy, R. A., Rumpho, M. E. & VanderZee, D. (1983). Germination of *Echinochloa crus-galli* (barnyard grass) seeds under anaerobic conditions: respiration and response to metabolic inhibitors. *Plant Physiology*, **72**, 787–794.

Michael, P. W. (1973). Barnyard grass (*Echinochloa*) in the Asian-Pacific region, with special reference to Australia. pp. 489–493. Proceedings of the 4th Asian-Pacific Weed Science Society Conference, Rotura.

Pradet, A. & Raymond, P. (1983). Adenine nucleotide ratios and adenylate energy charge in energy metabolism. *Annual Review of Plant Physiology*, **34**, 199–24.

Rumpho, M. E. (1982). *Anaerbobic metabolism in germinating seeds of* Echinochloa crus-galli *var.* oryzicola (*barnyard grass*). Ph.D. thesis, Washington State University.

Rumpho, M. E. & Kennedy, R. A. (1983). Anaerobiosis in *Echinochloa crus-galli* (barnyard grass) seedlings: Intermediary metabolism and ethanol tolerance. *Plant Physiology*, **72**, 44–49.

Rumpho, M. E., Pradet, A., Khalik, AA. & Kennedy, R. A. (1984). Energy charge and emergence of the coleoptile and radicle at varying oxygen levels in *Echinochloa crus-galli*. *Physiologia Plantarum*, **62**, 133–138.

Mitochondrial fine structure in imbibing seeds and seedlings of *Zea mays* L. under anoxia

B. B. VARTAPETIAN, H. H. SNKHCHIAN AND
I. P. GENEROZOVA

Timiriazev Institute of Plant Physiology, Academy of Sciences, Moscow 127276, U.S.S.R.

SUMMARY

1 Mitochondrial ultrastructure and respiratory capacity were studied both in anaerobic imbibing seeds (primary anoxia) and after aerobically grown seedlings were transferred into oxygen-free medium (secondary anoxia).

2 Under anoxic imbibition no signs of germination were observed and seeds eventually died. Nevertheless the assemblage of inner mitochondrial membranes (cristae) of seed embryos took place, being completed, as at aerobic imbibition, after 48 h.

3 Anaerobic formation of mitochondrial membranes could not be arrested by inhibition of either mitochondrial or cytoplasmic protein synthesis systems.

4 Alongside the anaerobic formation of mitochondrial membranes the respiratory function of seeds and mitochondria was concurrently activated.

5 At secondary anoxia, the signs of destructive changes of mitochondrial ultrastructure in seedlings' organs were observed even after only 1·5 h anoxia. These changes were reversed, after the seedlings had been transferred from anaerobic to aerobic medium, and also when they were continuously kept under strict anoxia.

6 Irreversible destructive changes in mitochondrial membranes of roots were observed after 24 h anoxia, while in coleoptiles and leaves the mitochondria remained intact for at least 3 days.

7 Despite these differences, common features of these organs' reaction to anoxia in terms of ultrastructural modification of mitochondrial membranes were established.

8 It is concluded that, even with no oxidative phosphorylation, formation of a functionally active inner mitochondrial membrane and its stabilization does take place, apparently through utilization of glycolytic ATP.

9 The observed phenomena are discussed and contrasted to the anaerobic degradation and aerobic formation of mitochondrial membranes in facultative anaerobic yeast.

INTRODUCTION

Higher plants, being aerobes, require a continuous supply of molecular oxygen from the environment. The deficiency, let alone complete absence, of oxygen, even for a short time, brings about considerable metabolic disturbances in most higher plants followed by the destruction of fine cell organization and, eventually, the plant's death.

It is known, however, both from scientific and common observation, that for many species of higher plants even prolonged anaerobic environmental conditions are not necessarily lethal. A great number of wild species are known to grow well on waterlogged and even marshy soils, which are absolutely unsuitable for very many other species, especially agricultural plants. Among cultivated plants a most remarkable example is rice, normally grown on regularly flooded anaerobic soils.

The two main strategies of the adaptation of higher plants to an anaerobic environment are for plants either (1) to avoid anaerobiosis by transporting O_2 from the aerated part to those localized in the oxygen-free medium, or (2) to be adapted on the molecular level through a substantial modification of cell metabolism at hypoxia and anoxia.

In the first case, plant cells, though localized in an anaerobic medium, show no resistance to anoxia. So we have suggested that such plants should be called 'apparently resistant to anoxia' (Vartapetian 1978). As a matter of fact, it has been experimentally shown that root cells of such plants (hydrophytes and hygrophytes) growing on anaerobic soils not only fail to exhibit higher resistance to anoxia, as compared to plants growing on well-aerated soils (mesophytes), but are, indeed, more sensitive to oxygen deficiency than mesophytes (Vartapetian 1973, 1982; Webb & Armstrong 1983). ap Rees and colleagues (Smith & ap Rees 1979a, b; ap Rees & Wilson 1984) have shown that, with regard to root carbohydrate metabolism under anaerobic conditions, plants growing in a waterlogged environment are quite similar to those growing on dry aerobic soils.

On the whole, the above authors have shown the cells and tissues of the first category plants to lack molecular mechanisms for adaptation to anoxia.

It has been suggested that the second category of plants be called 'truly resistant to anoxia' (Vartapetian 1978). Such plants are not only capable of surviving long-term anoxia with the cell ultrastructure remaining intact, but in fact grow vigorously with no O_2 in the medium. This category includes, in particular, rhizomatous shoots in some wild species (Barclay and Crawford 1982) and germinating seeds of rice (Ueda & Tsuji 1971; Vartapetian, Andreeva & Maslova 1971; Kordan 1972; Tsuji 1972; Vartapetian, Maslova & Andreeva 1972; Opik 1973), and *Echinochloa crus-galli* L. (Kennedy *et al.* 1980). Like facultative yeast, seeds and shoots of the above plants actively germinate or grow both in the presence of molecular oxygen and under its complete absence in the medium. Biochemical, physiological and cytological features of these plants, particularly the mitochondrial ultrastructure and function in an oxygen-free medium, have by now been thoroughly studied.

Unlike these plants, most other species, particularly agricultural plants, are intolerant of an oxygen-free medium and can be categorized as plants 'not resistant to anoxia'. Lacking the adaptation mechanisms at the molecular and organism level, or possessing them to a far lesser extent than the plants of the first two groups, the third category plants show, in the absence of O_2, disorganization of metabolism and degradation of cell ultrastructure, which is followed by the death of the whole organism.

The imbibing maize seeds and seedlings, used in the experiments described in the present study, belong to this third category. Even under a thin layer of water, i.e. at hypoxia, maize seed germination is fully inhibited, and the seed embryo eventually dies.

The mitochondrial ultrastructure of maize seed embryos during dry seed-imbibition under strict anoxia (primary anaerobiosis) was investigated. The respiratory function of seeds and embryos, and of mitochondria isolated from embryos, was also studied immediately after the transfer of seeds from the anaerobic to aerobic medium. Lastly, a study was made of the changes in the cell mitochondrial ultrastructure of the meristematic tissue in maize seedling roots, leaves and coleoptiles under short and long secondary anaerobiosis, i.e. anaerobiosis created after a 4 day-long preliminary seed germination under normal aerobic conditions.

MATERIALS AND METHODS

Seeds and seedlings of *Zea mays* L. ('Zherebkovskaya' and 'Pioneer' cultivars) were used in the experiments. To investigate the effect of primary anaerobiosis, twenty-five dry seeds were placed in 2-litre flasks with $1 \cdot 5$ l sterilized distilled water through which air (control) or gaseous nitrogen (experiment) were continuously passed. The traces of O_2 were removed from the commercial nitrogen ($O_2 10^{-3}\%$) by a biological cleaning technique (Vartapetian & Bobylev 1981). The experiments were performed in darkness at 25°C. In the measurement of respiration rate seeds of the same weight were selected. The embryo coleoptile cell ultrastructure was studied after 24 and 48 h of aerobic and 24, 48, 72, 96 and 120 h of anaerobic imbibition. The respiration of embryos was measured after 1, 3, 6, 24, 48 and 72 and 96 h of anaerobic and 1, 3, 6, 24, 48 h of aerobic imbibition.

To investigate the effect of secondary anoxia 4-day-old etiolated maize seedlings were placed into 1-litre flasks containing 800 ml of sterile water through which air or nitrogen stream was also bubbled. Specimens for electron microscopy were periodically taken over 24 h (root) and 5·5 days (coleoptile, leaf). The meristematic tissue of the leaf, coleoptile (the base part of these organs) and of the root (0·5–1 mm from the tip) was studied. The material for electron microscopy was fixed according to the method of Karnovsky (1965), and embedded in EPON; the sections were obtained using ultratomes LKB-4800 and OMU-3, stained with uranyl-acetate and lead citrate (Reynolds 1963) and examined under JEM-100 B and TESLA-613 electron microscopes. Mitochondria were counted within the microscope field of vision or on micrographs with a × 1700 magnification. The percentage of mitochondria having different structures was calculated after observing 70–100 organelles.

To study the action of chloramphenicol and cycloheximide on an embryo's mitochondrial formation, seeds were anaerobically imbibed in the presence of protein synthesis inhibitors.

Respiration rate of embryos detached from the seeds, as well as of mitochondria isolated from the seed embryos, was measured in a polarographic cell, using a platinum Clark-type, membrane-coated electrode, with fifteen detached embryos placed into the polarographic cell. Mitochondria were isolated from the embryos of anaerobically imbibed seeds according to the method of Day & Hansen (1977) with our slight modifications; 10 mM of succinate, malate or α-ketoglutaric acid were used as substrate. Low non-substrate concentrations of glutamic acid were added to eliminate the inhibitory action of the oxaloacetic acid formed.

RESULTS

Primary anaerobiosis

Embryos of dry and anaerobically imbibed seeds were examined to reveal changes in the cell ultrastructure, particularly in the mitochondrial structure.

No regular-shaped mitochondria were observed in the initial dry seed embryo cells. Instead, there were observed round bodies, typical of dry seeds, with a double membrane, but without any cristae, and similar in structure to yeast 'pro-mitochondria' (Fig. 1a). Yet, after 1 day of the seeds' anaerobic imbibition, a part of the mitochondria have distinct cristae, though there were also organelles without cristae (Fig. 1b). By 48 h of soaking nearly all mitochondria are fully formed. They have an oval shape, electron-dense matrix and elongated cristae randomly distributed inside the mitochondria (Fig. 1d).

The mitochondria formed under conditions of strict anoxia, after 2 days were similar in morphology and ultrastructure to the seed embryos' mitochondria which had imbibed under aerobic conditions for the same time (Fig. 1 d, e).

The embryos' mitochondria retained their ultrastructure under a further 4 and 5 days exposure to seeds to strict anoxia (Fig. 1f, g, h).

To examine whether the anoxic mitochondria formation in maize embryos is associated with *de novo* synthesis of its membrane proteins, the anaerobic imbibition of seeds was carried out in the presence of inhibitors of protein synthesis. As the proteins of mitochondrial cristae are synthesized both on the cytoplasmic and mitochondrial ribosomes, inhibitors were used that block the two protein synthesis systems, cycloheximide and chloramphenicol.

FIG 1. Mitochondria of embryo in dry seed and after anaerobic and aerobic imbibition. (a) Dry seed. Embryo coleoptile's 'promitochondrion' without any cristae; (b)–(h) Embryo coleoptile's mitochondria after imbibition of seeds; (b) 1 day N_2, mitochondria with deluted matrix and a few bubble-like cristae; (c) 1 day O_2; mitochondria with electron opaque matrix and a few narrow cristae; (d) 2 days N_2; mitochondria with electron-dense matrix and narrow cristae; (e) 2 days O_2, mitochondria as in (d); (f) 3 days N_2, mitochondria as in (d); (g) 4 days N_2, mitochondria as in (d); (h) 5 days N_2; mitochondria as in (d). Bar = 0·5 μm.

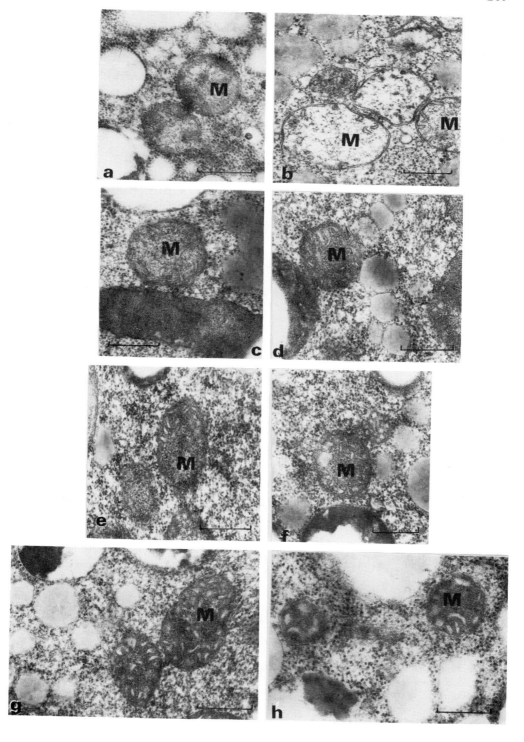

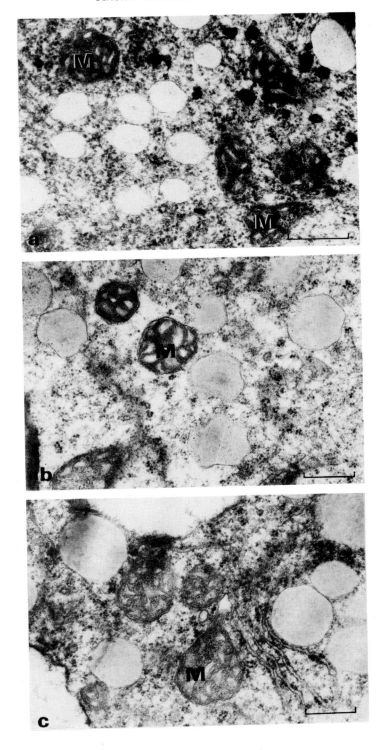

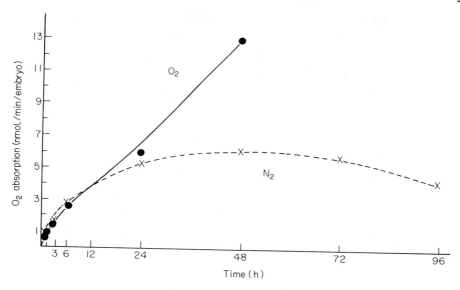

FIG. 3. Maize embryo respiration after anaerobic and aerobic seed imbibition.

As can be seen from Fig. 2, no substantial differences were found in the formation of mitochondrial membranes in the presence of the two inhibitors, as compared with the seeds imbibed without inhibitors.

Alongside the anaerobic formation of the mitochondrial ultrastructure, the seed embryos' respiratory activity was markedly enhanced, as measured immediately after the transfer of seeds from anaerobic to aerobic medium (Fig. 3). After 24 h anaerobic incubation the seed embryos' respiration intensity was virtually equal to that of the seeds that imbibed in air during the same period of time.

Significant differences between these seeds was observed as imbibition continued in the presence and absence of O_2. The respiration intensity of the embryo of aerobically imbibed seeds naturally grew in intensity, whereas in their anaerobic counterparts the respiration intensity, having reached its maximum by 24 h, no longer rose during the subsequent days and tended to decrease.

The mitochondria isolated from embryos of anaerobically imbibed seeds markedly oxidized the substrates of Krebs cycle-succinate, malate and α-ketoglutarate. The oxidation of succinate was inhibited by cyanide by 75%, indicating that the cytochrome pathway of respiratory chain is functioning in the mitochondria (data will be published elsewhere).

FIG. 2. Mitochondria of embryo coleoptiles after anaerobic imbibition of seeds in water or in solutions of inhibitors of protein synthesis. (a) Control, 2 days N_2 in water, mitochondria with developed cristae; (b) 2 days N_2 in chloramphenicol solution ($2 \cdot 10^{-3}$ M), mitochondria with developed cristae; (c) 2 days N_2 in cycloheximide solution (10^{-6} M), mitochondria with developed cristae. Bar = 0·5 μm.

Secondary anaerobiosis

The aim of these experiments was to study mitochondria in meristematic tissue cells of roots, coleoptiles and leaves in 4-day-old maize seedlings grown aerobically and then transferred to a strictly anoxic medium.

Besides prolonged exposure to anaerobiosis this study focused on changes in mitochondrial ultrastructure brought about by a short-term (1·5–3 h) exposure of seedlings to anoxia, which had not been investigated in earlier experiments.

Root

The root mitochondria in the seedlings after 4 days of aerobic germination had orthodox conformation. Kept for 24 h more in water and in an air stream, the seedlings did not show significant changes in the mitochondrial ultrastructure, though there were several cases of organelles of condensed conformation with electron-dense matrix and slightly dilated cristae (Fig. 4a).

By contrast, exposure to anoxia led to fairly significant rearrangement in mitochondrial ultrastructure. The latter was found to be heterogenous, i.e. the same conditions produced different organelle conformations—condensed, with parallel cristae, or swollen. Thus, after 1·5 h anoxia, the proportion of swollen mitochondria is only 14% of the total (Fig. 4b), whereas the other organelles are either of condensed conformation (44%) or with parallel cristae (41%). The swollen mitochondria, increasing in size, often form intricate ring-like profiles (Fig. 4b). After 3 h anoxia as much as about 50% of mitochondria acquire a swollen structure (Fig. 4c).

Nevertheless a further exposure to oxygen-free conditions does not increase the number of swollen mitochondria but, on the contrary, drastically reduces their number, with no such mitochondria observed after 7·5 h of anaerobiosis. By this time, there is an increased number of mitochondria with cristae stacked in parallel, whose proportion reaches 85% by 12 h of anoxia (Fig. 4d). This is followed by a decrease in their number and an increase in the proportion of the swollen mitochondria (Fig. 4f). Finally, after a brief appearance (by 19·5 h) of a considerable number (80%) of mitochondria with parallel cristae, irreversible degradation processes start in mitochondrial membranes from 21 h which lead, by 24 h of anoxia, to an almost complete breakdown of root cell mitochondria. Alongside the changes in the mitochondria's ultrastructure, their external morphology also undergoes continuous changes in anoxia: intricate ring-like mitochondrial profiles appear (Fig. 4e), which combine into a network of such organelles in the cytoplasm.

Coleoptiles

Mitochondrial ultrastructural modification pattern in coleoptiles is, on the whole, similar to that in roots, though there are significant distinctions between the two, e.g. with regard to the timing of modifications observed. Thus, already after 1·5 h

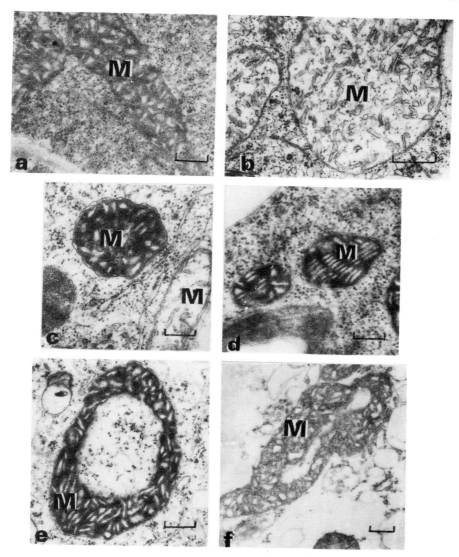

FIG. 4. Meristem mitochondria of roots of initiated seedlings and after anoxic exposure: (a) initiated seedling; (b) swollen mitochondria after 1·5 h of anoxia; (c) condensed and swollen mitochondria after 3 h of anoxia; (d) mitochondria with parallel cristae after 12 h of anoxia; (e) ring-like mitochondrial profiles after 13·5 h of anoxia; (f) degrading mitochondria after 21 h of anoxia. Bar = 0·5 μm.

of the coleoptile's exposure to anoxia, the destruction of its mitochondria can be seen, caused by the swelling of the organelles (Fig. 5b), although by this time over half of the mitochondria still retain a structure close to that of the control, and there are no mitochondria with parallel cristae.

Unlike roots, after 3 h of anoxia coleoptiles had no swollen organelles; the mitochondria are by this time either of condensed conformation or with parallel

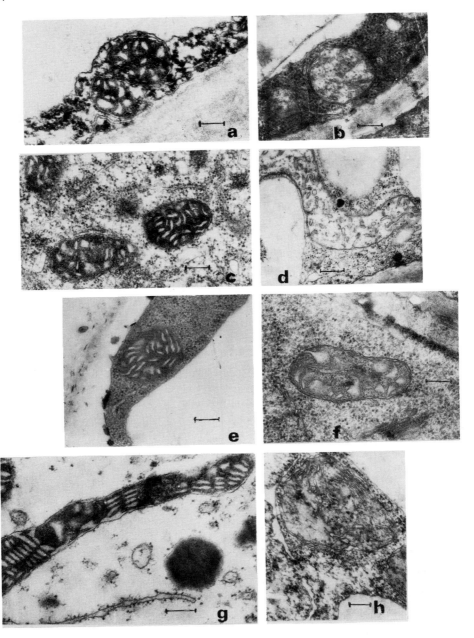

FIG. 5. Mitochondria of meristematic tissue of coleoptiles initiated seedling and after anoxic exposure: (a) initiated seedling 3; (b) swollen mitochondria after 1·5 h of anoxia; (c) mitochondria with parallel cristae after 4·5 h of anoxia; (d) swollen mitochondria after 12 h of anoxia; (e) mitochondria with parallel cristae after 18 h of anoxia; (f) condensed mitochondria after 24 h of anoxia; (g) mitochondria with parallel cristae, first signs of organellae's and cytoplasm's destruction after 3·5 days of anoxia; (h) degrading mitochondrion after 4 days anoxia. Bar = 0·5 μm.

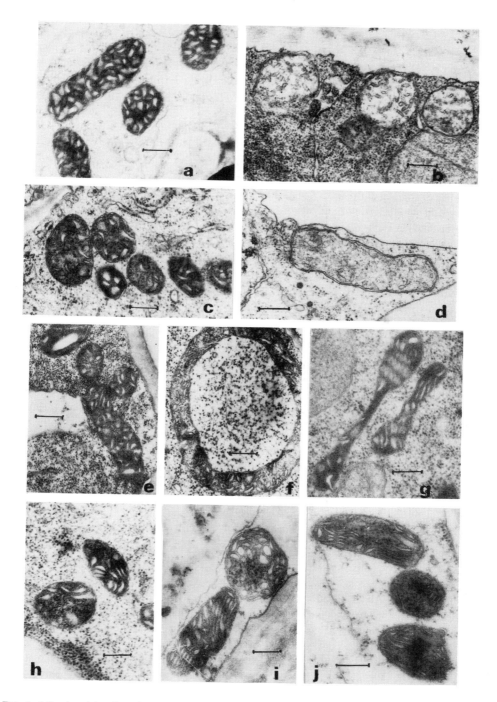

FIG. 6. Mitochondria of meristematic tissue of leaf's initiated seedling and after anoxic exposure: (a) initiated seedling; (b) swollen mitochondria after 1·5 h of anoxia; (c) recovery of mitochondrial ultrastructure after 3 h of anoxia; (d) swollen mitochondria after 4·5 h of anoxia; (e) recovery of mitochondrial ultrastructure after 12 h of anoxia; (f) ring-like mitochondrial profiles after 24 h of anoxia; (g, h) mitochondria with parallel cristae (g) after 36 h of anoxia and (h) after 72 h anoxia; (i) vacuolization and (j) destruction of cristae after 4·5 days of anoxia. Bar = 0·5 μm.

cristae (77%), the number of the latter soon sharply declining again. Longer anoxia resulted in recurrent alternations of the number of mitochondria with parallel cristae and of swollen ones (Fig. 5d, e), though condensed-type mitochondria were constantly present (Fig. 5c, f).

It should be noted that as the above cyclic changes in mitochondrial structure occurred, each structural type—swollen, condensed, and with parallel cristae—had its respective maximum ratio. For mitochondria with parallel cristae this maximum reached 80%, for swollen ones, 40%, for condensed, 70–80% of the total.

Under anoxic conditions, in coleoptile cells, as in roots, ring-like organelles appeared, which coincided either with an increase in the number of swollen mitochondria, or of those with parallel cristae.

Unlike roots, the destruction of individual mitochondria and other subcellular structures becomes distinct in coleoptiles only after 3·5–4 days of anoxia (Fig. 5g, h). By this time, mitochondrial cristae are swollen, and the matrix is distinctly electron-transparent; the cell cytoplasm is also clarified and the endoplasmic reticulum forms cyclic rows (no photo given).

After 5 days anoxia the cells appeared to be heavily damaged, the cytoplasm coagulated, with membranes in the organelles negatively contrasted (no photo given).

Leaf

In leaf cells, as in roots and coleoptiles, already after 1·5 h of anaerobiosis a part of the mitochondria acquires characteristic swollen shape of different patterns (Fig. 6b), while adjacent cells may retain mitochondria with the orthodox conformation. Swollen mitochondria were constantly observed in the cells (Fig. 6d) up to 9 h of exposure to anaerobiosis, when their number reached 50% of the total. By 12 h of anoxia a new significant structural change occurs in mitochondria, nearly all of which are now condensed (60%) or have parallel cristae (40%) (Fig. 6e). The ratio of mitochondria with parallel cristae later decreases, yet by 48 h and 72 h it again increases (up to 93% and 57% respectively) (Fig. 6g, h).

Thus, in the leaf, as in the root and coleoptile, the ratio of mitochondria of different structure shows wave-like changes, with periods of the ratio growth alternating with those of decrease for each organelle type. In the leaf, anoxic conditions also produce periodic changes not only in mitochondrial ultrastructure, but also in the shape of mitochondria acquiring oblong and ring-like profiles (Fig. 6f).

Irreversible destructive changes of mitochondrial membranes in part of leaf organelles start after 3 days of anoxia; as the matrix becomes clarified, the cristae swell, which is coupled with a progressive loss of their parallel arrangement, though even after 4·5 days of anoxia, single mitochondria with parallel cristae are retained (Fig. 5i, j). At the same time swollen ring-like mitochondria were observed in other cells. Complete degradation of the cell and destruction of all the membranes which became negatively contrasted, were observed after 5 days of anoxia.

DISCUSSION

Primary anaerobiosis

The results of the above-described investigations show that maize seeds, for whose germination the presence of O_2 in the medium is necessary, when imbibing under strict anaerobic conditions, nevertheless form structurally intact and functionally active mitochondria. Their formation is completed, as under conditions of normal aeration, after about 48 h of seed soaking.

There is an apparent time correlation between the self-assembly of the inner membrane of mitochondrial cristae and activation of the seed's respiratory function.

Since further anaerobic incubation is accompanied neither by significant ultrastructural modifications in mitochondria, nor by their enhanced respiratory activity, it may be suggested that anaerobic formation of functionally adequate mitochondria is effected through utilization of ready-made components of membrane precursors, as is the case with pea seeds germinating under conditions of normal aeration (Nawa & Asahi, 1973a, b, c).

It is the depletion of the reserves of these compounds—proteins and lipids—after 48 h imbibition which is most likely to limit the possibility of a further increase in the activity of the respiratory apparatus of the organelle in an anaerobic medium. By contrast, there is a rapid increase in the respiratory activity of the control seeds under aerobic conditions, apparently through the synthesis *de novo* of respiratory enzymes and the lipid components of membranes.

It is indeed surprising that after the seeds had completed mitochondrial formation and attained maximum respiration level, within the subsequent fairly long period (5 days) of strict anoxic conditions the embryo's mitochondrial structure did not differ much from regular organelles. This may explain why, after transfer from anaerobic to aerobic medium, there was a rapid rise in the seeds' respiratory activity with the seeds retaining the ability to germinate.

It is also to be noted that the present study deals with a plant structure which is quite sensitive to the presence of O_2 in the environment. As noted above, maize seeds, unlike seeds of rice (Ueda & Tsuji 1971; Vartapetian *et al.* 1971, 1972; Kordan 1972; Tsuji 1972) and *Echinochloa crus-galli* L. (Kennedy *et al.* 1980), die under prolonged anoxia. Nevertheless, even under conditions of strict exclusion of O_2 from the environment, the maize seed embryo's respiratory apparatus forms in the normal way, and it is ready to start active functioning on contact with molecular oxygen. This feature of seeds may have a survival value and may have been formed through selection, early spring being the sowing and germination season, with the seeds often in waterlogged anaerobic soil.

Secondary anaerobiosis

In secondary anaerobiosis the most conspicuous feature is the difference in tolerance to anoxia of the ultrastructure of a maize seedling's different organs. In leaves and coleoptiles irreversible destruction of single mitochondria occurs only after 3–3·5 days anaerobiosis, and complete degradation of all the cells of these organs, after 5 days, whereas in roots, in fact, the same complete degradation of the cells is brought about already after 24 h in an oxygen-free medium. The reason for this difference is not yet clear, and it was not specifically investigated in the present study. It may be associated with a more rapid depletion of the reserves of utilizable carbohydrates in the root meristem's cells compared to leaf and coleoptile cells, or with a lesser size of these reserves, or with a more vigorous metabolism. Pradet and co-workers (Saglio & Pradet 1980; Saglio, Raymond & Pradet 1980) have shown detached tips of maize roots to consume rapidly the reserves of carbohydrates available within a short period of time under both aerobic and anaerobic conditions. Though the root system was not, in our case, detached from the whole seedling, nevertheless, under conditions of strict anoxia, the inflow of carbohydrates from other parts of the plant to the root meristem may have been strongly inhibited. As a result, the expenditure of utilizable carbohydrates in the maize seedling meristem during anaerobic incubation could not be compensated sufficiently by their inflow from other parts.

Despite the difference in sensitivity to oxygen, all the seedling organs were found to have common characteristics with regard to the reaction of the meristematic cells' ultrastructure to anoxia, the most salient characteristic being an early reaction of mitochondria to anoxia observed in our experiments after only 1·5 h of anoxia, when a swelling of mitochondria is observed, bordering on their destruction. Such swelling was often accompanied by the formation of intricate ring-like and oblong mitochondrial profiles.

However, these destructive mitochondrial modifications appeared to be reversible. Paradoxically, further exposure of seedlings not only to aerobic medium but also to strict anoxia did intensify restoration of mitochondria and they retained the restored structure for quite a long period of time in the complete absence of O_2.

Another charactersitic similar to all the organs is the sequence of changes of mitochondrial ultrastructure under continued anaerobiosis: swollen mitochondria (including ring-like ones) → condensed mitochondria → mitochondria with parallel cristae (including ring-like ones) → swollen mitochondria → irreversibly degraded mitochondria. Although these processes are, on the whole, similar for all the seedling organs, there are distinctions, depending upon the organ. This applies, specifically, to the timing and duration of a specific structural condition of mitochondria. It is also of note that recurrent cyclical rearrangements of mitochondrial ultrastructure and morphology were observed under continued anaerobiosis.

Attention is also called to the heterogeneity of the mitochondrial ultrastructure which was observed at all anaerobic exposures in all organs studied. This is clearly demonstrated in Table 1 which was compiled using 2000 photos. One can see that at any anaerobic exposure—in the same or in different cells of the meristematic tissue of the same organ—mitochondria of different structures are

Table 1. Percent distribution of mitochondria of different ultrastructure correlated with time of measurement

Time	Orthodox	Condensed	Swollen	With parallel cristae
Root
1·5	No	44·70 ± 7·89	14·30 ± 2·36	41·00 ± 8·7
3	No	52·70 ± 12·21	47·30 ± 11·90	No
4·5	No	55·95 ± 1·23	22·79 ± 1·90	21·26 ± 1·45
6	No	60·90 ± 4·18	19·96 ± 4·14	19·14 ± 0.39
7·5	No	51·12 ± 5·69	No	48·87 ± 5·69
9	No	47·51 ± 5·19	No	52·49 ± 5·61
10·5	No	37·68 ± 4·99	No	62·32 ± 4·99
12	No	14·51 ± 1·52	No	85·29 ± 1·61
13·5	No	40·47 ± 7·74	No	59·53 ± 4·47
15	No	60·00 ± 8·56	No	40·00 ± 10·52
16·5	No	46·92 ± 4·75	22·24 ± 2·61	27·84 ± 2·80
18	No	32·73 ± 0·58	No	67·27 ± 0·57
19·5	No	20·13 ± 2·77	No	79·87 ± 2·78
21	No	7·93 ± 3·60	22·60 ± 4·3	69·47 ± 8·33
Coleoptile
1·5	45·5 ± 1·43	16·30 ± 5·90	38·10 ± 3·99	No
3	No	55·70 ± 2·97	No	44·30 ± 2·97
4·5	No	23·00 ± 2·90	No	77·00 ± 3·20
6	No	58·37 ± 16·50	30·13 ± 8·80	11·50 ± 3·70
9	No	73·50 ± 3·80	26·50 ± 3·80	No
12	No	60·10 ± 1·96	39·90 ± 1·96	No
18	No	28·30 ± 5·21	No	71·70 ± 5·21
24	No	48·33 ± 3·13	12·90 ± 1·16	38·77 ± 1·97
36	No	9·10 ± 0·80	No	90·90 ± 2·20
48	No	19·60 ± 1·45	19·50 ± 0·64	60·90 ± 1·34
Leaf
1·5	76·6 ± 6·1	No	23·30 ± 6·2	No
3	41·7 ± 5·3	18·40 ± 4·22	23·30 ± 4·0	No
4·5	64·3 ± 7·1	No	35·70 ± 9·7	No
6	65·3 ± 7·7		34·70 ± 7·7	No
9	53·3 ± 3·3		46·70 ± 4·1	No
12	No	60·20 ± 5·1	No	39·80 ± 5·0
18	No	73·50 ± 1·5	No	26·50 ± 1·5
24	No	62·30 ± 3·3	26·20 ± 2·0	1·70 ± 2·3
36	No	19·00 ± 2·9	9·30 ± 1·2	71·70 ± 3·3
48	No	7·00 ± 0·09	No	93·00 ± 0·7

observed. The table shows that mitochondria of condensed conformation occur with those having parallel cristae. These conformations appear to be easily reversible under anoxic conditions and this suggestion is supported by the existence of the intermediate mitochondrial conformation. Both forms are invariably observed in root and coleoptile specimens, whereas in leaf specimens the massive appearance of mitochondria of condensed conformation and those with parallel cristae is observed only at 12 h of anoxia.

Although we lack experimental data which could reveal the physiological nature of the phenomena responsible for the regular sequence of mitochondrial structural modifications observed during anaerobic incubation, it seems reasonable to suggest that it is the glycolytic energy provision to the cell which is responsible. This hypothesis is supported by earlier observations that the integrity of mitochondria under anoxia depends on their provision with exogenous sugar maintaining anaerobic glycolysis in cells (Vartapetian *et al*. 1977). It may be suggested that in the case of maize seedlings it is the shortage of ATP, which may occur during a changeover from aerobic to anaerobic metabolism, that causes a reversible destruction of the mitochondria which is then restored through intensified glycolysis providing an adequate amount of ATP.

It follows from this interpretation that the beginning of anoxic exposure (the first hours) is an early critical period of anaerobic impact, when ATP shortage in the meristematic tissue was most markedly manifested. The tissue's further survival under conditions of anoxia appears to depend upon the effectiveness and intensity of glycolysis. Specifically, the appearance and growth of the number of mitochondria with parallel cristae would coincide with or immediately follow the period of the most intensive glycolysis, and consequently of the greatest ATP production.

CONCLUDING COMMENTS

The formation, described above, of functionally adequate mitochondrial membranes (cristae) in maize embryos during seed imbibition at strict anoxia is of considerable interest by itself. It shows that not only stabilization of the inner mitochondrial membrane, but even its reassemblage, may occur under conditions of the complete blocking of oxidative phosphorylation, apparently through the utilization of the energy (ATP) produced by glycolysis. This is in sharp contrast to mitochondria which change into promitochondria (Meysel *et al*. 1964; Wallace & Linnane 1964; Wallace, Huang & Linnane 1968; Yellow, Kellerman & Linnane 1968; Watson, Haslam & Linnane 1970). In yeasts, promitochondria begin to be restored into functionally active mitochondria only after a repeated aeration of the medium (Meysel *et al*. 1964; Wallace & Linnane 1964). These observations, as well as the results of experiments obtained in Luzikov's laboratory (Luzikov *et al*. 1971; Luzikov, Zubatov & Rainina 1973; Luzikov 1985) in simulating anaerobiosis in the presence of O_2 by inhibiting the electron transport in yeast mitochondria by KCN accompanied by a similar structural and functional degradation of mitochondria, led the authors to make a basic generalization on the need for a constant electron transport in the respiratory chain of these organelles for membrane stabilization. This implies that the mitochondria can remain intact only through utilization of energy produced endogenously in mitochondria in the process of oxidative phosphorylation.

As can be seen from the present study and from earlier experiments with rice coleoptiles and pumpkin roots fed with exogenous glucose (Vartapetian, Andreeva & Kozlova 1976; Vartapetian *et al.* 1977), higher plant mitochondria drastically differ from those of yeast.

Not only prolonged stabilization of the mitochondria's inner membrane, but their formation as well, are possible in plants under complete arrest of electron transport in the respiratory chain of these organelles, apparently through the utilization of ATP of extramitochondrial, glycolytic origin.

How can these substantial differences between higher plants and yeast, with regard to the biogenesis and degradation of mitochondrial membranes, be accounted for? The present investigation of maize seedlings, in which we observed the destruction of seedling mitochondria under short-term anaerobiosis and the restoration of a normal ultrastructure in the same mitochondria when strictly anaerobic conditions were prolonged, seems to shed some light on the reasons for this destruction and permits a consistent explanation.

As was mentioned above, the explanation for the destruction of mitochondrial membranes in the roots, coleoptiles and leaves of a maize seedling, which was distinct in our experiments after 1·5 h of anoxia, is most likely to be when after transfer from aerobic to anaerobic medium, seedling cells begin soon to experience energy starvation due to a drastic decrease in ATP generation caused by the arrest of oxidative phosphorylation. This immediately begins to tell on the ultrastructure of mitochondrial membranes whose destruction is distinct as early as 1·5 h of anoxia. Under continued anoxia glycolysis appears to become intensified, creating an additional supply of ATP to the cell and a real possibility for swollen mitochondria to restore their normal ultrastructure.

In yeast, after a changeover from aerobic to anaerobic metabolism, apparently due to energy starvation, destruction of mitochondrial membranes is soon observed, as in the case of maize seedlings. But, unlike plants, subsequent intensification of glycolysis under anaerobic conditions in yeast is not accompanied by the restoration of the cristae in promitochondria. The reason is that yeast cells, unlike higher plants, are capable of vigorous metabolism and of reproduction in an oxygen-free medium. Therefore, under anaerobiosis, although promitochondria do form, there can be no question of mature, fully formed mitochondria, since the oxygen-free conditions preclude the synthesis of some enzymes, including cytochromes, and polyunsaturated fatty acids of phospholipids, which are part of the inner membrane.

Due to the shortage of some essential components in an anaerobic medium, the inner membrane (cristae) of mitochondria in newly forming promitochondria cannot be completed until the yeast cell suspension has been aerated. Thus, the degradation of mitochondria upon the transfer of yeast to anaerobic medium is not accounted for by the need for a constant electron transport so as to maintain the integrity of this membrane system, but by the deficiency of utilizable energy (ATP) in the early period of anaerobic incubation, leading to the destruction of membranes.

The case with higher plants is somewhat different because, as in yeast, the initial period of anaerobic incubation energy starvation causes destructive changes in mitochondria. However, under their continued exposure to anaerobiosis no propagation of plant cells or, apparently, mitochondria themselves takes place. Such being the case, as glycolysis is further intensified, the now favourable conditions with regard to energy provision in the cell create a real possibility for the mitochondrial inner membranes to restore a normal ultrastructure again through the utilization of pre-existing membrane components. In other words, by the time the cell is adequately provided with energy, it has at its disposal plastic compounds as well (proteins, co-enzymes, lipids), needed for the mitochondrial inner membrane (cristae). Self-assembly of mitochondrial membranes in anaerobically imbibing maize seed embryos could be interpreted on the same basis. It is the availability of structural components, which can be used to build complete mitochondrial membranes, that is a determining factor of mitochondrial formation in anaerobically imbibing seeds. The depletion of these components after 2 d of anoxia appears to be a decisive factor limiting further formation of mitochondria as well as further increase in the respiratory rate.

The hypothesis proposed, though requiring more corroborative evidence, can be used to explain the totality of the structural and functional variations of mitochondrial behaviour under anaerobic incubation both in plant and yeast.

REFERENCES

ap Rees, T. & Wilson, P. M. (1984). Effects of reduced supply of oxygen on the metabolism of roots of *Glyceria maxima* and *Pisum sativum*. *Zeitschrift für Pflanzenphysiologie*, **114**, 493–503.

Barclay, A. M. & Crawford, R. M. M. (1982). Plant growth and survival under strict anaerobiosis. *Journal of Experimental Botany*, **33**, 541–549.

Day, D. A. & Hanson, J. B. (1977). On methods for the isolation of mitochondria from etiolated corn shoots. *Plant Science Letters*, **11**, 99–104.

Karnovsky, M. J. (1965). A formaldehyde-glutaraldehyde fixative of high osmolatity for use in electron microscopy. *Journal of Cell Biology*, **27**, 137A–138A.

Kennedy, R. A., Barrett, S. C. H., VanderZee, D. A. & Rumpho, M. E. (1980). Germination and seedling growth under anaerobic conditions in *Echinochloa crus-galli* (barnyard grass). *Plant, Cell and Environment*, **3**, 243–248.

Kordan, H. A. (1972). Rice seedling germinated in water with normal and impeded environmental gas exchange. *Journal of Applied Ecology*, **9**, 527–533.

Luzikov, V. N. (1985). *The Mitochondrial Biogenesis and Breakdown*. 362 pp. Plenum Press, New York & London.

Luzikov, V. N., Zubatov, A. S., Rainina, E. J. & Bakeeva, L. E. (1971). Degradation and restoration of mitochondria upon deaeration and subsequent aeration of aerobically grown *Saccharomyces cerevisiae* cells. *Biochemica et Biophysica Acta*, **225**, 321–334.

Luzikov, V. N., Zubatov, A. S. & Rainina, E. J. (1973). Formation and degradation of mitochondria in the cell. 1. Increasing stability of mitochondria during aerobic growth of *Saccharomyces cerevisiae*. *Bioenergetics*, **5**, 129–149.

Meysel, M. N., Biruzova, B. I., Volkova, T. M., Malatian, M. N. & Medvedeva, G. A. (1964). Functional morphology and cytochemistry of microorganism's mitochondrial apparatus. In *Electron and Fluorescent Microscopy of Cells*. AS USSR, Nauka, M.-L.: 3–15 (in Russian).

Nawa, Y. & Asahi, T. (1973a). Relationship between the water content of pea cotyledons and mitochondrial development during the early stage of germination. *Plant Cell Physiology*, **14**, 607–610.

Nawa, Y. & Asahi, T. (1973b). Effect of cycloheximide on development of mitochondria in germinating pea cotyledons. *Agr. Biol. Chem.*, **37**, 937–939.

Nawa, Y. & Asahi, T. (1973c). Biochemical studies on development of mitochondria in pea cotyledons during the early stage of germination. Effects of antibiotics on the development. *Plant Physiology*, **51**, 833–838.

Opik, H. (1973). Effect of anaerobiosis on respiratory rate, cytochrome oxidase activity and mitochondrial structure in coleoptiles or rice (*Oryza sativa*). *Journal of Cell Science*, **12**, 725–736.

Reynolds, E. S. (1963). The use of lead citrate at high pH as an electron-opaque stain in electron microscopy. *Journal of Cell Biology*, **17**, 208–212.

Saglio, P. H. & Pradet, A. (1980). Soluble sugars, respiration and energy charge during aging of excised maize root tips. *Plant Physiology*, **66**, 516–519.

Saglio, P. H., Raymond, B. & Pradet, A. (1980). Metabolic activity and energy charge of excised maize root tips under anoxia. *Plant Physiology*, **66**, 1053–1057.

Smith, A. M. & ap Rees, T. (1979a). Pathways of carbohydrate fermentation in the roots of marsh plants. *Planta*, **146**, 327–334.

Smith, A. M. & ap Rees, T. (1979b). Effect of anaerobiosis on carbohydrate oxidation by roots of *Pisum sativum*. *Phytochemistry*, **18**, 1453–1458.

Tsuji, H. (1972). Respiratory activity in rice seedling germinated under strictly anaerobic conditions. *Botanical Magazine (Tokyo)*, **85**, 207–218.

Ueda, K. & Tsuji, H. (1971). Ultrastructural changes of organelles in coleoptile cells during anaerobic germination of rice seeds. *Protoplasma*, **73**, 203–215.

Vartapetian, B. B. (1973). Aeration of roots in relation to molecular oxygen transport in plants. In *Plant Responses to Climatic Factors*. Proc. Uppsala Symp., Paris, UNESCO, 1970, Ecology and conservation, **5**, 259–264.

Vartapetian, B. B. (1978). Life without oxygen. In *Plant Life in Anaerobic Environment*. Ed. by D. Hook & R. M. Crawford, Ann Arbor Science, Michigan, 1–13.

Vartapetian, B. B. (1982). Pasteur effect visualization by electron microscopy. *Naturwissenschaften*, **69**, 99.

Vartapetian, B. B., Andreeva, I. N. & Maslova, I. P. (1971). Ultrastructure of rice coleoptile's cells in aerobic and anaerobic conditions. Dokl. AN USSR, 166, 1231–1233 (in Russian).

Vartapetian, B. B., Maslova, I. P. & Andreeva, I. N. (1972). Mitochondria of anaerobic rice coleoptiles. *Phisiologia rasteniy*, **19**, 106–108 (in Russian).

Vartapetian, B. B. & Bobylev, Y. S. (1981). Biological method for fine purification of nitrogen and inert gases from O_2 traces. *Naturwissenschaften*, **68**, 329.

Vartapetian, B. B., Andreeva, I. N. & Kozlova, G. I. (1976). The resistance to anoxia and the mitochondrial fine structure of rice seedlings. *Protoplasma*, **88**, 215–224.

Vartapetian, B. B., Andreeva, I. N., Kozlova, G. I. & Agapova, L. P. (1977). Mitochondrial ultrastructure in roots of mesophyte and hydrophyte at anoxia and after glucose feeding. *Protoplasma*, **91**, 243–256.

Wallace, P. G. & Linnane, A. W. (1964). Oxygen-induced synthesis of yeast mitochondria. *Nature*, **201**, 1191–1194.

Wallace, P. G., Huang, M. & Linnane, A. W. (1968). The biogenesis of mitochondria. II. The influence of medium composition on the cytology of anaerobically grown *Saccharomyces cerevisiae*. *Journal of Cellular Biology*, **37**, 207–220.

Watson, K., Haslam, J. M. & Linnane, A. W. (1970). Biogenesis of mitochondria. XIII. The isolation of mitochondrial structures from anaerobically grown *Saccharomyces cerevisiae*. *Journal of Cellular Biology*, **46**, 88–96.

Webb, T. & Armstrong, W. (1983). The effects of anoxia and carbohydrates on the growth and viability of rice, pea and pumpkin roots. *Journal of Experimental Botany*, **34**, 579–603.

Yellow, D., Kellerman, G. M. & Linnane, A. W. (1968). The biogenesis of mitochondria. III. The lipid composition of aerobically and anaerobically grown *Saccharomyces cerevisiae* as related to the membrane systems of the cells. *Journal of Cell Biology*, **37**, 221–230.

B. Root Physiology under
Oxygen Stress

The metabolism of flood-tolerant plants

TOM AP REES, LOVEDAY E. T. JENKIN*, ALISON M.
SMITH† AND PATRICIA M. WILSON

Botany School, University of Cambridge, Downing Street, Cambridge CB2 3EA

SUMMARY

1   The pathways of fermentation in anoxic excised roots of the flood-intolerant *Pisum sativum* and the flood-tolerant *Ranunculus sceleratus, Senecio aquaticus* and *Glyceria maxima* were found to be similar. All depended primarily on the ethanolic fermentation and in none was malate a significant product.

2   The maximum catalytic activity of alcohol dehydrogenase in excised roots of *Pisum sativum* and *Glyceria maxima* did not alter greatly during 48 h anoxia.

3   The manner in which excised roots of *Pisum sativum* and *Glyceria maxima* metabolized [^{14}C]sucrose was determined at different concentrations of oxygen. No differences between the species were found: in both, hypoxia reduced, and anoxia almost abolished, incorporation of ^{14}C into the major polymers.

4   Excised shoots of *Typha angustifolia*, and excised roots of *Rhizophora stylosa* and *Avicennia marina*, all flood tolerant, showed an appreciable ability to convert labelled sugars to polymers, including protein, in anoxia.

INTRODUCTION

Plants vary widely in their ability to grow, or survive, with their underground parts in a flooded environment (Crawford 1982). We refer to this ability as flood tolerance, a term that we consider to be a useful and practicable description of an ecologically important property. In this article we consider, primarily from the results of our own work, whether flood tolerance in higher plants is associated with the possession of a distinctive primary metabolism.

One of the main characteristics of flooded soils is lack of oxygen (Armstrong 1978) and this is probably the most important factor limiting the growth and survival of higher plants in such habitats. The major effect of depriving non-photosynthetic cells of plants of oxygen is to prevent respiratory chain phosphorylation and thus stop the mechanism whereby the tissue makes the vast majority of its ATP. Such ATP that is made in anoxia results from fermentation. This poses two further problems. First, fermentation yields a maximum of 3 moles of ATP per mole of hexose equivalent whereas respiration gives a maximum of 39. Thus much more substrate has to be consumed to produce a given amount of ATP

Present addresses: *Cornwall Trust for Nature Conservation, Dairy Cottage, Trelissick, Truro, Cornwall;
†John Innes Institute, Colney Lane, Norwich NR4 7UH.

in anoxia than in air. Second, fermentative pathways use their own intermediates, for example pyruvate in glycolysis, as hydrogen acceptors. Thus breakdown of the substrate is incomplete, and the products of fermentation include organic compounds such as ethanol and lactate. In general, multicellular organisms are much less able to dispose of the products of fermentation than of CO_2, the product of respiration. Thus there is a danger that such products will accumulate to toxic concentrations. Although lack of ATP is probably the major hazard of anoxia, this is followed closely by inefficient use of substrate, and poisoning by the products of fermentation.

PATHWAYS OF FERMENTATION

The mechanism of fermentation is central to anaerobiosis. Variations that result in increased yield of ATP, or production of more manageable end-products, could be of tremendous significance. Therefore, in investigating whether flood-tolerant plants have a distinctive metabolism, it was logical to concentrate initially on their pathways of fermentation. The pioneering work, without which it is doubtful if the subject would have advanced sufficiently for this conference to have been held at this time, was done by Crawford and his colleagues (Crawford 1969; McManmon & Crawford 1971; Crawford 1972). They suggested that, in hypoxia and anoxia, roots of plants intolerant of flooding depended primarily on the alcoholic fermentation, i.e. glycolysis followed by conversion of pyruvate to ethanol and CO_2. In contrast roots of flood-tolerant plants were held to use fermentative pathways that produced compounds other than ethanol, notably malate, although shikimate and glycerol were also suggested. Malate production was seen as occuring via the carboxylation of glycolytic phospholenolpyruvate followed by the reduction of the oxaloacetate to malate via malate dehydrogenase. An immediate difficulty with this scheme is that it bypasses pyruvate kinase. Thus if free hexose was the substrate, and phosphoenolpyruvate carboxylase formed the oxaloacetate, there would be no net yield of ATP. The latter would require the breakdown of starch or sucrose to yield phosphorylated, rather than free, hexoses as the glycolytic substrate.

Products of fermentation

Against the above background we decided to compare the pathways of fermentation in anoxic excised roots of one flood-intolerant plant, *Pisum sativum* L., and three tolerant plants: the annual *Ranunculus sceleratus* L., the biennial *Senecio aquaticus* Hill, and the perennial *Glyceria maxima* (Hartm.) Holmberg (Smith & ap Rees 1979a,b). First, we identified the principal products of fermentation by incubating root apices in [U-^{14}C]sucrose under nitrogen for 4–5 h, and then determining the detailed distribution of label in both the roots and the surrounding medium. The latter is essential as our results showed that a high proportion of the products of [^{14}C]sucrose metabolism had leaked into the medium. Our analyses were completed without appreciable loss of ^{14}C so we did not overlook any major product of fermentation. The latter were identified by their accumulation of ^{14}C. We found no significant difference between the four species. In each there was appreciable label in ethanol and CO_2, significant label in alanine and slight labelling of lactate. Labelling of malate and other organic acids was minimal in all species.

The extent to which $[^{14}C]$sucrose labelled the products of fermentation only indicates the amounts of products formed. This is because differences in pool sizes of intermediates specific to different pathways could differentially affect the flow of label through the pathway. Accordingly, we measured the relative amounts of ethanol, lactate and malate that accumulated when the excised roots were subjected to anoxia. Both tissue and medium were analysed as we had already shown that tissue content was no guide to amount of product formed in fermentation. In addition we demonstrated that our method of stopping metabolism, and of extracting and measuring the metabolites, was reliable. The need for such a demonstration is paramount (Smith & ap Rees 1979b; ap Rees 1980). Metabolism was stopped by rapidly freezing to the temperature of liquid nitrogen. Reliability was checked by preparing duplicate samples of tissue. One was killed and extracted in the usual way; the other was treated similarly except that a measured amount of the metabolite to be assayed was added to the frozen sample before it was killed. Comparison of the amounts found in extracts of the two samples was taken as a measure of the extent to which the compound had survived the process of killing and extraction. All estimates of such recovery of the added compound were within 15% of the expected values.

Our analyses (Table 1) showed that all four species behaved in the same way. Ethanol showed the major increase, lactate a small increase, and there was no evidence of any increase in malate. If these measurements are considered with the results of the labelling experiments it becomes clear that there are no obvious differences between the four species in their pathways of fermentation. In all species the major products are ethanol and CO_2 with a demonstrable contribution from alanine and a small contribution from lactate.

TABLE 1. Changes in glycolytic and related metabolites after subjecting excised roots to 240 min anoxia (from Smith & ap Rees 1979a,b)

		Species			
Measurement	Compound	*Pisum sativum*	*Ranunculus sceleratus*	*Senecio aquaticus*	*Glyceria maxima*
Amount present	Ethanol	$38\cdot0 \pm 6\cdot9$	$2\cdot81 \pm 0\cdot90$	$1\cdot48 \pm 0\cdot61$	$1\cdot51 \pm 0\cdot79$
before transfer to	Alanine	$5\cdot4 \pm 0\cdot6$	—	—	—
anoxia (μmol g^{-1}	Lactate	$1\cdot7 \pm 0\cdot3$	$0\cdot98 \pm 0\cdot14$	$1\cdot09 \pm 0\cdot19$	$0\cdot68 \pm 0\cdot13$
fresh weight)	Malate	$2\cdot7 \pm 0\cdot2$	$1\cdot69 \pm 0\cdot41$	$0\cdot70 \pm 0\cdot15$	$11\cdot91 \pm 0\cdot37$
Change in content	Ethanol	$37\cdot0 \pm 3\cdot1$	$4\cdot51 \pm 1\cdot18$	$6\cdot51 \pm 1\cdot11$	$17\cdot68 \pm 1\cdot00$
after 240 min	Alanine	$5\cdot0 \pm 0\cdot4$	—	—	—
anoxia (μmol g^{-1}	Lactate	$1\cdot5 \pm 0\cdot3$	$0\cdot96 \pm 0\cdot18$	$1\cdot87 \pm 0\cdot37$	$3\cdot18 \pm 0\cdot16$
fresh weight)	Malate	$-2\cdot3 \pm 0\cdot2$	$-0\cdot21 \pm 0\cdot21$	$-0\cdot20 \pm 0\cdot12$	$-0\cdot91 \pm 0\cdot22$

Values are means $\pm$ S.E. from at least five different samples of root apices.

Alcohol dehydrogenase

The discrepancy between our results and the metabolic theory of flood tolerance led us to examine another aspect of the theory. This was that flooding leads to marked increases in the maximum catalytic activity of alcohol dehydrogenase in the roots of intolerant plants, but not in those of tolerant plants (Crawford 1969; McManmon & Crawford 1971). We determined the effects of anoxia on the maximum catalytic activity of alcohol dehydrogenase in excised roots, and of flooding on the activity of roots still attached to plants of *Pisum sativum* and *Glyceria maxima* (Jenkin & ap Rees 1983). Reliable estimation of the maximum catalytic activities of enzymes in plant tissues is attended by at least as many pitfalls as the measurement of substrates (ap Rees 1980). It is essential to optimize the assays and to provide evidence that significant losses of activity did not occur during extraction and assay of the enzyme. The latter may be achieved by demonstrating adequate recovery of pure enzymes in recovery experiments comparable to those we described for substrates. Unless these criteria are met, little faith may be placed in the results, as differential inactivation of the enzyme may have occurred in differently treated samples. Our assays were optimized and our estimates of recovery were within 12% of those expected.

We excised the apical 4–6 cm of the roots of pea and *Glyceria maxima*, incubated them under nitrogen, and estimated the maximum catalytic activity of alcohol dehydrogenase over the next 48 h (Table 2). We detected no clear difference between the behaviour of the roots of the two species. No significant change was detected in pea roots, and there was evidence of a slight rise in activity in *Glyceria* roots. On a fresh weight basis pea roots had by far the higher activity. This was due largely to differences in protein content. The values for freshly excised roots on a protein basis were 3·9 and 1·6 nkat mg^{-1} protein for pea and *Glyceria*, respectively. Comparable experiments with flooded plants showed that flooding doubled the activity of alcohol dehydrogenase in roots of *Glyceria maxima*, and led to a somewhat greater increase in pea roots (Jenkin & ap Rees 1983).

The above results provide no firm evidence for the view that flood tolerance is associated with a much smaller increase in alcohol dehydrogenase on flooding than

TABLE 2. Effect of incubation under nitrogen on maximum catalytic activity of alcohol dehydrogenase in excised roots of *Glyceria maxima* and *Pisum sativum* (from Jenkin & ap Rees 1983)

Time after transfer to nitrogen (h)	Enzyme activity (nkat g^{-1} fresh weight)	
	Glyceria maxima	*Pisum sativum*
0	7·1 ± 0·5	41·7 ± 7·7
2	8·6 ± 1·0	41·2 ± 7·2
12	10·1 ± 1·2	49·2 ± 12·7
18	10·4 ± 1·6	47·1 ± 11·5
24	12·3 ± 1·1	38·9 ± 13·9
42	—	52·5 ± 25·5
48	9·4 ± 4·2	—

Values are means ± S.E. of estimates from five separate samples.

is found in roots of intolerant species. There is ample evidence that lack of oxygen leads to synthesis of alcohol dehydrogenase in a range of other plants, mentioned by Hanson & Brown (1984): detailed studies have been made with maize (Freeling & Birchler 1981) and barley (Hanson, Jacobsen & Zwar 1984). However, too few measurements of the maximum catalytic activity of the enzyme have been made to allow us to decide whether the changes in other tissues are greater or smaller than those that we found on flooding the roots of pea and *Glyceria maxima*, and whether any such differences are associated with flood tolerance. To answer this question it will be necessary to measure the maximum catalytic activities of the different isoenzymes of alcohol dehydrogenase. When the presently available data on alcohol dehydrogenase are considered as a whole, they provide no convincing evidence for the view that flood tolerance is associated with minimal induction. Indeed the evidence points strongly in the opposite direction, as mutants of maize (Roberts *et al.* 1984b) and barley (Harberd & Edwards 1982), which lack the major isoenzyme of alcohol dehydrogenase, are markedly less capable of withstanding lack of oxygen than are plants with the normal complement of enzyme.

GENERAL METABOLISM

Sucrose metabolism by roots of Glyceria maxima *and* Pisum sativum

Flood-tolerant plants may differ from intolerant plants in aspects of metabolism that lie outside the principal pathways of fermentation. To investigate this possibility we studied the effects of a range of oxygen concentrations on the metabolism of [U-^{14}C]sucrose by excised roots of pea and *Glyceria maxima* (ap Rees & Wilson 1984). We argue that, in biochemical terms, the essential feature of root growth is the conversion of sucrose, translocated to the root, into the constituent polymers of the root; protein, nucleic acid and structural polysaccharides. Thus if the two species of root differ in the ability of their overall metabolism to withstand anoxia, such a difference should be reflected in differential sensitivity to lack of oxygen of the movement of label from [^{14}C]sucrose into the water-and-ethanol-insoluble fraction of the roots. This fraction would contain the vast majority of the polymers made during growth. We investigated whether this was so.

We excised the apical 6–8 cm of the roots, incubated them with [U-^{14}C]sucrose at different oxygen concentrations for 4 h, and then determined the detailed distribution of label in the medium and roots. To allow comparison of samples that had metabolized different amounts of label, the ^{14}C recovered per fraction is expressed as a percentage of the total ^{14}C metabolized by that sample (Table 3). This total fell with the oxygen concentration. The extent of this fall was similar in both species. The reduction between air and nitrogen was 50% for pea and 66% for *Glyceria*. The percentages of metabolized [^{14}C]sucrose that were converted to polymers in general, i.e. the insoluble fraction, and to protein in particular, were

TABLE 3. Effects of oxygen concentration on metabolism of [U-^{14}C]sucrose by excised roots of *Pisum sativum* and *Glyceria maxima*

Oxygen concentration (%, v/v)	*Pisum sativum*			*Glyceria maxima*		
	^{14}C metabolized (kBq)	% metabolized ^{14}C found in: Insoluble fraction	Protein	^{14}C metabolized (kBq)	% metabolized ^{14}C found in: Insoluble fraction	Protein
21·0	6·06	17	9	3·87	17	5
14·3	7·25	20	9	2·99	18	5
11·3	7·84	20	8	3·16	17	6
8·0	6·12	20	7	2·68	13	5
6·1	6·34	17	7	1·65	15	4
4·1	5·72	11	5	2·08	8	3
2·0	6·48	10	5	1·84	6	2
1·0	4·87	6	4	1·42	1	n.d.
0·0	3·31	3	1	1·35	1	n.d.

Roots were incubated for 4 h and then killed and analysed as described by ap Rees & Wilson (1984).
n.d. = none detected.

not much affected until the oxygen concentration fell below 6·1%. Thereafter a very sharp decline occurred so that polymer synthesis, particularly of proteins, from sucrose in anoxia was slight or negligible compared to that in air. The roots of the two species behaved very similarly. There is no evidence of the metabolism of the roots of *Glyceria maxima* being any more tolerant of lack of oxygen than that of pea roots. In both species, one of the most spectacular effects of hypoxia is a massive reduction in polymer synthesis from sucrose.

The apparent contradiction between the inability of the roots of *Glyceria maxima* to convert sucrose to polymers in hypoxia, and the knowledge that this plant thrives in soils that contain little oxygen (Lambert 1947), implies that, although the soil is anoxic, the roots are not. Examination of the roots of *Glyceria maxima* revealed extensive development of aerenchyma that reached right to the root apex (ap Rees & Wilson 1984). Evidence that this aerenchyma does supply oxygen to the root was obtained by sealing the roots of plants into boiling tubes that contained [U-14C]sucrose through which nitrogen was bubbled. This simulated a plant growing in flooded soil and permitted analysis of the fate of the [14C]sucrose under these conditions. The control plants were left intact with their shoots in air; the shoots were cut off the remaining plants and the cut ends sealed so that the aerenchyma that led to the roots had no direct access to the atmosphere. The roots of the latter plants behaved similarly to hypoxic excised roots in that there was little labelling of polymers and appreciable fermentation (Table 4). The roots of the intact plants readily incorporated label into polymers. The extent of this incorporation was slightly less than when excised roots were incubated in air, or when intact plants were incubated in aerated [14C]sucrose. These results suggest very strongly that the aerenchyma provided the roots with sufficient oxygen to permit appreciable, but perhaps not optimal, growth.

Two further points emerge from our analysis of the intact plants fed [14C]sucrose in the presence of nitrogen (Table 4). First, less than 1% of the metabolized label was found in the shoot. This argues against the view that the products of fermentation are transferred from roots to shoots for subsequent aerobic metabolism. Second, of the label metabolized by the roots, less than 4% was found in malate. Thus our inability to demonstrate that malate is a significant product of fermentation in excised roots of *Glyceria maxima* is not an artefact caused by the excision of the roots.

TABLE 4. Effects of reduced supply of oxygen on metabolism of [U-14C]sucrose by the roots of plants of *Glyceria maxima* (from ap Rees & Wilson 1984)

Treatment	[14C]sucrose metabolized (kBq)	Percentage of metabolized 14C recovered as:	
		Volatile compounds	Water and ethanol-insoluble compounds
Complete plant	6·23	5	14
	5·41	2	16
Plant with shoots removed and stem sealed	3·32	12	2
	3·54	9	4

Collectively our studies demonstrate a remarkable similarity in the responses of the flood-tolerant and -intolerant plants to hypoxia and anoxia. We argue that for the flood-tolerant species that we studied, the tolerance is not due to any obvious difference in primary metabolism. Our more detailed studies with *Glyceria maxima* strongly suggest that the ability to form aerenchyma is probably the major characteristic that distinguishes this plant from flood-tolerant species.

We stress, in particular, our complete inability to find any evidence that malate is a significant product of fermentation. This is despite claims for both *Glyceria maxima* and *Senecio aquaticus* that malate accumulates in excised roots during short periods of anoxia (Crawford 1967), and in the roots of flooded plants (Crawford & Tyler 1969). We may not say that malate is not a significant product of fermentation: the range of different plants and conditions of growth defeats any attempt to prove such a negative. However, we do argue that at present there is no convincing proof that malate is an appreciable product of fermentation in higher plants. As far as we are aware, claims for such formation of malate, even the more recent ones (Mendelssohn, McKee & Patrick 1981), fail to satisfy one or more of the following criteria. These are: authentication of measurements of enzymes and metabolites, appreciation that much of the products of fermentation may leak out of the tissue, proof of net accumulation of the compound as a product of fermentation, and a need to demonstrate that any accumulation of a compound in anoxia is significant in relation to the total amount of substrate fermented. We also note that it has not been possible to substantiate the claim that glycerol is formed as an alternative to ethanol in anoxia (Smith, Kalsi & Woolhouse 1984).

Our work adds appreciably to the growing evidence (Davies 1980) that ethanol and CO_2 are generally the major products of fermentation in higher plants. We also detected a much smaller but nonetheless significant accumulation of lactate. This is consistent with Davies' (1980) hypothesis that subjection of plants to anoxia leads to an initial accumulation of lactate that lowers the cytosolic pH so that the activity of lactate dehydrogenase decreases and that of pyruvate decarboxylase increases. The net result is an early switch from lactate to ethanol and CO_2 as the major products of fermentation. The fall in cytosolic pH has now been demonstrated and shown to coincide with ethanol production (Roberts *et al.* 1984b). Thus the weight of the available evidence suggests that in anoxia, except for a very brief initial period, plants depend primarily upon alcoholic fermentation. However, this dependence is not complete. We have demonstrated the formation of alanine as a product of fermentation in anoxia and there is evidence (Davies 1980) that both succinate and 4-amino butyrate accumulate in anoxia. The precise significance of this is not known, mainly because we do not know the pathways whereby these compounds are formed.

The view of anaerobic metabolism obtained from the relatively few species that have been studied in detail is one of dominance, but not exclusively so, of the ethanolic fermentation, and a marked inability to convert sucrose to polymers. The extent to which this view is general is not known. We thought that the inability to make polymers from sucrose might not be found in plants that inhabit very hypoxic environments. The following experiments were done to see if this was so.

T. AP REES *et al.* 235

[^{14}C]*Glucose metabolism by mangrove roots*

The mangroves *Rhizophora stylosa* Griff. and *Avicennia marina* (Forsk.) Vierh. grow in flooded soils deficient in oxygen (Dowling & McDonald 1982). We supplied [^{14}C]glucose, under nitrogen, to samples of excised roots that had been freshly harvested from plants growing naturally in flooded soil on the shores (*Rhizophora*) or inlets (*Avicennia*) at Cape Ferguson, Queensland. The samples were analysed as described by ap Rees & Wilson (1984) for roots of *Glyceria maxima*. Two features of the results (Table 5) are stressed.

TABLE 5. Metabolism of specifically labelled [^{14}C]glucose by excised roots of mangroves in anoxia

| | Percentage of supplied ^{14}C recovered per fraction | | |
| | *Rhizophora stylosa* | | *Avicennia marina* |
Position of label in [^{14}C]glucose	CO_2	Insoluble fraction	Insoluble fraction
C-1	0·90	3·3	18·2
	1·19	—	31·8
C-3,4	8·10	6·4	8·8
C-6	0·25	10·8	27·7
	0·27	13·6	35·5

Samples of freshly excised, mainly fibrous roots, fresh weight 1·7 g (*Rhizophora*) and 1·45 g (*Avicennia*) were incubated under nitrogen in 10 ml 0·35 mM [^{14}C]glucose (*Rhizophora*) in sterilized seawater or 0·7 mM [^{14}C]glucose (*Avicennia*) in sterilized 75% seawater for 17 h at 22°C. ^{14}C added to each sample of *Rhizophora* was: C-1, 15·3 kBq; C-3,4, 9·2 kBq; C-6, 16·0 kBq. Double these amounts were added to *Avicennia*, respectively. Samples were killed and extracted with 8% (v/v) HClO$_4$ and further extracted with water. Protein was isolated as described by ap Rees & Wilson (1984).

Roots of *Rhizophora stylosa* released both C-1 and C-6 of glucose as CO_2 in anoxia. Release from C-1 was significant in comparison to that from C-3,4. Ethanolic fermentation releases C-3 and 4, but none of the other carbons, of hexose as CO_2. The relatively high yield from [3,4-^{14}C]glucose suggests a dominance of the ethanolic fermentation. However, the yield from C-1 strongly suggests the operation of an additional pathway that is capable of releasing carbon dioxide. The fact that release from C-1 greatly exceeded that from C-6 is consistent with this additional pathway being the oxidative pentose phosphate pathway. Further evidence from a detailed knowledge of the distribution of ^{14}C amongst the other metabolites labelled is needed to substantiate our suggestions. Nonetheless the data in Table 5 strongly suggest that, whilst pyruvate decarboxylase may be the main means whereby hexose carbon is converted to CO_2 in anoxic plants, it is not the only one.

The second striking feature of the data in Table 5 is that, in both species, there was appreciable movement of label from [^{14}C]glucose into the insoluble fractions. At least some of this was in protein. Further analysis of the samples fed [6-^{14}C]glucose showed that 4·7 and 14·0% of the supplied ^{14}C was recovered in protein in roots of *Rhizophora stylosa* and *Avicennia marine*, respectively. Thus these roots differ from those of peas and *Glyceria maxima* in being able to catalyse appreciable polymer synthesis from respiratory intermediates in anoxia.

Metabolism of [¹⁴C]sucrose by shoots of Typha angustifolia

Further evidence that almost complete cessation of polymer synthesis from the sugars is not an inevitable consequence of anoxia in all higher plants is provided by studies of the effect of diminished oxygen concentration on the metabolism of [U-¹⁴C]sucrose by excised young shoots of *Typha angustifolia* (Jenkin & ap Rees 1986). Entire young shoots, about 5 cm long, were excised from their parent rhizomes, which were in their natural flooded habitat where the soil redox potential was below +250 mV. The experiments were similar to those described for *Glyceria maxima* in Table 3, except that the incubation in [U-¹⁴C]sucrose was for 5 h. The results are shown in Table 6.

TABLE 6. Effects of oxygen concentration on metabolism of [U-¹⁴C]sucrose by excised shoots of *Typha angustifolia* (from Jenkin & ap Rees 1986)

| | | Percentage of ¹⁴C metabolized recovered in: | |
Oxygen concentration (%, v/v)	¹⁴C metabolized (kBq g⁻¹ fresh weight)	Water and ethanol-insoluble fraction	Protein
21·0	0·103	22·8	5·6
14·3	0·148	17·1	6·9
11·3	0·070	20·2	7·5
8·0	0·073	10·1	3·8
6·1	0·101	14·6	5·2
3·9	0·124	19·0	4·5
2·0	0·067	14·2	6·8
1·0	0·091	9·2	2·8
0·0	0·103	7·0	0·9

We found important differences between the two flood-tolerant species, *Typha angustifolia* and *Glyceria maxima*, in the ways in which lack of oxygen affected sucrose metabolism. First, lowering the concentration of oxygen did not significantly affect the amount of [¹⁴C]sucrose metabolized by the *Typha* shoot (Table 6); it reduced it by two-thirds in the *Glyceria* roots (Table 3). Second, although lowering the oxygen concentration caused a decline in incorporation into polymers as a whole and protein in particular in the *Typha* shoots, this decline was less marked and occurred at a lower concentration of oxygen than in the *Glyceria* roots. For example, the ratio of the incorporation into the insoluble fraction in air to that in 1% oxygen was 17 for *Glyceria* roots but only 2·5 for *Typha* shoots. Thus the latter are characterized by an ability to convert a much higher proportion of their metabolized sucrose to polymers in hypoxia and anoxia than are the roots of the flood-tolerant *Glyceria maxima*. To this extent, the flood tolerance of *Typha angustifolia* is associated with differences in metabolism. We have not yet identified precisely what these differences are. The use of malate as a major product of fermentation does not appear to be one of them. The proportion of metabolized label from [¹⁴C]sucrose that was recovered in malate was under 3%.

FINAL COMMENTS

Generalizations about the metabolism of flood-tolerant plants are probably best confined to the statement that too few species have been studied in sufficient depth to reveal any distinct pattern. However, progress requires a working hypothesis and we, very tentatively, suggest the following. Tissues of a substantial proportion, possibly the vast majority, of flood-tolerant plants are not capable of significant growth in anoxia or severe hypoxia. The metabolism of such species probably does not differ in any major way from that of intolerant plants. Nonetheless, relatively minor differences may well occur and be of considerable ecological significance. For example, Roberts *et al.* (1984a) have suggested that the greater intolerance of peas *vis-à-vis* maize, to anoxia may be due to earlier onset of cytoplasmic acidosis in anoxia in peas, caused, at least in part, by a greater permeability of the pea tonoplast to protons. Flood tolerance, in most plants, is therefore likely to depend primarily upon transport of oxygen to the submerged parts. In some flood tolerant plants such transport of oxygen may be inadequate, or impossible due to lack of connexion between the submerged tissue and a source of oxygen. The metabolism of at least parts of such plants, for example, the young shoots of *Typha angustifolia*, may be characterized by an ability to synthesize from sucrose the polymers required for at least some growth. It seems likely that a major key to our understanding of the metabolism of flood tolerant plants lies in discovering the mechanisms that regulate polymer synthesis in hypoxia and anoxia.

ACKNOWLEDGMENTS

L.E.T.J. and A.M.S. thank the S.E.R.C. for research studentships. T. ap R. thanks Dr John Bunt, Director of the Australian Institute of Marine Science for appointing him a Visiting Investigator. Part of this work was carried out at the Australian Institute of Marine Science, Townsville, Queensland, Australia.

REFERENCES

ap Rees, T. (1980). Integration of pathways of synthesis and degradation of hexose phosphates. *The Biochemistry of Plants. Vol. 3. Carbohydrates: Structure and Function* (Ed. by J. Preiss), pp. 1–42. Academic Press, London and New York.

ap Rees, T. & Wilson, P. M. (1984). Effects of reduced supply of oxygen on the metabolism of roots of *Glyceria maxima* and *Pisum sativum*. *Zeitschrift für Pflanzenphysiologie*, **114**, 493–503.

Armstrong, W. (1978). Root aeration in the wetland condition. *Plant Life in Anaerobic Environments* (Ed. by D. D. Hook & R. M. M. Crawford), pp. 269–297. Ann Arbor Science, Ann Arbor.

Crawford, R. M. M. (1967). Alcohol dehydrogenase activity in relation to flooding tolerance in roots. *Journal of Experimental Botany*, **18**, 458–464.

Crawford, R. M. M. (1969). The physiological basis of flooding tolerance. *Berichte der Deutschen Botanischen Gesellschaft*, **82**, 111–114.

Crawford, R. M. M. (1972). Some metabolic aspects of ecology. *Botanical Society of Edinburgh Transactions*, **41**, 309–322.

Crawford, R. M. M. (1982). Physiological response to flooding. *Encyclopaedia of Plant Physiology. Vol. 12B. Water Relations and Carbon Assimilation* (Ed. by O. L. Lange, P. S. Nobel & C. B. Osmond), pp. 453–477. Springer-Verlag, Berlin and Heidelberg.

Crawford, R. M. M. & Tyler, P. D. (1969). Organic acid metabolism in relation to flooding tolerance in roots. *Journal of Ecology*, 57, 235–244.

Davies, D. D. (1980). Anaerobic metabolism and the production of organic acids. *The Biochemistry of Plants. Vol. 2. Metabolism and Respiration* (Ed. by D. D. Davies), pp. 581–611. Academic Press, London and New York.

Dowling, R. M. & McDonald, T. J. (1982). Mangrove communities of Queensland. *Mangrove Ecosystems of Australia* (Ed. by B. F. Clough), pp. 79–93. Australian National University Press, Canberra.

Freeling, M. & Birchler, J. A. (1981). Mutants and variants of the alcohol dehydrogenase-1 gene in maize. *Genetic Engineering, Principles and Methods. Vol. 3.* (Ed. by J. K. Setlow & A. Hollaender), pp. 223–264. Plenum, New York.

Hanson, A. D. & Brown, A. H. D. (1984). Three alcohol dehydrogenase genes in wild and cultivated barley: characterization of the products of variant alleles. *Biochemical Genetics*, 22, 495–515.

Hanson, A. D., Jacobsen, J. V. & Zwar, J. A. (1984). Regulated expression of three alcohol dehydrogenase genes in barley aleurone layers. *Plant Physiology*, 75, 573–581.

Harberd, N. P. & Edwards, K. J. R. (1982). The effect of a mutation causing alcohol dehydrogenase deficiency on flooding tolerance in barley. *New Phytologist*, 90, 631–644.

Jenkin, L. E. T. & ap Rees, T. (1983). Effects of anoxia and flooding on alcohol dehydrogenase in roots of *Glyceria maxima* and *Pisum sativum*. *Phytochemistry*, 22, 2389–2393.

Jenkin, L. E. T. & ap Rees, T. (1986). Effects of lack of oxygen on the metabolism of shoots of *Typha angustifolia*. *Phytochemistry*, 25, 823–827.

Lambert, J. M. (1947). Biological flora of the British Isles. *Glyceria maxima. Journal of Ecology*, 34, 310–344.

McManmon, M. & Crawford, R. M. M. (1971). A metabolic theory of flooding tolerance: the significance of enzyme distribution and behaviour. *New Phytologist*, 70, 299–306.

Mendelssohn, I. A., McKee, K. L. & Patrick, W. H. (1981). Oxygen deficiency in *Spartina alterniflora* roots: metabolic adaptation to anoxia. *Science*, 214, 439–441.

Roberts, J. K. M., Callis, J., Jardetzy, O., Walbot, V. & Freeling, M. (1984a). Cytoplasmic acidosis as a determinant of flooding intolerance in plants. *Proceedings of the National Academy of Sciences of the United States of America*, 81, 6029–6033.

Roberts, J. K. M., Callis, J., Wemmer, D., Walbot, V. & Jardetzy, O. (1984b). Mechanism of cytoplasmic pH regulation in hypoxic maize roots tips and its role in survival under hypoxia. *Proceedings of the National Academy of Sciences of the United States of America*, 81, 3379–3383.

Smith, A. M. & ap Rees, T. (1979a). Effects of anaerobiosis on carbohydrate oxidation by roots of *Pisum sativum*. *Phytochemistry*, 18, 1453–1458.

Smith, A. M. & ap Rees, T. (1979b). Pathways of carbohydrate fermentation in the roots of marsh plants. *Planta*, 146, 327–334.

Smith, A. M., Kalsi, G. & Woolhouse, H. W. (1984). Products of fermentation in the roots of alders (*Alnus* Mill.) *Planta*, 160, 272–275.

Root metabolic response of *Spartina alterniflora* to hypoxia

IRVING A. MENDELSSOHN AND KAREN L. MCKEE

Laboratory for Wetland Soils and Sediments, Center for Wetland Resources, Louisiana State University, Baton Rouge, Louisiana 70803, U.S.A.

SUMMARY

Root hypoxia resulting from nitrogen gas bubbling through a hydroponic rooting environment or from waterlogging of a soil system caused a signifcant increase in root alcohol dehydrogenase activity (ADH) in *Spartina alterniflora*. The well-developed aerenchyma system of this plant does not in itself provide sufficient root ventilation to allow for complete aerobic respiration. Although root malate concentrations increase due to hypoxia in *S. alterniflora*, the role, if any, of malate in the flood tolerance of this species has not yet been determined. Root ethanol accumulation in *S. alterniflora* during hypoxia is low even when ADH activity is induced, which suggests diffusion of ethanol from the root tissue. Root energy status of *S. alterniflora* during hypoxia is maintained at levels similar to that during root normoxia.

INTRODUCTION

Spartina alterniflora is the dominant intertidal angiosperm in salt marshes along the Atlantic and Gulf coasts of the United States. The productivity of this species rivals that of fertilized crops such as sugar-cane (Dawes 1981). Although *Spartina* is a highly productive plant, its productivity varies within and between marshes (Turner 1976); it is tallest and most productive along tidal creeks (Turner 1976) where soil drainage is maximal (Odum & Fanning 1973; Mendelssohn & Seneca 1980; Howes *et al.* 1981). With increasing distance inland from the tidal creeks, the productivity and height of *Spartina* decreases as subsurface drainage becomes negligible. Thus, gradients in primary productivity which typically occur within *Spartina*-dominated marshes are directly related to the degree of soil waterlogging (see Mendelssohn, McKee & Postek 1982; King *et al.* 1982; Wiegert, Chalmers & Randerson 1983). This differential soil drainage results in significant differences in soil oxygen demand as measured by soil redox potential (Mendelssohn & Seneca 1980; Howes *et al.* 1981). The fact that *Spartina* can grow in zones of different intensities of soil waterlogging has stimulated an interest in how this plant adapts metabolically to varying degrees of root hypoxia.

239

Although the aerenchyma system of *Spartina* is important in the transport of atmospheric oxygen to the roots (Teal & Kanwisher 1966; Carlson 1980; Morris & Dacey 1984), this transport is not always sufficient to maintain aerobic respiration. Morris & Dacey (1984) and Gleason & Dunn (1982) have demonstrated a reduction in below-ground respiration during hypoxia. Carbon dioxide release from intact roots of *S. alterniflora* decreased from approximately 13 $\mu l\ g^{-1}\ min^{-1}$ at 12% O_2 (the concentration of O_2 at which respiration was maximal) to 5 $\mu l\ g^{-1}\ min^{-1}$ at 0% oxygen (Morris & Dacey 1984). Oxygen deficiencies within the root apparently inhibit aerobic respiration even when the above ground tissue is exposed to the atmosphere. We would hypothesize that as the oxygen demand of the substrate increases, the degree of root oxygen deficit should also increase. Gleason & Zieman (1981) have shown that the oxygen concentration in the root tissue of *S. alterniflora* can decrease to as low as 3% during tidal flooding. At night, when photosynthetic O_2 production is absent, root oxygen concentrations can approach 0%. The purpose of this paper is to describe the root metabolic response of *S. alterniflora* to hypoxia. The specific objectives were to determine if:

(i) Hypoxic root environments stimulate anaerobic metabolism in the roots of flood-tolerant *S. alterniflora*.

(ii) Ethanol accumulates in the roots of *S. alterniflora* during hypoxia.

(iii) Malate accumulates in the roots of *S. alterniflora* during hypoxia.

(iv) Root energy status is maintained in the roots of *S. alterniflora* during hypoxia.

MATERIALS AND METHODS

The following experiments were conducted in a greenhouse where the light intensity averaged 1500 $\mu mol\ m^{-2}\ s^{-1}$ at midday and the temperature ranged from 30°C (day) to 20°C (night). Transplants of *S. alterniflora* were collected from the field in April 1983. Clones of these transplants were propagated vegetatively, established in 21 × 21 cm plastic pots containing a 1:1 soil mixture of Jiffy Mix (commercial potting soil, Jiffy Products of America, West Chicago, Illinois) and marsh sediment and allowed to grow for 1 month. Five pots were then permanently flooded with a nutrient solution (Mendelssohn & McKee 1985) by placing them in larger buckets; flooding depth was maintained at *c.* 3 cm above the pot rims and *c.* 5 cm above the soil surface. Five additional pots were maintained in a wet, but drained, condition by daily flooding from and drainage into a common reservoir containing the same nutrient solution as in the permanently-flooded treatment. This experiment was terminated after 2 months.

A second flooding experiment was conducted with *S. alterniflora* in natural marsh sediment. Ten marsh cores containing *c.* ten to twenty culms were collected from the field in June 1983 and placed in 21 × 21 cm plastic pots. Five pots were flooded by placing them in larger buckets and five others were maintained in a

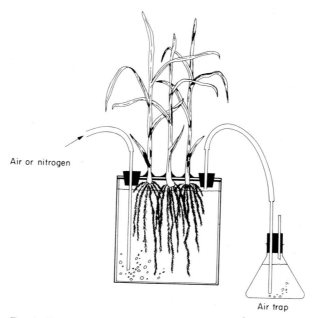

Air or nitrogen

Air trap

FIG. 1. Experimental apparatus for the control of root aeration.

drained condition as in the flooding experiment described above. This experiment was also terminated after 2 months.

Clones of *S. alterniflora* were also grown hydroponically in plastic pots (10 cm diameter × 15 cm high) which contained the same nutrient solution as above. A single culm was threaded through a 1 cm diameter hole in a plastic top which fitted snugly inside the rim of the pot (Fig. 1). The top of the pot through which the plant culm was growing contained three additional openings, two for the aeration tubing and one for an oxygen probe. Each pot was vigorously aerated with an aquarium pump while the plants were allowed to grow for 2 months. The nutrient solution was changed weekly. Five pots were randomly designated as controls and maintained as described above in the aerated nutrient solution. The tops of the other five pots were sealed with a non-toxic silicone sealant (General Electric Company) and the stems of the plants likewise sealed so that there was no leakage of atmospheric air into the pots. Nitrogen was purged through each treatment pot at a rate of *c*. 100 cm^3 min^{-1}. The experiment was terminated after 120 h.

Measurement of soil redox potential

Soil redox potentials (Eh) were measured in the flooded soil experiments at 1 and 15 cm depths with brightened platinum electrodes; the potential of a calomel electrode against a standard hydrogen electrode (+244 mV) was added to the measured potential to calculate Eh.

Measurement of oxygen concentration

Oxygen concentrations in the hydroponic culture solutions were measured with a YSI oxygen meter and probe which was inserted into each pot through an opening in the top, and sealed with silicon sealant.

Root collection

Plants from the flooded soil experiments were washed in tap water to remove all sediment from the root mat. Roots from the flooded natural sediment experiment were divided into two groups: those growing in the upper 5 cm of the pot and those in the lower part. Roots from the other flooded soil experiment and the hydroponic experiment were not divided, but selected from the entire root mat. Living roots, i.e. turgid and structurally intact, from all experiments were quickly separated from the dead roots and debris, rinsed in deionized water and divided into sample: (i) roots for enzyme analysis were placed in plastic bags, frozen in liquid nitrogen ($-197°C$), and stored on dry ice ($-79°C$) until analysis (within 24 h), (ii) roots for metabolite analysis were cut into 1–2 mm pieces which were immediately dropped into liquid nitrogen, fixed with 8% perchloric acid, and refrozen until analysis (within 24 h); (iii) roots for adenine nucleotide analysis were placed in plastic bags with c. 20 ml of deionized water, frozen in liquid nitrogen (the water forms a protective layer of ice which prevents tissue meltback during handling), and freeze-dried (3–4 days).

Adenylate analysis

The root samples were freeze-dried in a LabConco Freeze-Dryer 5, ground in a Wiley-Mill (No. 60 mesh sieve), and extracted in a 1 mM EDTA plus 5% PVPP solution, pH 7·4. Cryodesiccation followed by boiling extraction had previously been shown to be a satisfactory method for the determination of adenylates in plant tissue (Mendelssohn & McKee 1981; Delistraty & Hershner 1983). A 0·1 g sample was extracted in 10 ml of boiling extraction solution for 30 s, cooled in ice, centrifuged at 20 000 g at 4°C, and the supernatant fraction immediately prepared for the assay of the adenine nucleotides. Adenosine mono-, di-, and triphosphates were measured using the ATP-dependent light yielding reaction of the firefly-lantern luciferin luciferase (FLE-50, Sigma Chemical Co.) complex with a Model LS100 Beckman liquid scintillation counter. ATP was determined directly, while ADP and AMP were converted enzymatically to ATP and determined by subtraction (see Mendelssohn & McKee 1981, for details). Recoveries of added ATP, ADP, and AMP standards were 92–100% with this method.

Enzyme analysis

Root tissue (0·50 g fresh wt) from the flooded soil experiments was homogenized with a Brinkman Polytron homogenizer in 5 ml of a 50 mM HEPES buffer (pH 8·0) which contained 2 mM cysteine-HCl, 5mM $MgCl_2$, and 2% PVPP for the assay of alcohol dehydrogenase (ADH) (EC 1.1.1.1).

Root tissue from the hydroponic experiment was freeze-dried for both adenylate analysis and enzyme extraction. Preliminary tests showed that activities of ADH extracted from freeze-dried *Spartina* roots were consistently equal to that extracted from fresh tissue (unpubl. results). Extraction of ADH in this case was accomplished by vortexing 25 mg dry wt of the dried, ground roots in 5 ml of a 100 mM Tris buffer (pH 7·3) which contained 5 mM $MgCl_2$, 20 mM dithiothreitol, and 2% (w/v) of PVPP.

All plant extracts were centrifuged at 20 000 g for 30 min and the supernatant assayed at once. All procedures were carried out at 2–4°C until the assay at 30°C in a reaction mixture described as follows: 2·8 ml total volume, 5·5 mM $MgCl_2$, 0·26 mM NADH, and 0·40 mM acetaldehyde in either 14 mM HEPES (flooded soil experiment) or 14 mM Tris (hydroponic gas experiment) buffers, pH 8·0. Enzyme activity was measured by following the oxidation of NADH in the reaction cuvette for 3–4 min against a reference containing all components except acetaldehyde at a wavelength of 340 nm in a Beckman Model 35 spectrophotometer.

Total soluble protein was determined according to Bradford (1976) on the supernatant fraction from each of the sample extractions within 1 h after the enzyme assay. Each sample was read against a blank containing the appropriate extraction buffer.

Measurement of root metabolites

Roots from both the flooded soil and hydroponic gas experiments were quickly excised, weighed (0·5 g fresh wt), cut into 1–2 mm pieces which were dropped into liquid nitrogen in plastic sample bottles, fixed with 8% perchloric acid, and refrozen for 24 h until analysis. The roots frozen in perchloric acid were thawed, pelleted, and the supernatant fraction neutralized with 5 M K_2CO_3. Malate and ethanol were determined enzymatically on the neutralized extract by measuring the reduction of NAD at 340 nm. The specific reaction mixtures were as follows: malate—430 mM glycine, 340 mM hydrazine hydrate, 2·75 mM NAD, c. 37 IU/ml MDH in a total volume of 2·92 ml (pH 9·0); ethanol—68 mM $Na_4P_2O_7$, 68 mM semicarbazide, 19 mM glycine, 0·73 mM NAD, c. 54 IU/ml ADH in a total volume of 3·32 ml (pH 8·7).

RESULTS AND DISCUSSION

We previously described some aspects of the root metabolic response of the highly flood-tolerant plant, *Spartina alterniflora*, along a gradient of increasing soil waterlogging in a Louisiana salt marsh (Mendelssohn, McKee & Patrick 1981). In that study, the degree of root anaerobic metabolism was environmentally controlled by the intensity of soil reduction (Mendelssohn, McKee & Patrick 1981). At soil redox potentials indicating moderate reduction (+200 mV), root

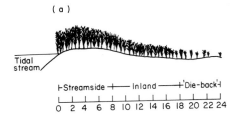

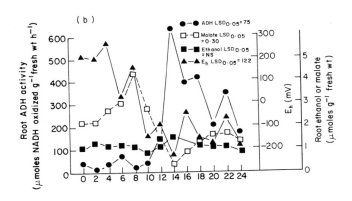

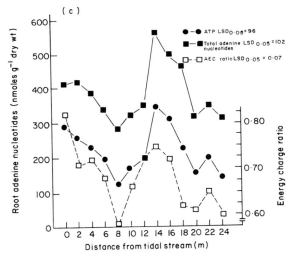

FIG. 2. (a) Sampling transect through streamside, inland, and dieback forms of *Spartina alterniflora* in a Louisiana salt marsh. (b) Variation in root alcohol dehydrogenase (ADH) activity and malate and ethanol concentrations in *S. alterniflora* and substrate Eh along the sampling transect. (c) Variations in root adenosine triphosphate (ATP) and total adenine nucleotide concentrations and adenylate energy charge (AEC) ratio in *S. alterniflora* along the sampling transect (from Mendelssohn, McKee & Patrick 1981).

respiration was primarily aerobic as evidenced by a lack of ADH activity and a high energy status (Fig. 2). However, where the soil redox potentials were low (-150 mV), ADH activity was induced to a high level and energy status was maintained at levels comparable to that in roots metabolizing primarily aerobically. At intermediate redox levels ($0 - {}^{+}100$ mV) ADH activity was not stimulated, malate accumulated, and energy status declined. This field study generated a number of hypotheses which were tested in both hydroponic and flooded soil experiments.

Hypothesis 1: Hypoxic root environments stimulate root ADH activity in flood-tolerant Spartina alterniflora

Spartina roots grown in a hypoxic ($[O_2] = 0.92 \pm 0.05$ mg l^{-1}) hydroponic culture for 5 days exhibited a significant six-fold increase in ADH activity over the aerated control (Fig. 3). This same species grown in a flooded soil system for 8 weeks also showed a significant increase in ADH activity (4.7-fold) (Fig. 4) although the increase was not as great as that observed in the hydroponic experiment. The absolute values for enzyme activity were also lower in the flooded soil experiment compared to the hydroponic experiment (Figs 3 & 4). The moderate soil reducing conditions (Eh $= + 14 \pm 15$ mV) developed in the flooded soil experiment may not have created an oxygen demand great enough to induce ADH activities as high as in the hydroponic experiment. In addition, the difference between short and long term response may have contributed to the dissimilarity in degree of ADH activity between the two experiments.

The results of both the short-term hydroponic experiment (Fig. 3) and the long-term flooded soil experiment (Fig. 4) support the ADH field data (Fig. 2).

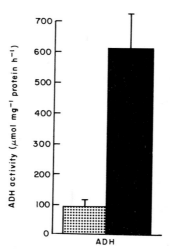

FIG. 3. Activities of alcohol dehydrogenase (ADH) measured in the roots of *Spartina alterniflora* growing in hydroponic culture after 120 h of hypoxia. Air (□); nitrogen (■).

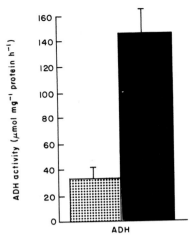

FIG. 4. Activities of alcohol dehydrogenase (ADH) measured in the roots of *Spartina alterniflora* growing in soil after 2 months of flooding. Drained (□); flooded (■).

Certainly, *Spartina* responds to a hypoxic root environment by stimulating relatively high ADH activity, which may indicate a stimulation in alcoholic fermentation (assuming that alcoholic fermentation is positively correlated with ADH activity (Keeley 1979)). The rate of stimulation is dependent upon the intensity of the soil oxygen demand (see below). This flood-tolerant plant does not adapt to flooding by reducing alcoholic fermentation, as has been suggested for some plants (Crawford 1967; Linhart & Baker 1973), but rather appears to accelerate this pathway of carbon metabolism, as has been found for rice and some flood-tolerant marsh plants (John & Greenway 1976; Smith & ap Rees 1979). The degree of stimulation of ADH activity caused by soil flooding or by flushing the root environment with nitrogen gas is extremely variable. For example, Jenkin & ap Rees (1983) demonstrated only a two-fold increase in root ADH activity in flooded *Glyceria maxima*. This small increase in ADH activity which was similar to that found for the flood-intolerant pea, *Pisum sativum*, led to the conclusion that the induction of alcohol dehydrogenase was of little significance in determining flood tolerance. However, some flood-tolerant species exhibit a large induction of ADH activity upon flooding. Smith & ap Rees (1979) demonstrated a five-fold increase in ADH activity in unaerated *Ranunculus sceleratus* roots and a four-fold increase for the roots of *Senecio aquaticus*. In rice seedlings, a five-fold increase in ADH activity was observed after flushing the culture solution with pure nitrogen for 15–18 h (Wignarajah, Greenaway & John 1976). River birch, *Betula nigra*, exhibited a twenty-five-fold increase in root ADH activity after 6 days of a nitrogen treatment to the roots compared to a 1·3-fold increase in the flood-intolerant European birch, *B. pendula* (Tripepi & Mitchell 1984). Monk, Crawford & Braendle (1984) demonstrated that the rhizomes of some marsh plant species showed an increase in ADH activity during a 16-day period in anoxia, e.g.

Schoenoplectus lacustris, while others did not, e.g. *Iris pseudacorus*. Some of these differences in ADH induction are certainly species-dependent, but we would suggest that the intensity and duration of exposure are also key factors in controlling the observed response, e.g. nitrogen gas purged through a culture solution would produce a different oxygen demand from that generated by flooding a sand substrate. Regardless of the above differences in ADH response to hypoxia, it is evident at least for *Spartina* that hypoxic root conditions stimulate significant ADH activity and that the well-developed aerenchyma system of this plant does not in itself provide sufficient root ventilation to allow for complete aerobic root respiration when soil oxygen demand is relatively high.

Hypothesis 2: Hypoxic root environments cause malate accumulation in Spartina alterniflora

Data collected in the field (Mendelssohn, McKee & Patrick 1981) indicated that root malate concentrations can increase in *Spartina* roots under moderately reducing soil conditions (Fig. 2). The results from both the hydroponic and flooded soil experiments support the conclusion that root malate concentrations increase during hypoxia (Figs 5 & 6). However, not all plant species appear to respond to hypoxia by increasing the concentration of malate. Several marsh plant species and two birch tree species responded to hypoxic or anaerobic conditions by decreasing root malate concentrations (Smith & ap Rees 1979; Tripepi & Mitchell 1984).

Although the accumulation of malate in the roots of *Spartina* both in the field (Mendelssohn, McKee & Patrick 1981) and in two separate greenhouse

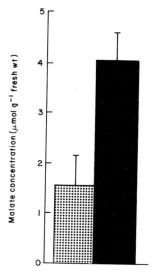

FIG. 5. Malate accumulation in the roots of *Spartina alterniflora* growing in hydroponic culture after 120 h of hypoxia. Air (□); nitrogen (■).

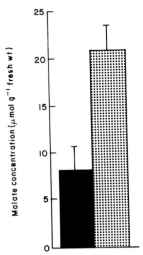

FIG. 6. Malate accumulation in the roots of *Spartina alterniflora* growing in soil after 2 months of flooding. Drained (□); nitrogen (■).

experiments (Figs 5 & 6) demonstrated that this response is not unusual in this species, the significance of this phenomenon is difficult to assess. Although McManmon & Crawford (1971) proposed that malate accumulates through an alternate anaerobic pathway which avoids alcoholic fermentation and the production of ethanol, others (Hiatt & Leggett 1974; Keeley 1979) have pointed out that malate may also accumulate in response to an ionic imbalance due to an excessive uptake of cations under flooded conditions. Mendelssohn, McKee & Patrick (1981) found that malate concentrations, which were relatively high in *Spartina* roots growing in moderately reduced soils, were significantly lower when ADH activity was stimulated to a high degree at sites where the soil was highly reduced. If malate were accumulating in response to an excessive uptake of cations, one would expect an increase in malate concentration to occur at the more reduced soil sites. The results of this field study suggest that malate might play an important role during flooding when substrate conditions are moderately, but not highly, reduced. The results of the present study demonstrated, however, that malate accumulation and alcoholic fermentation are not mutually exclusive (Figs 3–6). Although these results cannot answer the question of whether malate was produced as an alternative to ethanol in *Spartina* or accumulated through some other mechanism, they do not eliminate the possibility that malate may play some role in flood adaptation in this species. Further investigation is required to answer this question.

Hypothesis 3: Root ethanol accumulation during hypoxia is minimal in Spartina alterniflora

Root ethanol concentrations for *Spartina* plants growing in hydroponic culture were not detectable even when ADH activity was as high as 600 μmol mg^{-1}

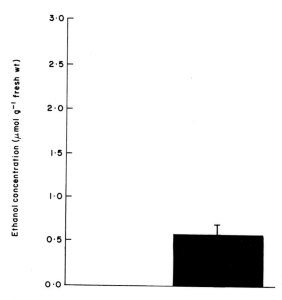

FIG. 7. Concentration of ethanol measured in the roots of *Spartina alterniflora* growing in soil after 2 months of flooding. Drained (□); flooded (■).

protein h^{-1}. These results suggested that *Spartina* has a high capacity for removing ethanol from its root tissue. Bertani, Brambilla & Menegus (1980) found that 98% of the total ethanol produced by rice seedlings was released into the rooting medium. Alpi & Beevers (1983) showed that ethanol readily diffused from rice seedlings and that only 21% remained within the plant tissue after a 24 h anaerobic incubation. The constant gas flow through the hydroponic system (Fig. 1) would contribute to the removal of ethanol. Roots collected both from the field and the flooded soil experiment also had exceptionally low ethanol concentrations (Figs 2 & 7). Hence, it would appear that *Spartina* easily rids itself of ethanol through diffusion into the surrounding medium and/or via internal oxidation.

Hypothesis 4: Root energy status is maintained during hypoxia in Spartina alterniflora

Root energy status, as indicated by the AEC ratio, was directly related to root ADH activity in *Spartina* growing along a flooding gradient in the field (Fig. 2). Short-term hydroponic culture of *Spartina* additionally demonstrated that the root energy status can be maintained under hypoxia at a level (0·85) equivalent to that for aerobically grown roots (Fig. 8). Rice also maintains a relatively high AEC when subjected to anoxia (Mocquot *et al.* 1981). After an initial decline in energy status following transfer to nitrogen, rice seedlings attained an AEC (within 24 h) similar to that maintained under aerobic conditions.

Although total adenine nucleotide concentrations were significantly lower in the hypoxic *Spartina* roots, no significant differences between treatment and

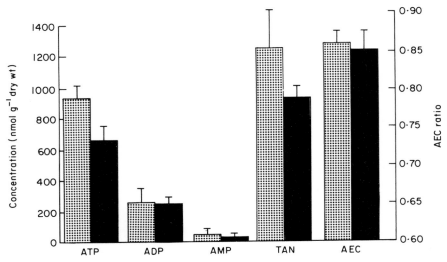

FIG. 8. Concentrations of adenine nucleotides and adenylate energy charge (AEC) ratio measured in the roots of *Spartina alterniflora* growing in hydroponic culture after 120 h of hypoxia. Air (□); nitrogen (■).

control root AEC values were observed (Fig. 8). The maintenance of low AMP concentrations in the hypoxic roots appears to have stabilized the AEC at a high level. The root AEC ratio of *Spartina* growing in a flooded soil system for 8 weeks was also maintained at a relatively high value (Fig. 9). In this experiment, the roots of the flooded treatment had a significantly greater AEC than that of the drained treatment. The high AMP concentrations in the roots of the drained treatment indicated that these roots may have been experiencing some type of stress (Fig. 9). *Spartina alterniflora* exhibits significantly less growth when grown under drained

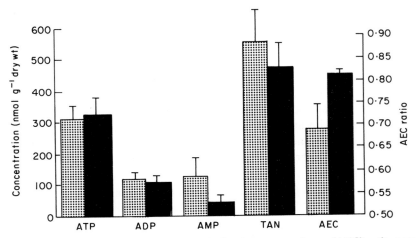

FIG. 9. Concentrations of adenine nucleotides and adenylate energy charge (AEC) ratio measured in the roots of *Spartina alterniflora* growing in soil after 2 months of flooding. Drained (□); nitrogen (■).

TABLE 1. The effect of drained and flooded soil treatments on alcohol dehydrogenase (ADH) activity and energy charge (AEC) ratio in *Spartina alterniflora* roots and soil redox potential (Eh)

| | Drained | | Flooded | |
	Roots at 0–5 cm	Roots at 10–15 cm	Roots at 0–5 cm	Roots at 10–15 cm
AEC ratio	0·80 ± 0·03	0·69 ± 0·03	0·66 ± 0·02	0·82 ± 0·02
ADH (μmol g^{-1} fresh wt h^{-1})	0 ± 0	15 ± 9	83 ± 15	295 ± 54
Eh (mV)	+457 ± 64	+274 ± 145	+12 ± 22	−166 ± 25

compared to flooded conditions (Parrondo, Gosselink & Hopkinson 1978). The lower AEC in the roots of drained *Spartina* may reflect the effect of this suboptimal growth environment.

Stratified sampling of *Spartina* roots growing in flooded and drained marsh soil more clearly demonstrated the relationship between soil reduction and root metabolism (Table 1). Initial soil Eh measurements indicated a difference in soil oxygen demand as a function of soil depth. Surface soil was more oxidized than subsurface soil, regardless of the treatment. Therefore, sampling was stratified so that surface roots in each treatment were analysed separately from subsurface roots.

In the drained treatment, the surface soil had a relatively high Eh of +457 mV (Table 1). The roots within this zone of the pot showed no detectable ADH activity and the AEC was relatively high at 0·80. The absence of ADH activity and the presence of a high energy status indicated that these roots were respiring aerobically. The soil Eh within the centre of the drained pots was +274 mV. Roots within this zone showed a small induction of ADH activity and a significantly lower AEC ratio than the roots at the surface (Table 1). Apparently, oxygen was limiting the production of ATP via aerobic respiration in these roots.

In the flooded pots, the soil surface Eh was significantly more reduced than in the drained treatment (Table 1). The flooded roots at the surface had higher ADH activities than the drained surface roots. The energy status of the flooded surface roots was relatively low. However, roots within the centre of the pots of the flooded treatment, where soil Eh indicated highly reducing conditions (−166 mV), showed a significant increase in ADH activity. The AEC ratio was high and equivalent to that of the surface roots from the drained treatment. The induction of high ADH activity may indicate a significant stimulation in alcoholic fermentation which apparently increased the anaerobic production of ATP, maintaining a high AEC ratio. *Spartina* appears to adapt to a more reduced soil environment by stimulating a high rate of alcoholic fermentation which provides the energy needed to support plant metabolism.

CONCLUSIONS

Spartina alterniflora is a plant highly adapted to an intertidal environment. Where soil reducing conditions are not severe the well-developed aerenchyma tissue of this species provides sufficient oxygen to support aerobic root respiration. However, where soil drainage is retarded and soil oxygen demand is increased, the aerenchyma tissue does not support complete aerobic respiration. We hypothesize that *Spartina* has adapted to this environmental variation, having the capacity to induce a high rate of alcoholic fermentation (as indicated by stimulated ADH activity) which maintains a high root energy status. Potentially toxic ethanol apparently diffuses from the root tissue. Although these adaptations allow *Spartina* to survive and grow in highly waterlogged conditions, it may be at the expense of considerable carbon loss to the plant via accelerated glucose consumption (to fuel alcoholic fermentation) and ethanol loss to the rooting medium. *Spartina alterniflora* thus responds to its flooded environment through a combination of anatomical and physiological adaptations.

REFERENCES

Alpi, A. & Beevers, H. (1983). Effects of O_2 concentration on rice seedlings. *Plant Physiology*, **71**, 30–34.

Bertani, A., Brambilla, I. & Menegus, F. (1980). Effect of anaerobiosis on rice seedlings: Growth, metabolic rate, and fate of fermentation products. *Journal of Experimental Botany*, **31**, 325–331.

Bradford, M. M. (1976). A rapid and sensitive method for the quantification of microgram quantities of protein utilizing the principle of protein-dye binding. *Analytical Biochemistry*, **72**, 248–254.

Carlson, P. R. (1980). *Oxygen diffusion from the roots of* Spartina alterniflora *and the role of* Spartina *in the sulfur cycle of salt marsh sediments*. Ph.D. dissertation, University of North Carolina. 188 pp.

Crawford, R. M. M. (1967). Alcohol dehydrogenase activity in relation to flooding tolerance in roots. *Journal of Experimental Botany*, **18**, 458–464.

Dawes, C. J. (1981). *Marine Botany*. John Wiley & Sons, New York. 628 pp.

Delistraty, D. A. & Hershner, C. (1983). Determination of adenine nucleotide levels in *Zostera marina* (eelgrass). *Journal of Applied Biochemistry*, **5**, 404–419.

Gleason, M. L. & Dunn, E. L. (1982). Effects of hypoxia on root and shoot respiration of *Spartina alterniflora*. *Estuarine Comparisons* (Ed. by V. S. Kennedy) pp. 243–253. Academic Press, New York.

Gleason, M. L. & Zieman, J. C. (1981). Influence of tidal inundation on internal oxygen supply of *Spartina alterniflora* and *Spartina patens*. *Estuarine and Coastal Shelf Science*, **13**, 47–57.

Hiatt, A. J. & Leggett, J. E. (1974). Ionic interactions and antagonisms in plants. *The Plant Root and its Environment*. (Ed. by E. W. Carson) University of Virginia Press, Charlottesville, Virginia.

Howes, B. L., Howarth, R. W., Teal, J. M. & Valiela, I. (1981). Oxidation-reduction potentials in a salt marsh: spatial patterns and interactions with primary production. *Limnology and Oceanography*, **26**, 350–360.

Jenkin, L. E. & ap Rees, T. (1983). Effects of anoxia and flooding on alcohol dehydrogenase in roots of *Glyceria maxima* and *Pisum sativum*. *Phytochemistry*, **22**, 2389–2393.

John, C. D. & Greenway, H. (1976). Alcoholic fermentation and activity of some enzymes in rice roots under anaerobiosis. *Australian Journal of Plant Physiology*, **3**, 325–336.

Keeley, J. E. (1979). Population differentiation along a flood frequency gradient: Physiological adaptations to flooding in *Nyssa sylvatica*. *Ecological Monographs*, **49**, 89–108.

King, G., Klug, M. J., Wiegert, R. G. & Chalmers, A. G. (1982). Relation of soil water movement and sulfide concentration to *Spartina alterniflora* production in a Georgia salt marsh. *Science*, **218**, 61–63.

Linhart, Y. B. & Baker, I. (1973). Intra-population differentiation of physiological response to flooding in a population of *Veronica peregrina* L. *Nature*, **242**, 275–276.

McManmon, M. & Crawford, R. M. M. (1971). A metabolic theory of flooding tolerance: the significance of enzyme distribution and behaviour. *New Phytology*, **70**, 299–306.

Mendelssohn, I. A. & McKee, K. L. (1981). Determination of adenine nucleotide levels and adenylate energy charge ratio in two *Spartina* species. *Aquatic Botany*, **11**, 37–55.

Mendelssohn, I. A. & McKee, K. L. (1985). The effect of nutrients on adenine nucleotide levels and the adenylate energy charge ratio in *Spartina alterniflora* and *Spartina patens*. *Plant, Cell and Environment*, **8**, 213–218.

Mendelssohn, I. A., McKee, K. L. & Patrick, Jr. W. H. (1981). Oxygen deficiency in *Spartina alterniflora* roots: metabolic adaptation to anoxia. *Science*, **214**, 439–441.

Mendelssohn, I. A., McKee, K. L. & Postek, M. T. (1982). Sublethal stresses controlling *Spartina alterniflora* productivity, (Ed. by B. Gopal, R. E. Turner, R. G. Wetzel & D. F. Whigham). *Wetlands: Ecology and Management*. pp. 223–242. National Institute of Ecology, Jaipur and International Scientific Publications.

Mendelssohn, I. A. & Seneca, E. D. (1980). The influence of soil drainage on the growth of salt marsh cordgrass *Spartina alterniflora* in North Carolina. *Estuarine and Coastal Marine Science*, **11**, 27–40.

Mocquot, B., Prat, C., Mouches, C. & Pradet, A. (1981). Effect of anoxia on energy charge and protein synthesis in rice embryo. *Plant Physiology*, **68**, 636–640.

Monk, L. S., Crawford, R. M. M. & Braendle, R. (1984). Fermentation rates and ethanol accumulation in relation to flooding tolerance in rhizomes of monocotyledonous species. *Journal of Experimental Botany*, **35**, 738–745.

Morris, J. T. & Dacey, J. W. H. (1984). Effects of O_2 on ammonium uptake and root respiration by *Spartina alterniflora*. *American Journal of Botany*, **71**, 979–985.

Odum, E. P. & Fanning, M. E. (1973). Comparison of the productivity of *Spartina alterniflora* and *Spartina cynosuroides* in Georgia coastal marshes. *Georgia Academy of Science Bulletin*, **31**, 1–12.

Parrondo, R. T., Gosselink, J. G. & Hopkinson, C. S. (1978). Effects of salinity and drainage on the growth of three saltmarsh grasses. *Botanical Gazette*, **139**, 102–107.

Smith, A. M. & ap Rees, T. (1979). Pathways of carbohydrate fermentation in the roots of marsh plants. *Planta*, **146**, 327–334.

Teal, J. M. & Kanwisher, J. W. (1966). Gas transport in the marsh grass, *Spartina alterniflora*. *Journal of Experimental Botany*, **17**, 355–361.

Tripepi, R. R. & Mitchell, C. A. (1984). Metabolic response of river birch and European birch roots to hypoxia. *Plant Physiology*, **76**, 31–35.

Turner, R. E. (1976). Geographic variations in salt marsh macrophyte production: a review. *Contributions to Marine Science*, **20**, 47–68.

Wiegert, R. G., Chalmers, A. G. & Randerson, P. F. (1983). Productivity gradients in salt marshes: the response of *Spartina alterniflora* to experimentally manipulated soil water movement. *Oikos*, **41**, 1–6.

Wignarajah, K., Greenway, H. & John, C. D. (1976). Effect of waterlogging on growth and activity of alcohol dehydrogenase in barley and rice. *New Phytology*, **77**, 585–592.

Effect of exogenous nitrate on anaerobic root metabolism

A. BERTANI, I. BRAMBILLA AND R. REGGIANI

Istituto Biosintesi Vegetale C.N.R., Via Bassini 15, 20133 Milano, Italy

SUMMARY

In the apical segment of sterile rice and wheat roots, reduction of nitrate in the absence of oxygen is promoted by the presence of exogenous nitrate. Also associated with this effect were modifications in the levels of NADH, catabolic reducing charge, CO_2 produced an adenylate energy charge. The comprehensive effect of nitrate on the anaerobic metabolism of the root is discussed in relation to root survival in anoxia.

INTRODUCTION

The NADH produced by the catabolism of lipids, polysaccharides and proteins is oxidized during aerobic plant growth through the mitochondrial electron transport chain with consequent ATP synthesis. In the absence of molecular oxygen, the cytochrome chain ceases to function and this leads to the accumulation of NADH. In this condition, the plant tissues face the problem of regenerating oxidized pyridine coenzymes in order to maintain energy production. The mechanism used by plants to re-oxidize the pyridine nucleotides involves the enzyme alcohol dehydrogenase and, to a lesser extent, lactate dehydrogenase (App & Meiss 1958; Hageman & Flesher 1960; Davies 1973). However, during anaerobic treatment the NADH concentration has been reported to increase (Roberts *et al.* 1984), indicating that in anoxia the synthesis of NADH always exceeds its re-oxidation.

In this respect, oxidized substances taken up from the environment and reduced by plant cells at the expense of NADH could play an important role in providing NAD^+ and nitrate could be one of these substances.

In previous work, barley and rice plants grown in liquid culture have been shown to withstand root anaerobiosis better when nitrate was supplied (Arnon 1937; Malavolta 1954). Moreover, nitrate-associated alleviation of symptoms of waterlogging damage have been reported in barley and wheat plants (Drew, Sisworo & Saker 1979; Trought & Drew 1981). On the other hand, metabolic investigations of nitrate assimilation have revealed that, in anoxia, this pathway is severely inhibited at the nitrate reductase step (Nance 1948; Ferrari & Varner 1971; Lee 1978).

The aim of this work is to analyse the effects of exogenous nitrate on the metabolism of excised rice and wheat roots in anoxia.

MATERIALS AND METHODS

Seeds of rice (*Oryza sativa* cv.Arborio) and wheat (*Triticum aestivum* cv.Flavio) were used. Dehulled seeds of rice were sterilized for 2 min with 70% (v/v) ethanol and for 30 min with 5% (w/v) calcium hypochlorite, each treatment being followed by rinsing with distilled water. For wheat seeds, the sterilization was carried out as described for rice seeds excluding the treatment with 70% ethanol. The seeds were then germinated for 3 days on sterile wet paper in Petri dishes at 25°C in the dark. The apical 10×10^{-3} m of the seedling roots was excised and grown in air at 25°C for 1 day on a modified Heller medium at pH 5·5 with the following composition: potassium, 10 mol m^{-3}; sodium, 0·9 mol m^{-3}; calcium, 0·5 mol m^{-3}; magnesium, 1 mol m^{-3}; nitrate, 7 mol m^{-3}, chloride, 4 mol m^{-3}; sulphate, 1 mol m^{-3}; phosphate 0·9 mol m^{-3}; iron, $3·7 \times 10^{-3}$ mol m^{-3}; iodine, $0·1 \times 10^{-3}$ mol m^{-3}; boron, 20×10^{-3} mol m^{-3}; zinc, $3·5 \times 10^{-3}$ mol m^{-3}; manganese, $0·4 \times 10^{-3}$ mol m^{-3}; copper, $0·1 \times 10^{-3}$ mol m^{-3}; thiamin, $0·3 \times 10^{-3}$ mol m^{-3}; sucrose, 2% (w/v). The excised roots (25 in each sample with a total fresh weight of 75×10^{-3} g) were then placed in a jar with 5×10^{-6} m^3 of modified Heller medium (with or without 7 mol m^{-3} NO$_3^-$) and made anaerobic by flushing nitrogen gas (99·999% nitrogen) through the growth medium. The time required to obtain the anaerobic condition in the jar was not more than 20 min, as checked by Clark oxygen electrode (YSI Model 53 Biological Oxygen Monitor) and GasPack anaerobic systems (BBL, U.S.A.). When NO$_3^-$ was omitted, Cl$^-$ was added to the medium to retain the osmolarity of the solution.

In all experiments bacterial contamination was checked by incubating 3×10^{-7} m^3 of the culture medium on a medium containing agar 15 kg m^{-3}, peptone 5 kg m^{-3} and beef extract 3 kg m^{-3}. Data from contaminated replicates were discarded.

Anaerobic treatments were applied for 3, 8 and 24 h to rice roots and 1·5, 8 and 24 h to wheat roots. After the treatments were completed, the medium was removed from the jars and further root metabolism immediately blocked with HClO$_4$ 6×10^2 mol m^{-3} at 2°C. All operations were carried out without opening the jar and under a continous stream of nitrogen gas. The excised roots were then disrupted by grinding and the homogenate cleared by centrifugation at 10 000 g for 15 min. The samples were neutralized with K$_2$CO$_3$ $2·4 \times 10^3$ mol m^{-3}. If the samples were to be used for NADH determinations, homogenation, centrifugation, neutralization and assay were carried out within 40 min to avoid acid degradation of reduced nucleotides. Experiments carried out as above, replacing the roots with NADH standards, showed recoveries exceeding 85%. When the excised roots were stained with tetrazolium salts, the method described by Bertani, Brambilla & Menegus (1981) was followed.

Carbon dioxide evolution was continuously monitored by Infra Red Gas Analyzer (Leybold-Heraeus GmbH, West Germany). Free amino acids were assayed as described previously (Bertani & Brambilla 1982) and nitrite was measured as reported by Hageman & Hucklesby (1971). ATP, ADP and AMP were assayed by bioluminescence with the luciferine-luciferase assay using the ATP

Monitoring Kit (LKB, Finland), according to the method described by Carver & Walker (1983). The pyridine nucleotides, either reduced or oxidized (after enzymatic reduction), were determined by bioluminescence with the luciferase assay using NADH Monitoring Kit (LKB, Finland). The emisson of light from the luciferase reaction was monitored by Biolumat LB 9500 (Berthold company, West Germany). The adenylate energy charge was defined as ATP + 0·5 ADP/ATP + ADP + AMP and the catabolic reducing charge as NADH/NADH + NAD$^+$.

RESULTS

Growth and survival

The excised rice and wheat roots previously grown in aerobic conditions on Heller medium (7 mol m^{-3} NO$_3^-$) were subjected to anaerobiosis in the presence or absence of exogenous nitrate. The imposition of anaerobiosis entirely blocked the growth and increase in fresh weight of roots incubated with or without nitrate. The viability of rice roots, measured as the capacity to be stained with tetrazolium salts (Bertani, Brambilla & Menegus 1981), was maintained both in the presence and absence of nitrate for the whole duration of the anaerobic treatment (24 h). Conversely, the viability of wheat roots was not entirely preserved in 24 h of anaerobic treatment and a difference of tolerance of oxygen deficiency was observed between roots grown with and without nitrate. In presence of exogenous nitrate, a greater capacity for root survival was, in fact, observed (Table 1) by staining with tetrazolium salts. Moreover, in order to assess cellular integrity the

TABLE 1. Effect of exogenous nitrate on the ability of the apical segment of wheat roots (0·2 × 10^{-3} m) to be stained with tetrazolium salts after 8 and 24 h of anaerobiosis. Three different experiments were carried out and the percentage values were calculated as (number of stained apex roots/total apex roots) × 100

	Hours of anaerobiosis	
	8	24
With nitrate	64%	5%
Without nitrate	12%	0%

TABLE 2. Release of amino acids into the culture medium from excised roots of wheat subjected to anaerobiosis in the presence or absence of nitrate

Hours of anaerobiosis	With nitrate	Without nitrate
0	—	—
1·5	0·56 ± 0·13	0·80 ± 0·16
8	3·06 ± 0·16	5·03 ± 0·26
24	10·96 ± 0·23	12·93 ± 0·39

The concentrations are reported as 10^{-6} mol g^{-1}. All values are mean (±S.E., n = 4).

amino acids lost in the growth medium were determined. The exogenous nitrate acted by limiting the release of these substances from the wheat roots (Table 2).

Nitrate reduction

Under anaerobic conditions the inhibition of nitrite reductase leads to an accumulation of nitrite which is released into the surrounding medium. The production of nitrite in such conditions can be used to characterize the nitrate reduction in plant tissue (Ferrari & Varner 1971). In the present work, in order to assess the anaerobic nitrate reduction carried out by rice and wheat roots incubated in the presence or absence of nitrate, the amounts of nitrite present inside the roots and excreted into the culture medium were measured. The behaviour of rice and wheat plants as regards anaerobic nitrate reduction was very similar and can be summarized as follows: the level of endogenous nitrite was negligible and showed no significant difference in anaerobically treated roots in the presence or absence of nitrate. Conversely, nitrite was released into the surrounding medium especially by roots supplied with nitrate (Table 3). Release of nitrite from roots of rice and wheat was observed in the first hours of anaerobiosis and then ceased as the treatment proceeded, as reported with regard to tobacco cells, barley aleurone layers and corn leaf section by Ferrari, Yoder & Filner (1973).

TABLE 3. Release of nitrite into the culture medium from excised roots of rice and wheat subjected to anaerobiosis in the presence or absence of nitrate

Hours of anaerobiosis	Rice		Wheat	
	With NO_3^-	Without NO_3^-	With NO_3^-	Without NO_3^-
0	15·1 ± 1·6	7·2 ± 0·8	9·5 ± 0·9	5·5 ± 0·6
1·5			308·5 ± 21·5	22·8 ± 3·2
3	457·3 ± 47·4	25·7 ± 4·0		
8	488·5 ± 30·6	23·0 ± 3·9	614·2 ± 78·9	61·1 ± 17·4
24	398·3 ± 27·0	18·9 ± 3·9	673·4 ± 85·7	92·3 ± 28·0

The concentrations are reported as 10^{-9} mol g^{-1}. All values are mean (±S.E., $n = 4$).

The body of observations suggests that exogenous nitrate stimulates nitrate reduction to nitrite in the first hours of anoxia. This reduction, fairly low in absolute value, was appreciably greater than the reduction carried out by rice and wheat roots anaerobically incubated in the absence of nitrate.

Pyridine nucleotide pools

The levels of NAD^+ and NADH in rice and wheat roots were considered in order to assess whether the anaerobic conversion of nitrate to nitrite carried out by these roots modified the pyridine nucleotide pools.

In Fig. 1 the concentrations of pyridine nucleotides present in rice and wheat roots incubated in the presence or absence of nitrate are reported. The effect of

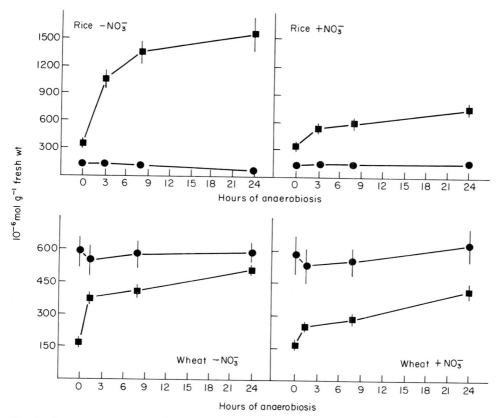

FIG. 1. Concentrations of NAD$^+$ and NADH in root tips of rice and wheat anaerobically grown in the presence or absence of exogenous nitrate. All values are mean ($\pm$ S.E., $n = 3$). ($\bullet$), NAD$^+$; ($\blacksquare$), NADH.

exogenous nitrate was fairly similar for both species considered. In excised roots incubated in nitrogen-free medium there was, in the first hours of anaerobiosis, a rapid and dramatic increase in NADH concentration. After this initial phase, the level showed only a slight further increase as the treatment proceeded. A decrease was observed in NAD$^+$ concentration. When nitrate was supplied to the growth medium, the dramatic increase of NADH was considerably diminished (Fig. 1). However, the increase in NADH concentration observed between the eighth and twenty-fourth hour of anaerobiosis was similar for roots fed or not fed with nitrate.

The level of NAD$^+$ was not affected by the presence of nitrate in wheat roots, while the nitrate-fed roots of rice showed a concentration of this nucleotide slightly higher than that present in roots incubated in the absence of nitrate.

In both species the great increase in NADH concentration could not be interpreted as a simple reduction of oxidized nucleotides because the pool of NAD$^+$ showed only small variation in the transfer from air to nitrogen. As a consequence of this, the total pool of oxidized and reduced pyridine nucleotides increased during the anaerobic treatment. However, since at least in rice the degradation of pyridine nucleotide phosphates (NADPH and NADP$^+$) does not contribute in a significant way to the anaerobic increase of pyridine nucleotides (Reggiani, Brambilla & Bertani 1985), a net synthesis of pyridine coenzyme must be invoked to explain this behaviour.

Appreciable differences in the ratio NAD$^+$/NADH were observed between rice and wheat roots. At present, although we still do not have a convincing explanation for this, it is possible that broad variations of this ratio depending on vegetative growth or species occur, as already described by Mukherji, Dey & Sircar (1968) and Quebedeaux (1981). Consequently, as shown in Fig. 2, the catabolic reducing charge was very different between the two species. The effect of anaerobiosis and

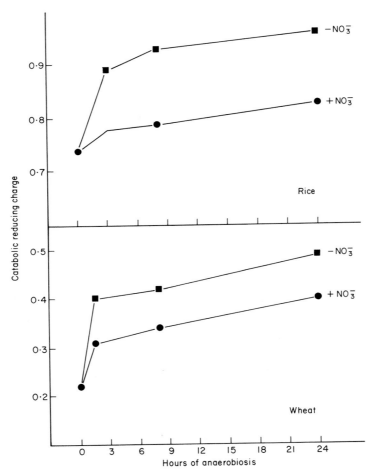

FIG. 2. Levels of catabolic reducing charge in excised roots of rice and wheat subjected to anaerobiosis in the presence or absence of exogenous nitrate. Each value is the mean of three determinations. The coefficient of variation was less than ± 10%. (●), with nitrate; (■), without nitrate.

exogenous nitrate on this parameter was instead similar in rice and wheat root tips. In particular as shown in Fig. 2, the presence of exogenous nitrate during the anaerobic treatment lowered the catabolic reducing charge. Such difference established in the first hours of anoxia was then maintained for all the treatment.

CO_2 evolution and adenylate energy charge

As shown in Fig. 3, the external nitrate stimulated in roots of rice and wheat the rate of CO_2 evolution in comparison with the rate of CO_2 production in roots grown on nitrogen-free medium. The entity of the effect was similar for both the species examined. Experiments carried out with both plants in the absence of external

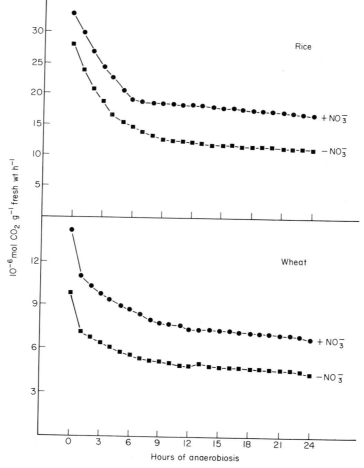

FIG. 3. Mean ($n = 3$) rate of CO_2 production from root tips of rice and wheat grown anaerobically in the presence or absence of exogenous nitrate. The coefficient of variation was less than $\pm 10\%$. Symbols as in Fig. 2.

TABLE 4. Effect of anaerobiosis and exogenous nitrate on the levels of adenylate energy charge in root tips of rice and wheat

Hours of anaerobiosis	Rice		Wheat	
	With NO_3^-	Without NO_3^-	With NO_3^-	Without NO_3^-
0	0·85	0·85	0·91	0·91
1·5			0·60	0·55
3	0·48	0·38		
8	0·41	0·34	0·48	0·43
24	0·37	0·30	0·31	0·31

All values are calculated from three different experiments. The coefficient of variation was less than ±10%.

sucrose revealed that the amount of CO_2 produced declined rapidly and became negligible after a few hours of anoxia (data not shown). This observation leads to the conclusion that the CO_2 produced by rice and wheat roots subjected to anaerobiosis is essentially due to carbohydrate breakdown. Since CO_2 is one of the main end-products of alcoholic fermentation and alcoholic fermentation is the main process for producing energy under anoxia, the adenylate energy charge was also considered.

In the rice roots, in the transfer from air to nitrogen, the adenylate energy charge dropped rapidly from 0·85 to 0·38–0·48 (Table 4). After this initial phase, the values showed only a slight further decrease as the treatment proceeded. In nitrate-fed rice roots a constantly higher level of adenylate energy charge was observed at all times during anaerobiosis. In the wheat roots, the adenylate energy charge dropped from the value of 0·91 in air to 0·55–0·60 at 1·5 h of anoxia (Table 4). As the anaerobic treatment proceeded the adenylate energy charge values constantly decreased. In nitrate-fed wheat roots a higher level of adenylate energy charge was observed only in the first 8 h of anaerobiosis.

DISCUSSION

Like other plant tissues (Townsend 1970; Jarwozski 1971; Klepper, Flesher & Hageman 1971; Kende, Hahn & Kays 1972; Streeter & Bosler 1972; Ferrari, Yoder & Filner 1973), rice and wheat roots subjected to anaerobiosis are able to reduce nitrate to nitrite (Table 3). As described by Ferrari, Yoder & Filner (1973), this reduction was, however, limited in time. These results extend the idea that nitrate, at least in the first hours of anoxia, plays an important role in plant tissues as an oxidizing agent, maintaining the oxido-reductive state at a more oxidized level. In this study, the catabolic reducing charge was used as an indicator of the oxido-reductive state. The data reported in Fig. 2 confirm the above idea and the fact that the reduction of nitrate, requiring NADH, is a powerful means of limiting the formation of a reduced environment in the cell in conditions of anoxia.

Previous works (Garcia-Novo & Crawford 1973; Crawford 1982), have suggested a possible effect of nitrate reduction on the metabolism of plant tissues subjected to anaerobiosis. However, the results reported in Fig. 3 and Table 4 are the first clear evidence of interaction between nitrate reduction and anaerobic metabolism. We have shown that in rice and wheat root tips exogenous nitrate stimulates the CO_2 evolution and maintains the adenylate energy charge at a higher level than in root tips not supplied with nitrate. Although a clear metabolic linkage between catabolic reducing charge, carbohydrate consumption and adenylate energy charge is difficult to establish, we suggest that the lower level of the catabolic reducing charge is a condition, in the absence of oxygen, which favours the fermentative activity of the roots. As a consequence of the greater consumption of sugars, there would be a greater availability of energy-rich compounds in roots anaerobically grown on nitrate. However, this hypothesis, which is attractive, calls for further verification.

The beneficial effect of exogenous nitrate on survival of wheat roots in anoxia (Tables 1 and 2) involved, in our experimental conditions, a greater consumption of sugars. The culture medium of the roots contained a high level of carbohydrates (2% sucrose) and phenomena of sugars starvation were thus avoided. However, in other conditions in which the availability of carbohydrates is limited (field condition), would the presence of nitrate affect the cell viability?

Waterlogging experiments carried out on young barley and wheat plants seem to indicate a positive role of nitrate in the alleviation of flooding injury (Drew, Sisworo & Saker 1979; Trought & Drew 1981). However, to clarify this question and, in general, the role of nitrate in the survival of intact roots in anoxia, further investigations are in course.

REFERENCES

App, A. A. & Meiss, A. N. (1958). Effect of aeration on rice alcohol dehydrogenase. *Archives of Biochemistry and Biophysics*, 77, 181–190.

Arnon, D. I. (1937). Ammonium and nitrate nitrogen nutrition of barley and rice at different seasons in relation to hydrogen-ion concentrations, manganese, copper and oxygen supplied. *Soil Science*, 44, 91–121.

Bertani, A. & Brambilla, I. (1982). Effect of decreasing oxygen concentration on some aspects of protein and amino-acid metabolism in rice roots. *Zeitschrift für Pflanzenphysiologie Band*, 107, 193–200.

Bertani, A., Brambilla, I. & Menegus, F. (1981). Effect of anaerobiosis on carbohydrate content in rice roots. *Biochemie Physiologie Pflänzen*, 176, 835–840.

Carver, K. A. & Walker, D. A. (1983). The effect of adenylate kinase activity in a reconstituted chloroplast system; determination of ATP, ADP and AMP levels with luciferin-luciferase. *Science Tools*, 30, 1–5.

Crawford, R. M. M. (1982). Physiological responses to flooding. *Encyclopedia of Plant Physiology Vol. 12B* (Ed. by O. L. Lange, P. S. Nobel, C. B. Osmond & H. Ziegler) pp. 453–477. Springer-Verlag, Berlin, Heidelberg and New York.

Davies, D. D. (1973). Control of and by pH. *Symposium of the Society of Experimental Biology*, 27, 513–529.

Drew, M. C., Sisworo, E. J. & Saker, L. R. (1979). Alleviation of waterlogging damage to young barley plants by application of nitrate and a synthetic cytokinin, and a comparison between the effects of waterlogging, nitrogen deficiency and root excision. *New Phytologist*, 82, 315–329.

Ferrari, T. E. & Varner, J. E. (1971). Intact tissue assay for nitrate reductase in barley aleurone layers. *Plant Physiology*, 47, 790–794.

Ferrari, T. E., Yoder, O. C. & Filner, P. (1973). Anaerobic nitrite production by plant cells and tissues: evidence for two nitrate pools. *Plant Physiology*, **51**, 423–431.

Garcia-Novo, F. & Crawford, R. M. M. (1973). Soil aeration, nitrate reduction and flooding tolerance in higher plants. *New Phytologist*, **72**, 1031–1039.

Hageman, R. H. & Flesher, D. (1960). The effect of anaerobic environment on the activity of alcohol dehydrogenase and other enzymes of corn seedlings. *Archives of Biochemistry and Biophysics*, **87**, 203–209.

Hageman, R. H. & Hucklesby, D. P. (1971). Nitrate reductase from higher plants. *Methods in Enzymology Vol. 23* (Ed. by S. P. Colowick & N. O. Kaplan) pp. 491–503. Academic Press, New York and London.

Jarwozski, E. G. (1971). Nitrate reductase assay in intact plant tissues. *Biochemical and Biophysical Research Communications*, **43**, 1274–1279.

Kende, H., Hahn, H. & Kays, S. E. (1972). Enhancement of nitrate reductase activity by benzyladenine in *Angrostemma githago*. *Plant Physiology*, **48**, 702–706.

Klepper, L., Flesher, D. & Hageman, R. H. (1971). Generation of reduced nicotinamide adenine nucleotide for nitrate reduction in green leaves. *Plant Physiology*, **48**, 580–590.

Lee, R. B. (1978). Inorganic nitrogen metabolism in barley roots under poorly aerated conditions. *Journal of Experimental Botany*, **110**, 693–708.

Malavolta, E. (1954). Studies on the nitrogenous nutrition of rice. *Plant Physiology*, **29**, 98–99.

Mukherji, S., Dey, B. & Sircar, S. M. (1968). Changes in nicotinic acid content and its nucleotides derivatives of rice and wheat seeds during germination. *Physiologia Plantarum*, **21**, 360–368.

Nance, J. F. (1948). The role of oxygen in nitrate assimilation by wheat roots. *American Journal of Botany*, **35**, 602–660.

Quebedeaux, B. (1981). Adenylate and nicotinamide nucleotides in developing soybean seeds during seed-fill. *Plant Physiology*, **68**, 23–27.

Reggiani, R., Brambilla, I. & Bertani, A. (1985). Effect of exogenous nitrate on anaerobic metabolism in excised rice roots I. Nitrate reduction and pyridine nucleotide pools. *Journal of Experimental Botany*, **36**, 1193–1199.

Roberts, J. K. M., Callis, J., Wemmer, D., Walbot, V. & Jardetzky, O. (1984). Mechanism of cytoplasmic pH regulation in hypoxic maize root tips and its role in survival under hypoxia. *Proceedings of the National Academy of Science, U.S.A.*, **81**, 3379–3383.

Streeter, J. G. & Bosler, M. E. (1972). Comparison of *in vitro* and *in vivo* assay for nitrate reductase in soybean leaves. *Plant Physiology*, **49**, 448–450.

Townsend, L. R. (1970). Effect of form of N and pH on nitrate reductase activity in lowbush blueberry leaves and roots. *Canadian Journal of Plant Science*, **50**, 603–605.

Trought, M. C. T. & Drew, M. C. (1981). Alleviation of injury to young wheat plants in anaerobic solution cultures in relation to the supply of nitrate and other inorganic nutrients. *Journal of Experimental Botany*, **32**, 509–522.

The effects of anoxia on the ultrastructure of pea roots

D. D. DAVIES, P. KENWORTHY AND B. MOCQUOT*

School of Biological Sciences, University of East Anglia, Norwich, NR4 7TJ

K. ROBERTS

John Innes Institute, Colney Lane, Norwich, NR4 7UH

SUMMARY

Pea seedlings were made anaerobic and the ultrastructure of the root apex examined by electron microscopy. Sections showed the presence of whorls of endoplasmic reticulum, formed in response to the anoxic conditions. Sectioning in different planes showed that the reticulum was arranged in a series of concentric spheres. The condensation of the reticulum occurs when the energy charge of the root tip falls below 0·6 due to the lack of oxygen or to the presence of uncouplers of oxidative phosphorylation. They can also be induced by centrifuging the root tip. It is proposed that the extended endoplasmic reticulum is an unstable energy-requiring structure, whereas the concentric spheres of reticulum represent a stable low-energy structure. The consequences of sequestering ribosomes within the spheres is briefly discussed in relation to the biochemical adaptations to flooding.

INTRODUCTION

Some of the fundamental concepts of ecology and physiology have a common origin. Thus the Theory of Tolerance (Good 1953) and the Law of Limiting Factors (Blackman 1905) are derivatives of Leibig's Law of the Limit (1843). These theories and Laws were not rigorously derived, they are based on arguments of analogy and contain an element of the obvious. If a plant cannot tolerate anoxic conditions it is unlikely that its geographical distribution will include marshes! Similarly, if the rate of photosynthesis is limited by light intensity, it is unlikely that increasing the CO_2 concentration will greatly increase photosynthesis. These self-evident examples conceal the over-simplification which is inherent in the concepts. Blackman's Law of Limiting Factors requires that the flux through a system be controlled by the pace of the slowest reaction—the pacemaker or master reaction. The flaws in this argument have been pointed out by several authors, but none more forcibly than by Kacser who has provided a rigorous treatment of the thesis that rate control is a system property—all reactions contributing to the overall flux (Kacser & Burns 1973). We do not have an equally rigorous treatment of the Theory of Tolerance, but we can qualitatively recognize that many

*Present address: Station de physiologie Végétale, INRA, Pont-de-la-Maye Bordeaux, France.

factors contribute to the overall tolerance. Investigators of flooding tolerance have tended to concentrate their studies on a single strategy of adaptation. Such single-mindedness has produced a penetrating analysis of the adaptations. However, it is unlikely that plants have adapted to their environment in such a single-minded way. Anatomical features can facilitate the diffusion of oxygen into roots (Armstrong 1979) or along the surface of leaves (Raskin & Kende 1983) and hence ameliorate the anoxic conditions. However, the oxygen demand of the growing root apex is so high that under anoxic conditions a switch to anaerobic metabolism is inevitable.

THE PASTEUR EFFECT

The conservation principle, variously defined as 'oxygen inhibits fermentation' or 'the conservation of carbon by oxygen' was first enunciated by Pasteur (1861) and in 1926 was termed the Pasteur effect by Warburg. In the absence of a Pasteur effect the ratio CO_2 produced in N_2/CO_2 produced in air would be 0·33 for a plant whose fermentation is represented by the equation.

$$\text{Glucose} \longrightarrow 2 \text{ ethanol} + 2 \text{ } CO_2$$

The measurement of the Pasteur effect is notoriously difficult: data suggesting the absence of a Pasteur effect in maize, sweet peas and buckwheat was reported by Leach (1936) although subsequent work has established the existence of the Pasteur effect in maize and sweet peas but not in buckwheat (Effer & Ranson 1967). Again, a large Pasteur effect was reported in rice by Taylor (1942) but a very small value was reported by Crawford (1967). Some of these discrepancies are undoubtedly due to technical problems but additionally there is evidence that glucose metabolism in the absence of oxygen involves reactions other than glycolysis, e.g. the pentose phosphate pathway (Kennedy, Rumpho & Vander Zee 1983) and in yet other tissues, lactate may be an end product of glycolysis either in short periods of anoxia (Davies, Grego & Kenworthy 1974) and over long periods of anoxia (Hanson & Jacobsen 1984). It is thus clear that the interpretation of the ratio CO_2 in N_2/CO_2 in air requires a detailed knowledge of the underlying biochemistry.

THE BIOCHEMISTRY OF THE ANAEROBIC RESPONSE

Glycolysis was one of the first metabolic pathways to be elucidated. In mammals glycolysis produces lactic acid and in yeast, ethanol. In plants it has been argued (Davies, Grego & Kenworthy 1974) that because pyruvic decarboxylase has an acid pH optimum, it cannot function at neutral or alkaline pH so that before the cell can decarboxylate pyruvate and produce ethanol, the pH of the cytoplasm must become acid. It was suggested that under anaerobic conditions lactic acid accumulates and lowers the pH of the cytosol to a point at which pyruvic decarboxylase is active and so allows the formation of ethanol. This sequence of

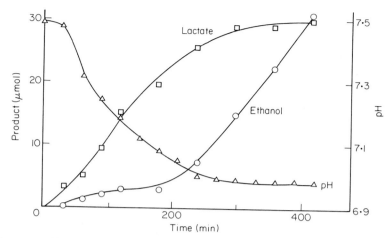

FIG. 1. Sequence of formation of lactate and ethanol in relation to pH during glycolysis by an extract of pea seeds (after Davies, Grego & Kenworthy 1974).

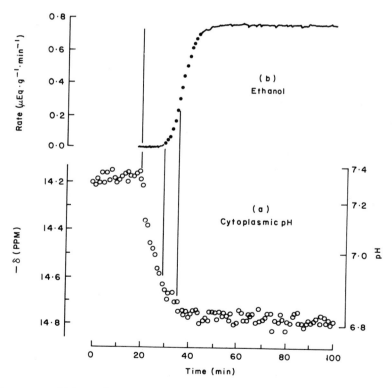

FIG. 2. Time course of (a) cytoplasmic pH and (b) rate of ethanol production in perfused maize root tips during hypoxia, determined *in vivo* (after Roberts *et al.* 1984).

events, which is depicted in Fig. 1, has now been supported by N.M.R. studies (Roberts *et al*. 1984) and an example of the evidence obtained is shown in Fig. 2. The cessation of lactate production could be due to the operation of an ATP controlled pH-stat (Davies 1973). At pH values above 7 lactate dehydrogenase is not inhibited to any significant extent by ATP, but as the pH falls below 7 the enzyme becomes increasingly sensitive to ATP. Recently it has been shown (Asker & Davies 1984), that the five isoenzymes of lactate dehydrogenase respond differently to ATP with isoenzyme 4 being particularly sensitive to ATP at low pH. These reactions are minor variants of the well established reactions of the Embden-Meyerhof pathway of glycolysis, which is probably the major metabolic pathway involved in the anaerobic response. The pathway produces 2 moles of ATP per mole of glucose compared with 36 moles of ATP per mole of glucose during oxidation to CO_2. Hence to meet the cells' demands for ATP, an efficient transport of glucose is required, and when this supply fails, the energy charge of the cell drops and this can lead to cell death. Whilst the Embden-Meyerhof pathway is probably the major pathway in roots, we should not close our minds to the possibility of other metabolic pathways being involved and data showing that large amounts of $^{14}CO_2$ can be anaerobically produced from glucose-1-^{14}C by some species, requires an alternative pathway.

GENE ACTIVATION AND THE ANAEROBIC RESPONSE

During the first hour or so of anoxia there is a cessation of protein synthesis which is in association with a near complete dissociation of polysomes, although the mRNA remains translatable for at least 5 h (Sachs, Freeling & Okimoto 1980). After the first hour or so of anoxia, the cells synthesize a group of transition polypeptides; their size is about 30 000 but their identity is unknown. After about 2 h, the cells synthesize a group of twenty or so anaerobic polypeptides, including the products of the Aahl 1 and AAh 2 genes (Sachs & Freeling 1978) and also glucose phosphateisomerase (Kelly & Freeling 1984). This pattern of response has been demonstrated in roots from a number of species. The synthesis of similar, but not identical, anaerobic polypeptides has been demonstrated in rice embryos (Mocquot *et al*. 1981) and *E. coli* (Smith & Niedhardt 1983).

ULTRASTRUCTURE CHANGES ASSOCIATED WITH
THE ANAEROBIC RESPONSE

Previous studies (Cooke, Roberts & Davies 1980) have suggested that stress, perhaps acting via hormones, affects the permeability of certain membranes, particularly the tonoplast, and allows the vacuolar proteolytic enzymes to interact with cytoplasmic proteins. We therefore examined the effect of anoxia on pea root tip cells, using two systems: (i) excised root-tips maintained under anaerobic conditions in the presence of glucose (0·2 M) as described by Saglio, Raymond & Pradet (1980) and (ii) intact pea seedlings with primary roots 1–2 cm in length

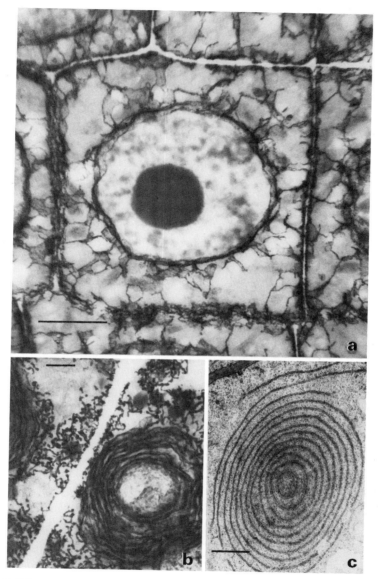

FIG. 3. (a) Cell from pea root meristem, stained with osmic acid soak method to reveal the normal arrangement of endoplasmic reticulum (ER). This 1 μm thick section, with no post staining, shows radial and peripheral cisternae (bar = 5 μm). (b) A thick section of a similar cell to (a) but from a root tip following 6 h anoxia. Tubular ER is seen, while the rough ER has formed concentric whorls (bar = 1 μm). (c) Anoxia-induced whorls of ER seen in a conventionally fixed and stained thin section (bar = 1 μm).

which were transferred to anaerobic jars. No obvious changes were observed in the tonoplast membrane. We then paid particular attention to the structure of the cortical cells up to 1 mm from the junction of the root cap and the meristem. The major changes which we have observed occur in the endoplasmic reticulum, although other studies have demonstrated changes in the fine structure of mitochondria (Vartapetian, Andreeva & Kozlova 1976). The endoplasmic reticulum in aerobic cells adopts an extended configuration, revealed much more clearly in thick (1 μm) sections of roots which had been specifically stained for endoplasmic reticulum using an osmic acid soak technique. The bulk of the cisternal endoplasmic reticulum is arranged close to, and parallel to, the cell wall together with radially arranged sheets centred on the cell nucleus (Fig. 3a). These sheets are attached to, and intermingled with, elaborate clusters of tubular endoplasmic reticulum. Clear effects of anoxia on the arrangement of the endoplasmic reticulum can be detected after 15 min, with cisternal endoplasmic reticulum becoming arranged in layered stacks near the cell periphery. After 1 h, many cells were in the characteristic final anoxic state in which almost all of the cisternal endoplasmic reticulum is arranged in large concentric spheres, with sometimes as many as a dozen layers (Fig. 3c). Similar effects have been reported in tomato roots by Morrisset (1983). Ribosomes are still attached to these membranes, but surface views reveal that they are not part of organized polysomes. The tubular endoplasmic reticulum is not so dramatically affected. After 6 h the effect is dramatic but after much longer times (16 h) the cells are beginning to show signs of damage and senescence. Roots that were placed back for 1 h in aerobic conditions, after 6 h anoxia, show a recovery of the normal structural arrangement of endoplasmic reticulum. The anoxia-induced rearrangements of endoplasmic reticulum are therefore both rapid and reversible. To eliminate the possibility that these arrangements were in some way a toxic symptom induced by common by-products of anoxia, we subjected aerobic roots for 6 h to 0·2% ethanol (ten times the level reached in anoxic roots). We also examined ethylene-treated roots. Neither showed any trace of abnormality in the arrangement of endoplasmic reticulum.

THE SYNTHESIS OF FATTY ACIDS UNDER ANAEROBIC CONDITIONS

Membrane lipid bilayers are mostly fluid at physiological temperatures. However, bacteria regulate the fluidity of their membrane phospholipds in response to temperature. The mechanism of regulation seems to occur via the incorporation of proportionally more unsaturated fatty acids into membrane lipids as the temperature decreases (de Mendoza & Cronan 1983). In higher plants marked changes in the degree of saturation of fatty acids do not appear to be associated with temperature changes. On the other hand, Harris & James (1969) concluded that oxygen availability is the rate limiting factor for desaturation and Rebeille, Bligny & Douce (1980) showed that the composition of the fatty acids of cells of

TABLE 1. Effect of oxygen on the fatty acid composition of fat isolated from cells of maple

Fatty acid	Oxygen concentration (μM)				
	12·5	25	37·7	62·5	253
C_{16}	24	24	24	26	27
C_{18}	5	3	1·4	1·5	1·5
$C_{18:1}$	44	28	23	4	3·5
$C_{18:2}$	15	30	33	45	45
$C_{18:3}$	12	15	18	23	23

After Rebeille, Bligny & Douce (1980). Results are presented as a % of total fatty acids.

TABLE 2. Fatty acid composition of rice coleoptiles

Age	Condition	% of total fatty acids							
		C_{12}	C_{14}	C_{16}	$C_{16:1}$	C_{18}	$C_{18:2}$	$C_{18:2}$	$C_{18:3}$
3 days	Air	0·4	0·8	11·6	2·0	2·5	18·2	48·4	16·0
	N_2	1·0	2·6	14·7	1·1	2·4	30·3	44	3·8

After Vartapetian, Mazliak & Lance (1978).

TABLE 3. Labelling of fatty acid in 4-day rice seedling coleoptiles after incubation in [^{14}C]

Incubation	Condition	Total ^{14}C (cpm/g tissue)	% of total ^{14}C in fatty acids								Unknown OH acid
			C_{12}	C_{14}	C_{16}	$C_{16:1}$	C_{18}	$C_{18:1}$	$C_{18:2}$	$C_{18:3}$	
4 h	Air	122 000	—	1·3	27·6	20·2	4·0	46·7	—	—	—
	N_2	633 000	2·6	19·3	49·6	—	10·5	—	—	—	17·9
24 h	Air	22 900	—	4·2	40·9	7·9	5·1	30·4	11·4	—	—
	N_2	1 189 000	22·5	28·3	37·2	—	12·0	—	—	—	—

After Vartapetian, Mazliak & Lance (1978).

the sugar maple (*Acer pseudo-platanus*) was affected by oxygen concentration. At oxygen concentrations below 60 μM the molar proportion of oleate increased whilst linoleate decreased (Table 1). Aeration resulted in a rapid formation of linoleate at the expense of oleate. However, it has been reported that the lipid composition of rice seedlings grown in anaerobic conditions was similar to that of control plants grown in the presence of oxygen (Costes, Bazier & Vartapetian 1975). The study of lipid biosynthesis in rice seedlings grown under anaerobic conditions produced a paradox (Vartapetian, Mazliak & Lance 1978). On the one hand, the fatty acid composition proved to be independent of the oxygen concentration (Table 2), suggesting that desaturation can occur in the absence of oxygen. On the other hand, the labelling of fatty acids from acetate ^{14}C, failed to detect the synthesis of any unsaturated fatty acids in the absence of oxygen (Table 3).

We have therefore examined the fatty acid composition of the membranes and the synthesis of fatty acids in pea roots to determine if the observed changes in the endoplasmic reticulum were correlated with changes in fatty acids.

TABLE 4. Effect of a 5 h incubation in air or nitrogen on the fatty acid composition of membranes isolated from pea root tips

Fraction	Condition	% of total fatty acids				
		C_{16}	C_{18}	$C_{18:1}$	$C_{18:2}$	$C_{18:3} + C_{20:1}$
4	Air	26·3	3·1	2·7	58·0	9·9
	N_2	23·8	2·9	3·2	61·1	9·0
3	Air	24·6	3·1	2·6	61·1	8·6
	N_2	25·3	3·1	3·2	59·8	8·6
2	Air	21·7	4·2	3·6	62·0	8·5
	N_2	23·4	4·1	3·6	61·0	7·9

Seeds were germinated for 86 h before the roots were excised.

Pea seeds were germinated for 86 h at 25°C and the root tips (3 cm) excised. Batches of root tips (100) were incubated for 5 h in the culture medium of Saglio & Pradet (1980) containing glucose (0·2 M) and [^{14}C]-acetate (3 μCi ml^{-1}) or [^{3}H]-acetate (15 μCi ml^{-1}) in nitrogen or air. Membrane fractions were isolated from the roots, essentially as described by Owens & Northcote (1981), fixed in boiling isopropanol and after esterification the fatty acids were analysed by gas–liquid chromatography as described by Mazliak *et al*. (1972).

The detailed analysis of these experiments will be published elsewhere but the main conclusions are apparent from the data presented in Tables 4 and 5.

Table 4 shows that over a 5 h period of anoxia there is no major change in the fatty acid composition of the pea root membranes including the endoplasmic reticulum, suggesting that the change in organization of the endoplasmic reticulum is unlikely to be due to gross changes in fatty acid composition. The incorporation of acetate into fatty acids, including the unsaturated fatty acids, occurs under nitrogen or air, albeit with a much reduced incorporation into the unsaturated acids (Table 5). The effect of nitrogen on fatty acid metabolism is thus to increase the synthesis of stearic acid at the expense of oleic and linoleic acids. It could be argued that this shift in balance between the newly formed saturated and unsaturated fatty acids might be sufficient to cause a lipid phase transition which is sometimes called the order–disorder transition. We are inclined not to accept this causal relationship

TABLE 5. Labelling of fatty acids in 86 h pea roots, after incubation in [^{14}C] or [^{3}H]-acetate for 5 h after excision (air + [^{3}H]-acetate 15 μCi ml^{-1}; nitrogen + [^{14}C]-acetate 3 μCi ml^{-1}).

Fraction	Condition	Radioactivity (cpm)	% of radioactivity in fatty acids				
			C_{16}	C_{18}	$C_{18:1}$	$C_{18:2}$	$C_{18:3}$
4	Air [^{3}H]	221 000	49·3	8·7	14·6	25·0	2·4
	N_2 [^{14}C]	47 500	53·1	32·5	3·1	8·5	2·8
3	Air [^{3}H]	253 000	48·9	7·7	12·5	29·8	1·1
	N_2 [^{14}C]	40 000	54·4	30·2	3·5	10·4	1·4
2	Air [^{3}H]	38 700	48·3	8·1	13·1	28·0	2·5
	N_2 [^{14}C]	6 300	53·7	30·2	5·0	7·6	3·5

and suggest that since oxygen is required for desaturation, changes in the ratio saturated/desaturated fatty acids are likely to reflect the availability of oxygen, but not to be germane to changes in organization of the endoplasmic reticulum. It is of interest to note that Hetherington, Hunter & Crawford (1982) have reported marked changes in the ratio saturated/desaturated fatty acids between rhizomes of *Iris pseudoacorns* kept in air or nitrogen, whereas there was little change in the ratio between rhizomes of *Iris germanica* kept in air or nitrogen. Further work is needed on the possible link between tolerance of anoxia and lipid composition.

THE EFFECT OF ANOXIA ON THE ENERGY CHARGE
OF PEA ROOTS

We have measured the changes in amounts of adenine nucleotides which occur when peas are kept in atmospheres containing varying amounts of oxygen. Some of our results are shown in Fig. 4 and a complete analysis for the 6 h period is presented in Table 6. We then attempted to relate the formation of the whorls of endoplasmic reticulum to the energy status of the cells. After a 6 h incubation at oxygen concentrations above 3% we could not detect whorls. At concentrations of oxygen below 2% whorls were present in most or all of the cells at the root tip. The transition corresponds to an energy charge of about 0·6. We then examined the effects of the uncoupler of oxidative phosphorylation, FCCP (M-fluoro-carbonylcyamide-phenylhydrazone), on the energy charge and on the forma-

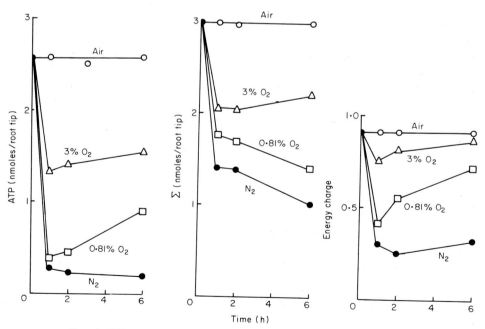

FIG. 4. Effect of anoxia on the level of adenine nucleotides in pea roots.

TABLE 6. Adenine nucleotides present in root tips of pea seedlings exposed to atmospheres containing various concentrations of oxygen for 6 h

(nmoles/root tip)

Atmosphere	ATP	ADP	AMP	ATP/ADP	Σ	E.C.
N_2	0·18	0·25	0·52	0·73	0·96	0·33
0·81% O_2	0·92	0·41	0·20	2·24	1·53	0·73
3% O_2	1·78	0·50	0·03	3·56	2·32	0·88
Air	2·45	0·55	Trace	4·45	3·00	0·91Est.

Root tips (1 cm) were excised after fixation in $-100°C$ ether as described by Raymond & Pradet (1980).

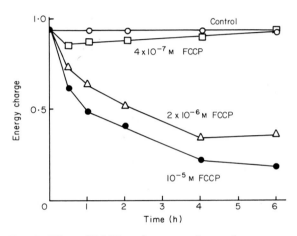

FIG. 5. Effect of FCCP on the energy charge of pea roots.

tion of whorls. The effect of FCCP on energy charge is shown in Fig. 5. At concentrations of FCCP greater than 10^{-6} M whorls are formed in the presence of air (Fig. 6c) whereas at concentrations less than 5×10^{-7} M whorls are not formed. Once again the critical energy charge for the formation of whorls appears to be about 0·6–0·7.

CONCLUSIONS

We propose that the extended form of the endoplasmic reticulum which is found in aerobic conditions is an unstable structure whose maintenance requires energy, so that when the supply of ATP is reduced by anoxia, or other means, the reticulum collapses to form a compact whorl which is a low energy state. Further support for our view that the whorls are formed by self-assembly to a low energy state is provided by the demonstration (Fig. 6 a and b) that whorls of endoplasmic reticulum are formed during centrifugation. It should be noted that precautions were taken to ensure that the roots did not go anaerobic during the centrifugation.

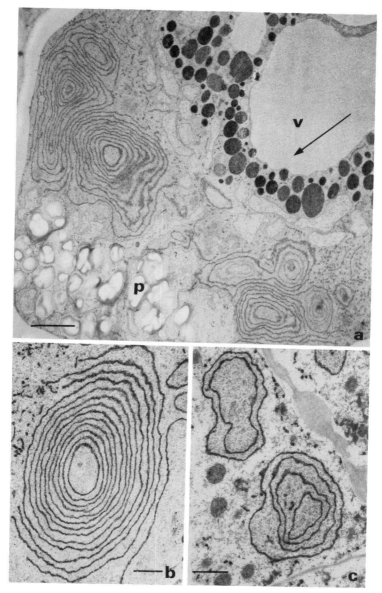

Fig. 6. (a) Thin section of a root meristem cell from a pea seedling which had been centrifuged for 6 h at 5000 g. The organelles are stratified between the vacuole (v) and the starch grain filled plastids (p). The rough ER has formed clear concentric whorls. The arrow indicates the direction of centrifugal force (bar = 2 μm). (b) A whorl of ER from a centrifuged root tip cell as in (a). The ER is stained by the osmic method (bar = 1 μm). (c) Whorls of ER found in cells of root tip cells treated for 6 h with 2×10^{-5} M FCCP (bar = 1 μm).

The reversal of the process is an energy-requiring step. When oxygen is introduced into a nitrogen atmosphere, there is a very rapid (2 min) increase in the energy charge and within 15–20 min the endoplasmic reticulum reverts to the extended form.

We do not wish to argue that it is the energy charge *per se* which controls the state of the reticulum, but rather the overall supply of energy in the form of ATP.

We have been concerned with the response of the pea root to anoxia over relatively short periods of time. The time scale of these changes in energy charge corresponds to the time scale of the gene activation involved in the anaerobic response. It is tempting to speculate that the formation of whorls of endoplasmic reticulum removes ribosomes, which could otherwise be involved in general protein synthesis. The free ribosomes are then involved in the selective translation of the m-RNAs corresponding to the anaerobic genes. Having been tempted by the speculation, it remains to test the proposal.

REFERENCES

Armstrong, W. (1979). Aeration in higher plants. *Adv. Botanical Research*, 7, 225–332.

Asker, H. & Davies, D. D. (1984). The physiological role of the isoenzymes of lactate dehydrogenase in potatoes. *Planta*, 161, 272–280.

Blackman, F. F. (1905). Optima and limiting factors. *Annals of Botany*, 19, 281–295.

Cooke, R. J., Roberts, K. & Davies, D. D. (1980). The mechanism of stress-induced protein degradation in *Lemna minor*. *Plant Physiology*, 66, 1119–1122.

Cooke, R. J., Grego, S., Roberts, K. & Davies, D. D. (1980). The mechanism of deuterium oxide-induced protein degradation in *Lemna minor*. *Planta*, 148, 374–380.

Costes, C., Bazier, R. & Vartapetian, B. B. (1975). Abstracts: XII International Botanical Congress, Leningrad, p. 353.

Crawford, R. M. M. (1967). Alcohol dehydrogenase activity in relation to flooding tolerance in roots. *Journal of Experimental Botany*, 18, 458–464.

Davies, D. D. (1973). Control of and by pH. *S.E.B. Symposium* 27, 513–529. Cambridge University Press.

Davies, D. D., Grego, S. & Kenworthy, P. (1974). The control of the production of lactate and ethanol by higher plants. *Planta*, 118, 297–310.

de Mendoza, D. & Cronan, J. E. (1983). Thermal regulation of membrane lipid fluidity in bacteria. *Trends in Biochemical Science*, 8, 49–56.

Effer, W. R. & Ranson, S. L. (1967). Respiratory metabolism in buckwheat seedlings. *Plant Physiology*, 42, 1042–1045.

Good, R. (1953). *The Geography of the Flowering Plant* 2nd ed. Longmans, Green & Co., London.

Hanson, A. D. & Jacobsen, J. V. (1984). Control of lactate dehydrogenase, lactate glycolysis and α amylase by O_2 deficit in barley aleurone layers. *Plant Physiology*, 75, 566–572.

Harris, P. & James, A. T. (1969). Effect of low temperature on fatty acid biosynthesis in seeds. *Biophysica et Biochimica Acta*, 187, 13–18.

Hetherington, A. M., Hunter, M. I. A. S. & Crawford, R. M. M. (1982). Contrasting effects of anoxia on rhizome lipids in *Iris* species. *Phytochemistry*, 21, 1275–1278.

Kacser, H. & Burns, J. A. (1973). The control of flux. *S.E.B. Symposium*, 27, 65–104. Cambridge University Press.

Kelly, P. M. & Freeling, M. (1984). Anaerobic expression of maize glucose phosphate isomerase. *Journal of Biological Chemistry*, 259, 673–677.

Kennedy, R. A., Rumpho, M. E. & Vander Zee, D. (1983). Germination of *Echinochloa crus-gali* seeds under anaerobic conditions. *Plant Physiology*, 72, 787–794.

Leach, W. (1936). Researches on plant respiration. *Proceedings of the Royal Society of London Series B*, **119**, 507.

Leibig, J. (1843). *Chemistry in its Application to Agriculture & Physiology.* 3rd Edn, Petersen, Philadelphia.

Mazliak, P., Oursel, A., Abdelkader, A. B. et Grosbois, M. (1972). Biosynthese des acides gras dans les mitochondries vegetales isolees. *European Journal of Biochemistry*, **28**, 399–411.

Mocquot, B., Prat, C., Mouches, C. & Pradet, A. (1981). Effect of anoxia on energy charge and protein synthesis in rice embryo. *Plant Physiology*, **68**, 636–640.

Morrisset, C. (1983). Effects of energetic shortage upon the ultrastructure of some organelles, in excised roots of *Lycopersicon esculentium* cultivation *in vitro*. *Cytologia*, **48**, 349–361.

Owens, R. J. & Northcote, D. H. (1981). The location of arabinosyl: hydroxyproline transferase in the membrane system of potato tissue culture cells. *Biochemical Journal*, **195**, 661–667.

Pasteur, L. (1861). Experiences et vues nouvelles sur la nature des fermentations. *Comptes Rendus*, **52**, 1260–1264.

Raskin, I. & Kende, H. (1983). How does deep water rice solve its aeration problem? *Plant Physiology*, **72**, 447–454.

Raymond, P. & Pradet, A. (1980). Stabilization of adenine nucleotide ratios at various values by an oxygen limitation of respiration in germinating lettuce (*Lactuca sativa*) seeds. *Biochemical Journal*, **190**, 39–44.

Rebeille, F., Bligny, B. & Douce, R. (1980). Role de l'oxygene et de la temperature sur la composition en acides gras des cellules isolées d'erable. *Biophysica et Biochimica Acta*, **620**, 1–9.

Roberts, J. K. M., Callis, J., Weinmer, D., Walbot, V. & Jardetzky, O. (1984). The mechanism of cytoplasmic pH regulation in hypoxic maize root tips and its role in survival under hypoxia. *Proceedings of the National Academy of Sciences, USA*, **81**, 3379–3383.

Sachs, M. M. & Freeling, M. (1978). Selective synthesis of alcohol dehydrogenase during anaerobic treatment of maize. *Mol. Gen. Genet.* **16**, 111–115.

Sachs, M. M., Freeling, M. & Okimoto, R. (1980). The anaerobic proteins of maize. *Cell*, **20**, 761–767.

Saglio, P. H. & Pradet, A. (1980). Soluble sugars, respiration and energy charge during aging of excised maize root tips. *Plant Physiology*, **66**, 516–519.

Saglio, P. H., Raymond, P. & Pradet, A. (1980). Metabolic activity and energy charge of excised maize root tips under anoxia. *Plant Physiology*, **66**, 1053–1057.

Smith, M. W. & Niedhardt, F. C. (1983). Proteins induced by anaerobiosis in *Escherichia coli*. *Journal of Bacteriology*, **154**, 336–343.

Taylor, D. L. (1942). Influence of oxygen tension on respiration, fermentation and growth in wheat and rice. *American Journal of Botany*, **29**, 721–738.

Vartapetian, B. B., Mazliak, P. & Lance, C. (1978). Lipid biosynthesis in rice coleoptiles grown in the presence or in the absence of oxygen. *Plant Science letters*, **13**, 321–328.

Vartapetian, B. B., Andreeva, I. N. & Kozlova, G. I. (1976). The resistance to anoxia and the mitochondrial fine structure of rice seedlings. *Protoplasma*, **88**, 215–224.

Warburg, O. (1926). Uber die Wirkung von Blausaureathylester (Athylcarbylamin) auf die Pasteursche Reaktion. *Biochem. Z.* **172**, 432–441.

C. Whole Plant Responses to Oxygen Stress

Metabolic responses of four families of loblolly pine to two flood regimes

DONAL D. HOOK AND STEWART DENSLOW

Department of Forestry, Clemson University, Clemson, South Carolina 29407, U.S.A.

SUMMARY

Comparisons of growth, root morphology and root metabolism were made between three open pollinated families (F_1, F_2 and F_3) from wet sites and a seed lot that included several genetically improved coastal plain families (G_1) of loblolly pine (*Pinus taeda* L.) grown under two flooded regimes. Although the genetically improved families were taller at planting time than the three wet-site families, their total height exceeded the shortest family on only two measurement dates and mean relative height growth-rate did not vary significantly among the families at any measurement date. The wet-site families showed a trend of increasing mean relative height growth-rate late in the growing season whereas G_1 levelled off or continued to decrease, but the values were not significant by measurement dates. On the other hand, metabolic responses, as measured by CO_2 production, ethanol and malate accumulation, and alcohol dehydrogenase activity, appeared to segregate family responses to flood regime fairly distinctly. Therefore use of anaerobic metabolism measurements seems to have potential for screening for flood-tolerant genotypes in loblolly pine but further research is needed to clarify relationships between metabolism and growth responses. Differences in anaerobic metabolism responses among the families between the cyclic and continuously flooded treatments point out the weakness of utilizing only flooded and unflooded treatments in evaluating adaptive traits of plants.

INTRODUCTION

Loblolly pine (*Pinus taeda* L.) is most prevalent on mesophytic sites throughout its range, but in the Atlantic and Gulf coastal plains individual trees frequently survive and grow well in shallow swamps, sloughs, and other marginal wet sites. Consequently a general belief has developed among some who are familiar with the species that one or more wet-site genotypes of loblolly pine exists. However, scientific evidence to support this belief is lacking. Williams & Bridgwater (1981), in comparing deep peat and coastal plain families of loblolly pine, found only a small response to water regimes and that except for number of lateral roots, morphological traits and phytomass were statistically nonsignificant. Topa (1984) found that a wet-site loblolly pine family outgrew a very dry site family under waterlogged conditions, had significantly greater oxidizing capacity around the root system and took up more phosphorus than did the dry-site family. Goddard, Zobel & Hollis (1976) found fertilizer × family interactions for loblolly pine families were usually insignificant, but the responses were more variable on wet sites. Usually on wet sites deficient in phosphorus the progeny that did poorest without fertilizers responded the most when it was applied. Also, they found one genotype to be generally responsive to fertilizers. Jahromi (1971) found evidence that some slash pine (*Pinus elliottii*) genotypes were very responsive to the addition of P and N on soils severely deficient in phosphorus.

Flood-tolerant species appear to have specific metabolic and physiological mechanisms that enable them to survive under root anaerobiosis (Crawford 1978; Davies 1980), so some of these traits have been used to screen for genetic variation within and between species for flood tolerance. Experiments with sweet potatoes showed variations in ethanol accumulation and alcohol dehydrogenase (ADH) activity between cultivars during anaerobic and post-anaerobic treatments (Chang, Hammett & Pharr 1982). The flood-tolerant cultivar showed higher ADH activities during post-anaerobic treatments but lower levels of ethanol accumulation than the less tolerant cultivars. Bromegrass and maize genotypes and *Senecio* species tolerant to wet sites had lower ADH activities than intolerant genotypes or species under waterlogging stress (Marshall, Brove & Pryor 1973; Brown, Marshall & Munday 1976; Crawford 1978). Nelson *et al.* (1983) had some success in screening *Phaseolus vulgaris* L. genotypes for flood tolerance by combining several physiological measurements and one visual method.

Roots of loblolly pine seedlings are known to produce ethanol and malate under anaerobic conditions in relation to age of seedlings, length of flooding treatment, and phosphorus application (Hook *et al.* 1983; DeBell *et al.* 1984). The species, therefore, expresses metabolic adaptation under flooding stress that may be of value in screening for flood-tolerant genotypes.

This paper reports an exploratory study of the effects on growth and metabolic traits of soil flooding and applied phosphorus in wind-pollinated progeny (families) from three parent trees growing on wet sites and a seed lot including several genetically improved coastal plain families of loblolly pine.

METHOD

Seed sources

During the fall of 1982, seeds were collected from headwater swamps in South Carolina. Two trees (F_2 and F_3) were selected in Reserve Swamp on Hobcaw Barony near Georgetown, South Carolina and one tree (F_1) in Blue Bird Swamp on the Francis Marion National Forest in Berkeley County, South Carolina. Because 1982 was a poor seed year on the wet sites in coastal South Carolina, no opportunity existed to select for the best phenotype; one Reserve Swamp tree was average in growth and form, the other was poor in both traits. and the Blue Bird Swamp phenotype was of poor form and exhibited only moderate growth. The seed lot of genetically improved families was obtained from Westvaco Seed Orchard near Summerville, South Carolina and contained germplasm from several parents from the lower Atlantic coastal plain. It is referred to herein as family G_1 for convenience.

After 45 days stratification, the seeds were placed in styrofoam trays containing 128 cavities filled with a 1:1:1 mixture of soil, vermiculite and peat. Trays were placed in an environmentally controlled room with 16 h light and a 29–21°C day/night temperature range. Seed germinated in late January and early February 1983. After germination, the trays were transferred to a greenhouse with a minimum temperature of 20°C. Seedlings were watered daily, drenched twice weekly with a half strength Hoagland's nutrient solution, sprayed periodically with Captan to prevent damping off and planted in the soil tanks on March 23–29, 1983.

Soil and water control facilities

The soil tanks (see Hook *et al.* 1970) were filled to 1·5 m deep with B2 horizon sandy loam taken from a Goldsboro series (Aquic Paleudult, fine loamy siliceous thermic). This was covered with 30 cm of A1 horizon soil collected from the Bethera series (Typic Paleaquult, clayey mixed thermic). Both soils are inherently low in phosphorus and were obtained from the lower coastal plain of a Pleistocene terrace.

Each of the six tanks (1·83 m × 1·83 m × 1·83 m) was divided in half, perpendicular to the flood reservoirs (which are located on opposite sides at the growth tanks) by 1·9 cm thick plywood to a depth of 106 cm. Three tanks were kept flooded to 0–2 cm deep (F) from April 28 to October 1983 by a float valve using deep well-water. Three tanks were flooded the same as the above for 4 weeks and the water table was dropped to 30 cm below the soil surface for 4 weeks (F/D). This cycle was repeated every 8 weeks through October 1983.

One half of each tank was randomly assigned to receive 100 μg kg^{-1} phosphorus in the surface soil. Phosphorus was applied as monocalcium phosphate granules and thoroughly mixed into the surface soil. An equal amount of calcium was applied to the non-fertilized half of each tank as calcium sulphate to ensure that calcium was constant (McKee *et al.* 1984). The study was established with three replications. Fifteen seedlings of each family were randomly assigned a position in each tank and were planted in three rows of five seedlings each with a 15 cm spacing parallel to the flood reservoirs.

Seedling height was measured after planting in March and every 4 weeks through October 1983. Mean relative height growth-rate was computed for each 4-week period by the equation:

$$\bar{R}_\text{H} = \frac{\ln H_2 - \ln H_1}{T_2 - T_1}$$

Where H is height in cm and T is time in weeks.

During November, seedlings from one of the cyclic treatment tanks and two of the continuously flooded tanks were harvested to obtain phytomass and morphological data. Height, the length of the five longest first-order lateral roots, the length of tap root, the number of live succent root tips, the number of first-order lateral roots at 0–3, 3–6, and 6 cm below the root collar, and fresh and dry weight were measured on each seedling.

Carbon dioxide, ethanol and malate

Harvested seedlings were stored with their roots in water and returned to the laboratory where about one-half of the active root tips were excised from each seedling or group of seedlings for this assay and the remainder were used for ADH assays. Group sampling was used for the treatments without phosphorus and continuously flooded treatments in order to obtain enough experimental materials. The excised tips from each seedling or group of seedlings were incubated separately in 15 ml flasks in a respirometer under nitrogen for 2 h to determine periodic CO_2 production. Ethanol and malate accumulation after 2 h were determined as in Hook & Brown (1973) and Gutmann & Wahlefeld (1974) with minor modifications.

Alcohol dehydrogenase extraction and activity

Chemicals

Hepes-buffer was purchased from Calbiochem-Behring (La Jolla, CA). Yeast-derived alcohol dehydrogenase, NAD, PVP-40, Tris and Bicine were purchased from Sigma Chemical Co. (St. Louis, MO). All other chemicals were analytical reagent grade. (Abbreviations used: Hepes, N-2-hydroxyethylpiperazine-N-2-ethanesulphonic acid; ADH, alcohol dehydrogenase; Bicine, N,N-bix(2-hydroxyethyl)glycine; NAD, nicotinamide-adenine dinueleotide; PVP-40, poly-vinylpyrrolidone (soluble PVP)).

Enzyme assay

Alcohol dehydrogenase (ADH) activity was measured at room temperature by following the increase in absorbance at 340 nm on a Bausch and Lomb Spectronic 710 spectrophotometer. The activity of a sample was taken as the difference in absorbance ($\times 1000$) between 1 and 2 min after addition of enzyme. The reaction was initiated by the addition of 0·5 ml of tissue extract to the cuvette containing the rest of the reaction mixture (2·5 ml). The final reaction mixture (after enzyme addition) contained 42 mM Bicine (pH 9·0), 100 mM ethanol, 1·2 mM NAD, 5 mM $MgCl_2$, 24 mM mercaptoethanol, 1% glycerol, and 0·625% PVP-40.

Extraction methods

Extraction buffer contained 100 mM Hepes (pH 7·5), 100 mM mercaptoethanol, 1 mM NAD, 5% glycerol, and 2·5% PVP-40. Root tips were excised from the same seedlings as used for CO_2 ethanol, and malate analysis and ADH extraction and analysis were run concurrently.

Excised roots were collected on the day of analysis and 50 to 100 mg lots of tissue were stored immediately in screen-top vials in liquid nitrogen until time for analysis. Each sample in turn was pulverized under liquid nitrogen in a porcelain mortar and pestle in the presence of 25 mg PVP-40. Pulverized tissue was transferred 'dry' to a 2·0 ml volume Duall glass-on-glass tissue grinder (Kontes Glass, Vineland, NJ) and homogenized for 30 s on ice in 1 ml of added extraction buffer. A second 1 ml aliquot of extraction buffer was used to rinse the mortar and was separately homogenized in the Duall grinder. The two homgenates were combined and centrifuged at 20 000 × g, 20 min at 0°C and assayed as above (Denslow & Hook 1986).

ANALYSIS OF DATA

Due to the limited number of actively growing root tips in the continuously flooded treatments without phosphorus, physiological data on root metabolism were too fragmented to analyse and were not included. Also, since periodic height growth showed no interactions between families and phosphorus, only the data from treatments with phosphorus are presented. Periodic height growth was analysed by SAS procedures for a randomized block design. Phytomass, morphology and metabolic data were analysed by the t-test using Satterthwaite's approximations and one-way analysis of variance.

RESULTS

Period height and mean relative height growth rate

The genetically improved family, G_1, was taller at germination and planting time when all treatment data were combined (not shown) but its height in the cyclic (F/D) treatment with phosphorus was significantly different from the three wet-site families only at the 12 and 16 week measurements (Table 1). There were no significant differences in height among the families in the continuously flooded treatment. Mean relative height growth rate ($\bar{R}_H$) for 4-week periods did not vary significantly among the families by measurement period (Table 2), but the trend in the cyclic flooded treatment was for $\bar{R}_H$ of wet-site families to increase and G_1 to decrease late in the season.

Dry weight and root morphology

Dry weights of needles, stem and roots showed somewhat similar patterns to height relationships among the families in the F/D and F treatments (Tables 1 and 3). Generally F_2 was lowest in weight with G_1 and F_1 averaging slightly higher than F_3. Length of the tap root was shorter in the F than in the F/D treatment (Table 3) and lateral root lengths (data not shown) were longer in the F/D than the F

TABLE 1. Mean heights (cm) of seedlings from all subplots with added phosphorus at 4-week intervals

		Cyclic			Flood regimes		Continuous			
Week	G_1	F_1	F_2	F_3	P^*	G_1	F_1	F_2	F_3	P^*
0	6·1	5·5	5·2	5·4	0·15	6·0	5·4	5·2	5·6	0·14
4	8·0	6·8	6·5	6·9	0·10	7·8	7·0	6·8	7·4	0·24
8	14·9	13·2	11·5	12·9	0·08	11·0	9·7	8·9	10·0	0·30
12	20·4[a]†	17·0[b]	15·8[b]	17·4[b]	0·03	13·6	11·5	10·6	11·9	0·18
16	21·4[a]	18·1[b]	16·4[b]	17·6[b]	0·02	14·6	11·6	10·9	12·2	0·16
20	22·4	19·4	17·5	18·5	0·08	14·9	11·3	10·7	11·9	0·12
24	23·4	20·8	18·9	19·8	0·24	14·8	11·3	11·1	12·0	0·13

* Probability of a greater F in a one-way analysis of variance.
† Means from sources in rows with the same letters are not significantly different ($P = 0.05$).

TABLE 2. Mean relative height growth rate ($\bar{R}_H$) by 4-week intervals and for the 24-week period*

Period		Cyclically flooded		
		Family		
Week	G_1	F_1	F_2	F_3
0–4	0·066	0·055	0·056	0·062
4–8	0·156	0·166	0·140	0·157
8–12	0·079	0·063	0·079	0·057
12–16	0·012	0·016	0·009	0·003
16–20	0·011	0·017	0·016	0·013
20–24	0·011	0·017	0·019	0·018
0–24	0·056	0·056	0·053	0·054

* Although $\bar{R}_H$ did not vary significantly among sources ($P = 0.05$) for any single measurement period, the trend for the Families to decrease at 12–16 weeks and increase thereafter appears to differ from the trend of G_1 which decreased or levelled off.

treatment but no family differences were evident. The number of live root tips below 6 cm soil depth ranged from 1·5 for F_1 to 5·4 for F_3 in the F/D treatment and from 0·7 to 1·6 in the F treatment (Table 3). The harvested data showed overall growth patterns similar to the periodic height growth data from all tanks combined. Large effects of flooding on dry weight and root morphology, except the number of lateral roots at 0–3 and 3–6 cm, were evident. Family differences were not pronounced but G_1 generally had the largest dry weight and F_2 had the least dry weight. Family$_2$ appeared to have as long or longer tap roots and as many or more live root tips deep in the soil than the other families (Table 3).

Metabolism

Patterns of CO_2 production and ADH activity of the four families fell into two categories. Family$_1$ and G_1 showed significant differences (t-test using Satterthwaite's approximation) in the values of these variables between flooding regimes; carbon dioxide was significantly higher for F_1 and G_1 ($P = 0.039$ and

TABLE 3. Dry weight components of seedlings and root morphology measurements by treatment

Treatment	Component	G_1	F_1	F_2	F_3
			Dry weight (g)		
Cyclically flooded	Needle	4·64 ± 1·36	4·38 ± 1·83	2·69 ± 1·28	4·00 ± 1·20
	Stem	2·52 ± 1·04	2·88 ± 1·61	1·41 ± 0·69	2·44 ± 0·76
	Root	2·48 ± 1·15	2·22 ± 0·78	1·40 ± 0·42	2·17 ± 0·77
Continuously flooded	Needle	0·84 ± 0·35	0·85 ± 0·95	0·51 ± 0·16	0·90 ± 0·27
	Stem	0·51 ± 0·16	0·42 ± 0·24	0·31 ± 0·09	0·51 ± 0·17
	Root	0·58 ± 0·28	0·51 ± 0·47	0·39 ± 0·14	0·55 ± 0·21
			Morphology		
Cyclically flooded	Tap root length (cm)	17·9 ± 4·2	16·6 ± 6·0	20·0 ± 4·7	19·8 ± 5·8
	Number live root tips	3·4 ± 3·4	1·5 ± 1·3	4·6 ± 4·0	5·4 ± 4·2
Continuously flooded	Tap root length (cm)	7·9 ± 1·3	7·2 ± 2·3	8·2 ± 1·3	8·7 ± 1·8
	Number live root tips	1·5 ± 1·7	0·7 ± 1·3	1·3 ± 1·8	1·6 ± 1·8

Means are from fifteen seedlings per replication fron the cyclically and two replications from the continuously flooded treatment. Standard deviations are shown following the mean.

0·024) respectively, and ADH activity was significantly lower for F_1 and G_1 ($P = 0.0001$ and 0.0209) respectively than F_2 and F_3 in the F/D as compared to the F treatment (Fig. 1). By the same tests, F_2 and F_3 showed no significant differences in ADH activity or CO_2 production between flooding levels or families. One-way analysis of variance of the four families at each flooding level showed a family effect for both variables. Carbon dioxide showed a family effect under continuous flooding ($P = 0.048$) and ADH under alternate flooding ($P = 0.0063$) and continuous flooding ($P = 0.0148$).

Malate levels were significantly higher ($P = 0.044$, t-test, Satterthwaite's approximation) in the continuously flooded condition than the cyclic flooding treatment when all data were combined. This difference showed up only for F_2 when using data from any single family. Analysis of variance of the families under either flooding schedule showed no significance.

Ethanol production values under continuous flooding did not vary significantly among families. Under alternate flooding there was a significant family effect ($P = 0.0015$, one-way analysis of variance) with F_1 and G_1 most similar, F_3 above them and F_2 below. Only F_3 approached significance in the difference between ethanol production levels under the different flooding levels ($P = 0.08$, t-test, Satterthwaite's approximation.)

DISCUSSION

The fact that the genetically improved family was taller than the wet-site families at germination and at planting probably accounts in part for its greater height during a

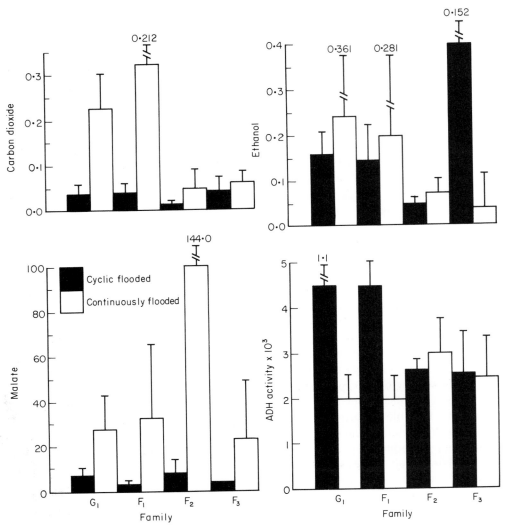

FIG. 1. Carbon dioxide, ethanol and malate in excised roots after 2 h incubation in prepurified nitrogen gas. ADH activity on extracts from live root tips excised and plunged into liquid nitrogen. CO_2 in μmol mg^{-1} freshweight (FW) h^{-1}. Ethanol in μmol mg^{-1} FW h^{-1}. Malate in nmol mg^{-1} FW h^{-1}. ADH = units per mg FW. 'T' bars represent standard deviations about the mean. Each mean contains three to five observations except for F_3 in the F/D treatment for malate where only one observation was available. Means for the cyclic flooded treatment (F/D) were from the roots of individual seedlings whereas the means from continuously flooded treatments (F) were pooled from groups of seedlings.

portion of the experiment. However, its mean relative height growth-rate during the first 16 weeks indicated that it performed as well in the flooded treatments early in the growing season as did the wet-site families. Later in the season, from 16th to the 24th week, the wet-site families seemed to outperform it. Since the study was conducted for only one growing season it is not possible to determine whether these apparent shifts in $\bar{R}_H$ late in the season were due to seasonal, long-term adaptations, or experimental error.

An inverse relationship between ethanol concentration and ADH activity has been reported for the roots of cultivars of sweet potatoes with different flood-tolerances under anaerobic and post-anaerobic conditions (Chang, Hammett & Pharr 1982). Higher ADH and pyruvate decarboxylase activities and lower ethanol concentrations were associated with a flood-tolerant cultivar and the reverse with flood-sensitive cultivars. During exposure to 100% CO_2 atmosphere and upon return to air, the flood-sensitive cultivars produced more ethanol and leaked more electrolytes than the flood-tolerant cultivar (Chang, Hammett & Pharr 1983). Although our ethanol and ADH data do not show clear reverse relationships between ADH activity and ethanol the same trends were apparent (Fig. 1) as reported by Chang, Hammett & Pharr (1982), but varied by genotype and whether the roots had been flooded continuously or intermittently.

Plants moderately tolerant to flooding may show increased alcoholic fermentation as a result of anaerobiosis, i.e. CO_2 and ethanol production increase (Crawford 1978). The roots measured in this experiment showed only a partial response in this direction when the two flood regimes were compared. Family$_1$ and G_1 maintained a constant ethanol production between the two flooding regimes, but changed CO_2 production and ADH activity. This pattern may reflect CO_2 production changes due to variation in the activity of the pentose phosphate pathway or due to changes in the amount of pyruvate decarboxylated by pyruvate dehydrogenase as a first step toward α-ketoglutarate and amino acid biosynthesis (DeBell et al. 1984). Increased malate in equilibrium with oxaloacetate to feed into this section of the Krebs cycle could be part of a mechanism such as the latter.

Family$_2$ had the shortest initial height of any family and its average height remained shorter throughout the experiment. Yet, its height was never significantly less than F_1 and F_3 and its $\bar{R}_H$ was equal to or greater than F_1 and F_3 in the F treatment. Also, F_2 was the only family when compared individually that showed an increase in malate accumulation in the F over the F/D treatment. This response is interesting in that the parent of F was located on the wettest site of the three wet-site family parents.

Family$_3$ had a growth pattern similar to F_1 and G_1 but had a different physiological pattern. No change was apparent in CO_2 production and ADH activity between flooding levels but ethanol production in the F/D versus the F treatment (Fig. 1) increased. As mentioned above the CO_2 produced by the pyruvate decarboxylase reaction may be partially consumed in the production of oxaloacetate and malate from phosphoenolpyruvate, so it could account in part for the lack of increased CO_2 production in the F/D in conjuction with the higher level of ethanol.

Continuous flooding effects as compared to cyclic flooding were expressed across all families by greater malate accumulation. Hence, continuous anaerobic conditions in the root environment appeared to stimulate alternate pathways, as described above, for carbohydrate breakdown in this species.

Although G_1, and F_1 to a lesser degree, appeared to be superior in growth and to have similar physiological responses, it seems premature to judge either or both to be better adapted to the two flood regimes for several reasons. First, more distinct differences in growth than occurred seem to be needed to declare a difference. Second, the apparent increases in $\bar{R}_H$ by all three wet-site families late in the season and a decline in $\bar{R}_H$ for G_1 during the same period, suggests that more time under these conditions is needed for adaptations to be fully expressed. Third, the relatively long tap roots, large number of live root tips below 6 cm soil depth in both treatments, and large accumulation of malate in the continuously flooded treatment by F_2 (the family with poorest growth) raises a caution flag against interpreting the relative merits of the adaptation mechanisms at this time.

The large variation among the families in physiological responses between the two flood regimes without corresponding growth responses was not anticipated. Much of the literature on flood tolerance is based on experiments that contrast plant responses between flooded and unflooded conditions (Hook & Crawford 1978, Kozlowski 1984) but little information is available about physiological responses among different flood regimes. DeBell *et al.* (1984) showed that CO_2 production and the accumulation of ethanol and malate in the roots of loblolly pine seedlings from seasonally flooded treatments were equal to or greater than drained treatments but less than continuously flooded treatments. Their data were collected early in the spring after seedlings in the seasonal and continuously flooded treatment had over-wintered (Nov.–April) under flooded conditions. In our experiment, the data were collected during a normal 4-week flooding phase of the F/D treatment in November. Keeley (1979) also showed variations in physiological responses of swamp tupelo seedlings in relation to their parent site flood regime and length of flooding. In summary, it can be concluded that the type of flood regime may have a major influence on the physiological response of a species and therefore interpreting flood tolerance mechanisms by contrasting only flooded and unflooded treatments may in some cases be misleading. This would be particularly so if the flood treatments from one experiment to another result in different hypoxic and reducing stresses on the plant. Some of the disagreements among researchers on flood tolerance mechanism may revolve around this problem. For instance, Smith, Kalsi & Woolhouse (1984) concluded that their work on alder was not consistent with Crawfords's 'metabolic theory' of flood tolerance, yet all we know about the prior stress under which the alder grew is that they came from marshes or a farm. In the future it would be most beneficial if researchers specified the degree of soil reduction by redox potential measurements or at least provided some information on length, time, and depth of flooding regimes as well as soil descriptions.

The data herein suggests that physiological responses to flooding may be detected before growth responses become clearly evident and that more than one physiological parameter will be needed for screening. This is consistent with the conclusions of Chang, Hammett & Pharr (1983) that no single biochemical parameter was suitable for screening for flood tolerance among sweet potato cultivars. The apparent sensitivity of physiological measurements for detecting differences in loblolly pine responses to different flood regimes is encouraging.

The large variation in data within treatments was probably due as much to genetic differences within family as to experimental error. More homogeneous genetic material would be desirable for segregating families but since loblolly pine is difficult to propagate by rooting or tissue culture it will be some time before a clonal bank of desirable stock can be developed to make appropriate test. Until such time, longer-term experiments of 2 years or more and the use of more diverse genetic stock (using stock from dry and wet sites and from improved stock bred for wet sites when available) should provide useful information for breeders and other biotechnological decisions.

ACKNOWLEDGMENTS

This research was supported in part by a cooperative grant from the USDA, Forest Service, Southeastern Forest Experimental Station Asheville, North Carolina. The assistance of William H. McKee jr. in planning and Martha McKevlin and Martha Singletary in harvesting the experiment was most helpful. We also thank Westvaco Corporation, Summerville, South Carolina, for providing the seed lot for family G_1.

REFERENCES

Brown, A. D., Marshall, D. R. & Munday, J. (1976). Adaptedness of variants of an alcohol dehydrogenase locu in *Bromus mollis* L. (Soft Bromegrass). *Australian Journal of Biological Science*, **29**, 389–396.

Chang, L. A., Hammett, L. K. & Pharr, P. M. (1982). Ethanol, alcohol dehydrogenase, and pyruvate decarboxylase in storage roots of four sweet potato cultivars during simulated flood-damage and storage. *Journal of American Society of Horticulture Science*, **107**, 674–677.

Chang, L. A., Hammett, L. K. & Pharr, P. M. (1983). Internal gas atmospheres, ethanol, and leakage of electolytes from a flood-tolerant and a flood-susceptible sweet potato cultivar as influenced by anaerobiosis. *Canadian Journal of Botany*, **61**, 3399–3404.

Crawford, R. M. M. (1978). Metabolic adaptations to anoxia. *Plant Life in Anaerobic Environments* (Ed. by D. D. Hook & R. M. M. Crawford), pp. 119–136. Ann Arbor Science, Ann Arbor, Michigan.

Davies, D. D. (1980). Anaerobic metabolism and the production of organic acids. *The Biochemistry of Plants* (Ed. by P. K. Stumph & E. E. Conn). *Vol 2. Metabolism and Respiration* (Ed. by D. D. Davies), pp. 581–611. Academic Press. New York.

DeBell, D. S., Hook, D. D., McKee, W. H., Jr. & Askew, J. L. (1984). Growth and physiology of loblolly pine roots under various water table level and phosphorus treatments. *Forest Science*, **30**, 705–714.

Denslow, S. & Hook, D. D. (1986). Extraction of alcohol dehydrogenase from fresh root tips of loblolly pine. *Canadian Journal of Forestry Research*, **16**, 146–48.

Goddard, R. E., Zobel, B. J. & Hollis, C. A. (1976). Responses of *Pinus taeda* and *Pinus elliottii* to varied nutrition. *Tree Physiology and Yield Improvement* (Ed. by M. G. R. Cannell & F. T. Last), pp. 449–462. Academic Press, New York.

Gutmann, I. & Wahlefeld, A. W. (1974). L-(–)-malate determination with malate dehydrogenase and NAD. *Methods of Enzymatic Analysis* (Ed. by H. U. Bergmeyer), 2nd edn, Vol. 3, pp.1585–1589. Academic Press, New York.

Hook, D. D., Langdon, O. G., Stubbs, J. & Brown, C. L. (1970). Effect of water regimes on the survival, growth, and morphology of tupelo seedlings. *Forest Science*, **16**, 304–311.

Hook, D. D. & Brown, C. L. (1973). Root adaptations and relative flood tolerance of five hardwood species. *Forest Science,* **19**, 225–229.

Hook, D. D. & R. M. M. Crawford (Eds) (1978). *Plant Life in Anaerobic Environments.* Ann Arbor Science, Ann Arbor, Michigan.

Hook, D. D., DeBell, D. S., McKee, W. H., Jr. & Askew, J. L. (1983). Responses of loblolly pine (mesophyte) and swamp tupelo (hydrophte) seedlings to soil flooding and phosphorus. *Plant and Soil,* **71**, 388–394.

Jahromi, S. (1971). *Genetic variation in nutrient absorption in slash pine.* Ph.D. thesis, University of Florida, Gainesville, Florida.

Keeley, J. (1979). Population differentiation along a flood frequency gradient: physiological adaptations to flooding in *Nyssa sylvatica. Ecological Monograph,* **49**, 89–108.

Kozlowski, T. T. (Ed.) (1984). *Flooding and Plant Growth.* Academic Press, New York.

Marshall, D. R., Brove, P. & Pryor, A. J. (1973). Adaptive significance of alcohol dehydrogenase isozymes in maize. *Nature New Biology,* **244**, 16–17.

McKee, W. H., Jr., Hook, D. D., DeBell, D. S. & Askew, J. L. (1984). Growth and nutrient status of loblolly pine seedlings in relation to flooding and phosphorus. *Soil Science Society of America Journal,* **48(6)**, 1438–1442.

Nelson, R. B., Davis, D. W., Palta, J. P. & Lang, D. R. (1983). Measurements of soil waterlogging tolerance in *Phaseolus vulgaris* L.: a comparison of screening techniques. *Scientia Horticulturae,* **20**, 303–313.

Smith, A. M., Kalsi, G. & Woolhouse, H. W. (1984). Products of fermentation in the roots of alder (*Alnus* Mill). *Planta,* **160**, 272–275.

Topa, M. A. (1984). *Effects of anaerobic growth conditions on biomass accumulation, root morphology, and efficiencies of nutrient uptake and utilization in seedlings of some southern coastal plain pine species.* Ph.D. thesis, Duke University, Durham, North Carolina, U.S.A.

Williams, C. & Bridgwater, F. (1981). Screening loblolly pine for adaptability to deep peat site: a seedling study of two edaphic seed sources from eastern North Carolina. Proceedings of the 16th Southern Forestry Tree Improvement Conference, Blacksburg, Virginia. May 26–28, 1981.

Ethanol, acetaldehyde, ethylene release and ACC concentration of rhizomes from marsh plants under normoxia, hypoxia and anoxia

C. STUDER AND R. BRAENDLE

Institut für Pflanzenphysiologie, Universität Bern, Altenbergrain 21, Bern, CH 3013, Switzerland

SUMMARY

Rhizomes of *Acorus calamus*, *Glyceria maxima*, *Phragmites australis*, *Schoenoplectus lacustris* and *Typha latifolia* were incubated for 72 h in a closed system initially containing 21% O_2, 1% O_2 or N_2. Gas samples from the head space were analysed daily by GC. ACC and M-ACC contents of the tissues were determined at the end of the incubation period. The rhizomes of all species produced ethanol and acetaldehyde under anoxia, to a lesser extent under hypoxia, and in *S. lacustris* even under normoxia. *Acorus calamus* and *G. maxima* showed a different time-course of ethanol release compared to *P. australis*, *S. lacustris* and *T. latifolia*. The differences were probably due to either non-limited (*A. calamus* and *G. maxima*) or limited fermentation processes. Ethylene release depended on O_2 availability and the species. The highest production rate under normoxia was detected in *A. calamus* (565 pM g^{-1} fw.d). During the first day, ethylene production rates were frequently higher than later on, because of additional ethylene induced by wounding. *Schoenoplectus lacustris* and *T. latifolia* produced little ethylene after the period of wound ethylene release (*S. lacustris* 22 pM g^{-1} fw.d and *T. latifolia* 7 pM g^{-1} fw.d). The rhizomes of all species accumulated ACC under an O_2 deficit stress. M-ACC concentrations were much higher than the ACC levels, independently of the incubation conditions. An exception was *A. calamus* with low M-ACC levels. The M-ACC concentrations in the various treatments were not significantly different, but once again with the exception of *A. calamus*.

INTRODUCTION

The emersed shoots of most of our wetland plants, being rhizome geophytes, die back periodically. The rhizomes, storage organs as well as bud-bearing organs, survive flooding and regenerate new shoots easily (Braendle 1985). Rhizomes are, therefore, of special interest for studies of adaptation mechanisms to stress from flooding or O_2 deficiency (Braendle 1980).

In the whole plant a well-formed aerenchyma normally allows efficient O_2 transport from the shoots to the rhizomes, avoiding O_2 deficit stress (Armstrong 1979; Studer & Braendle 1984). In case of a blockage of O_2 transport the internal

293

O_2 concentration of the rhizomes drops sharply, as the aerenchymatous tissues do not function as an O_2 reservoir (Crawford 1982).

Rhizomes of a number of wetland plants have been shown to endure strict anoxia for a long time. They even partly form new shoots (Barclay & Crawford 1982). The main source of energy in rhizomes under anoxia is alcoholic fermentation (Braendle 1983; Monk, Crawford & Braendle 1984). Thus a big stock of fermentable compounds is necessary (Sale & Wetzel 1983; Steinmann & Braendle 1984). Rhizomes of flood-tolerant species did not accumulate high amounts of ethanol when gassed with N_2 in an open system (Monk, Crawford & Braendle 1984). The ability to eliminate ethanol easily may be the explanation. However, nothing can be concluded concerning the method of release and the total production of ethanol by rhizomes in an open system. Investigation of the head space atmosphere by GC may overcome this problem. At the same time it allows the detection of other important volatile compounds, e.g. acetaldehyde and ethylene (Kimmerer & Kozlowski 1981).

Ethylene is able to stimulate shoot extension in wetland species (Malone & Ridge 1983; Kende, Metraux & Rastin 1984). Moreover it is implicated in the aerenchyma formation (Jackson & Drew 1984). Ethylene is formed from its precursor 1-amino-cyclopropane-1-carboxylic acid (ACC, Adams & Yang 1979). The step from ACC to ethylene requires O_2 (Yang 1984). Another pathway for use of ACC is malonylation to malonyl-ACC. This reduces the level of free ACC available for ethylene formation (Amrhein *et al.* 1981; Kionka & Amrhein 1984). An increased ethylene release and an accumulation of M-ACC was reported as an answer to different stresses (Kimmerer & Kozlowski 1981; Liu, Hoffman & Yang 1983; Fuhrer & Fuhrer-Fries 1985). An enhanced CO_2 concentration, especially appearing in flooded rhizomes (Laing 1940; Steinmann & Braendle 1981), may act as a positive effector of ethylene formation (Yang 1984). Hence it is important to know to what extent and under what conditions the investigated rhizomes produce ethylene and/or accumulate ACC or M-ACC.

MATERIALS AND METHODS

Rhizomes of *Acorus calamus*, *Glyceria maxima*, *Phragmites australis*, *Schoenoplectus lacustris* and *Typha latifolia* were collected during summer and early autumn from our freeland cultures (Studer & Braendle 1984;. The roots were cut off and pieces of rhizome were surface-sterilized in NaOCl (2%) and chloramphenicol (50 μg ml^{-1}) for 15 min. The incubation containers (infusion flasks of 150 ml) were filled with 10 ml of slightly acidified water, containing 50 μg ml^{-1} chloramphenicol and a few glass balls, to avoid direct contact between water and rhizomes. The containers were closed and made airtight with a rubber septum and were connected with U-tubes, which allowed pressure compensation before taking the samples. At the beginning of the experiments the containers were flushed with air, 1%/99% O_2 or N_2 for 1 h. All experiments were carried out at 25°C. At the end the rhizomes were deep-frozen for further analysis.

For the detection of ethanol, acetaldehyde and ethylene gas, samples of 250 μl were withdrawn with a gas-tight syringe through the rubber septum. The samples were injected in a GC (Perkin-Elmer, Dual FID, SIGMA 300, Poropak Q 100–120 M, oven 130°C, N_2 carrier gas flow 30 ml min^{-1}). The GC response was compared with calibration curves (Kimmerer & Kozlowski 1981).

The deep-frozen rhizomes were pulverized at -18°C to detect ACC (1-amino-cyclopropane-1-carboxylic acid) and Malonyl-ACC. The powder was homogenized in 80% ethanol (v/v) and extracted according to Bufler *et al.* (1980) for 16 h at 4°C. The resulting extract was centrifuged at 10 000 *g* for 10 min. The pellet was resuspended in 80% ethanol (v/v) and recentrifuged. The combined supernatant fractions were vacuum-dried at 40°C. The remaining residues were dissolved in 2 ml H_2O and centrifuged at 10 000 *g* for 10 min. M-ACC was hydrolysed according to Liu, Hoffman & Yang (1983) with 2 M HCl at 100°C for 10 min. After neutralization with 10 M NaOH, the solution was again centrifuged and assayed for free ACC. Free ACC was determined according to Lizada & Yang (1979), but with 16 μmol $HgCl_2$ in the non-hydrolysed and 8 μmol $HgCl_2$ in the hydrolysed extracts. A mixture of 14% NaOCl and 10 M NaOH (2:1, v/v) was injected to start the ethylene-forming reaction in closed flasks. The injected mixture was 1/10 of the volume of the ACC-containing solution. The GC response was compared with a calibration curve produced from ACC (Sigma, St Louis, U.S.A.).

The O_2 and CO_2 concentrations in the incubation containers were measured according to Steinmann & Braendle (1981). The preparation of the rhizome for SEM photography was carried out as described earlier (Studer & Braendle 1984).

RESULTS

Figure 1 shows the ethanol release of the rhizomes. All species release more ethanol under N_2 than when initially exposed to 1% O_2/99% N_2. Under both

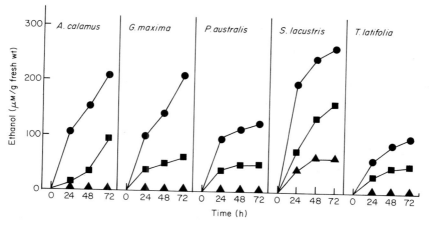

FIG. 1. Ethanol release of rhizomes under normoxia (▲), severe hypoxia (■), and anoxia (●).

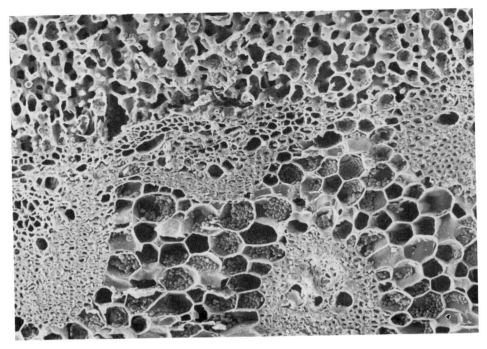

FIG. 2. SEM picture of the rhizome from *Schoenoplectus lacustris* showing the aerenchymatous cortex and the denser vascular cylinder with starch-filled cells. Enlargement ×250.

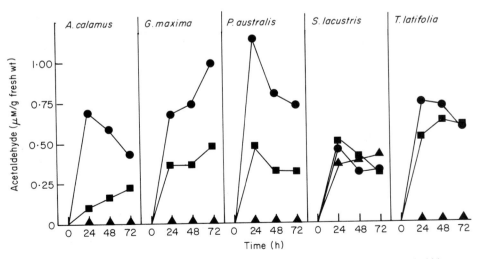

FIG. 3. Acetaldehyde content in the surrounding atmosphere of rhizomes under normoxia (▲), severe hypoxia (■), and anoxia (●).

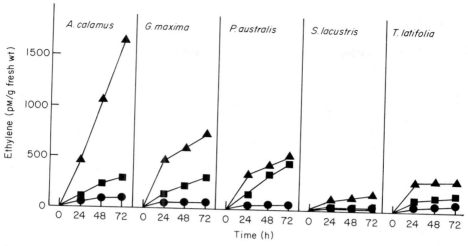

FIG. 4. Ethylene release of rhizomes under normoxia (▲), severe hypoxia (■), and anoxia (●).

treatments the rhizomes of *P. australis, S. lacustris* and *T. latifolia* show a decreasing ethanol production rate during the incubation period. This effect is less distinct in *A. calamus*, and *G. maxima* under the anoxic treatment. *Schoenoplectus lacustris* releases ethanol even under high O_2 concentrations.

This species has a vascular cylinder almost free of air spaces (Fig. 2). O_2 diffusion is hindered and is not sufficient to sustain normal respiration, causing anaerobic zones (Haldemann 1983).

Ethanol release is always accompanied by a release of acetaldehyde (Fig. 3). The concentrations of acetaldehyde in the containers are much lower and tend to achieve an equilibrium depending on re-metabolization and ethanol production rate.

Figure 4 demonstrates ethylene release. It depends on the species investigated and on the conditions applied. Control rhizomes of *A. calamus* produce the most and do not show any decline after the first day as the other species do. Generally, release is highest in the controls, much lower under severe hypoxia and practically nil under anoxia. *Schoenoplectus lacustris* and *T. latifolia* release only a little ethylene after the first day. CO_2, which may favour ethylene release (Osborne 1984), increases in all containers up to the end of the incubation period. The values achieved are: 2–4% CO_2 under anoxia, 2–5% under hypoxia with traces of O_2 and 4–6% CO_2 in the controls with 10–15% O_2.

Figure 5 shows the ACC accumulation after an O_2 deficit stress of 72 h. It is highest in *A. calamus* and correlates as well as in *P. australis* and *T. latifolia* with the ethylene release of the controls. This is not the case in the other species. In comparison with ACC contents, the M-ACC levels are normally much higher (Fig. 6). Only *A. calamus* contains very low amounts of M-ACC. There are no significant differences in the levels of M-ACC between the various treatments. The only

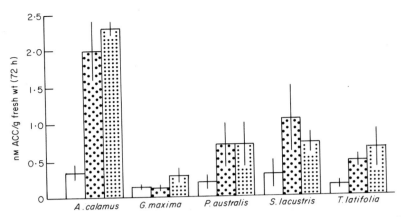

FIG. 5. ACC contents of rhizomes after 72 h of incubation under normoxia (□), severe hypoxia (▣) and anoxia (▦). Mean values of five samples ±S.D.

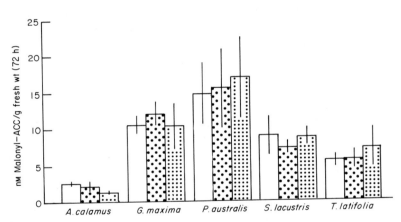

FIG. 6. Malonyl-ACC content of rhizomes after 72 h of incubation under normoxia (□), severe hypoxia (▣) and anoxia (▦). Mean values of five samples ±S.D.

exception is *A. calamus*. The unusual behaviour of this species cannot yet be explained.

DISCUSSION

As well as acetaldehyde, rhizomes of all species investigated release considerable amounts of ethanol under an O_2 deficit stress. The time courses of the release in *P. australis*, *S. lacustris* and *T. latifolia* under anoxia and severe hypoxia decrease with increasing duration of O_2 deficit stress. This may reflect a limitation of the fermentation process (Crawford 1978). The inhibition of the fermentation processes is more pronounced under severe hypoxic conditions, which correspond better to natural conditions (Crawford 1982). The O_2 concentration of 1% in the surrounding atmosphere is too low to sustain respiration for prolonged periods. The limitation is absent in *A. calamus* and, under anoxia only, in *G. maxima*. *Acorus calamus* and, perhaps to a lesser extent, *G. maxima* carry green shoots or part of shoots during the whole year. These species may not develop this kind of limitation, because they are able to release ethanol more easily or they are able to re-metabolize ethanol in the green parts standing in contact with oxygenated water (Cossins 1978).

Ethylene release also differs between the species and depends on O_2 availability. The higher ethylene production at the beginning of the incubation period under normoxia may be due to additional ethylene induced by wounding. An exception is *A. calamus*. Rhizomes of this species are the best ethylene producers and perhaps the wound release is too low to be shown clearly. The rhizomes of this species are not really buried at the natural site and grow very fast. Perhaps a connection exists between this behaviour and the ethylene-enhanced supergrowth (Osborne 1984).

All of the species investigated accumulate ACC under an O_2 deficit stress. But there exist some remarkable differences. The ACC accumulation in *A. calamus*. *P. australis* and *T. latifolia* correlates, as expected, more or less with ethylene release in presence of oxygen. The ACC accumulation is too low in *G. maxima* and too high in *S. lacustris*. This may be an indication of other factors regulating ACC accumulation, besides the limited conversion of ACC to ethylene. An enhanced ACC synthesis under low O_2 concentration (Reid & Bradford 1984) or a regulation of the M-ACC forming enzymes by D-amino acids (Kionka & Amrhein 1984) could be taken into consideration. In this context the high amounts of free ACC and the lower amounts of M-ACC in *A. calamus* may be of interest. To resolve this problem requires further experimentation.

The question arises as to how ethylene is involved in the flooding response in buried and regenerating rhizomes with anaerobic metabolism and high internal CO_2 concentrations. There is evidence that CO_2 promotes growth in amphibious plants (Osborne 1984).

We believe that the critical stage for rhizome survival is the timespan needed to build up a small new shoot under anoxia. *Schoenoplectus lacustris* is able to do this as long as the reserves are sufficient (Steinmann & Braendle 1984). Afterwards ACC is transported eventually from the rhizome into the shoot (Reid & Bradford 1984). When in contact with oxygenated water, growth may be accelerated. The latter response has been observed in rhizomes of *S. lacustris*, regenerating to whole plants.

ACKNOWLEDGMENT

We thank Dr J. C. Rutter for improving the style of the manuscript.

REFERENCES

Adams, D. O. & Yang, S. F. (1979). Ethylene biosynthesis: identification of 1-aminocyclo-propane-1-carboxylic acid as an intermediate in the conversion of methionine to ethylene. *Proceedings of the National Academy of Sciences U.S.A.*, **76**, 170–174.

Amrhein, N., Schneebeck, D., Skorupka, H. & Tophof, S. (1981). Identification of a major metabolite of the ethylene precursor 1-aminocyclopropane-1-carboxylic acid in higher plants. *Naturwissenshaften*, **68**, 619–620.

Armstrong, W. (1979). Aeration in higher plants. *Advances in Botanical Research*, 7, 225–332.

Barclay, A. M. & Crawford, R. M. M. (1982). Plant growth and survival under strict anaerobiosis. *Journal of Experimental Botany*, **134**, 541–549.

Braendle, R. (1980). Die Ueberflutungstoleranz der Seebinse (*Schoenoplectus lacustris* (L.) Palla: II. Uebersicht über die verschiedenen Anpassungsstrategien. *Vierteljahresschrift der Naturforschenden Gesellschaft Zürich*, **125**, 177–185.

Braendle, R. (1983). Evolution der Gärungskapazität in den flut- und anoxiatoleranten Rhizomen von *Phalaris arundinacea*, *Phragmites communis*, *Schoenoplectus lacustris* und *Typha latifolia*. *Botanica Helvetica*, **93**, 39–45.

Braendle, R. (1985). Kohlenhydrategehalte und Vitalität isolierter Rhizome von *Phragmites australis*, *Schoenoplectus lacustris* und *Typha latifolia* nach mehrwöchigem O_2-Mangelstress. *Flora*, **177**, 317–321.

Bufler, G., Mor, Y., Reid, M. S. & Yang, S. F. (1980). Changes in 1-aminocyclopropane-1-carboxylic acid content of cut carnation flowers in relation to their senescence. *Planta*, **150**, 439–442.

Cossins, E. A. (1978). Ethanol metabolism in plants. *Plant Life in Anaerobic Environments* (Ed. by D. D. Hook & R. M. M. Crawford), pp. 169–202. Ann Arbor, Michigan.

Crawford, R. M. M. (1978). Metabolic adaptation to anoxia. *Encyclopedia of Plant Physiology* (Ed. by O. L. Lange, P. S. Nobel, C. B. Osmond & H. Ziegler), pp. 453–477. Springer, Heidelberg.

Crawford, R. M. M. (1982). Physiological responses to flooding. *Encyclopedia of Plant Physiology*, 12B (Ed. by O. L. Lange, P. S. Nobel, C. B. Osmond & H. Ziegler), pp. 453–477. Springer, Heidelberg.

Fuhrer, J. & Fuhrer-Fries, C. B. (1985). Formation and transport of 1-aminocyclopropane-1-carboxylic acid in pea plants. *Phytochemistry*, **24**, 19–22.

Haldemann, C. (1983). *Die Ueberflutungstoleranz der Teichbinse* (Schoenoplectus lacustris *(L.) Palla): Umgehung des Sauerstoff-Defizits und Sauerstoffabgabe behalmter Rhizome*. Thesis, University of Berne.

Jackson, M. B. & Drew, M. C. (1984). Effects of flooding on growth and metabolism of herbaceous plants. *Flooding and Plant Growth* (Ed. by T. T. Kozlowski), pp. 47–128. Academic Press, Inc., New York.

Kende, H., Metraux, J. P. & Raskin, I. (1984). Ethylene-mediated growth response in submerged deep-water rice. *Ethylene* (Ed. by Y. Fuchs & E. Chalutz), pp. 121–128. M. Nijkoff & Dr W. Junk, The Hague.

Kimmerer, T. W. & Kozlowski, T. T. (1981). Ethylene, ethane, acetaldehyde and ethanol production by plants under stress. *Plant Physiology*, **69**, 840–847.

Kionka, C. & Amrhein, N. (1984). The enzymatic malonylation of 1-aminocyclopropane-1-carboxylic acid in homogenates of mung-bean hypocotyls. *Planta*, **162**, 226–235.

Laing, H. E. (1940). The composition of the internal atmosphere of *Nuphar advenum* and other water plants. *American Journal of Botany*, **27**, 861–868.

Liu, Y., Hoffman, N. E. & Yang, S. F. (1983). Relationship between the malonylation of 1-aminocyclopropane-1-carboxylic acid and D-amino acids in mung-bean hypocotyls. *Planta*, **158**, 437–441.

Lizada, M. C. C. & Yang, S. F. (1979). A simple and sensitive assay for 1-aminocyclo-propane-1-carboxylic acid. *Analytical Biochemistry*, **100**, 140–145.

Malone, M. & Ridge, I. (1983). Ethylene-induced growth and proton excretion in the aquatic plant *Nymphoides peltata*. *Planta*, **157**, 71–73.

Monk, L. S., Crawford, R. M. M. & Braendle, R. (1984). Fermentation rates and ethanol accumulation in relation to flooding tolerance in rhizomes of monocotyledonous species. *Journal of Experimental Botany*, **35**, 738–745.

Osborne, D. J. (1984). Ethylene and plants of aquatic and semi-aquatic environments: a review. *Plant Growth Regulation*, 2, 167–185.

Reid, D. M. & Bradford, K. J. (1984). Effects of flooding on hormone relations. *Flooding and Plant Growth* (Ed. by T. T. Kozlowski), pp. 195–219. Academic Press, Inc., New York.

Sale, P. J. M. & Wetzel, R. G. (1983). Growth and metabolism of *Typha* species in relation to cutting treatments. *Aquatic Botany*, 15, 321–334.

Steinmann, F. & Braendle, R. (1981). Die Ueberflutungstoleranz der Teichbinse (*Schoenoplectus lacustris* (L.) Palla): III. Beziehungen zwischen der Sauerstoffversorgung und der 'Adenylate Energy Charge' der Rhizome in Abhängigkeit von der Sauerstoffkonzentration in der Umgebung. *Flora*, 171, 307–314.

Steinmann, F. & Braendle, R. (1984). Auswirkungen von Halmverlusten auf den Kohlehydratstoffwechsel überfluteter Seebinsenrhizome *Schoenoplectus lacustris* (L.) Palla. *Flora*, 175, 295–299.

Studer, C. & Brändle, R. (1984). Sauerstoffkonsum und Versorgung der Rhizome von *Acorus Calamus* L., *Glyceria maxima* (Hartmann) Holmberg, *Menyanthes trifoliata* L., *Phalaris arundinaceca* L., *Phragmites communis* Trin. und *Typha latifolia* L. *Botanica Helvetica*, 94, 23–31.

Yang, S. F. (1984). The formation of ethylene from 1-aminocyclopropane-1-carboxylic acid. *Ethylene* (Ed. by Y. Fuchs & E. Chalutz), pp. 1–8. M. Nijkoff & Dr W. Junk, The Hague.

Some aspects of internal plant aeration in amphibious habitats

T. J. GAYNARD AND W. ARMSTRONG

Department of Plant Biology and Genetics, The University, Hull HU6 7RX, U.K.

SUMMARY

1   The effects of total and partial submergence on the internal oxygen regime of an amphibious species, *Eriophorum angustifolium*, are described and discussed.

2   The oxygen regime within the plants was monitored chiefly using polarographic sensors around the root apices, although occasionally sensors were applied to the leaves.

3   In darkness any degree of submergence might adversely affect the oxygen supply to the roots, but in the absence of short 'free' leaves no significant effect occurs unless water levels rise above the leaf sheath region of the major leaf group. Oxygen flux from root apices ceased only at full submergence.

4   Photosynthetic activity enhances the root's internal oxygen supply in the unsubmerged condition and at all levels of submergence since the sheath region and water layers help to prevent the ready escape of endogenously produced oxygen.

5   Oxygen levels were raised both by increasing radiant flux density and by concentrations of free dissolved carbon dioxide $>0.1-0.3$ mM in the submerging fluids. The bicarbonate ion had no influence on the aeration process.

6   It is suggested that under certain circumstances net gains in root aeration and rhizosphere oxygenation might accrue from the photosynthetic activities.

7   Total submergence experiments suggested that the oxygen storage potential of the aerenchyma in *E. angustifolium* is fairly limited and that critical oxygen pressures for respiration are low.

8   The interpretation of results was facilitated by the use of a functional electrical analogue of the leaf–stem–root system.

INTRODUCTION

Soil saturation and soil anaerobiosis usually go hand in hand, so if normal root activities of plants are to be sustained there must be adequate provision for internal oxygen transport from the above-ground parts of the plant. That such provision amply exists in many wetland species is readily demonstrated by examining the relationships between internal oxygen concentrations and such activities as root respiration and root extension. If roots are made anoxic, growth is usually halted very quickly but normally internal oxygen transport can sustain root extension 'indefinitely', even in excised roots, provided that a sufficiency of respirable substrate is provided (Webb & Armstrong 1983). Furthermore, such activities can

303

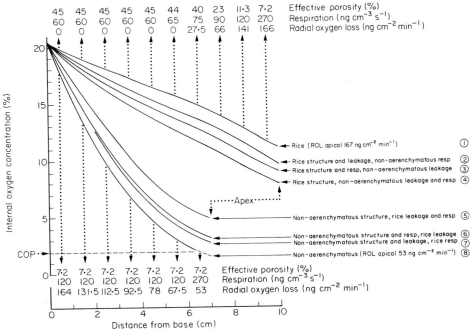

FIG. 1. An analogue analysis showing how the various characteristics of wetland (aerenchymatous rice root) and non-wetland (non-aerenchymatous) roots contribute to the oxygen status of the root in the wetland condition. The data was compiled on the assumption that the wetland soil, where aerated, consumes oxygen at the uniform rate of $5 \cdot 27 \times 10^{-8}$ g cm^{-3} s^{-1}, and that oxygen diffusivity in the soil was a uniform 1×10^{-5} cm^2 s^{-1}. It was assumed also that wall permeability to oxygen of the rice root declined from a maximum of 100% at the apex, to zero at 5 cm from the apex; in the non-aerenchymatous root the minimum value (60%) was attained at 6 cm. (After Armstrong 1979.)

be sustained at constant rates until very low oxygen levels are reached (Greenwood 1968; Armstrong & Gaynard 1976; Armstrong & Webb in press). Even in non-aerenchymatous plants internal oxygen transport is significant in short roots (Armstrong & Healy 1984) and in aerated soils will supplement the oxygen supply from the soil (Armstrong 1979). In the wetland condition the internal oxygen supply can also serve to alter the conditions in the rhizosphere by outward diffusion and the oxygen regime in the root is a function of interactions between root respiratory demand, pore-space resistance, and radial leakage to the soil. The consequence of these interactions is an acropetal oxygen concentration gradient along the root and as the root grows the effects will magnify (Fig. 1(8)). In lysigenous aerenchymatous systems, however, the pattern is much modified (Fig. 1(1)) since oxygen demands, diffusive resistance and leakage are reduced substantially in basal regions and oxygen gradients in the roots are therefore much less steep (Armstrong 1979).

The foregoing assumes that oxygen enters the plant very close to the root base (i.e. immediately above the root-shoot junction). With amphibious species, however, a good deal of the above-ground parts may become immersed in free-standing water. Very little has been published concerning the effects of partial or total flooding on the internal aeration of emergent species. In recent years the functioning of aerenchyma as an oxygen reservoir has attracted some attention (Armstrong & Gaynard 1976; Crawford & Smirnoff, see Crawford 1982; Studer & Braendle 1984), whilst oxygen fluctuations in *Spartina alterniflora* and *S. patens* in relation to aerenchyma and in response to photosynthesis and tidal flooding have also been examined (Gleason 1980; Gleason & Zieman 1981); a number of particularly thought-provoking papers have also appeared: Dacey (1980) and Dacey & Klug (1982) on mass-flow 'winds' in the water lily, *Nuphar*, and Raskin & Kende (1983) on a mechanism of mass-flow involved in the aeration of deep water rice. Oxygen relations and carbon dioxide fixation in totally submerged aquatics (and an unusual terrestrial relative) have been examined recently (Keeley 1981; Sand-Jensen & Prahl 1982; Sand-Jensen, Prahl & Stokholm 1982; Keeley, Osmond & Raven 1984).

One wetland species which is very widespread in Britain (and indeed throughout the northern hemisphere), which prolifically multiplies vegetatively and which is very easily handled in the laboratory, is the cotton grass *Eriophorum angustifolium* Honck., a member of the Cyperaceae. Although not amphibious in the same sense as the *Spartinas* it is frequently to be found growing half-submerged and undoubtedly at times experiences total immersion. Some time ago we embarked on a study of internal aeration in this and other species and in this paper we report on some aspects of that work which seemed of relevance to the amphibious theme of the conference. We would stress, however, that the study was not specifically undertaken with an amphibious theme in mind.

THE PLANT

Much of the morphology and anatomy of *E. angustifolium* has been described by Phillips (1953, 1954a,b) and Metcalfe (1971). The species is perennial, spreading by means of rapidly growing rhizomes. The rhizome bears scale leaves, buds and roots, save at the end, where it turns upwards to produce a swollen stem bearing leaves, roots and buds (Fig. 2). These axillary buds elongate to produce daughter rhizomes and ultimately a sympodial system.

The foliage leaves are normally divisible into three distinct regions: a narrow, pointed tip of triangular cross-section often red in colour, a green channelled blade of approximately V-shaped cross-section, and a sheathing base, green to pink to colourless, and crescentric in section. The adaxial surface of the leaf bases bears a large membraneous 'stipule'-like outgrowth which extends laterally from the leaf margin and tightly envelops the inner leaves, effectively extending the sheath region some distance along them. In smaller plants the sheath can extend almost the whole length of the shoot. In addition to the normal foliage leaves, shorter outer leaves occur which are not included in the main leaf group. These 'free' leaves are less distinctly divisible into tip, blade and base and may wither at an early stage

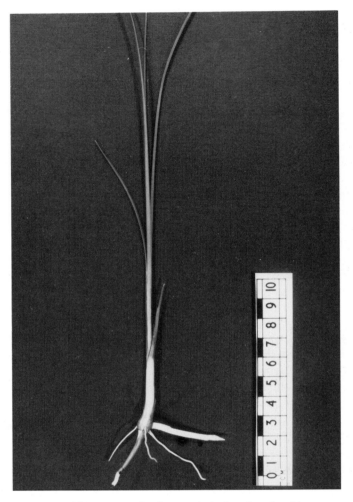

FIG. 2. *Eriophorum angustifolium*; vegetative features including short basal leaves, strongly sheathed main leaf group and young daughter rhizome.

of growth. They rarely extend beyond the sheath region but as will be seen can influence the aeration of the root system. Stomata are found on the abaxial surface only.

The plant is highly aerenchymatous and the root, whilst non-aerenchymatous at the apex, has spaces reaching to within three cell layers of the root–cap junction (Fig. 3). Sub-apically aerenchyma develops in the root in two stages (see Fig. 4a and Fig. 5). The root-wall also becomes relatively impermeable to oxygen in the sub-apical parts, (Fig. 4b); it may also afford greater than expected resistance in apical parts (Gaynard 1979). The leaf is aerenchymatous throughout with *c*. 14 parallel channels interrupted by transverse diaphragms (*c*. 3 per cm) (Fig. 6a). Variations in porosity along a leaf are shown in Fig. 4c. Porosities in stems are

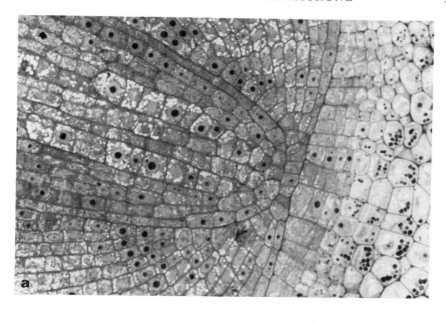

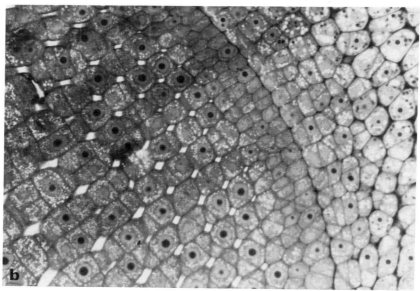

FIG. 3. (a) *E. angustifolium*: Radial longitudinal section through root apex in the region of the root–root cap junction (×725). Gas spaces present in cortex but not visible due to the particular plane of the section. (b) Tangential section of the root apex. Cortical intercellular gas space is visible within three cells of the junction.

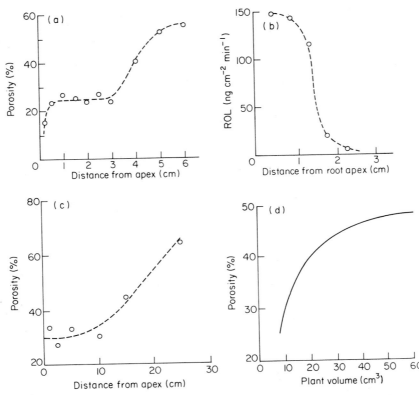

Fig. 4. (a) Changes in gas-filled porosity along a root of *E. angustifolium*. (b) Variation in ROL along a root of *E. angustifolium* reflecting basipetal reduction in wall permeability to oxygen. (c) Gas-filled porosity along a leaf of *E. angustifolium*. (d) Variation in total porosity with plant volume.

lower and the gas-space is largely attributed to aerenchyma in the outer cortex. In the root–shoot junction there is still gaseous connexion (Fig. 6b), but porosities are lower still (Gaynard 1979). Whole-plant porosities vary with plant volume as shown in Fig. 4d.

PARTIAL SUBMERGENCE AND ROOT AERATION

Whole plants or single leaf–stem–root units were used and experiments performed both in light and darkness and with and without added carbon dioxide in the flooding medium. The plants were established as illustrated (Fig. 7), with anaerobic agar-water (0·05% w/v) medium (plus supporting electrolyte KCl $6·94 \times 10^{-4}$ g cm^{-3} for the polarographic operations) in the lower part of the chamber and just covering the root–shoot junction. The oxygen regime at the root-wall 2–7 mm behind the apex was monitored polarographically as radial oxygen loss (ROL) from the root using sleeving Pt-cathodes (Armstrong 1979). Aerial parts of the plants

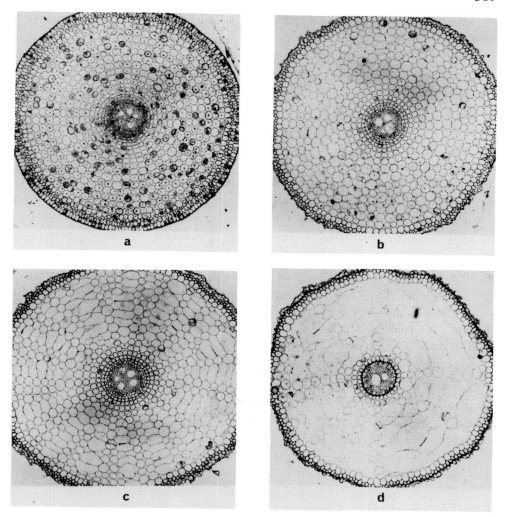

FIG. 5. Basipetal development of lysigenous gas spaces in the root of *E. angustifolium*. (a)–(d): transverse sections at 0·2, 2·0, 4·0, 6·0 cm from the root tip (×100).

were submerged as required using the same agar-water medium; for details see Gaynard (1979).

RESULTS AND DISCUSSION

Sequential submergence: dark and illuminated

Plants were submerged in stages until completely immersed. At each stage the equilibrated value of ROL was recorded and it should be noted that this value can

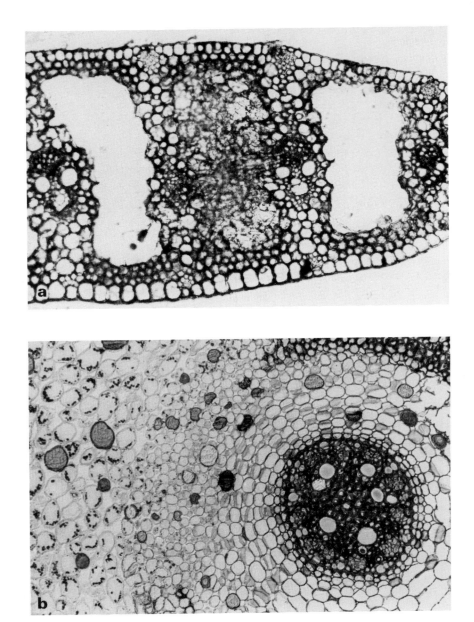

FIG. 6. (a) *E. angustifolium*: fresh transverse section through leaf; two outer inter-vascular channels show the open lacunate condition; the central channel is occluded at this point by a low porosity tissue diaphragm. (b) Section through the root–shoot junction of *E. angustifolium* showing the continuity of the gas spaces in the cortices of root and stem (×270).

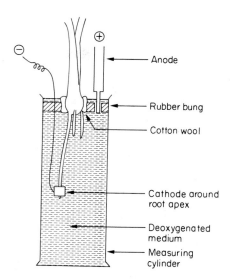

FIG. 7. Basic experimental assembly for the measurement of oxygen loss from roots. A taller vessel with a movable bung assembly was employed for the submergence experiments.

be translated into an oxygen concentration at the root surface (Armstrong 1979). In most cases before submergence this amounted to *c.* 7–10% oxygen.

Single leaf–stem–root preparations: dark

ROL declined with submergence as illustrated in Fig. 8a and there was a marked correlation between the fall in ROL and the distribution of stomata: submergence of the sheath region where stomata are absent or infrequent had little effect; submergence in the regions of high stomatal density markedly reduced ROL.

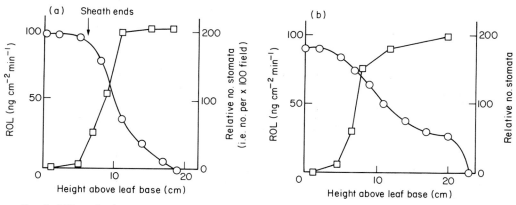

FIG. 8. Effect of submergence on ROL: single leaf–stem–root units; ROL (○); stomata (□); (a) in darkness; leaf length, 19·0 cm. (b) in low light; leaf length, 23·0 cm.

Clearly the major cause of the decline was the gradual accumulation of sink activity and diffusive resistance brought about by the blocking of stomatal entry points which routed oxygen through an ever increasing length of leaf. It should be noted that although the change in slope of the oxygen-submergence curve at *c.* 10 cm corresponds with a flattening of the stomatal curve, it should not be assumed that there is a causal relationship since the stomatal number required for aeration is very low (Gaynard 1979). Also to be noted is the apparent state of anoxia at full flooding and this will be commented on later. The oxygen regime (2·5%) previously identified by Armstrong & Gaynard (1976) as corresponding with the critical oxygen pressure for respiration (COPR) occurred at the ROL of 34·5 ng cm^{-2} min^{-2} when the degree of flooding was *c.* 76%.

Single leaf–stem–root preparations: light
Laboratory lighting (15 μE m^{-2} s^{-1}) brought about a major departure from the flux pattern obtained in darkness (see Fig. 8b): in the latter stages of submergence the decline in ROL was very much reduced and something of a plateau formed before the final and abrupt decline to anoxia at full flooding. In view of this the effect appears to be a photosynthetic one dependent upon a supply of carbon dioxide via the exposed leaf apex and the restraining effects of blocked stomata and internal diffusive resistance on oxygen escape. Critical oxygen pressures were thus avoided until full flooding.

Whole plants: dark
Several examples are shown (Fig. 9a): one, (A), in which there were none of the short free leaves mentioned earlier, and two others in which free leaves were present, (B) and (C).

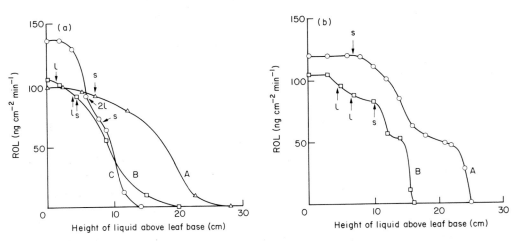

Fig. 9. Effect of submergence on ROL: intact plants. Length of 'free' leaves shown by (l); end of sheath indicated by (s). (a) darkness; longest leaves, (A) 28 cm, (B) 20 cm, (C) 14 cm. (b) low light; longest leaves, (A) 25 cm, (B) 16 cm.

The basic pattern noted with the single leaf preparations was modified by the free leaves. In (B) submergence of the shorter of two free leaves masks the initial plateau whilst submergence of the longer one probably sustains the relatively steep decline in ROL. In (C) the initial plateau is present but simultaneous submergence of two free leaves extending to 6 cm is reflected as a sharp fall in ROL. Further submergence to 8 cm slightly decreases the slope, but inundation past the sheath causes the ROL to fall steeply again.

Whole plants: light
Two examples are illustrated in Fig. 9b. The effect of free leaves is apparent and again a photosynthetic plateau is to be seen.

Submergence levels: light and dissolved carbon dioxide effects

Experiments were performed in a darkroom with light provided unidirectionally as required using a 500 W incandescent lamp (Phillips 'Arga-photo'). Incident radiation was measured as flux density in the photosynthetically active spectral region (400–700 nm). Radiation at the vessel surface was varied simply by altering the position of the light source or by using a paper 'shutter'. A cold-water heat shield was placed between the lamp and plant to absorb the infra-red radiation; the ambient temperature was kept constant by using fans to circulate the air.

Effects of submergence and light intensity: no added carbon dioxide

The relationship between ROL and illumination for each of five levels of submergence is shown in Fig. 10. A photosynthetic effect related to the level of submergence can be observed in each case, the order being:

$$1/2 \geq 1/4 > 0 > 3/4 > 1$$

If the submerged parts were kept in the dark illuminating the unsubmerged parts had no affect on the oxygen regime.

To explain these observations it should first be noted that oxygen leaking from roots can be derived from two sources: the atmosphere and photosynthesis. The effectiveness of the atmospheric source will be at a maximum at zero submergence (i.e. when the diffusive resistance between the atmosphere and root apices is minimized). The atmosphere, however, will also buffer the system against photosynthetic oxygen losses. The photosynthetic source will be enhanced by submergence since the escape of oxygen and respiratory carbon dioxide from the submerged leaves will be reduced. At the same time carbon dioxide supply from the atmosphere will be more restricted and this will tend to reduce oxygen production. The resulting ROL is due, therefore, to the interaction of these factors, and maximum ROL will occur when the degree of submergence is such that the balance

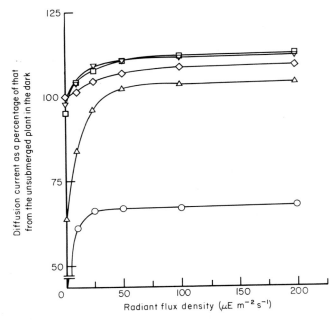

FIG. 10. Diffusion current as a function of incident radiant flux. No added CO_2. Unsubmerged (◊); ¼ submerged (△); ½ submerged (□); ¾ submerged (△); fully submerged (○).

between 'jacketing' and carbon dioxide entry is optimal. This occurred in the half-submerged plant when ROL was boosted to 111% of the dark unsubmerged value. Even in the unsubmerged plant, however, the ROL was raised to the 108% level and this is attributable to photosynthesis in the astomatal leaf bases; when the basal regions of the leaves were blackened illumination did not have an effect. In the fully submerged plant the ROL was raised from a dark value of zero to 66% of the darkened, unsubmerged reading. In this case the atmospheric carbon dioxide source was absent as was the oxygen buffering influence of the atmosphere. Photosynthesis must have been dependent chiefly upon the carbon dioxide derived from respiration though it is possible that a small amount of the gas may have been derived from the surrounding medium.

Effects of submergence, light and carbon dioxide
Carbon dioxide was supplied either as the free dissolved gas or the bicarbonate ion, either to the roots or leaves, and at one or both of two light intensities (100 and 500 $\mu E\ m^{-2}\ s^{-1}$). In the experiments using free carbon dioxide the pH of the medium was adjusted to 4·0 with hydrochloric acid. Carbon dioxide-saturated medium was prepared by adjusting the pH of the liquid to 4·0 and bubbling overnight with carbon dioxide gas. The desired carbon dioxide concentration in the bathing medium was obtained by adding an appropriate volume of this saturated solution. When bicarbonate was used the pH of the medium was adjusted to 8·3

with KOH solution. At this pH the concentration of bicarbonate is a hundred times that of free carbon dioxide (Wium-Andersen 1971). A stock solution of 100 mM bicarbonate (as $KHCO_3$) was freshly prepared before each experiment and was added to the bathing medium to obtain the required concentration. Since the dissolution of $KHCO_3$ is an endothermic process care was taken to allow the stock solution to come to laboratory temperature before adding.

Test whether a pH change from 4·0 to 8·3 could *per se* alter ROL the oxygen flux was monitored from a plant 3/4 submerged in medium at pH 4·0 and illuminated at 100 μE m^{-2}s^{-1}. The pH was raised as indicated and the ROL after equilibration was identical to that at pH 4·0, indicating that the pH change did not affect photosynthetic rate. Wium-Andersen (1971) obtained a similar result with *Lobelia*.

Bicarbonate. Bicarbonate was supplied to both roots and leaves and was used at concentrations of 2, 5, and 20 mM with 3/4 submergence and 100 μE light flux. Regardless of concentration, however, the oxygen regime in the roots never varied from the original value but addition of HCl always produced an almost immediate rise in oxygen levels. The results indicate that *E. angustifolium* is unable to utilize the bicarbonate. The waxy cuticle probably prevents the uptake of bicarbonate through the leaf surface, whilst any bicarbonate absorbed by the roots is clearly insufficient to produce a measureable effect. However, the rapid increase in ROL on addition of HCl (which would have liberated carbon dioxide *in situ*) is strong evidence that freed carbon dioxide entered the plant from the bathing medium and caused an elevation in oxygen pressure by increasing the photosynthetic rate.

Free carbon dioxide—roots. Wium-Andersen (1971) found that carbon dioxide supplied to the roots of *Lobelia* illuminated at 13 mW cm^{-2} was limiting only up to 1 mM. In the present experiments the carbon dioxide concentration was raised to double this value but at no time did an increase in ROL occur. This indicates that any carbon dioxide absorbed by the roots of *Eriophorum* is not sufficient to produce any significant effect on aeration if utilized in photosynthesis, a result which is in agreement with that of Yoshida *et al*. (1974) for rice.

Free carbon dioxide—leaves. Results for a radiation flux of 100 μE m^{-2} s^{-1} are shown in Fig. 11a and for 500 μE m^{-2} s^{-1} in Fig. 11b. In general ROL at 100 μE increased with carbon dioxide concentration until a saturation level was approached.

In the totally submerged plant there was little effect of increased concentration below 0·2 mM; above this value the increase in ROL was hyperbolic with saturation occurring at c. 1·1 mM. A similar result was obtained for the 3/4 submerged plant; in this case ROL increased hyperbolically above 0·1 mM. These solution concentrations are equivalent to gas-phase levels of c. 0·5 and 0·25% carbon dioxide respectively. One explanation for these results might be that in the submerged aerial parts respiratory carbon dioxide had only accumulated to these levels before the addition of carbon dioxide to the bathing medium. Consequently

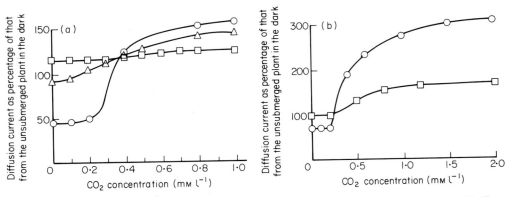

FIG. 11. Oxygen diffusion current as a function of dissolved carbon dioxide concentration (a) at 100 μE m^{-2} s^{-1}; (b) at 500 μE m^{-2} s^{-1}. $\frac{1}{2}$ submerged ($\square$); $\frac{3}{4}$ submerged ($\triangle$); fully submerged ($\bigcirc$).

solution concentrations below 0·1–0·2 mM would not be expected to enhance the photosynthetic effect. A second possible reason for the ineffectiveness of low concentrations may have been synergism between carbon dioxide uptake in the guard cells and a high boundary-layer resistance to carbon dioxide diffusion into the plant in the unstirred medium. This may have necessitated threshold concentrations in the bulk solution in excess of 0·2 mM before the concentration gradient between the solution and internal atmosphere was sufficient to allow diffusion of the gas through the stomata.

In the half-submerged plant ROL increased only above 0·3 mM. In this case there was less surface exposed to the carbon dioxide, and this together with a readier leakage of oxygen to the atmosphere may have been such that any enhancement of photosynthesis by carbon dioxide concentrations below 0·3 mM was made ineffective.

At 500 μE a similar pattern emerges but at carbon dioxide concentrations above the threshold value ROL becomes much higher than previously and the saturation concentration is almost doubled.

To place these results in a field context measurements were made of solar radiation and of carbon dioxide concentrations in pools where *E. angustifolium* was abundant. Even visually dull days gave a radiant flux density of 800 μE m^{-2}s^{-1} and the free carbon dioxide concentrations in the pool waters approached 0·8 mM (method of Mackereth 1963). The implications would seem to be that partial submergence in the field may lead to a considerable enhancement of oxygen levels during daylight and a consequence of this might be that in long days there might be in terms of aeration a net advantage to be gained both in root growth and rhizosphere oxygenation.

Where there is overlap the results are generally in agreement with those obtained by Gleason (1980) who found for example that shoot base lacunae in *S. alterniflora* become virtually anoxic during midsummer high tides in darkness and that photosynthesis contributed significantly to oxygen supply during flooding.

Conclusions

1 In darkness any degree of submergence might adversely affect the oxygen supply to the below-ground parts of *E. angustifolium*. In the absence of short free leaves, however, no significant effect occurs unless water levels rise above the leaf-sheath region.

2 In darkness progressive submergence beyond the sheath region can cause a steep decline in the oxygen supply to the root system but oxygen levels, at least in short roots (≤ 7 cm), apparently fall to zero at room temperatures only at full flooding. Stomatal resistance is not thought to play a significant role in submergence effects.

3 Photosynthetic activity enhanced the root's internal oxygen supply in the unsubmerged condition and at all levels of submergence; the sheath region and water layers acted as jackets hindering the escape of endogenously produced oxygen.

4 Neither the bicarbonate ion in solution around the leaves nor carbon dioxide in solution around the roots caused any enhancement of the photosynthetic effect but around the leaves dissolved carbon dioxide levels up to $1 \cdot 0$ mM (r.flux 100 μE) and $2 \cdot 0$ mM (r.flux 500 μE) caused oxygen levels in fully submerged plants to rise to 160% and 310% of the levels in darkened unsubmerged plants.

5 It is suggested that under certain circumstances (e.g. long days) net gains in root aeration may accrue from the photosynthetic activities of the submerged portions.

TOTAL SUBMERGENCE AND AERATION

The primary aim in terms of root aeration was to determine to what extent aerenchyma might serve as an oxygen reservoir but attention was focused later on the detailed dynamics of oxygen depletion and on respiratory critical oxygen pressures. The following is a very abbreviated account of the findings.

Oxygen flux from the roots always declined rapidly soon after submergence but the exact form of the time dependent plot was related to the size of plant (Fig. 12). Irrespective of size, however, flux did not decline immediately; there was first a lag phase. Subsequently in the smaller plants there was a linear decline apparently indicative of a constant rate of oxygen depletion. Eventually linearity was again lost and the curves 'tailed' to a zero point. Initially it was assumed that the point of inflexion before the 'tail' was caused by the reaching of a critical oxygen pressure within the plant (e.g. the root) and this assumption was later supported by electrically modelling the system. The mean critical oxygen pressure determined from the submergence curves was $0 \cdot 02$ atmos ($+/- 0 \cdot 007$, 25 observations). The time from flooding to zero flux was 50–70 min (Fig. 13) and this was apparently independent of plant volume. If the time taken to reach the COPR is plotted against plant volume, however, the result is somewhat different and the difference has been explained in terms of shoot:root ratios (see Gaynard 1979).

With the larger plants the initial lag phase precedes a second stage in which a lag component may still be operating since this is followed by a sharper drop than the

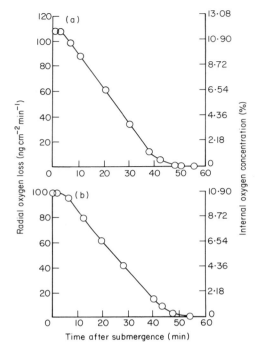

FIG. 12. Fall in radial oxygen loss and internal apical oxygen concentration following submergence in anaerobic medium at 23°C. (a) Small plant; (b) larger plant.

subsequent linear phase. With these plants the 'tail' portion of the curves was shorter. The explanation for these stages was elucidated from modelling studies beyond the scope of this paper.

The results imply that aerenchyma may not provide a very adequate buffer against root anoxia under full-flooding although at field temperatures aerobic conditions might persist for much longer periods. At summer temperatures, however, the period may not much exceed 2 h. (Preliminary trials involving flooding with stirred aerated waters have given similar results.)

Subsequent experimental studies showed that leaf oxygen concentrations declined along a somewhat similar time curve to the roots but the 'tail' portion extended further, a feature which could be explained in terms of root-wall resistances and oxygen consumption within the wall which, when internal oxygen levels are very low, might be sufficient to prevent throughflow across the wall. Electrical modelling confirmed that root and leaf would come to extinction at around the same time. Leaf electrodes suggested that the time to extinction is *c*. 100 min at laboratory temperatures.

At 20°C oxygen depletion curves similar to those in Fig. 12 were obtained by Studer & Braendle (1984) in their studies of excised rhizome aeration in several species. At 5°C, however, the concentration against time relationship tended to be curvilinear throughout and the time to oxygen extinction was approximately

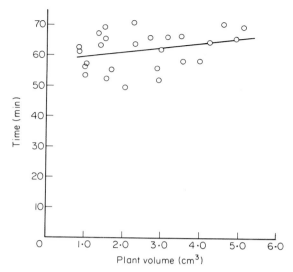

FIG. 13. Time taken to exhaust the internal oxygen reservoir as a function of plant volume. $y = 1.26x + 57.7, r = 0.271$.

trebled in three species, more than quadrupled in two more, and in *Glyceria maxima* was increased sevenfold. The excised rhizomes had remained exposed to air and the curvilinear decline probably indicates a major concentration-dependent leakage of oxygen into the system and was revealed at the lower temperature because of much lower respiratory rates.

Conclusions

1 The oxygen depletion curve in the root indicates that respiratory critical oxygen pressures in the plant are very low.

2 The relationship between respiratory activity and oxygen concentration in *E. angustifolium* is such that the oxygen storage capacity of the aerenchyma system is fairly rapidly exhausted at root temperatures. At summer field temperatures the maximum period of aerobiosis may extend to *c*. 3–4 h.

FINAL COMMENTS

Eriophorum angustifolium is a species which can experience a wide range of flooding regimes and the results have shown that unless fully submerged in darkness the likelihood is that an oxygen presence will be maintained even in root apices. There was no evidence of any significant mass flow component in the aeration of the plant and in this as in other related species aerenchyma extends to the extremities of the leaves and this must play a major role in the plant's flood tolerance. Photosynthetic activity can clearly enhance root aeration in this species and it would seem that partial submergence might be advantageous under certain circumstances. It would be interesting in further studies to investigate more thoroughly the implications that there may be optimal flooding depths in amphibious habitats.

REFERENCES

Armstrong, W. (1979). Aeration in higher plants. *Advances in Botanical Research* (Ed. by H. W. Woolhouse), pp. 226–332. Academic Press, London.

Armstrong, W. & Gaynard, T. (1976). The critical oxygen pressures for respiration in intact plants. *Physiologia Plantarum*, **37**, 200–206.

Armstrong, W. & Healy, M. T. (1984). Oxygen diffusion in pea. III. Changes in the oxygen status of the primary root relative to seedling age. *New Phytologist*, **96**, 179–185.

Armstrong, W. & Webb, T. (1986). In press.

Crawford, R. M. M. (1982). Physiological responses to flooding. *Encyclopedia of Plant Physiology* (Ed. by O. L. Lange, P. S. Nobel, C. B. Osmond & H. Ziegler). pp. 453–477. Springer-Verlag, Berlin.

Dacey, J. H. W. (1980). Internal winds in water lilies: an adaptation for life in anaerobic sediments. *Science*, **210**, 1017–1019.

Dacey, J. H. W. & Klug, M. J. (1982). Ventilation by floating leaves in *Nuphar*. *American Journal of Botany*, **69**, 999–1003.

Gaynard, T. J. (1979). *Some aspects of internal aeration in wetland plants*. Ph.D. thesis. University of Hull, U.K.

Gleason, M. L. (1980). *Influence of tidal inundation on internal oxygen supply of* Spartina alterniflora *and* Spartina patens. Ph.D thesis. University of Virginia, U.S.A.

Gleason, M. L. & Zieman, J. C. (1981). Influence of tidal inundation on internal oxygen supply of *Spartina alterniflora* and *Spartina patens*. *Estuarine, Coastal and Shelf Science*, **13**, 45–57.

Greenwood, D. J. (1968). Root growth and oxygen distribution in soil. *Transactions of the 9th International Congress of Soil Science*, **1**, 823–832.

Keeley, J. E. (1981). *Isoetes Howellii*: a submerged aquatic CAM plant? *American Journal of Botany*, **68**, 420–424.

Keeley, J. E., Osmond, C. B. & Raven, J. A. (1984). Stylites, a vascular land plant without stomata absorbs CO_2 via its roots. *Nature, London*, **310**, 694–695.

Mackereth, F. J. H. (1963). *Water Analysis for Limnologists*. Freshwater Biological Association, Publication No. 21.

Metcalfe, C. R. (1971). *Anatomy of Monocotyledons, Vol. V. Cyperaceae.* Oxford University Press, Oxford.

Phillips, M. E. (1953). Studies in the quantitative morphology and ecology of *Eriophorum augustifolium* Roth. I. The rhizome system. *Journal of Ecology*, **41**, 295–318.

Phillips, M. E. (1954a). Studies in the quantitative morphology and ecology of *Eriophorum angustifolium* Roth. II. The leafy shoot. *New Phytologist*, **53**, 312–343.

Phillips, M. E. (1954b). *Eriophorum angustifolium. Journal of Ecology*, **42**, 612–622.

Raskin, I. & Kende, H. (1983). How does deep-water rice solve its aeration problem? *Plant Physiology*, **72**, 447–454.

Sand-Jensen, K. & Prahl, C. (1982). Oxygen exchange with the lacunae and across the leaves and roots of the submerged vascular macrophyte *Lobelia dortmanna* L. *New Phytologist*, **91**, 103–120.

Sand-Jensen, K., Prahl, C. & Stokholm, H. (1982). Oxygen release from roots of submerged aquatic macrophytes. *Oikos*, **38**, 349–354.

Studer, C. & Braendle, R. (1984). Sauerstoffkonsum und Versorgung der Rhizome von *Acorus Calamus* l., *Glyceria maxima* (Hartmann) Holmberg, *Menyanthes trifoliata* L., *Phalaris arundinacea* L., *Phragmites communis* Trin. und *Typha latifolia* L. *Botanica Helvetica*, **94**, 23–31.

Webb, T. & Armstrong, W. (1983). The effects of anoxia and carbohydrates on the growth and viability of rice, pea and pumpkin roots. *Journal of Experimental Botany*, **34**, 579–603.

Wium-Andersen, S. (1971). Photosynthetic uptake of free CO_2 by the roots of *Lobelia dortmanna*. *Physiologia Plantarum*, **25**, 245–248.

Yoshida, S., Coronel, V., Parao, T. & Reyes de los, E. (1974). Soil carbon dioxide flux and rice photosynthesis. *Soil Science and Plant Nutrition*, **20**, 381–386.

Mechanisms of acclimation to flooding and oxygen shortage in non-wetland species

M. C. DREW

Department of Horticultural Sciences, Texas A & M University, College Station, Texas 77843-2133, U.S.A.

SUMMARY

This paper reviews physiological mechanisms by which non-wetland species acclimate to transient flooding and root hypoxia or anoxia. Particular emphasis is given to the induction and role of aerenchymatous adventitious roots that can extend into anaerobic surroundings. Recent studies of energy metabolism indicate internal oxygen transport from the shoot to the root tip when the root environment lacks oxygen. Tracer work shows that aerenchymatous roots maintain radial transport of nutrient ions to the xylem, despite the lysis of much of the root cortex.

INTRODUCTION

In non-wetland habitats, poor soil aeration and oxygen shortage in the rooting zone can develop either through restricted soil drainage, or because of a bulk movement of water into the soil profile, producing a rising water table. Poor drainage of water through the soil profile results in an inadequate system of continuous, gas-filled soil pores for rapid gas exchange with the air, so that products of plant and microbial metabolism like carbon dioxide and other gases became entrapped, while the oxygen required for oxidative metabolism is partially excluded from the soil (Gambrell & Patrick 1978; Ponnamperuma 1984).

Some soils are naturally poorly draining, but on agricultural land additional problems arise from farming operations (Cannell 1977). Wheeling by farm traffic compresses and smears the wet soil, and so too on a fine scale does the passage of the plough-sole, so that the macropores that enable water to drain quickly, and subsequently also conduct the major portion of gas exchange with the soil, are eliminated. Microvolumes of the soil that are surrounded by water-saturated micropores soon become anaerobic (oxygen-free), although these may coexist with other fully aerobic volumes, so that the aeration status of the soil is highly heterogeneous (Greenwood 1968). Under extreme conditions impeded drainage causes the entire soil profile to become water-saturated or flooded, stagnant water can be seen ponded at the surface, and if soil temperatures are sufficiently high to encourage rapid respiration large portions of the soil profile can become anaerobic.

The challenge to physiologists and ecologists is to understand the means by which wetland species appear to thrive under such conditions of excess water while non-wetland species clearly do not. However, the range of response of non-wetland species to flooding and oxygen shortage is considerable, from a mere check to growth, to characteristic symptoms of flooding injury to shoots, to rapid death (Jackson & Drew 1984). Yield-losses from winter wheat crops subjected to prolonged mid-winter waterlogging may be only 15% or less (Cannell *et al.* 1980) and a similar yield depression occurred with winter oilseed rape (*Brassica napus*) following waterlogging between January and March in Britain (Cannell & Belford 1980). By contrast, seed yield losses of pea (*Pisum sativum*) at a susceptible preflowering growth stage was 42% after only 5 days waterlogging (Belford *et al.* 1980).

Clearly, some species are better adapted to flooding than others, which suggests that genetically determined characteristics confer a degree of resistance to the stresses associated with flooding.

Removal of oxygen from the soil atmosphere results in a sequence of biological and chemical changes in the root environment (see Gambrell & Patrick 1978; Ponnamperuma 1984). Lack of oxygen can by itself prove injurious to root metabolism and indirectly to the shoot (Drew 1983). The gradual accumulation to toxic concentrations of organic metabolites and reduced inorganic ions in the soil water are an additional hazard to plant survival (Drew & Lynch 1980). Earlier reviews have summarized evidence that flooding injury to the shoots of susceptible species is associated with root dysfunction and consequent abrupt changes in mineral nutrient status, water relations, level of plant growth substances and toxins. Here, we are concerned with physiological adaptations in non-wetland plants that appear to minimize these changes, and confer an improved resistance to flooding.

FORMATION OF AERENCHYMA IN ADVENTITIOUS ROOTS

None of the vegetative tissues of the dryland crop plants grown to temperate climates are truly anoxia tolerant, in contrast to the ability of the organs of some wetland species to survive exposure to oxygen-free environments for weeks or months (Crawford 1982; Barclay & Crawford 1982). Thus, exposure of the root systems of crop plants to transient flooding or oxygen deprivation leads to the death of the initial root system if oxygen is not reintroduced within 3–70 h (Jackson & Drew 1984).

Survival of herbaceous plants under these conditions has long been known to be closely associated with the formation of a new, adventitious root system from root primordia in the basal part of the stem (Kramer 1951; Jackson 1955). These roots presumably continue the function of the earlier, now moribund, tissues but their physiology has been little explored. The adventitious roots of a wide range of dryland crop species (wheat, barley, maize, sunflower, tomato) develop an aerenchyma (or an increased gas-filled volume) in response to flooding or when the nutrient solution is made oxygen-deficient (see Table 2 in Jackson & Drew 1984). Aerenchyma formation takes place by the lysis of cortical cells, leaving gas-filled cavities or lacunae (Fig. 1). The effectiveness with which these cavities convey oxygen from the aerial tissues to the root tip is discussed in a later section.

In the root tips of adventitious roots of maize, an early stage in aerenchyma

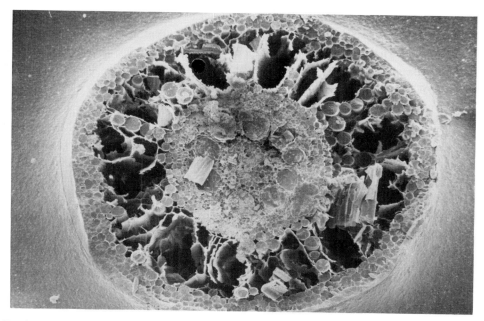

Fig. 1. Maize root aerenchyma. The scanning electron micrograph is of a frozen, hydrated, transverse fracture, embedded in an inert supporting medium. Note the cortical lacunae, wall residues, and three files of intact cortical cells. The zone was 50 mm from the root tip. ×80 (micrograph courtesy of Dr J. Sargent, W.R.O., Begbroke).

formation induced by oxygen shortage is detected as a collapse of individual cells in the mid-cortex, in the zone 10 mm from the apex, where cell expansion is just complete (Campbell & Drew 1983). Affected cells showed a loss of turgor and inward collapse, abnormally dense cytoplasm and loss of tonoplast, while surrounding cells retained turgidity and a normal ultrastructure. Loss of tonoplast integrity would quickly lead to cell death, through cytoplasmic acidosis brought about by leakage of protons from the vacuole (Roberts *et al.* 1984). Nuclear magnetic resonance of ^{31}P showed that in heatlhy cells in the root tip of maize the pH of the vacuolar contents (5·6) contrasted sharply with that of the cytoplasm (7·3). Loss of viability under anoxia was closely associated with a decrease in cytoplasmic pH. In lysing cells of maize signs of loss of tonoplast integrity preceded any other change in subcellular structure that could be detected in the transmission electron microscope (Campbell & Drew 1983). We noted that the plasma membrane and mitochondria appeared intact at this stage, which suggests that the tonoplast is specifically affected by the mechanisms of cell lysis.

Cell lysis during aerenchyma formation was attributed by McPherson (1939), and many since, to the direct effects of anoxia, since increased formation of aerenchyma was apparent in roots extending in an oxygen-deficient environment. However, several lines of evidence point to endogenous ethylene rather than to the direct effect of anoxia as the agent of induction of cell lysis.

Kawase (1974, 1981) was the first to discover that exogenous applications of ethephon (a synthetic compound that forms ethylene in plant cells) or ethylene in air induced cavity formation in the hypocotyl of sunflower. When the stem base and emerging adventitious (nodal) roots of maize plants growing in nutrient solution were bubbled with low concentrations of ethylene ($0.1–1.0$ μl 1^{-1} in air), the pattern of gas space formation appeared indistinguishable from that brought about by treating roots with oxygen-deficient media (Drew, Jackson & Giffard 1979). The induction of gas spaces by exogenous ethylene, or by exposure to sub-ambient oxygen, could be blocked either by low concentrations of Ag^+, an inhibitor of ethylene action (Drew *et al.* 1981) or by inhibitors of ethylene biosynthesis, like aminoethoxyvinylglycine (AVG) or amino-oxyacetate (Konings 1982; Konings & de Wolf 1984; Jackson *et al.* 1985). An important confirmatory test was to treat roots with a solution of AVG together with the ethylene precursor, 1-amino-1-cylopropanecarboxylic acid (ACC). The release within the tissue of ethylene from the precursor bypassed the inhibitory effect of AVG and induced aerenchyma in the growing roots, thus demonstrating that AVG was not blocking ethylene biosynthesis by acting as a general, non-specific inhibitor (Jackson *et al.* 1985).

The induction of aerenchyma by exposing roots to low oxygen partial pressures appears to take place by stimulating ethylene biosynthesis. Extraction gases from intercellular spaces shows an augmentation of ethylene when oxygen is in short supply (Drew, Jackson & Giffard 1979) and the rate of ethylene formation by excised segments previously treated with different partial pressures of oxygen showed a marked rise at oxygen partial pressures of $0.03–1.12$ atm (Jackson *et al.* 1985). However, in contrast to the hypoxic stimulation of ethylene biosynthesis, strict anoxia arrests both the synthesis of ethylene and the formation of aerenchyma. Interference in ethylene synthesis under anoxia is not surprising in view of the ATP-dependent biosynthetic pathway and the requirement for molecular oxygen in converting ACC to ethylene (Young & Hoffman 1984) but anoxic roots fail to respond to exogenous ethylene.

It is clear, therefore, that aerenchyma formation is not a general degeneration response to lack of oxygen, but a selective, physiologically controlled induction of collapse and lysis specifically in cortical cells. Not all the cortical cells in transection are seen to respond to ethylene, and it would be interesting to know the biomolecular basis for their resistance, as well as the mechanism of stimulation of the biosynthetic pathway.

Cell lysis and aerenchyma formation are not uniquely induced by lack of oxygen. Shortage of a nitrogen source (nitrate or ammonium ions) or of phosphate in the nutrient solution induces a similar response (Konings & Verschuren 1980; Drew & Saker 1983). Future work should determine whether ethylene has a role to play in bringing about this phenomenon.

FUNCTION OF AERENCHYMA IN INTERNAL TRANSPORT OF OXYGEN

Characteristically the roots of non-wetland plants differ from wetland species in having a smaller gas-filled volume and a greater permeability to the outward diffusion of oxygen (Armstrong 1979). Non-wetland roots are thus less able to conserve oxygen and conduct it towards the dividing and expanding cells of the root apex. In a wetland species like rice, gas-filled lacunae are prominent and intercellular spaces extend into the meristematic zone, while the presence of suberized and lignified layers of cells beneath the epidermis presumably contribute to the low permeability to oxygen (Clark & Harris 1981). These anatomical features are less developed, or absent, in non-wetland plants. Experiments with roots of various dryland crop species held in anaerobic media demonstrate that oxygen from the shoot can diffuse through intracellular spaces along roots over distances of several cm (Greenwood 1967; Greenwood & Goodman 1971). How effective this internal transport of oxygen may be in maintaining aerobic respiration is less well understood, since it is difficult to determine whether oxygen diffusing to the outer medium is that in excess of respiratory requirements, or oxygen that has failed to reach the sites of consumption.

One approach to resolving this question is to determine the effectiveness of internally transported oxygen by measuring its utilization, from estimates of energy metabolism at different distances along the root from the oxygen source. In seminal roots of maize, Saglio, Raymond & Pradet (1983) deduced from measurements of the adenylate energy charge (AEC) that the contribution to respiration from internal movement of oxygen was already very small at distances of 8 cm from the oxygen source (the shoot). Their results would seem to minimize the importance of internal movement of oxygen in a non-wetland root, but the roots were non-aerenchymatous.

Recently, we measured energy metabolism in roots of maize in an anaerobic medium in the presence and absence of an induced aerenchyma (Drew, Saglio & Pradet 1985). Aerenchyma was induced by pretreating the emerging adventitious roots with 5% oxygen in nitrogen for 11 days while non-aerenchymatous control roots were treated with 40% oxygen in nitrogen. After the pretreatment, roots of intact plants were placed in nutrient solution containing glucose as an additional respiratory substrate, and NaF to block fermentation, and the solution was vigorously bubbled with oxygen-free nitrogen gas so that the medium was anaerobic for about 1 h. Adventitious roots, up to 21 cm long, were then excised at the junction with the shoot and immediately plunged into liquid nitrogen so that exposure to air was minimal. The frozen roots were divided into segments corresponding to different distances from the shoot (up to 21 cm) and extracted for nucleotides. As a check on the influence of fully aerobic or anaerobic conditions on energy metabolism, corresponding segments from excised roots that had been incubated in anaerobic atmosphere were also extracted. Analysis of these excised

TABLE 1. Adenine-nucleotide content of excised maize root segments

	Length[*] (mm)	ATP	ATP/ADP	AEC[†]
Aerobic treatment	10–20	1220	4·68	0·89
	100–110	586	4·85	0·91
	195–205	747	5·33	0·88
	205–210	1120	3·42	0·74
Anerobic treatment	10–20	46	0·05	0·11
	100–110	28	0·06	0·13
	195–205	21	0·04	0·09
	205–210	59	0·05	0·12

In the aerobic treatment, the atmosphere was 40% (v/v) oxygen in nitrogen gas. In the anaerobic treatment the atmosphere was oxygen-free nitrogen gas. The amounts of nucleotides are in pmol per segment.

[*]Distance along the root (i.e. from the junction with the shoot) before excision.

[†]Adenylate energy charge $= \dfrac{[ATP] + 0·5 \, [ADP]}{[ATP] + [ADP] + [AMP]}$.

From Drew, Saglio & Pradet (1985).

segments showed that anoxia led to a very low concentration of ATP and to small values for ATP/ADP and the AEC (Table 1). Essentially the same pattern was found when segments were incubated after excision from roots pretreated with 5% (v/v) oxygen in nitrogen, or with NaF. In the intact roots (Table 2), dependent on a supply of oxygen from the shoot, the ATP concentrations, the ratio ATP/ADP and the AEC were all greater in the distal segments of aerenchymatous than non-aerenchymatous roots.

From these results it seems reasonable to conclude that the formation of an aerenchyma in maize had assisted internal aeration, and that there was an effective utilization of oxygen to boost energy metabolism. However, it is also apparent that

TABLE 2. Adenine-nucleotide content of segments along intact maize roots in an anaerobic environment

	Length (mm)	ATP	ATP/ADP	AEC
Aerobic treatment	10–20	1020	3·82	0·83
	100–110	227	1·05	0·50
	195–205	104	0·53	0·23
	205–210	287	0·71	0·38
5% oxygen pretreatment (aerenchyma present)	10–20	604	3·58	0·77
	100–110	1020	1·46	0·56
	195–205	647	1·74	0·67
	205–210	860	1·48	0·62

Plants had been pretreated for 11 days to induce aerenchyma, or to prevent aerenchyma formation. The amounts of nucleotides are in pmol per segment.
From Drew, Saglio & Pradet (1985).

internal transport was insufficient to maintain aerobic respiration at the rate that occurred in fully aerobic tissue. An ATP/ADP ratio of 1·5 in the apical segment of aerenchymatous roots (Table 2) would correspond to about 30% of the maximum respiration rate (compare Saglio, Raymond & Pradet 1983, Fig. 6), but this is appreciably greater than the ATP/ADP ratio of 0·5 in non-aerenchymatous root apices, corresponding to a respiration rate of only 10% of the maximum. At present we do not know whether such intermediate levels of energy metabolism apply uniformly across the root tissue, or whether the more closely-packed stelar tissues are at a much lower energy status. Such considerations are relevant to questions as to the effectiveness of aerenchymatous roots in ion transport, since the activity of ion pumps involved in loading of the xylem will be sensitive to the energy status and oxygen supply to the xylem parenchyma cells.

FUNCTION OF AERENCHYMATOUS ROOTS IN ION TRANSPORT

The formation of a lysigenous aerenchyma in maize and other cereals takes place by extensive loss of cells in the root cortex, external to the endodermis with cell collapse becoming first visible about 10 mm behind the root tip. Degradation of the root cortex in this matter might be expected to greatly inhibit radial ion transport to the xylem through loss of apoplastic and symplastic pathways. Ion uptake might then become confined to the apical zones of roots, but little transport to the xylem would be expected to take place there.

To assess the effect of cortical degeneration on radial ion transport we have recently compared the properties of non-aerenchymatous maize roots with aerenchymatous ones (Drew & Saker 1985). Aerenchyma was induced, as described above, by treatment with an atmosphere of 5% (v/v) oxygen in nitrogen. Adventitious roots of intact plants were inserted into an apparatus that allowed a 5–12 mm long segment, located 80–100 mm from the root tip, to be continuously supplied with nutrient solution containing radioactively labelled nutrient ions. During the uptake process the roots were fully exposed to aerated nutrient solution, whatever their pretreatment had been, so that oxygen supply was not limiting to respiration. After 24 h exposure to the labelled nutrient solution, plants were divided into treated segments, roots and shoots, which were counted separately for radioactivity. The transport of label out of the treated segment to the remainder of the plant was calculated from the radioactive counts and the specific activity of the labelled solution.

The results of a series of experiments to measure K^+ uptake and transport are given in Figs 2 and 3. When uptake by the labelled segment is expressed per unit root volume (Fig. 2), there is a trend for all the parameters to be greater in aerenchymatous roots, although only the greater accumulation of K^+ from the lower potassium concentration by the labelled segment (Seg) is statistically significant. However, the diameter of aerenchymatous roots is usually greater than non-aerenchymatous ones. When the same uptake data are expressed per unit root

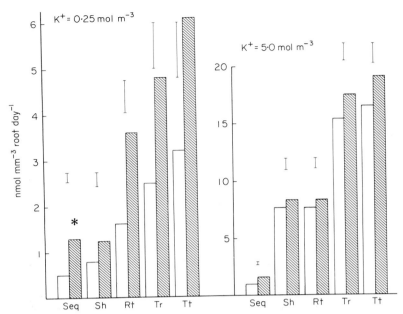

FIG. 2. Uptake and transport of K^+ by root segments of intact maize plants, with and without an induced aerenchyma, in 24 h, expressed per unit root volume. Seg, labelled root segment; Sh, transport to shoot; Rt, transported to other roots; Tr, total transported: Tt, total uptake. Bars indicate LSD ($P < 0.05$). * indicates statistically significant differences.

length (Fig. 3), it is apparent that the aerenchymatous roots are more effective in terms of transport to the xylem (Tr) and total uptake (Tt) and some other parameters. In parallel work we have also found transport of phosphate and of chloride to take place in aerenchymatous roots at least as effectively as in non-aerenchymatous ones.

Possible reasons for the apparent lack of effect of cortical degeneration on radial ion transport are discussed in detail elsewhere (Drew & Saker 1985). Our observations undoubtedly raise questions about the role of the cortical cells in radial ion transport. These cells have long been regarded as forming a 'collecting system', so that ion entry into the symplast is accelerated by transport across the greatly enlarged surface area of plasma membranes of cortical cells, accessible to ions in the outer solution via the free space. Entry of ions into the symplast is thus assumed to be rate limiting to ion transport to the xylem. However, our results suggest, on the contrary, that movement of ions across the cortex is not rate limiting, even in aerenchymatous roots with a degraded cortex. The corollary must be that the more important step is loading of ions from the xylem parenchyma to the xylem. In aerenchymatous roots, some radial cell walls remain after cell lysis and cavity formation, and there is evidence that they contribute an apoplastic pathway for ion movement towards the endodermis (Drew & Fourcy, unpubl.).

The anatomy of aerenchymatous maize roots resembles in some major features that of rice, with prominent lacunae, radial files of intact cells, and radial wall

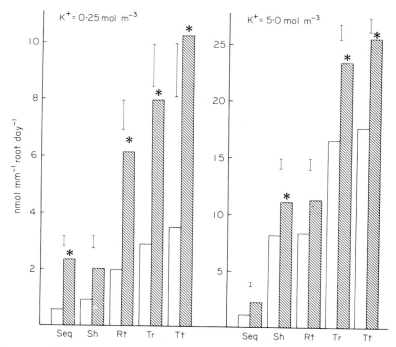

FIG. 3. Uptake and transport of K$^+$ by root segment of intact maize plants, with and without an induced aerenchyma, in 24 h, expressed per unit root length. Symbols as in Fig. 2.

residues. In rice the suberized hypodermis and the layer of sclerenchyma beneath are additional structures that may inhibit ion movement across the root. In young roots, a symplasmic pathway across these potential barriers is provided by plasmodesmata, but these are infrequent, and in more mature root zones (150 mm from the root tip) the sclerenchyma cell develop thick secondary walls and lose their cytoplasm. Clearly, such changes would be expected to arrest ion transport and restrict ion uptake to the younger zones of the root (Clark & Harris 1981). A similar conclusion concerning ion uptake properties in *Carex arenaria* was reached by Robards, Clarkson & Sanderson (1979), on the basis of the formation of a suberized hypodermis lacking plasmodesmata. Such roots were shown to be unable to transport water, phosphate or calcium ions. Perhaps the ability to conserve internal oxygen carries a penalty: interference in ion transport. However, to date there have been no reports concerning the ion transport properties of rice roots as a function of age or anatomical development, and a direct examination of this question is needed.

CONCLUSION

The possible importance of the adventitious root system has been emphasized in this paper in relation to acclimation to flooding. The aerenchymatous roots are able to grow into anaerobic media, the formation of lacunae can be shown to assist internal oxygenation and energy metabolism, and aerenchymatous roots function well in the transport of ions to the xylem. Perhaps the speed with which dryland cereal species acclimate to flooding is restricted by the number of adventitious roots initiated at the base of the shoot and the delay in developing a new root system of a size to replace fully the original one. It may also be erroneous to assume that the formation of an aerenchymatous root system is the only means by which species acclimate to flooding. Among five flood-tolerant tree species, two did not develop aerenchyma (Joly & Crawford 1982) so that their resistance mechanisms must include some other physiological properties. Additionally, the recognition that truly anoxia-tolerant species can survive, or even show limited growth, in the complete absence of oxygen (Barclay & Crawford 1982) supports the long-held view that some species acclimate metabolically, their energy status maintained at a sufficient level by anaerobic metabolism.

ACKNOWLEDGMENTS

I am grateful to L. R. Saker for reading a draft of the manuscript, and to Joan Llewellyn for typing it.

REFERENCES

Armstrong, W. (1979). Aeration in higher plants. *Advances in Botanical Research*, 7, 225–331.

Barclay, A. M. & Crawford, R. M. M. (1982). Plant growth and survival under strict anaerobiosis. *Journal of Experimental Botany*, 33, 541–549.

Belford, R. K., Cannell, R. Q., Thomson, R. J. & Dennis, C. W. (1980). Effects of waterlogging at different stages of development on the growth and yield of peas (*Pisum sativum* L.) *Journal of the Science of Food and Agriculture*, 31, 857–869.

Campbell, R. & Drew, M. C. (1983). Electron microscopy of gas space (aerenchyma) formation in adventitious roots of *Zea mays* L. subjected to oxygen shortage. *Planta*, 157, 350–357.

Cannell, R. Q. (1977). Soil aeration and compaction in relation to root growth and soil management. *Applied Biology*, 2, 1–86.

Cannell, R. Q. & Belford, R. K. (1980). Effects of waterlogging at different stages of development on the growth and yield of winter oilseed rape (*Brassica napus* L.). *Journal of the Science of Food and Agriculture*, 31, 963–965.

Cannell, R. Q., Belford, R. K., Gales, K., Dennis, C. W. & Prew, R. D. (1980). Effects of waterlogging at different stages of development on the growth and yield of winter wheat. *Journal of the Science of Food and Agriculture*, 31, 117–132.

Clark, L. H. & Harris, W. H. (1981). Observations on the root anatomy of rice (*Oryza sativa* L.). *American Journal of Botany*, 68, 154–161.

Crawford, R. M. M. (1982). Physiological responses to flooding. In *Physiological Plant Ecology 2, Encyclopaedia of Plant Physiology N.S. Vol. 12B.* (Ed by O. L. Large, P. S. Nobel, C. B. Osmond & H. Ziegler) pp. 453–477. Springer-Verlag, Berlin.

Drew, M. C. (1983). Plant injury and adaptation to oxygen deficiency in the root environment: a review. *Plant and Soil*, 75, 179–199.

Drew, M. C., Jackson, M. B. & Giffard, S. (1979). Ethylene-promoted adventitious rooting and development of cortical airspaces (aerenchyma) in roots may be adaptive responses to flooding in *Zea mays* L. *Planta*, 147, 83–88.

Drew, M. C., Jackson, M. B., Giffard, S. C. & Campbell, R. (1981). Inhibition by silver ions of gas space (aerenchyma) formation in adventitious roots of *Zea mays* L. subjected to exogenous ethylene or oxygen deficiency. *Planta*, 153, 217–224.

Drew, M. C. & Lynch, J. M. (1980). Soil anaerobiosis, micro-organisms, and root function. *Annual Review of Phytopathology*, **18**, 37–66.

Drew, M. C., Saglio, P. H. & Pradet, A. (1985). Larger adenylate energy charge and ATP/ADP ratios in aerenchymatous roots of *Zea mays* in anaerobic media as a consequence of improved internal oxygen transport. *Planta*, in press.

Drew, M. C. & Saker, L. R. (1983). Induction of aerenchyma formation by nutrient deficiency in well aerated maize roots. *Agricultural Research Council Letcombe Laboratory Annual Report*, **1982**, 41–42.

Drew, M.C. & Saker, L. R. (1985). Ion transport to the xylem in aerenchymatous roots of *Zea mays* L. *Journal of Experimental Botany*, in press.

Gambrell, R. P. & Patrick, W. H. (1978). Chemical and microbiological properties of anaerobic soils and sediments. In *Plant Life in Anaerobic Environments* (Ed. by D. D. Hoak & R. M. M. Crawford) pp. 375–423. Ann Arbor Science, Ann Arbor, Michigan.

Greenwood, D. J. (1967). Studies on the transport of oxygen through the stems and roots of vegetable seedlings. *New Phytologist*, **66**, 337–347.

Greenwood, D. J. (1968). Effects of oxygen distribution in the soil on plant growth. In *Root Growth* (Ed. by W. J. Whittington) pp. 202–221. Butterworth, London.

Greenwood, D. J. & Goodman, D. (1971). Studies on the supply of oxygen to the roots of mustard seedlings (*Sinapsis alba* L.). *New Phytologist*, **70**, 85–96.

Jackson, M. B. & Drew, M. C. (1984). Effects of flooding on growth and metabolism of herbaceous plants. In *Flooding and Plant Growth*. (Ed. by T. T. Kozlowski) pp. 47–128. Academic Press, New York.

Jackson, M. B., Fenning, T. M., Drew, M. C. & Saker, L. R. (1985). Stimulation of ethylene production and gas space (aerenchyma) formation in adventitious roots of *Zea mays* L. by small partial pressures of oxygen. *Planta*, in press.

Jackson, W. T. (1955). The role of adventitious roots in recovery of shoots following flooding of the original root systems. *American Journal of Botany*, **42**, 816–819.

Joly, C. A. & Crawford, R. M. M. (1982). Variation in tolerance and metabolic responses to flooding in some tropical trees. *Journal of Experimental Botany*, **33**, 799–809.

Kawase, M. (1974). Role of ethylene in induction of flooding damage in sunflower. *Physiologia Plantarum*, **31**, 29–38.

Kawase, M. (1981). Anatomical and morphological adaptations of plants to waterlogging. *HortScience*, **16**, 30–34.

Konings, H. (1982). Ethylene-promoted formation of aerenchyma in seedling roots of *Zea mays* L. under aerated and non-aerated conditions. *Physiologia Plantarum*, **54**, 119–124.

Konings, H. & Verschuren, G. (1980). Formation of aerenchyma in roots of *Zea mays* in aerated solutions, and its relation to nutrient supply. *Physiologia Plantarum*, **49**, 265–270.

Konings, H. & de Wolf, A. (1984). Promotion and inhibition by plant growth regulators of aerenchyma formation in seedling roots of *Zea mays*. *Physiologia Plantarum*, **60**, 309–314.

Kramer, P. J. (1951). Causes of injury to plants resulting from flooding of the soil. *Plant Physiology*, **29**, 241–245.

McPherson, D. C. (1939). Cortical air spaces in the roots of *Zea mays* L. *New Phytologist*, **39**, 190–202.

Ponnamperuma, F. N. (1984). Effects of flooding on soils. In *Flooding and Plant Growth* (Ed. by T. T. Kozlowski), pp. 9–45. Academic Press, New York.

Robards, A. W., Clarkson, D. T. & Sanderson, J. (1979). Structure and permeability of the epidermal/hypodermal layers of the sand sedge (*Carex arenaria* L.). *Protoplasma*, **101**, 331–347.

Roberts, J. K. M., Callis, J., Jardetzky, O., Walbat, V. & Freeling, M. (1984). Cytoplasmic acidosis as a determinant of flooding intolerance in plants. *Proceedings of the National Academy of Sciences, U.S.A*, **81**, 6029–6033.

Saglio, P. H., Raymond, P. & Pradet, A. (1983). Oxygen transport and root respiration of maize seedlings. A quantitative approach using the correlation between ATP/ADP and the respiration rate controlled by oxygen tension. *Plant Physiology*, **72**, 1035–1039.

Young, S. F. & Hoffman, N. E. (1984). Ethylene biosynthesis and its regulation in higher plants. *Annual Review of Plant Physiology*, **35**, 155–189.

The ecophysiology of mangroves

G. R. STEWART*

Department of Botany, Birkbeck College, University of London, Malet Street, London, WC1E 7HX

M. POPP

Institut für Pflanzenphysiologie, Universität Wien, Althanstrasse 14, Postfach 285, Vienna, A 1091, Austria

SUMMARY

Mangrove vegetation occurs by all tropical seas, particularly on lagoons, inlets and estuaries. The soil is flooded with water either permanently or at high tide. Salinity can range from above that of sea water to almost fresh water. This paper considers (a) the characteristics of carbon and nitrogen assimilation exhibited by mangroves, (b) their response to flooding, long and short term, (c) their mechanisms of salt tolerance and the role played by compatible solutes, and (d) the possible involvement of compatible solutes in the temperature adaptation of mangroves. The physiological characteristics of these tropical woody halophytes is compared with those of temperate herbaceous halophytes.

INTRODUCTION

The term mangrove is generally used in two senses: firstly to describe the woody plant communities distributed in the tropical and sub-tropical tidal zone between the lowest and highest tidal limits and secondly for the plant species found in these communities (Barth 1982). Over fifty species of mangrove have been described, the majority of which are dicotyledonous. Typical mangroves are *Avicennia*, *Brugiera*, *Ceriops*, *Conocarpus*, *Kandelia*, *Laguncularia*, *Lumnitzera*, *Rhizophora* and *Sonneratia*. Mention should also be made of the halophytic fern, *Acrostichum aureum*, which is found in nearly all mangrove vegetation.

The dominance of tree species in tropical coastal vegetation presents a marked contrast to the saltmarsh vegetation of temperate areas which is composed largely of small herbaceous species. Mary Kingsley, the Victorian naturalist, vividly describes the appearance of mangrove vegetation, 'In every other direction you will see the apparently endless walls of mangrove, unvarying in colour, unvarying in form, unvarying in height' (*Travels in West Africa*, 1897). This is a description of the ideal, uniform material for physiological studies.

Mangroves are characterized by their ability to grow in saline soils which are almost invariably anoxic and where climatic conditions provide a combination of

---

*Present address: Department of Botany and Microbiology, University College London, Gower Street, London, WC1E 6BT.

high temperatures and high irradiance. In this paper we will consider the characteristics of carbon and nitrogen assimilation, responses to flooding, mechanisms of salt tolerance and tolerance of high temperatures.

CARBON AND NITROGEN ASSIMILATION

Photosynthesis

Studies of photosynthesis have been carried out on a number of mangrove species. The light response curves indicate saturation of net photosynthesis at photon flux densities of around 700–800 mol m^{-2} s^{-1} (Moore *et al.* 1972, 1973; Attiwill & Clough 1980; Andrews, Clough & Miller 1984) and the light compensation points are around 30 mol photons m^{-2} s^{-1} (Andrews *et al.* 1984). The temperature optimum for photosynthesis of *Rhizophora apiculata* leaves was found to be 30°C and a large reduction in rate occurred at temperatures a few degrees above the optimum (Andrews *et al.* 1984). The CO_2 compensation point for *R. apiculata* was about 60 bars (Andrews *et al.* 1984). The physiological characteristics reported for mangroves appear then to be those typical of plant species exhibiting C$_3$ photosynthesis.

Joshi and his co-workers have studied some biochemical features of photosynthesis in a number of mangrove species. Phosphoenol pyruvate carboxylase was found to be more active than ribulose bisphosphate carboxylase (Joshi *et al.* 1975, 1984). Aspartate and alanine were found to be heavily labelled following short-term exposure to $^{14}CO_2$. The activity of pyruvate phosphate dikinase of several mangroves was found to be similar to that of plants with C$_4$ photosynthesis. *Rhizophora mucronata* was reported to exhibit high levels of NAD-malic enzyme in the light and negligible activity in darkness, another characteristic of some C$_4$-plants. On the basis of these observations Joshi and co-workers have proposed the operation of a modified C$_4$ pathway in mangroves.

Other contributors to this volume have drawn attention to aquatic plants exhibiting C$_4$ physiological characteristics and C$_3$ biochemical characteristics; can it really be that mangroves have a C$_3$ physiology and a C$_4$ biochemistry? The differences in results obtained by Joshi *et al.* (1984) and Andrews *et al.* (1984) cannot be ascribed to the use of different species since several have been worked on by both groups. However, the $\delta^{13}C$ values reported for sixteen species of Australian mangroves range from $-32\cdot2‰$ to $-24\cdot6‰$ (Andrews *et al.* 1984). These are values fairly typical of C$_3$ plants and the most valid conclusion regarding photosynthesis would appear to be that they, like other tree species, photosynthesize via the C$_3$ pathway.

Nitrogen assimilation

The soil of mangrove vegetation is generally very anoxic, the result of prolonged waterlogging. These are conditions which inhibit or restrict the activities of soil-nitrifying bacteria. It is hardly surprising then that in a study of nitrate reductase activity by species of mangrove vegetation in West Africa, little evidence was found for nitrate utilization (Stewart & Orebamjo 1984). All of these mangrove species were found able to utilize nitrate as a nitrogen source and to exhibit nitrate reductase activity when grown in nitrate (Stewart & Orebamjo 1984). *Avicennia nitida (A. germinas)* was found to be active in root assimilation of nitrate and to transport nitrogen from root to shoot in the form of the amide asparagine and the ureides allantoin and allantoic acid (Stewart & Orebamjo 1984). Two other species, *Aegiceras* and *Aegialitis*, also appear to be root assimilators of nitrate (Popp & Stewart unpubl. data). Investigations of ammonia assimilation in *A. germinans* indicate that high activities of the enzymes of the glutamate synthase cycle, glutamine synthetase and glutamate synthase are present. The addition of an inhibitor of glutamine synthetase, methionine sulphoximine, to roots of *A. germinans* was found to bring about a rapid accumulation of ammonia (Abbey *et al.* 1986). This suggests that ammonia assimilation is via the glutamate synthetase cycle rather than by the possible alternative route catalysed by glutamate dehydrogenase (see Miflin & Lea 1981 for further discussion of the pathways of ammonia assimilation).

It is now recognized that in addition to its role in primary ammonia assimilation glutamine synthetase is also active in the recycling of ammonia released in photorespiration (Keys *et al.* 1978; Woo & Osmond 1982). The leaves of many plants exhibit two forms of glutamine synthetase, one located in the chloroplasts and the other in the cytosol (see McNally & Hirel 1983). Leaves of *A. germinans* exhibit very small amounts of the chloroplastic form of glutamine synthetase, over 95% of the activity being associated with the cytosolic isoforms (Abbey *et al.* 1986). Low activities of chloroplastic glutamine synthetase are associated with low rates of photorespiration (see e.g. Chandler *et al.* 1985) and these are reported in mangroves by Joshi *et al.* (1984). In our studies of several woody plants low chloroplastic glutamine synthetase activities is a common feature of plants which preferentially assimilate nitrate in their roots.

Responses to flooding

There are relatively few tree species which are able to tolerate permanently flooded soil but among the best examples are mangroves. All mangroves have a system of laterally spreading cable roots and smaller vertically-descending anchor roots, which although relatively shallow is quite extensive and results in high root/shoot ratios (see Saenger 1982). Several species exhibit above-ground roots of one form or another. In *Avicennia* and *Xylocarpus* there are pneumatophores, which are unbranched, negatively geotropic, are covered in lenticells and have a well developed aerenchyma; *Sonneratia* has similar but branched pneumatophores (Gill & Tomlinson 1975). Stilt roots are characteristic of *Rhizophora* and *Ceriops* and like pneumatophores these are rich in lenticels (Saenger 1982). The stilt roots of *Rhizophora* have a high oxygen content (15–18%) and if the lenticels are blocked

TABLE 1. Response of *Avicennia nitida* to flooding

Treatment	Growth* Shoot extension (cm^{-1})	% Mortality	Respiration μl O_2M^{-1} gfw		Alcohol dehydrogenase nmol min^{-1} mg^{-1} protein		Glutamine synthetase nmol min^{-1} mg^{-1} protein	
			Primary roots	Laterals	Primary roots	Laterals	Primary roots	Laterals
Drained	14·6	8·3	2·8	1·1	96	118	88	60
Drained + 50 mM NaCl	14·0	46	4·7	1·7	117	199	94	90
Flooded	11·3	42	3·1	1·9	132	465	39	54
Flooded + 50 mM NaCl	14·4	17	4·3	3·3	95	262	30	60

*All measurements were made after 6 months' growth. Flooding was achieved by immersing pots in water or 50 mM NaCl.

there is a rapid depletion of oxygen (Scholander, van Dam & Scholander 1955). It is assumed that those above-ground roots function in aeration and Canoy (1975) has shown an increase in the number of stilt roots produced by *Rhizophora* at reduced oxygen concentrations. Laboratory studies indicate that the formation of pneumatophores is an adaptive response in *Avicennia*, prolonged flooding being necessary for their development (Smirnoff & Stewart, unpubl. data). *Aegialitis* and *Excoecaria* have no specialized aerial roots and tend to be found on less waterlogged sediments. The growth of young seedlings of *Avicennia* is reduced and mortality increased by flooding (Table 1). The addition of 50 mM NaCl increased growth of flooded seedlings and decreased their mortality. Stimulation of growth under poor aeration by salt has been reported for *Puccinellia peisonis* (Stelzer & Lauchli 1977). There is little indication that flooding alters the respiration of primary root tips or lateral roots of young *Avicennia* seedings (Table 1). Alcohol dehydrogenase activity increases somewhat in lateral roots when flooded although the increase is less when NaCl is present. The increase in alcohol dehydrogenase is greatest in secondary roots, suggesting these may have a less efficient aeration system than primary roots. Glutamine synthetase activity was relatively unaffected by flooding. These results indicate adjustment to flooding in the metabolism and growth of *Avicennia*. Short-term experiments show much greater perturbations of metabolism, there being a sixfold to tenfold increase in alcohol dehydrogenase and a 50–60% decrease in glutamine synthetase over the first few days of flooding (Smirnoff & Stewart, unpubl. data). The longer term adaptation to flooding is likely to involve morphological as well as biochemical changes in root characteristics.

Mechanisms of salt regulation

Various physiological strategies have been suggested to control the uptake, accumulation and secretion of salt, and various authors have proposed that mangroves can be classified into two or three categories which reflects the primary mechanism of salt regulation. Thus we have salt secretors, salt excluders and salt accumulators (Joshi *et al.* 1975; Lear & Turner 1977; Saenger 1982). In our view these categorizations are somewhat misleading, particularly that of salt excluders, since all mangrove species accumulate large amounts of salt in their leaves, irrespective of the mechanisms employed to regulate salt content (Table 2; Popp 1984a,b). In this respect mangroves resemble many other halophytic angiosperms where osmotic adjustment to high external salt concentrations is achieved (in large part at least) by the uptake and accumulation of sodium and chloride ions. In culture solution the leaf water potential of *A. germinans* was found to decrease more or less linearly with external water potential, and leaf solute potential to decrease in parallel with leaf water potential (Stewart & Ahmad 1983). Leaf analysis showed that in this species Na and Cl account for over 60% of the leaf solute potential (Stewart & Ahmad 1983). It is clear then from laboratory and field studies that NaCl accumulation is the key feature of osmotic adjustment in mangroves as in many other halophytes.

TABLE 2. Sodium and chloride contents of mangrove leaf tissue

Species	Na$^+$	Cl^f	Total inorganic ions (Na$^+$, K$^+$, Ca^{2+}, Mg^{2+}, Cl$^-$, SO$_4^{2-}$
	mol m^{-3} plant water		mol m^{-3} plant water
Acanthus ilicifolius	480	550	1203
Aegialitis annulata	520	555	1415
Aegiceras corniculatum	520	560	1340
Avicennia marina	500	560	1303
Bruguiera gymnorhiza	350	580	1292
Ceriops tagal	300	550	1339
Rhizophora apiculata	400	500	1176
Rhizophora lamarckii	370	520	1206
Rhizophora stylosa	400	530	1312
Sonneratia alba	400	490	1073

Data is calculated from Popp (1984a, b). Young leaves harvested from field-grown plants were studied.

The salt concentrations found in the leaves of various field-grown mangrove species are over 500 mM, irrespective of their ability to secrete or exclude salt (Table 2). Studies of the salt sensitivity of a range of enzymes from halophytic plants have shown that they are markedly inhibited (50–90%) by high concentrations of NaCl (Flowers, Troke & Yeo 1977). High concentrations of NaCl, however, are reported to stimulate photosystems I and II in *Avicennia marina* and other salt-tolerant species (Critchley 1982; Critchley *et al.* 1982). However the results obtained by Ball, Taylor & Terry (1984) give no evidence that the photosynthetic membranes of salt-tolerant species are adapted to function at salt concentrations found in leaves of salt-grown plants. The requirement for high chloride concentration for maximum rates of photosynthetic O$_2$ evolution in *A. marina* appears to be an artefact arising from the loss of the 23 kD polypeptide during isolation of the thylakoids (Anderson *et al.* 1984). There is no reason then to suppose that enzymes from mangroves are any more salt-resistant than those of herbaceous halophytes. Ion localization studies (Harvey *et al.* 1981) show very marked differences in Na and Cl concentrations in different compartments of the cell. Vacuolar Na and Cl concentrations in the mesophyll cells of the herbaceous halophyte *Suaeda maritima* are at least five times those of the cytoplasm and chloroplast. Cytoplasmic ion concentrations considerably lower than those of the vacuole necessitate a lowering of cytoplasmic water potential in order to maintain water potential equilibrium across the tonoplast. Osmotic adjustment in the cytoplasm of halophytes is thought to be achieved by the synthesis of compatible organic solutes. Many angiosperm halophytes are characterized by the presence of large amounts of the amino acid proline or the methylated onium compound, glycine betaine, and both have been suggested to function as compatible cytoplasmic solutes (Stewart & Lee 1974; Storey & Wyn Jones 1975).

Glycine betaine is present in large amounts in the mangroves *Avicennia marina* (Wyn Jones & Storey 1981) and *A. nitida* (Stewart & Ahmad 1983). Popp has carried out an extensive analysis of the soluble organic compounds present in

TABLE 3. Concentrations of low molecular weight solute present in mangroves

Species	Solute mol m^{-3} plant water
Acanthus ilicifolius	Glycine betaine 39
Aegialitis annulata	Pinitol 105; chiroinositol 150. proline 28
Aegiceras corniculatum	Mannitol 230
Avicennia marina	Glycine betaine 61
Bruguiera gymnorhiza	Pinitol 80
Ceriops tagal	Pinitol 150
Excoecaria agallocha	Quebrachitol 90
Heritiera littoralis	Glycine betaine 72
Hibiscus tiliaceus	Glycine betaine 102
Lumnitzera littorea	Mannitol 110
Osbornea octodenta	?
Rhizophora apiculata	Pinitol 220
Scyphiphora hydrophylacea	Mannitol 225
Sonneratia alba	Mannitol 200
Xylocarpus mekongensis	Proline 57

Data is calculated from Popp (1984a, b). Young leaves harvested from field-grown plants were studied.

several species of Australian mangroves. The principal compounds present which could serve as compatible solutes are glycine betaine (*Avicennia* sp., *Hibiscus tilaeous*), proline (*Aegialitis* and *Xylocarpus*) and the sugar alcohols, chiro-inositol (*Aegialitis*), mannitol (*Aegiceras*, *Lumnitzera*, *Sonneratia* and *Scyphiphora*), pinitol (all Rhizophoraceae and the fern *Acrostichum*) and quebrachitol (*Excoecaria*). See Popp (1984b), Popp, Larher & Weigel (1984) and Table 4.

The occurrence of large amounts of sugar alcohols in many mangrove species contrasts with the very restricted occurrence of these compounds in temperate herbaceous halophytes (see Stewart & Ahmad 1983).

Salt exclusion

Based on observations of very low Na and Cl concentrations in the xylem sap of various mangrove species, Scholander (1968) proposed a non-metabolic ultrafiltration process in mangrove roots. Salt exclusion from the xylem sap is reported to be more pronounced in non-secreting species such as *Rhizophora mucronata* than in secreting species such as *Aegialitis annulata* (Atkinson *et al.* 1967). Recent studies with a secreting species *Avicennia marina* (Field 1984), which is described by Scholander (1962, 1966) as being a partial excluder but capable of excreting large amounts of salt from its leaves, cast doubt on Scholander's results. Exudate was always found to be at least 50% of the salinity of the seawater bathing the roots.

Although non-secreting species may exclude 90–95% of the NaCl in the soil solution calculations based on xylem concentrations, transpiration rates and salt content of leaves suggest the amount of salt delivered to the leaf could be as much as thirty times that found in the leaf (Clough 1984). One possibility is that NaCl is removed from the xylem fluid during movement up to the leaves. The woody parts of mangroves have a high salt content (Spain & Holt 1980) and stem tissue of *A. germinans* accumulates similar concentrations of salt as found in leaf tissue (Stewart & Ahmad, unpubl. data).

Another possibility is that of retranslocation of salt (Clough, Andrews & Cowan 1982) via the phloem. The salt-tolerant herbaceous species *Trifolium alexandrium* (Winter 1982) has been shown to retranslocate Na and Cl from leaves to other parts of the plant. Popp (1984a) observed that while the concentrations of Ma^{2+} and SO^{2-}_4 increased in older leaves of the Rhizophoreaceae, Na^+ and Cl^- concentrations did not and it was suggested that this occurred in part by transport out of leaves of Na^+ and Cl^-.

Succulence

The influence of sodium on succulence is regarded as a general phenomenon in plants (Jennings 1968). The potential importance of changes in leaf succulence in regulating salt concentrations is shown for the mangrove *Laguncularia racemosa* (Biebl & Kinzel 1965). The chloride content of leaves of increasing age was found to rise when calculated on a leaf area basis but remained more or less constant when calculated on the basis of plant water per leaf area. Leaf succulence appears to alter less in salt-secreting species than in non-secreting species (Table 4).

TABLE 4. Degree of succulence in various mangrove leaves and leaves from *Kalanchoe blossfeldiana*

	g plant water . m^{-2} leaf area	g plant water . g^{-1} dry substance	Average leaf area mm^{-2}
Aegialitis annulata	239·6	2·66	1248
Aegialitis annulata	231·1	1·39	1709
Aegiceras corniculatum	215·8	2·20	373
Aegiceras corniculatum	220·7	1·11	1628
Avicennia marina	336·8	2·21	502
Avicenna marina	370·3	1·36	1704
Ceriops tagal	265·8	1·69	285
Ceriops tagal	425·6	2·51	1713
Bruguiera exaristata	264·1	1·83	1007
Bruguiera exaristata	510·2	2·30	1583
Rhizophora stylosa	296·4	2·09	939
Rhizophora stylosa	* 679·1	2·42	2638
*Laguncularia racemosa**	368·2	—	1635
Laguncularia racemosa	1077·5	—	1380
Laguncularia racemosa	1140·4	—	1845
Kalanchoe blossfeldiana Short day (CAM mode)	848·2	14·25	—
Kalanchoe blossfeldiana Short day (CAM mode)	1008·4	27·34	—
Long day (no CAM)	526·9	7·20	—
Long day (no CAM)	552·3	7·77	—

*Calculated from Biebl & Kinzel (1965).

Salt secretion

Salt secretion via salt glands occurs in only six genera of mangroves, *Acanthus*, *Aegialitis*, *Aegiceres*, *Avicennia*, *Laguncularia* and *Sonneratia*, and the structure of these glands has been reviewed by Saenger (1982). The rates of salt secretion exhibited by some mangrove species is markedly higher than those reported for herbaceous salt-secreting halophytes (Table 5). In *Avicennia* the salt glands are on the underside of the leaves and are active only during the day.

Other ions are secreted along with Na^+ and Cl^-, in particular K^+, and Mg^{2+} and some SO^{2-}_4 are present in the secretion (Popp 1984a). Oxalate has been found in the secretion from salt gland of *Avicennia marina* (Popp & Gordon, unpubl. data). A similar secretion of oxalate has been reported for *Atriplex hymenelytra* (Bennert & Schmidt 1983). Ball (1981) has shown that the salt glands of *Aegiceras corniculatum* and *Avicennia marina* can secrete 47% and 33% respectively of the total sodium and chloride absorbed by plants grown on 50% seawater. In these and other salt-secreting species the salt glands must play a major role in the regulation of salt content.

TABLE 5. Salt secretion rates in mangroves and herbaceous halophytes

Species	Ion	Secretion rate $\mu mol\ m^{-2}\ s^{-1}$	Reference
Mangroves			
Aegialitis annulata	NaCl	0·1–0·6	Scholander et al. (1962)
Aegialitis annulata	Cl^-	0·93	Atkinson et al. (1967)
Aegiceras corniculatum	NaCl	1·1–2·2	Scholander et al. (1962)
Avicennia marina	Cl^-	0·1–0·2	Scholander et al. (1962)
Avicennia marina	Cl^-	0·53	Boon & Allaway (1982)
Avicennia marina	Na^+	0·40	Boon & Allaway (1982)
	Cl^-	0·046	
Avicennia nitida (50 mM)	Na^+	≥·053	Ahmad & Stewart (unpubl. data)
Avicennia nitida (400 mM)	Cl^-	0·30	Ahmad & Stewart (unpubl. data)
	Na^+	0·33	
Herbaceous halophytes			
Armeria maritima	Na^+	≥·002	Rozema, Gude & Pollak (1981)
Glaux maritima	Na^+	0·006	Rozema, gude & Pollak (1981)
Limonium vulgare	Na^+	0·020	Rozema, Gude & Pollak (1981)
Spartina anglica	Na^+	0·044	Rozema, Gude & Pollak (1981)

Temperature adaptation

The greatest diversity of mangroves and the best developed are found near the equator in Indonesia, New Guinea and the Philippines, where solar radiation and air temperatures are generally high. Various workers have shown that leaf temperature is a critical factor determining photosynthetic rates under natural conditions (Miller 1972; Moore et al. 1973; Ball 1981; Andrews et al. 1984). The rate of assimilation is relatively temperature insensitive over the range 17–30°C but falls off sharply above 30°C. Andrews *et al.* (1984) have shown that the nearly vertical orientation of sun leaves in the Rhizophoraceae is critical as an adaptation which minimizes leaf temperature. The stomata of Rhizophora are also very sensitive to small increases in temperature above 30°C (Andrews *et al.* 1984). In mesophytic plants transpiration can account for up to 50% of heat loss from leaves (Gates 1968). Closure of stomata of mangroves could lead to increased leaf temperatures, particularly in the more succulent species. Leaf temperatures of some mangroves exceed ambient air temperature by as much as 15°C (Clough, Andrews & Cowan 1982).

It is interesting that several of the compatible solutes accumulated in mangroves increase the heat stability of a variety of enzymes from plants and animals (Paleg *et al.* 1981). More recently it has been shown that mannitol and sorbitol also enhance the heat stability of glutamine sythetase and other enzymes (Smirnoff & Stewart 1985). Nash, Paleg & Wiskich (1982) have shown that proline and betaine also increase the heat stability of enzymes in isolated intact mitochondria and chloroplasts. Glycine betaine and other compatible solutes have been shown to protect against oxalate destabilization of membranes (Jolivet, Hamelin & Larher 1983); this may be important in those mangroves which are salt secretors since they contain free oxalate (Popp 1984a).

It may be significant in relation to the wide occurrence of polysols in mangroves that sorbitol and mannitol appear more effective than proline or glycine betaine in preventing heat damage of enzymes (Smirnoff & Stewart 1985).

Thus the accumulation of compatible solutes in mangroves may, in addition to playing a role in osmoregulation, be significant in protecting enzymes and membranes against heat and chemical characterization.

CONCLUDING REMARKS

The morphological, physiological and biochemical characteristics of mangroves exhibit many features suggestive of their adaptation to growth in saline, anoxic soils in habitats with high air temperatures and irradiance. Although the mangrove environment is a harsh one with respect to the amplitude and fluctuations of its physicochemical characteristics, it is one in which competition is low. On the basis of their morphological, and in part also their reproductive, features Saenger (1982) has grouped mangroves into colonizing, seral and climax species. *Rhizophora* and *Avincennia* are regarded as colonizers, *Ceriops* and *Aegiceras* as seral species and *Lumnitzera*, *Aegialitis* and *Osbornia* as climax species. The extent to which these groups also exhibit common physiological attributes deserves further consideration.

The mechanism by which mangroves adjust to low external water potentials, avoid intracellular salt toxicity and regulate their salt content are similar to those seen in temperate herbaceous halophytes. The role played by the regulation of salt uptake and secretion in controlling concentrations is well established. The physiological significance of other potential mechanisms which control tissue salt concentrations, such as retranslocation and storage in non-metabolic tissues, requires to be determined by further experimentation.

Mangrove vegetation is confined to the tropical and sub-tropical intertidal zone. Many of the morphological, physiological and biochemical features which seem to allow them to grow and reproduce in this very harsh environment are also characteristic of temperate herbaceous halophytes. It is striking therefore that woody halophytes are more or less absent from temperate and cooler saline environments. Is it that the energetic and nutritional demands of halophilism preclude the growth and reproduction of the woody life form in anything other than the sub-tropical—tropical environment? The would-be-researcher tackling this and other problems of mangrove ecology would be well advised to heed Mary Kingsley's words 'and why, having reached this point of absurdity, you need have gone and painted the lily and adorned the rose, by being such a colossal ass as to come fooling about in mangrove swamps' (*Travels in West Africa*, 1897).

REFERENCES

Abbey, K., Patel, M., Stewart, G. R. & Sumar, N. (1986). Features of ammonia metabolism in woody plants. *Annals of Botany*, (in press).

Anderson, B., Critchley, C., Ryme, I. J., Jansson, C., Larsson, E. & Anderson, J. M. (1984). Modification of the chloride requirements for photosynthetic O_2 evolution: the role of the 23kD polypeptide. *FEBS Letters*, **168**, 113–117.

Andrews, T. J., Clough, B. F. & Miller, G. J. (1984). Photosynthetic gas exchange properties and carbon isotope ratios of some mangroves in North Queensland. In *Physiology and Management of Mangroves* (Ed. by H. J. Teas), pp. 15–24. Dr W. Junk. The Hague.

Atkinson, M. R., Findlay, G. P., Hope, A. B., Pitman, M. G., Saddler, H. D. W. & West, K. R. (1967). Salt regulation in the mangroves *Rhisophora mucronata* Lam and *Àegialitis annulata* R. Bv. *Australian Journal of Biological Sciences*, **20**, 588–599.

Attiwill, P. M. & Clough, B. F. (1980). Carbon dioxide and water vapour exchange in the white mangrove. *Photosynthetica*, **14**, 40–47.

Ball, M. C. (1981). *Physiology of photosynthesis in two mangrove species: responses to salinity and other environmental factors*. Ph.D. thesis, Australian National University.

Ball, M. C., Taylor, S. E. & Terry, N. (1984). Properties of thylakoid membranes of the mangroves *Avicennia germinans* and *Avicennia marina* and the sugar beet, *Beta vulgaris*, grown under different salinity conditions. *Plant Physiology*, **76**, 531–535.

Barth, H. (1982). The biogeography of mangroves. In *Tasks for Vegetation Science Vol. 2*, (Ed. by D. N. Sen & K. S. Majpurohit) pp. 35–60. Dr W. Junk, The Hague.

Bennert, H. W. & Schmidt, B. (1983). Untersachungen zer salgebscheidung bei Atriplex hymenelytra (Tork) Wats (Chenopodiaceae). *Flora*, **174**, 341–355.

Biebl, R. & Kinzel, H. (1965). Blatbau und Salzhaushalt von *Laguncularia racemosa* (L.) Gaertn, f. und anderer mangrovenblaume auf Puerto Rica. *Osterreichische Botanische Zeitung*, **112**, 56–93.

Boon, P. I. & Allaway, G. W. (1982). Assessment of leaf-washing techniques for measuring salt secretion in *Avicennia marina* (Forsk) Vieh. *Australian Journal of Plant Physiology*, **9**, 725–734.

Canoy, M. J. (1975). Diversity and stability in a Puerto Rican *Rhizophora mangle* L. forest. In *Proceedings of the International Symposium on Biology and Management of Mangroves Vol. 1*, (Ed. by G. E. Walsh, S. C. Snedaker & H. J. Teas). pp. 344–356. Institute of Food and Agricultural Sciences, University of Florida, Gainesville, Florida.

Chandler, G., Ladley, P., McNally, S. F., Patel, M., Stewart, G. R. & Sumar, N. (1985). The activity and isoform complement of glutamine synthetase in *Panicum* species differing in photosynthetic pathways. *Journal of Plant Physiology*, **121**, 13–21.

Clough, B. F. (1984). Mangroves. In *Control of Crop Productivity* (Ed. by C. J. Pearson), pp. 253–268. Academic Press, Australia.

Clough, B. F., Andrews, T. J. & Cowan, I. R. (1982). Physiological processes in mangroves. In *Mangrove Ecosystems in Australia: Structure, Function and Management*, (Ed. by B. F. Clough) pp. 193–210, Australian National University Press, Canberra.

Critchley, C. (1982). Stimulation of photosynthetic electron transport on a salt tolerant plant by high chloride concentrations. *Nature*, **298**, 483–485.

Critchley, C., Baianu, I. C., Govindjee, H. S. & Gutowsky, H. S. (1982). The role of chloride in O_2 evolution by thylakoids from salt tolerant higher plants. *Biochemica et Biophysica Acta*, **682**, 436–445.

Field, C. D. (1984). Movement of ions and water into the xylem sap of tropical mangroves. In *Physiology and Management of Mangroves* (Ed. by H. J. Teas), pp. 49–52. Dr W. Junk, The Hague.

Flowers, T. J., Troke, P. F. & Yeo, A. R. (1977). The mechanism of salt tolerance in halophytes. *Annual Review of Plant Physiology*, **28**, 89–121.

Gates, D. M. (1968). Transpiration and leaf temperature. *Annual Review of Plant Physiology*, **19**, 211–238.

Gill, A. M. & Tomlinson, P. B. (1975). Aerial roots; an array of forms and functions. In *The Development and Function of Roots* (Ed. by J. G. Torrey & D. T. Clarkson) pp. 237–260. Academic Press, London.

Harvey, D. M. R., Hall, J. L., Flowers, T. J. & Kent, B. (1981). Quantitative ion localization with *Suaeda maritima* leaf mesophyll cells. *Planta*, **151**, 555–560.

Jennings, D. H. (1968). Halophytes, succulence and sodium in plants—a unified theory. *New Phytologist*, **67**, 899–911.

Jolivet, Y., Hamelin, J. & Larher, F. (1983). Osmoregulation in halophytic higher plants; the protective effects of glycine betaine and other related solutes against oxalate destabilization of membranes in belt root cells. *Zeitchrift für Planzen Physiologie*, **109**, 171–180.

Joshi, G. V., Bhosale, L., Jamale, B. B. & Karadge, B. A. (1975). Photosynthetic carbon metabolism in plant. In *Proceedings of the International Symposium on Biology and Management of Mangroves Vol. II*, (Ed. by G. E. Walsh & H. J. Teas), pp. 595–607. University of Florida, Gainesville.

Joshi, G. V., Sontakke, S., Bhosale, L. & Waghmode, A. P. (1984). Photosynthesis and photorespiration in mangroves. In *Physiology and Management of Mangroves* (Ed. by H. J. Teas) Dr W. Junk, The Hague.

Keys, A. J., Bird, I. F., Cornelius, M. F., Lea, P. J., Wallsgrove, R. M. & Miflin, B. J. (1978). The photorespiratory nitrogen cycle. *Nature*, **275**, 741–742.

Lear, R. & Turner, T. (1977). *Mangroves of Australia*, 84 pp. University of Queensland Press, St Lucia.

McNally, S. F. & Hirel, B. (1983). Glutamine synthetase isoforms in higher plants. *Physiologie Vegetale*, **21**, 761–774.

Miflin, B. J. & Lea, P. J. (1980). Ammonia assimilation. In *Biochemistry of Plants Vol. 5*, (Ed. by P. J. Stumpf & E. E. Conn), pp. 169–202. Academic Press, London.

Miller, P. C. (1972). Bioclimate, leaf temperature and primary production in red mangrove canopies in South Florida. *Ecology*, **53**, 22–45.

Moore, R. T., Miller, P. C., Albright, D. & Tieszen, L. L. (1972). Comparative gas exchange characteristics of three mangrove species in winter. *Photosynthetica*, **6**, 3897–393.

Moore, R. T., Miller, P. C., Ehleringer, J. & Lawrence, W. (1973). Seasonal trends in gas exchange characteristics of three mangrove species. *Photosynthetica*, **7**, 387–394.

Nash, D., Paleg, L. G. & Wiskich, J. T. (1982). Effect of proline, betaine and some other solutes on the heat stability of mitochondrial enzymes. *Australian Journal of Plant Physiology*, **9**, 47–57.

Paleg, L. G., Douglas, T. J., von Daal, A. & Keech, D. B. (1981). Proline and betaine protect enzymes against heat inaction. *Australian Journal of Plant Physiology*, **8**, 107–114.

Popp, M. (1984a). Chemical composition of Australian mangroves. I. Inorganic ions and organic acids. *Zeitschrift für Pflanzenphysiologie*, **113**, 411–421.

Popp, M. (1984b). Chemical composition of Australian mangroves. II. Low molecular weight carbohydrates. *Zeitschrift für Pflanzenphysiologie*, **113**, 411–421.

Popp, M., Larher, F. & Weigel, P. (1984). Chemical composition of Australian mangroves. III. Free amino acids, total methylated onium compounds and total nitrogen. *Zeitschrift für Pflanzenphysiologie*, **113**, 422–432.

Rozema, J., Gude, H. & Pollak, G. (1981). An ecophysiological study of the salt secretion of four halophytes. *New Phytologist*, **81**, 201–217.

Saenger, P. (1982). Morphological, anatomical and reproductive adaptations of Australian mangroves. In *Mangrove Ecosystems in Australia* (Ed. by B. F. Clough), pp. 151–189. Australian National University, Canberra.

Scholander, P. F. (1968). How mangroves desalinate seawater. *Physiologica Plantarum*, **21**, 251–261.

Scholander P. F., Hammel, H. T., Hemmingsen, E. & Gorey, W. (1962). Salt balance in mangroves. *Plant Physiology*, **37**, 722–729.

Scholander, P. F., van Dam, L. & Scholander, S. I. (1955). Gas exchange in the roots of mangroves. *American Journal of Botany*, **42**, 92–98.

Smirnoff, N. & Stewart, G. R. (1985). Stress metabolites and their role in coastal plants. *Vegetatio*, **62**, 273–278.

Spain, A. V. & Holt, J. A. (1980). *The Elemental Status of the Foliage and Branchwood of Seven Mangrove Species from Northern Queensland*. Commonwealth Scientific and Industrial Research Organization, Australian Division of Soils, Divisional Report No. 49.

Stelzer, R. & Lauchli, A. (1977). Salz-und uberflutungstoleranz von *Puccinellia peisonis* I. Der einfluss von NaCl- und KCl-salinitat auf das Wachstum bei variierter suaerstoffversorgung der wurzel. *Zeitschrift für Planzenphysiologie*, **83**, 35–42.

Stewart, G. R. & Ahmad, I. (1983). Adaptation to salinity in angiosperm halophytes. In *Metals and Micronutrients, Uptake and Utilization by Plants* (Ed. by D. A. Robb & W. S. Pierpoint) pp. 33–50. Academic Press, London.

Stewart, G. R. & Lee, J. A. (1974). The role of proline accumulation in halophytes. *Planta*, **120**, 279–289.

Stewart, G. R. & Orebamjo, T. O. (1984). Studies of nitrate utilization by the dominant species of regrowth vegetation of tropical West Africa: a Nigerian example. In *Nitrogen as an Ecological Factor* (Ed. by J. A. Lee, S. McNeill & I. H. Rorison), pp. 167–188. Blackwell Scientific Publications, Oxford.

Storey, R. & Wyn Jones, R. G. (1975). Betaine and chlorine levels in plants and their relationship to NaCl stress. *Plant Science Letters*, **4**, 161–168.

Winter, E. (1982). Salt tolerance of *Trifolium alexandriaum*. *Australian Journal of Plant Physiology*, **9**, 227–237.

Woo, K. C. & Osmond, B. B. (1982). Stimulation of ammonia and 2-oxoglutarate dependent O_2 evolution in isolated chloroplasts by dicarboxylate and the role of the chloroplast in photorespiratory nitrogen recycling. *Plant Physiology*, **69**, 591–596.

Wyn Jones, R. G. & Storey, R. (1981). Betaines. In *Physiology and Biochemistry of Drought Resistance in Plants*. (Ed. by L. G. Paleg & D. Aspinall), pp. 171–204. Academic Press, Sydney.

Pelvetia canaliculata, a high-shore seaweed that shuns the sea

DORNFORD A. RUGG AND TREVOR A. NORTON

Department of Marine Biology, University of Liverpool, Port Erin, Isle of Man

SUMMARY

1 *Pelvetia canaliculata* (L.) Dcne et Thur. Inhabits higher levels on rocky shores than any other perennial macroalga. The plants are unable to curb water loss or to store water effectively and after several days of continuous exposure on the shore the thallus may lose 96% of its initial water content. *Pelvetia* simply tolerates such stress rather than avoiding it. It also surpasses other intertidal fucoids in the extent to which its photosynthesis recovers on resubmergence after desiccation.

2 *Pelvetia* is confined to the high shore by its inability to compete with the seaweeds dominant at lower levels. Their removal enables it to colonize below its usual zone. However, although transplants to the mid-shore initially thrive, they decay over winter. In culture, too, *Pelvetia* plants decay if kept submerged for 6 or more hours out of every 12.

3 Two distinct forms of extremely destructive decay are described for the first time, although the causal organisms have not yet been identified. The decaying lesions may represent pathogenic infections, but the symptoms are not transmitted to healthy *Pelvetia* either by contact with infected plants or by transference of rotting tissue from lesions.

4 *Pelvetia* does not exhibit any obvious physiological decline during periods of prolonged submersion. Indeed, it can be growing rapidly immediately prior to the outbreak of decay.

5 A further possibility discussed is that endobiotic fungi invariably found in *Pelvetia*'s thallus may become parasitic when the seaweed is submerged for long periods.

INTRODUCTION

Pelvetia canaliculata Dcne et Thur. is a quite remarkable seaweed. Its habitat is more maritime than marine and, even more surprisingly, it appears to be incapable of a truly marine existence. The evidence for these two claims is presented below.

THE EVIDENCE

Habitat and tolerance to drought

Although *Pelvetia* is a fucoid alga it ventures far higher on the shore than any *Fucus* would dare. Not only is *Pelvetia* exposed by every tide, but during neap tides in calm weather it may not be wetted by the sea for several consecutive days (Schonbeck & Norton 1978).

Clearly *Pelvetia* extends to the very limits of existence for a marine plant. It lives where desiccation is an almost constant companion. Surprisingly, there is little about the plant's anatomy, its shape or its growth habit that enables it to curb water loss. Nor does it possess a significant internal reservoir of water or cling tenaciously to water bound into its cells (Schonbeck & Norton 1979a). Indeed, it actually loses water slightly faster than intertidal *Fucus* plants do. In consequence, during prolonged exposure to air on the shore the plants may lose up to 96% of their water content and become dry and brittle (Schonbeck & Norton 1979a). Crushed in the hand, they crumble into fragments. In the laboratory, plants losing 90–96% of their tissue water subsequently suffered little or no damage on resubmergence (Schonbeck & Norton 1979a). The secret of *Pelvetia*'s success is a biochemical toughness that confers the ability to tolerate the effects of prolonged desiccation. This intrinsic tolerance is allied to a resilience which enables the plant when resubmerged to recover far more completely than is the case for *Fucus* Dring & Brown 1982). *Pelvetia* therefore often survives on the shore when *Fucus* plants at lower levels on the shore are killed off (Schonbeck & Norton 1978). A further asset may be the ability to imbibe nutrients very rapidly during very brief, intermittent periods of submergence (Schonbeck & Norton 1979b).

Absence from lower on the shore

Most intertidal seaweeds form distinct zones at characteristic levels on the shore, but not infrequently some plants are found far out of their zone. For example, if the shore is disturbed either naturally or experimentally, a high-shore dweller such as *Fucus spiralis* L. may establish itself all over the shore, even becoming abundant near to low water mark (Burrows & Lodge 1951). In contrast, *Pelvetia* rarely becomes established below its usual zone. Juvenile plants make a transient seasonal appearance in the *F. spiralis* zone, but we have never seen attached adult plants of *Pelvetia* on either the middle or the lower shore, even when the usual dominant seaweeds have been removed.

What forces restrict *Pelvetia* to such a narrow zone so high on the intertidal? Is it simply unable to disperse to the lower reaches of the shore? Although there is no information on the dissemination of *Pelvetia*, some seaweeds do have very restricted dispersal ranges, with the majority of their propagules falling within a few metres of the parent plants (see Deysher & Norton 1982). However, the eggs and zygotes of *Pelvetia* are similar to those of *Fucus* species which have been shown to recruit in quantity up to 60 m from the nearest adult plants (Burrows & Lodge 1950). What is more, *Pelvetia* releases its eggs on spring tides (Subrahmanyan 1957) when the parent plants are certain to be submerged twice daily and the disseminules are most likely to be broadcast by tidal currents. Eggs or zygotes washed downshore may need to travel only a few metres in order to colonize the lower shore.

Undoubtedly space becomes available lower on the shore only sporadically and as there is a great abundance and variety of propagules available to colonize it, *Pelvetia*'s possible contribution may be swamped. Until the plants become quite large they might easily be overlooked. However, almost every year space predictably appears immediately below the *Pelvetia* zone when the plants in the upper *F. spiralis* zone are killed off during hot summer days (Schonbeck & Norton 1978). This space is invariably colonized by a new generation of *Pelvetia* plants (Schonbeck & Norton 1980). This attests to the effectiveness of the plants' reproduction and their ability to disperse propagules beyond the confines of the parental zone. Equally predictably, however, recruitment does not lead to eventual dominance by *Pelvetia*, for *F. spiralis* reproduces at about the same time and it simultaneously recolonizes its own lost ground. Invariably, it succeeds at the expense of *Pelvetia*.

With the help of an experimenter, however, *Pelvetia* can successfully colonize below its own zone. Starting with either naturally or artificially cleared rock, repeatedly removing *Fucus* plants allows *Pelvetia* to grow to maturity (Schonbeck & Norton 1980). In Fig. 1 selective weeding out of *Fucus* from every alternate quadrat on an experimental plot in the *F. spiralis* zone resulted in a chequerboard of *Pelvetia* and *Fucus* dominated squares. In the absence of *Fucus*, *Pelvetia* thrives, but if *Fucus* is present *Pelvetia* germlings are rapidly overgrown and eventually die out.

FIG. 1. An experimental 0.25×0.25 m quadrat in the *Fucus spiralis* zone at Port Loy, Isle of Cumbrae 18 months after being first colonized by *Pelvetia* and *F. spiralis*. For the last 6 of these months all the plants of *F. spiralis* were selectively removed from each alternate sub-square allowing *Pelvetia canaliculata* to become dominant. In the intervening unweeded sub-squares *F. spiralis* dominates.

Clearly, living so high on the shore, hardiness is the main key to success for *Pelvetia*. Natural selection has encouraged *Pelvetia* to survive rather than to compete. Indeed competitiveness is a very minor asset to a plant living beyond the range of all of its potential competitors. The acquisition of hardiness many even necessitate compromises that significantly reduce the plants' competitiveness. Certainly, *Pelvetia* plants have a higher proportion of dry matter than do other fucoids, and in individual plants this correlates strongly with drought tolerance and is inversely proportional to growth rate (Schonbeck & Norton 1979c). It seems that rapid growth produces tissue which contains less dry matter and is less drought tolerant. Higher dry matter content may improve tolerance to desiccation by simply retarding cell wall collapse and thereby minimizing damage to the cell membrane.

The rivals of *Pelvetia* are less hardy but grow far faster and rapidly produce a dense blanketing canopy. Young adult plants of *Pelvetia* increase in length by only about 3–5 cm per year (Fig. 2, and also Subrahmanyan, 1960; Schonbeck & Norton 1980). *Pelvetia* can achieve only a quarter of the rate of growth exhibited by *F. spiralis* when the two species are growing side by side (Schonbeck & Norton 1980). Even more significantly, tiny juvenile *Pelvetia* also grow very slowly. Even under favourable growing conditions the germlings take a month to reach a length of only 200 μm. In the same time *F. spiralis* germlings reach 750 μm (Schonbeck & Norton 1980). For *Pelvetia* to reach 750 μm would take at least 4 or 5 months (Subrahmanyan 1960; T. A. Norton unpubl. results).

The magnitude of the competitive superiority of *F. spiralis* was strikingly demonstrated when hot weather denuded much of the upper reaches of the *F. spiralis* zone. The area was colonized densely by germlings of both *Pelvetia* and *F. spiralis*, but within 10 months *F. spiralis* was completely dominant. However, a subsequent hot spell wiped out the *Fucus* plants, leaving the unharmed *Pelvetia* beneath to grow unhampered until *Fucus* recolonized the area. Nonetheless, 8

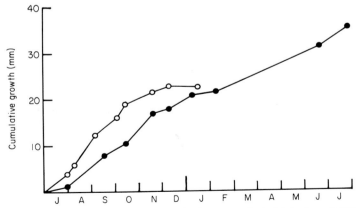

FIG. 2. Cumulative linear growth of *Pelvetia canaliculata* transplanted into its own zone (●) and the mid-shore (○) in July. The growth of the mid-shore transplants ceased when they began to decay in the following December.

months later the new generation of *F. spiralis* had again superseded the established *Pelvetia* plants (Schonbeck & Norton 1980). It is noteworthy that lower shore *Fucus* species are even faster-growing than *F. spiralis*.

Whether established adult plants of *Pelvetia* could survive in the mid-shore was tested by transplantation. For a time, during the spring and summer, the transplants grew almost twice as fast in the midshore region as control plants did in the *Pelvetia* zone (Fig. 2). However, during the autumn and winter the mid-shore transplants deteriorated and died whereas the plants in the *Pelvetia* zone, in spite of some losses, thrived. The results of this experiment conducted in the Firth of Clyde, Scotland were so striking that the procedure was repeated at St Michael's Island on the Isle of Man. The results were very similar. Transplants were made in February to three levels on the shore. Those in the *Pelvetia* zone survived in good condition for at least 9 months, until December; those in the mid *Ascophyllum* zone were fine until September, but then rapidly became moribund; those even lower, in the mid *F. serratus* zone, decayed within 2 months.

What fate befalls these plants? Do they succumb to adverse physical conditions or to biotic factors? They did not seem to be dislodged by wave action or the whiplash from adjacent plants. Nor were they excessively damaged by either abrasion or grazers. In fact the plants in question were neither detached nor destroyed, they quite simply decayed.

Several factors might cause *Pelvetia* to deteriorate on the lower shore. For example it is a light-loving plant and in dim light conditions (such as beneath a canopy of *F. spiralis*, or on the frequently submerged lower shore) it may be unable to exceed the compensation point for sufficiently long periods, with obviously deleterious consequences (Schonbeck & Norton 1980). However, *Pelvetia* seems well suited to dim light conditions. Its germlings survive 3 months even if cultured in complete darkness and there is some evidence that for adult plants the minimum irradiance required for growth is very low indeed (Schonbeck & Norton 1980). Chemical interference also seems unlikely, for in culture the growth of neither germlings nor adult *Pelvetia* is retarded by the presence of *Fucus* plants (Schonbeck & Norton 1980).

Submersion-induced decay

Even intertidal seaweeds are essentially aquatic plants and thrive when cultured under permanent submersion. This is true of all the fucoid algae tested and claims to the contrary (e.g. Fischer 1929; Fulcher & McCully 1969; Rusanowski & Vadas 1973) are based on very dubious evidence. However, *Pelvetia* is an exception. When cultured in a simulated tidal regime it grows well for long periods even if submerged for only 1 h out of every 12 (Schonbeck & Norton 1979c), but if the plants are immersed for 6 or more hours in every 12 they decay (Schonbeck & Norton 1979b).

The decay that occurs in culture closely resembles the brown rot seen in transplants on the shore. Initially superficial pale brown moribund patches appear

FIG. 3. (a) Well advanced brown soft rot on a *Pelvetia canaliculata* plant from St Michael's Isle, Isle of Man. (b) White tip rot on a *Pelvetia canaliculata* plant in culture.

scattered over the surface of the plant. They spread very rapidly and can affect the entire plant (Fig. 3a). Often the rot erodes completely through the thallus and all branches distal to the decay become detached, even if they are healthy. Once decay has begun the decline of the plant may be quite rapid and the loss of tissue from the combined effects of decay and detachment is total.

This 'brown soft rot' is previously undescribed and is quite distinct from the tissue destruction that may follow extreme desiccation, for example (Schonbeck & Norton 1978). Such damage discolours the thallus a dark rusty-brown, the injured tissue often remains firm for some while after the discolouration appears and it is usually sharply demarcated from the adjacent healthy tissue. In contrast, the 'brown soft rot' is a paler colour, the tissue rapidly softens and at the edges gradually grades into healthy thallus.

In culture 'brown soft rot' is often followed by a 'white tip rot' only rarely observed in nature. 'White tip rot' appears to begin in the medulla, but the first externally visible symptoms are small white or cream patches of soft tissue that appear only at the distal apices of branches (Fig. 3b). The patches subsequently coalesce to cover the entire tip and then spread progressively down the branch. Within days of the first appearance of 'white tip rot' the apical cell of the infected tip becomes inactive.

There are two likely explanations for the apparent scarcity of white tip rot in *Pelvetia* transplanted downshore. Firstly, the blanched tip tissue is so soft that it could probably only persist in exceptionally calm conditions such as in culture. Secondly, white tip rot invariably appears after the brown soft rot is well established. As the brown soft rot is potentially so destructive, especially on the shore where weakened branches are readily removed by wave action, the plant may be largely demolished before white tip rot has a chance to manifest itself.

An intriguing question presents itself. Are these rots pathological infections to which *Pelvetia* succumbs, or is *Pelvetia* unable to live under water and the rots are merely the first visible symptoms of its decline and death?

Culture experiments have revealed much about the rots. They develop when the *Pelvetia* plants are grown in a variety of culture media, even in unenriched flowing seawater. The initial symptoms of brown rot usually appear after about four weeks of continuous submersion, but both the onset of the infection and its subsequent rate of spread can be accelerated or retarded by the culture conditions (Table 1). The same is true for white tip rot (Table 1).

Although many of the factors tested significantly influence the rate of decay, no clear overall pattern emerges. For example, dim light accelerates brown soft rot (although not white tip rot), but so does a longer photoperiod which effectively subjects the plants to *more* light not less. Enhanced nutrient concentration also exacerbates decay. In nature *Pelvetia* lives so high on the shore that nutrients are only available during very brief, intermittent periods of submergence and the plants are probably adept at the rapid uptake of nutrients when they are available (Schonbeck & Norton 1979b). Could *Pelvetia* perhaps poison itself by excessive nutrient uptake during prolonged periods of immersion? This is a tempting theory, for *Pelvetia* decays when transplanted to the lower shore where it is submerged for long periods. Also the transplants die in autumn and winter when the concentration of nutrients in the sea is at its greatest. However, even at their maximum the nutrient concentrations in the sea remain quite low. Furthermore, although the growth of *Pelvetia* may be inhibited by high concentrations of nutrients, decay appears even in cultures where the plants are clearly nutrient limited, and it even spreads more rapidly than in plants that were grown in slightly enriched medium.

TABLE 1. The influence of various environmental factors on the onset and progression of the brown and white rots of *Pelvetia canaliculata* in culture

Accelerate	Retard
Prolonged or continuous submersion	Brief submersion in a daily 'tidal' cycle
High nutrient concentrations	Low nutrient concentrations
Continuous flow culture	Static culture
Continuous illumination	12:12 photoperiod
Low photon flux densities*	High photon flux densities*
Damaged or grazed plants	Undamaged plants
Winter-collected plants	Summer-collected plants

*Brown soft rot only.

Perhaps the most telling evidence against the idea that the decay follows a physiological decline of the plant, is that the time taken for the rot to appear bears no relation to the growth rate of the *Pelvetia* plant. There is often no obvious prior decline in the plant. It may be growing rapidly immediately before decay commences.

A striking phenomenon is that plants collected from the same area at the same time and then cultured submerged, tend to decay simultaneously. In contrast, plants collected from different shores even if collected on the same day and subsequently cultured together, do not exhibit this synchrony. It appears that the plant's previous experience may predispose it to decay.

Another contributory factor may be injury to the plant. Certainly, the first signs of decay often appear at the site of obvious prior damage to the thallus. Experimentally inflicted injuries also precipitate decay although, surprisingly, seemingly slight puncture wounds are more likely to accelerate decay than is extensive surface abrasion or cuts. In nature, thallus damage is commonplace. The frond may be snapped by wave action, abraded against the underlying rock, or punctured and scraped by grazers. The effects of *Hyale nilssoni* (Rathke), a herbivore known to eat *Pelvetia* (Moore 1977), were investigated in the laboratory. Athough the damage inflicted by the grazers appeared to be quite slight, their presence significantly accelerated the onset of decay. In the absence of *Hyale*, the presence of its faeces (a potential source of microbial infection) had no effect. It is assumed that breaches in the thallus may allow the ingress of pathogens.

The surface of *Pelvetia* abounds with micro-organisms, although they are less abundant in summer when no doubt their growth is curbed by frequent severe desiccation during exposure to air. In culture, the microbial flora on the plant's surface proliferates. Prolonged submergence might encourage the growth of an aquatic pathogen that is kept in check if *Pelvetia* is intermittently desiccated, or perhaps aerial exposure allows some volatile toxin to evaporate.

Micro-organisms are certainly especially abundant in the lesions on the surface of decaying *Pelvetia*. Isolates from the lesions are being cultured for identification and for use in an attempt to infect healthy plants. However, it is not yet clear that any of these organisms are pathogenic rather than invaders that decompose the thallus once it has died. If *Pelvetia* does indeed become diseased, the condition does not appear to be very infectious. Growing healthy plants in the same culture dish as decaying plants does not precipitate their decay, nor does growing them in medium in which plants have previously rotted. Even transferring decaying tissue from lesions onto healthy plants has little effect.

The fungal connection

Pelvetia harbours within its tissues several fungi, of which an ascomycete, *Mycosphaerella ascophylli* Cotton, is particularly abundant. This fungus has never been found free-living and is a constant companion of adult *Pelvetia* (Sutherland 1915; Webber 1967). The nature of the relationship is not clear, but the extreme difficulty experienced in attempting to culture the fungus divorced from the alga (Fries 1979), even in the presence of extracts of *Pelvetia* (D. A. Rugg, unpubl. results), hints at an intimate and obligate association.

The fungal hyphae ramify extensively throughout the mucilage-filled intercellular spaces of *Pelvetia*. Haustoria insinuate between the cell walls of the 'host' plant, but have not been observed to penetrate into the cells (Webber 1967; Kingham 1976). There is no evidence that the fungus harms *Pelvetia*, although the seaweed's exceedingly slow growth rate may reflect the drain exerted on its resources by the fungus. Certainly *in vitro*, *Mycosphaerella ascophylli* exhibits an efficient uptake of laminarin and mannitol (Kingham & Evans 1977; Fries 1979) which are abundant reserve products of brown algae such as *Pelvetia*.

Mycosphaerella ascophylli is probably potentially a dangerous organism. Certainly, most other members of the genus are pathogens causing leaf spot and stem spot diseases of angiosperms. Moreover, the symptoms of brown soft rot in *Pelvetia* are not unlike those induced by other ascomycete pathogens of seaweeds (Kohlmeyer 1974). It is therefore at least possible that *Mycosphaerella* is implicated in the fate of submersed *Pelvetia*. It may be significant that juvenile plants of *Pelvetia* are not usually infected with *Mycosphaerella* for the first year of their life (Webber 1959, 1967; Kingham 1976) and grow perfectly well without decaying even if kept permanently submerged for 50 days (Schonbeck & Norton 1980). It is only plants containing fungus that are adverse to prolonged submersion.

Many symbiotic fungi are potentially harmful. They are capable of *in vitro* utilization of some of the wall polymers of their partner, but usually do not do so in nature (Lewis 1975). Presumably when the fungus is part of a symbiotic relationship the production of cytolitic enzymes is suppressed or their activity is neutralized by the partner. There is some evidence that the availability of soluble carbohydrates represses the synthesis of such enzymes, but in adverse conditions their re-synthesis can result in necrotrophic parasitism (Lewis 1975).

In spite of the rather different relationship between fungus and alga in the *Mycosphaerella/Pelvetia* liaison, and that in the lichen symbiosis, it is tempting to draw parallels. Even in lichens the relationship wavers uneasily between mutualism and parasitism, and the possession of haustoria (as in *Mycosphaerella*) has been interpreted as a vestigial echo of parasitic intent (Ahmadjian 1982). The delicate balance between the two partners in lichens can quite easily be fatally disrupted as a result of the presence of excessive nutrients or moisture (Scott 1960; Hill & Ahmadjian 1972). In some lichens alternating periods of wetting and drying are essential to maintain a healthy plant (Farrar 1976). Under conditions of excessive moisture the alga may be killed by the respiration of the fungus, which is more restrained under drier conditions (Harris & Kershaw 1971). Perhaps in *Pelvetia* too, periodic desiccation is necessary to curb the excesses of its endobiotic fungus. We are at present attempting to eliminate the fungus from *Pelvetia* to investigate whether this might improve *Pelvetia*'s tolerance of submersion.

However, if the fungus is implicated in *Pelvetia*'s inability to tolerate long submersion, how can *Ascophyllum nodosum* (L.) Le Jol., which is equally infested with *Mycosphaerella ascophylli*, thrive both continuously immersed in culture, and also on the frequently submerged mid-shore region? Could there be some unique aspect of the *Pelvetia/Mycosphaerella* association that results in lethal consequences under prolonged submersion? Perhaps a clue is to be gleaned from the reserve products manufactured by the alga. Alone among the brown algae *Pelvetia canaliculata* produces, in addition to mannitol, a further alditol, the C_7 compound volemitol (Lindberg 1954; Quillet 1957). Volemitol is produced in large amounts (Young 1972) and is rapidly ^{14}C-labelled during photosynthesis (Kremer 1973). It may therefore be considered a typical assimilation product of *Pelvetia canaliculata*. Although the production of volemitol stems from the photosynthesis of the alga itself, it appears to be stimulated by the presence of *Mycosphaerella*. In infected *Pelvetia* plants volemitol production exceeds that of mannitol, whereas in uninfected germlings and in adult plants treated with a fungicide, volemitol production is reduced to almost half that of mannitol (Kingham & Evans 1977).

It is interesting to note that sugar alcohols including mannitol and sometimes volemitol, are also the principal storage compounds of lichen fungi and they are often produced in quantities vastly exceeding those needed for fungal growth. It has been postulated (J. F. Farrar pers. comm.) that they may be required to enable the fungus to cope with wetting and drying cycles.

Another member of the genus *Pelvetia fastigiata* (J. Ag.) De. Toni inhabits the upper shore on the Pacific coast of North America. It does not harbour *Mycosphaerella*, nor can it synthesize volemitol (Kremer 1976). Whether or not it can tolerate prolonged submersion is unknown.

CONCLUSIONS

Pelvetia canaliculata is essentially an aquatic plant. It is clearly closely related taxonomically to other fucoid algae and it resembles them in quite fundamental ways. *Fucus* and *Pelvetia* share a similar anatomy, pattern of development, reproduction and life-history. It is in their ecology that they differ most markedly.

Pelvetia's ability to survive far higher on the shore than any other perennial seaweed is well studied. It seems that only its extreme biochemical and physiological resilience enable it to survive. On the other hand its singular intolerance to prolonged submersion remains a puzzling enigma. It is not even clear whether it is some aspect of submersion that is lethal, or periodic immersion that is vital to satisfy some quirk of *Pelvetia*'s metabolism or to curb the excesses of potential pathogens or parasites. Whatever the cause, *Pelvetia* is almost certainly the only seaweed that cannot survive in the sea.

ACKNOWLEDGMENTS

We would like to thank Professor M. B. Wilkins and his staff for the help and facilities we enjoyed at the Botany Department, Glasgow University during the initial stages of the investigation.

REFERENCES

Ahmadjian, V. (1982). Algal/fungal symbioses. *Progress in Phycological Research* (Ed. by F. E. Round & D. J. Chapman), pp. 179–233 Elsevier Biomedical Press, Amsterdam, New York, Oxford.

Burrows, E. M. & Lodge, S. M. (1950). Note on the interrelationships of *Patella, Balanus* and *Fucus* on a semi-exposed coast. *Report of the Marine Biological Station, Port Erin*, **62**, 30–34.

Burrows, E. M. & Lodge, S. M. (1951). Autecology and the species problem of *Fucus. Journal of the Marine Biological Association of the United Kingdom*, **30**, 161–176.

Deysher, L. & Norton, T. A. (1982). Dispersal and colonization in *Sargassum muticum* (Yendo) Fensholt. *Journal of Experimental Marine Biology and Ecology*, **56**, 179–195.

Dring, M. J. & Brown, F. A. (1982). Photosynthesis of intertidal brown algae during and after periods of emersion: a renewed search for physiological causes of zonation. *Marine Ecology Progress Series*, **8**, 301–308.

Farrar, J. F. (1976). Ecological physiology of the lichen *Hypogymnia physodes*. II. Effects of wetting and drying cycles, and the concept of physiological buffering. *New Phytologist*, **77**, 105–113.

Fischer, E. (1929). Recherches de bionomie et de l'oceanographie littorales sur la Rance et le littoral de la Manche. *Annales de l'Institut Oceanographique, Monaco*, **5**, 201–429.

Fries, N. (1979). Physiological characteristics of *Mycosphaerella ascophylli*; a fungal endophyte of the marine brown alga *Ascophyllum nodosum. Physiologica Plantarum*, **45**, 117–121.

Fulcher, R. G. & McCully, M. E. (1969). Laboratory culture of the intertidal brown alga *Fucus vesiculosus. Canadian Journal of Botany*, **47**, 219–222.

Harris, G. P. & Kershaw, K. A. (1971). Thallus growth and the distribution of stored metabolites in the phycobionts of the lichens *Parmelia sulcata & P. physodes. Canadian Journal of Botany*, **49**, 1367–1372.

Hill, D. J. & Ahmadjian, V. (1972). Relationships between carbohydrate movement and the symbiosis in lichens with green algae. *Planta*, **103**, 267–277.

Kingham, D. L. (1976). *Studies relating to the fucacean endomycobiont* Mycosphaerella ascophylli. Ph.D. thesis, University of Leeds.

Kingham, D. L. & Evans, L. V. (1977). The *Pelvetia/Ascophyllum–Mycosphaerella* inter-relationship. *British Phycological Journal*, **12**, 120.

Kohlmeyer, J. (1974). Higher fungi as parasites and symbionts of algae. *Veröffenlichungen des Instituts für Meeresforschung in Bremerhaven, Supplement*, **5**, 339–356.

Kremer, B. P. (1973). Untersuchungen zur Physiologie von Volemit in der marinen Braunalge *Pelvetia canaliculata. Marine Biology*, **22**, 31–35.

Kremer, B. P. (1976). Distribution and biochemistry of alditols in the genus *Pelvetia* (Phaeophyceae, Fucales.) *British Phycological Journal*, **11**, 239–243.

Lewis, D. H. (1975). Comparative aspects of the carbon nutrition of mycorrhizas. *Endomycorrhizas* (Ed. by F. E. Sanders, B. Mosse & P. B. Tinker) pp. 119–148. Academic Press, London & New York.

Lindberg, B. (1954). Low-molecular carbohydrates in algae. IV. Investigation of *Pelvetia canaliculata. Acta Chemical Scandinavica*, **8**, 817–820.

Moore, P. G. (1977). Organization in simple communities: observations on the natural history of *Hyale nilssoni* (Amphipoda) in littoral seaweeds. *Biology of Benthic Organisms* (Ed. by B. F. Keagan, P. O. Ceidligh & P. J. S. Boaden) pp. 443–451. Pergamon Press, Oxford.

Quillet, M. (1957). Volemitol et mannitol chez les Phéophycées. *Bulletin du Laboratoire Maritime, Dinard*, **43**, 119–124.

Rusanowski, P. C. & Vadas, R. L. (1973). A tide-simulating apparatus for the study of intertidal marine algae. *Bulletin of the Ecological Society of America*, **54**, 35.

Schonbeck, M. W. & Norton, T. A. (1978). Factors controlling the upper limits of fucoid algae on the shore. *Journal of Experimental Marine Biology and Ecology*, **31**, 303–313.

Schonbeck, M. W. & Norton, T. A. (1979a). An investigation of drought avoidance in intertidal fucoid algae. *Botanica Marina*, **22**, 133–144.

Schonbeck, M. W. & Norton, T. A. (1979b). The effects of brief periodic submergence on intertidal algae. *Estuarine and Coastal Marine Science*, **8**, 205–211.

Schonbeck, M. W. & Norton, T. A. (1979c). Drought-hardening in the upper-shore seaweeds *Fucus spiralis* and *Pelvetia canaliculata. Journal of Ecology*, **67**, 687–696.

Schonbeck, M. W. & Norton, T. A. (1980). Factors controlling the lower limits of fucoid algae on the shore. *Journal of Experimental Marine Biology and Ecology*, **43**, 131–150.

Scott, G. D. (1960). Lichen symbiosis I. The relationship between nutrition and moisture content in the maintenance of the symbiotic state. *New Phytologist*, **59**, 374–381.

Subrahmanyan, R. (1957). Observations on the anatomy, cytology, development of the reproductive structures, fertilization and embryology of *Pelvetia canaliculata* Dcne et Thur. Pt 3. The liberation of reproductive bodies, fertilization and embryology. *Journal of the Indian Botanical Society*, **36**, 373–395.

Subrahmanyan, R. (1960). Ecological studies on the Fucales I. *Pelvetia canaliculata* Dcne et Thur. *Journal of the Indian Botanical Society*, **398**, 614–630.

Sutherland, G. K. (1915). New marine fungi on *Pelvetia*. *New Phytologist*, **14**, 33–42.

Webber, F. C. (1959). *Marine fungi*. Ph.D. thesis. University College of Wales, Aberystwyth.

Webber, F. C. (1967). Observations on the structure, life history and biology on *Mycosphaerella ascophylli*. *Transactions of the British Mycological Society*, **50**, 583–601.

Young, M. (1972). Seven-carbon sugars and alcohols in the algae. *British Phycological Journal*, **7**, 285.

Environmental control of macroalgal phenology

C. A. MAGGS

Department of Botany, Queen's University, Belfast BT7 1NN, Northern Ireland

M. D. GUIRY

Department of Botany, University College, Galway, Ireland

SUMMARY

1 Environmental control of macroalgal phenology was studied by laboratory culture in conjunction with observation of field populations.
2 Irish populations of three subtidal red algae, *Halarachnion*, *Atractophora* and *Halymenia*, were examined regularly for development and reproduction of each life history phase, and isolates were cultured in a variety of temperature and daylength regimes.
3 Short-day photoperiodic responses were observed in some phases, but were generally expressed over fairly narrow temperature ranges.
4 Temperature and daylength regulate growth and development in these algae by specific effects on each life history phase, and their annual cycles produce a characteristic phenological pattern for each species.

INTRODUCTION

The ecological aspects of photoperiodic and temperature control of reproduction in marine algae are difficult to pursue, but the rewards in the form of a better understanding of the factors controlling development and distribution are probably greater than from strictly laboratory-orientated physiological studies. In the past, the tendency has been either to describe phenological phenomena in the wild without reference to critical environmental variables or to carry out studies in controlled conditions with no comparative field work, so that correlations have often been largely speculative.

Making sense of the phenology of a plant in the field and its development under controlled conditions requires a detailed knowledge of the sequence of phases in its life history. In many species some of the most important events take place in small cryptic phases; frequently the relationship between such phases and the erect stage may not be known and initial laboratory studies are required to establish the sequence of events. The study of algal species growing on bedrock is hampered by the difficulty of locating microscopic stages and it may be necessary to clear the substratum completely or to plant out artificial substrata (e.g. microscope slides) which are examined periodically for algal development. As an alternative, mobile substrata such as gravel, shells and pebbles, which are sufficiently stable to support macroalgae but small enough to be collected entire for laboratory examination, are ideal for detailed studies of field phenology. The potential of these communities has often been ignored due to the difficulties involved in identification of the high proportion of crustose algae (Irvine & Maggs 1983).

Temperature, photoperiod, light quality, irradiance (and total light dosage) are the most important environmental factors regulating phenology of macroalgae, although nutrient levels have been implicated for the large perennial kelps (Chapman & Craigie 1977). The possible role of variation in grazing pressure as a determinant of algal phenology has recently come under scrutiny (e.g. Lubchenco & Cubit 1980; Dethier 1981); seasonal variation in wave action and sand-scour are also likely to be important in some communities (Dethier 1982; Littler & Littler 1983). There may be a threshold for any one or a combination of these factors at any life history stage but the most obvious effects of environmental conditions are on the reproduction of the different phases. In consequence many phenological studies are confined to the recording of reproductive structures (Kain 1982) even though their occurrence does not necessarily indicate that viable gametes or spores are being released. Temperature controls the onset of reproduction in many algae (Lüning 1980) and in an increasing number of instances photoperiodic responses have been found (Dring 1984). Several sublittoral species (e.g. *Acrosymphyton purpuriferum*, Breeman & ten Hoopen 1981) show enhanced short-day induction of sporangia as daylengths are decreased to less than 8 h and this may be related to the reduction of submarine daylength relative to the surface by poor water quality (Lüning & Dring 1979). Even in the intertidal zone, poor light transmission can have profound results for algal photoperiodic responses (Breeman *et al.* 1984). In a few algae, temperature and photoperiodic effects combine to create reproductive 'windows', confining reproduction to short periods and acting as barriers to the geographical spread of these species (*Trailliella*: Lüning 1980; *Gigartina acicularis*: Guiry & Cunningham 1984). Evidence of latitudinal ecotypic variation in temperature and photoperiodic responses has recently been obtained for the brown alga *Scytosiphon* (Lüning 1980) and two red algae (*Dumontia*: Rietema & Breeman 1982; *Rhodochorton*: Dring & West 1983).

Appropriate timing of life history events may be critical for algae on mobile substrata as storm disturbance, particularly severe in winter, can destroy their erect phases. These algae generally have heteromorphic life histories, in which spores of a seasonally-conspicuous, often ephemeral, erect phase grow into a more resistant perennating crustose phase. The majority of reported photoperiodic responses involve species with a heteromorphic life history and it is not surprising that mobile substrata are rich in such species.

The data presented here concern three red algal species, typical of these communities, which exhibit contrasting field phenologies and a variety of responses in culture to temperature and daylength.

MATERIALS AND METHODS

Field methods

Observations on field populations were made at approximately monthly intervals during the study periods by collecting samples of the mobile substrata and attached algae. Samples were examined under a dissecting microscope and all recognizable life history stages of the algae were recorded. Bottom temperature was taken on each sampling date. Daylength was obtained from tables for Shannon airport, near Galway, and calculations using equations in Dring (1984) showed that, during spring and autumn, there is only 10 min difference in daylength between the two sites.

The Galway site, a bed of free-living coralline algae (maerl) at 10 m depth, located 1 km offshore at Finavarra, Co. Clare, Ireland (53°0·5'N, 9°5'W), was studied from April 1981 to April 1982. A cobble bottom at 5–7 m depth at Strangford Lough Narrows, Co. Down, N. Ireland (54°6'N, 5°29'W), was examined throughout 1984.

Laboratory studies

Unialgal cultures were isolated from spores released by field-collected specimens, and maintained as described previously (Maggs & Guiry 1982; Guiry & Cunningham 1984).

Studies of temperature and photoperiodic responses were made on algae cultured at a photon irradiance of 5–7 μmol m^{-2} s^{-1} (Philips Cool-White) in a range of temperatures and daylengths as shown in the results. Unless otherwise stated, algae were grown initially at 15°C, 16:8 h (light:dark). Results for two of the species are qualitative, since quantitative data are not readily obtainable for these algae. Responses in *Halymenia* were quantified as described by Maggs & Guiry (1982).

For growth measurements single plants were grown in ten separate Repli-dish compartments at two daylengths, 16:8 h and 8:$\overline{16}$ h, at a range of temperatures. In order to distinguish photosynthetic effects from possible photoperiodic effects on growth, cultures at 16:8 h were grown at 5 μmol m^{-2} s^{-1} whilst those under 8:$\overline{16}$ h were kept at 10 μmol m^{-2} s^{-1}. Plant length, initially 1–3 mm, was measured weekly with a micrometer eyepiece. Plants of this size exhibit little lateral growth, so length was considered to be adequate for growth rate determination. Relative growth rates were calculated according to Hunt (1978). Error bars on all graphs are 95% confidence limits.

RESULTS

Halarachnion ligulatum

Growth and development in culture

Halarachnion ligulatum (Furcellariaceae, Giartinales) exhibits a heteromorphic life history (Boillot 1965; Kornmann & Sahling 1977) in which tetraspores produced by a small crustose sporophyte (previously known as *Cruoria rosea*) grow into blade-like gametophyte thalli which eventually bear carposporophytes.

In culture, sporophytes formed tetrasporangia under all environmental conditions tested (Table 1), although the numbers of sporangia per plant appeared to be greater under short days (SD) than in long days (LD). Developing tetraspores ̄gave rise initially to discs similar to the sporophyte. Erect axes were subsequently developed by the discs under LD at 15°C (Table 1), whereas at 10°C they were observed only in the SD treatment, a night-break apparently preventing upright initiation. Growth responses of the erect axes to temperature (at two daylengths at 15°C) are shown in Fig. 1. The optimum temperature for growth was 10°C; growth declined rapidly at 15°C under both LD and SD, although no reproduction was observed. Initially, the growth rate at 5°C was slow, but by week 4 was similar to that at 10°C although the plants were then smaller.

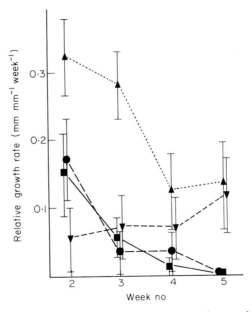

FIG. 1. Long-term effects of temperature and photoperiod on growth rate of *Halarachnion ligulatum* gametophyte blades (■—■ 15°C, 16:8 h light:dark; ●– –● 15°C 8:16 h; ▲. . . . .▲ 10°C 16:8 h; ▼– – – –▼ 5°C 16:8 h).

TABLE 1. Effects of environmental conditions on the development of *Halarachnion ligulatum*

Culture conditions	Tetrasporangia formed by sporophyte	Erect axes developed by gametophyte
15°C, 16:8̄ h	+	+
10°C, 16:8̄ h	+	−
10°C, 8:1̄6̄ h	+	+
10°C, 8:7̄·5̄:1:7̄·5̄ h	+	−

Field phenology

At Galway tetrasporophyte crusts reproduced throughout the year (Fig. 2). Small (<1 mm) erect axes were also present all year but developed into large blades only during spring and summer. Growth peaked in May and reproduction started before the beginning of June. The blades showed necrosis and grazing damage early in August and large thalli were absent by the end of August.

Atractophora hypnoides

Growth and development in culture

The life history of *Atractophora hypnoides* (Naccariaceae, ?Nemaliales) (Maggs, Guiry & Irvine 1983) follows a similar pattern to that of *Halarachnion*, but the

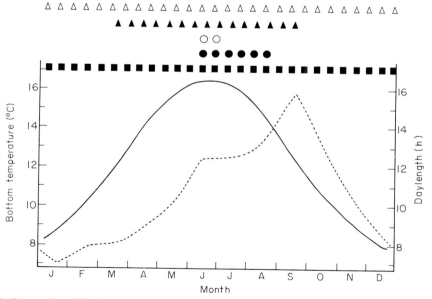

FIG. 2. Seasonality of *Halarachnion ligulatum* (Rhodophyta) at Galway Bay, Ireland during 1981–82. (△) small blades <1 mm high; (▲) blades 1–4 mm high; (○) gametangia; (●) carposporophytes; (■) tetrasporangia; (——) daylength; (.....) temperature.

TABLE 2. Effect of environmental conditions on the development of *Atractophora hypnoides*

Culture conditions	Tetrasporangia formed by sporophyte	Erect axes developed by gametophyte
15°C, 16:$\overline{8}$ h	−	+
15°C, 8:$\overline{16}$ h	+*	+
15°C, 8:$\overline{7\cdot5}$:1:$\overline{7\cdot5}$ h	−	No data
10°C, 16:$\overline{8}$ h	−	+
10°C, 8:$\overline{16}$ h	+	+
10°C, 4:$\overline{20}$ h	+	No data
10°C, 8:$\overline{7\cdot5}$:1:$\overline{7\cdot5}$ h	−	No data
5°C, 8:$\overline{16}$ h	−	No data

*Only after 4–6 months exposure.

erect gametophyte phase is filamentous (Dixon & Irvine 1977) rather than foliose. The sequence of phases is not obligatory as in *Halarachnion*: spores from the crustose phase (previously known as *Rhododiscus pulcherrimus*) can recycle the parent crust.

Sporophytes were large enough to reproduce in culture when about 4 months old. These crusts showed an absolute SD photoperiodic response (Table 2), forming tetrasporangia only at ≤8 h day at 10 and 15°C. A one-hour night-break was effective in preventing sporogenesis at both temperatures. At 15°C prolonged exposure was necessary to induce reproduction, and the largest numbers of spores were produced when crusts were transferred from 15°C 16:$\overline{8}$ h via 15°C 8:$\overline{16}$ h to 10°C 8:$\overline{16}$ h. Germinating tetraspores showed no photoperiodic effects (Table 2) and rapidly (8–20 days) initiated erect axes under all conditions tested. Growth versus irradiance experiments at 15°C (Fig. 3) showed that plants can grow at very

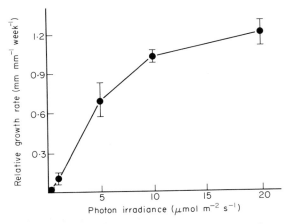

FIG. 3. Effect of photon irradiance on growth rate of *Atractophora hypnoides* gametophyte axes at 15°C, 16:$\overline{8}$ h.

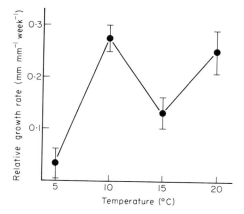

FIG. 4. Effect of temperature on initial growth rate of *Atractophora hypnoides* gametophyte axes at 16:8 h photoregime. Results were taken 1 week after transfer to experimental conditions.

low photon irradiances ($0.18–1$ μmol m^{-2} s^{-1}), and that saturation occurs at about 10 μmol m^{-2} s^{-1}. Photon irradiances above 30 μmol are inhibitory, and 50 μmol is lethal.

The effect of temperature on initial growth rates of gametophytes is shown in Fig. 4. Growth at 5°C is negligible, but all plants survived 5 weeks' treatment. The poor growth at 15°C can be attributed to immediate diversion of resources to reproduction at 15°C. Figure 5 shows that under 15°C LD conditions reproduction

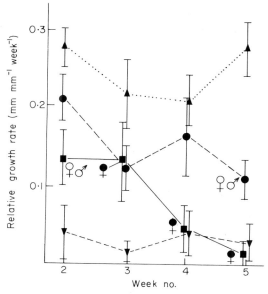

FIG. 5. Long-term effects of temperature on growth rate and reproduction of *Atractophora hypnoides* gametophyte axes. (■——■ 15°C 16:8 h; ●---● 15°C 8:16 h; ▲....▲ 10°C 16:8 h; 16:8 h; ▼----▼ 5°C 16:8 h; ♀♂ gametangia; ● carposporophytes)

had already started by week 2 (when plants were only 3–4 mm high) so that growth had effectively ceased by week 4; similar results were obtained at 20°C. At 15° SD reproduction was delayed until weeks 4–5, an apparent SD inhibition of reproduction, so that rapid growth continued until then. Gametangia were not formed at 10°C, and a high growth rate was maintained at this temperature.

Field phenology

Crustose tetrasporophytes in both study populations (Fig. 6) were reproductive from October/November until May/June, only sterile crusts being found in the intervening period. Tetrasporophytes probably live for several years: under optimal conditions in culture they take 12 months to reach the largest field plants observed. Erect gametophytes, which bear gametangia and carposporophytes, are rarely found and have been collected only in July/August, although specimens collected in mid-July were already 10 cm high, with extensive lateral development and cortication. Growth rates in culture suggest that these plants must have germinated at least 10 weeks earlier, i.e. in or before May. The great preponderance of tetrasporophytes at both sites indicates that the population is maintained almost entirely by tetrasporophyte recycling.

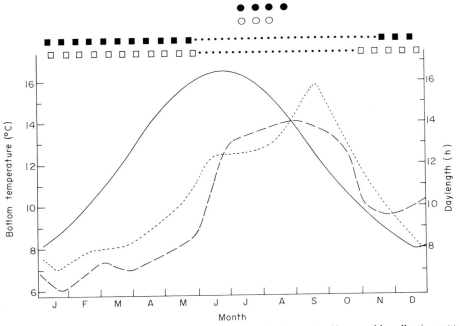

FIG. 6. Seasonality of *Atractophora hypnoides* (Rhodophyta), determined by monthly collections at two subtidal sites in Ireland during 1981–82 (Galway Bay) and 1984 (Strangford Lough). (●) gametophyte, bearing gametangia and carposporophytes: Galway; (○) gametophyte: Strangford (· · · ·) non-reproductive sporophyte; (■) tetrasporangia: Galway; (□) tetrasporangia: Strangford; (———) daylength; (.....) temperature: Galway; (– – –) temperature: Strangford).

Halymenia latifolia

Growth and development in culture

Halymenia latifolia has an isomorphic life history (Maggs & Guiry 1982) but both carpospores and tetraspores give rise initially to branching, *Conchocelis*-like, filaments which penetrate soft calcareous substrata such as old shells and dead coralline algae. The blade-like sporangial and gametophytic thalli arise from these boring filaments.

Initiation of blades from the shell-boring filaments in culture was not influenced by temperature or daylength; instead the determining factor is mechanical disturbance. Blades always developed within 1·5 months of removal of all emergent filaments from the bored calcareous substrata. A linear relationship between growth rate and temperatures of 5–20°C was observed in blades of a *Halymenia* isolate from Brittany (Fig. 7), as no reproduction occurs in thalli of the size studied. The Galway strain showed negligible initial growth at 5°C, and blades died after 2–4 weeks at this temperature, although the filamentous growths were still healthy after 3 months. At 10°C and 15°C the slight reduction of growth rate at SD relative to LD was due to rapid reproduction under 8-h daylength (see Fig. 9). However, at 15°C and 20°C some gamete formation occurred within 2 weeks even at 16-h daylength (Fig. 9) and this affected the observed growth rates.

Both the sporophyte and gametophyte phases showed SD photoperiodic enhancement of reproduction during 12-day exposures to the daylengths shown in

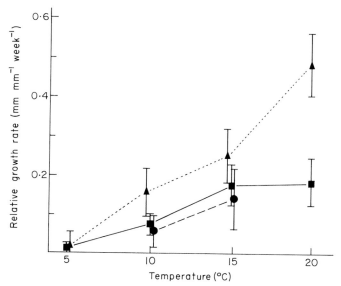

FIG. 7. Effect of temperature on growth rate of gametophyte blades of two isolates of *Halymenia latifolia*, showing growth rates for week 2 after transfer to experimental conditions. Blades of the Galway strain died after 2–4 weeks at 5°C. (▲...▲ Brittany strain, 16:8 h; ■——■ Galway strain, 16:8 h; ●----● Galway strain, 8:$\overline{16}$ h.)

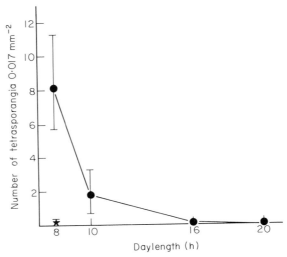

FIG. 8. Effect of photoperiod on tetrasporangial production by *Halymenia latifolia* sporophyte blades (Galway isolate) at 15°C. Tetrasporangia were counted after 12 days exposure to the experimental daylengths (see Maggs & Guiry 1982 for details). (★ night-break treatment)

Figs 8 and 9. A similar level of reproduction was observed after only 6 days at 8:$\overline{16}$ h (Maggs & Guiry 1982). The sporophyte (Fig. 8) showed a marked drop in sporangial production between 8-h and 10-h days. Very few sporangia were formed at 16 h, none at 20 h, and a 1-h night-break regime reduced sporogenesis to the 16-h daylength level. Production of female gametes (Fig. 10) did not differ significantly between 8 to 12-h daylengths, but these promoted gamete formation relative to 16-h days. The much reduced level of gametogenesis in the night-break

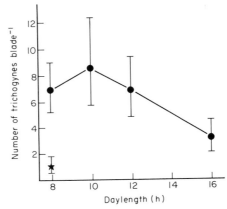

FIG. 9. Effect of photoperiod on female gamete production by *Halymenia latifolia* gametophyte blades (Galway isolate) at 15°C. Numbers of trichogynes (extensions of the sessile female gamete, the carpogonium) were counted on twenty blades in each treatment. (★ night-break treatment).

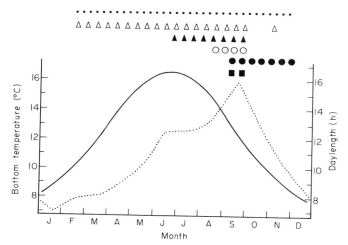

FIG. 10. Seasonality of *Halymenia latifolia* (Rhodophyta) at Galway Bay, Ireland, during 1981–82. (△) small gametophyte and sporophyte blades <1 mm high, (••••) filaments boring in calcareous substrata; (▲) blades 1–4 mm high; (○) gametangia; (●) carposporophyte development; (■) tetrasporangia; (——) daylength; (. . . .) temperature.

treatment indicates that further inhibition of gamete production would probably be found at daylengths longer than 16 h.

Field phenology

Small blades (<1 mm) were found in November and from February onwards (Fig. 10), and were apparently absent during December and January. Larger, non-reproductive, blades were observed only from June to September, and gametangia were produced during August/September, mature carposporophytes developing during September. Plants collected in December were decaying at the tips and had been partially destroyed by grazing molluscs. Tetrasporangial plants were found in September only. The reproductive phenology of this species at Roscoff, Brittany (Feldmann 1954) is remarkably similar: gametangia are found between July and September, carposporophytes during September/October and tetrasporangia in August.

DISCUSSION

These results of field and laboratory studies on red algal phenology demonstrate that environmental influences may have important effects at several points in the life history of each species.

In *Halarachnion*, the major factor determining the marked seasonality of gametophytes appears to be the influence of water temperature on gametophyte growth, although this is not clearly separable from daylength effects. The tetraspores produced throughout the year develop into discs which quickly give rise to erect axes in response to short days (in winter) or high temperature (in summer).

Little growth of the blades takes place during January and February (possibly due to a combination of low temperature and short daylength), and the May peak in growth rate seems to be correlated with the rise in water temperature to about 10°C, although the near-maximum daylength may also be important. The blades rapidly reach reproductive maturity, release carpospores, and die before the temperature is maximal. The carpospores give rise to sporophyte crusts which in culture became fertile within 1·5 months. Since the growth rate of gametophytes at 15°C in culture was slow and declined rapidly, it is likely that tetraspores germinating during the late summer do not form large blades, but over-winter as holdfast discs with small erect axes. Any large blades which develop during late autumn, when the temperature is again optimal for growth, are probably removed by equinoctial storms. The gametophyte phenology resembles the seasonality of *Callithamnion byssoides* (Kapraun 1978), which is also controlled primarily by temperature. *Callithamnion byssoides* is a spring–early summer annual in North Carolina (U.S.A.), disappearing after reproduction when the water temperature rises above its upper tolerance limit. Some latitudinal ecotypic variation in temperature response has been reported in this species (Kapraun 1978). Isolates from Texas, where *C. byssoides* is a summer–autumn annual, show a high temperature tolerance. The reproductive periodicity of *Halarachnion* gametophytes is constant over a wide latitudinal span (Norway to Morocco: Dixon & Irvine 1977) which suggests that temperature ecotypes may exist in this species also.

The commencement of *Atractophora* tetrasporophyte reproduction in October/November, at daylength 9–10 h, accords well with the SD response of this phase in culture, although further experiments at intermediate daylengths are necessary to determine the maximum daylength at which sporogenesis can occur. Water temperature is also an important influence on sporogenesis and the SD response in culture was expressed over a narrow temperature range. Sporulation started at both sites when the water temperature dropped to below 10·5°C and terminated when it rose to 10·5–11°C, so that the reproductive season of the Strangford population was extended by 3–4 weeks at either end relative to the Galway population. In Ireland, inductive daylengths coincide with favourable temperatures, resulting in a long reproductive season. This contrasts with the results of temperature–daylength interaction in some other species in which the SD response is confined to a short seasonal window (van den Hoek 1982; Dring 1984).

Tetraspores released in winter at <10°C develop slowly, as indicated by the temperature response in culture, and gametophytes would grow little until the temperature approached 10°C in April/May. The ephemerality of the gametophyte reflects the rapid growth and development of this phase in culture. Under ideal conditions, tetraspores can grow into small gametophytes bearing mature carposporophytes within 4–5 weeks. The apparent long-day effect on gametangial and carposporophyte development merits further investigation, as LD responses have rarely been demonstrated in macroalgae (Dring 1984). Mature field-collected plants show an unusual range of size, from 0·5–1 cm to 10 cm. The ultimate size of the plant relates to water temperature during development, since at high temperatures plants rapidly become reproductive, while at lower temperatures only vegetative growth takes place. The saturation of growth at low photon irradiances and the low light tolerance appear to be typical of sublittoral red algae (Lüning 1980).

Atractophora hypnoides is known from a small geographical range, from Northern Ireland to the northern Bay of Biscay. Unless latitudinal ecotypic responses to temperature and daylength exist, the species will be restricted to a small area by its requirement for winter temperatures <10°C and summer temperatures of *c* 15°C or above. Its southern limit in fact corresponds quite closely to the position of the 10°C February isotherm (van den Hoek 1975, 1982). The distribution of another heteromorphic red alga, the Mediterranean–Atlantic species *Acrosymphyton purpuriferum*, is also thought to be limited by a summer minimum and a winter maximum temperature (Breeman & ten Hoopen 1981).

The occurrence of small *Halymenia* blades during 10 months of the year probably results from the activities of grazing molluscs which provide the necessary mechanical disturbance. The principal grazer on maerl is *Acmaea virginea* which feeds on shell-boring and coralline algae (Farrow & Clokie 1979) and keeps the shells and maerl clear of superficial microalgae and algal sporelings. The radula action also wears away shell nacre, creating a more favourable surface for the settlement of algal spores and the development of shell-boring algae. Several examples of grazer-dependent algae have recently been reported (Slocum 1980; Dethier 1981; Steneck 1982), but these are usually crustose species or phases which are overgrown by other algae in the absence of grazing. Small erect blades do not survive the two coldest months (they die at 5°C in culture), and the population is maintained at this site by the *Conchocelis*-like filaments, which can survive prolonged exposure to low temperature. The delay (until June) in appearance of slightly larger blades was probably due to water temperature, which did not reach 10°C until May. Daylength is unlikely to be the major factor in the development of these blades as it is already decreasing by the time they are evident, and during August plants were still only 4–5 mm long. Rapid growth occurred only during late August and September, when the temperature approached 16°C: plants collected in September were up to 70 mm long.

Reproduction is stimulated by decreasing daylength from July onwards. The earlier appearance of gametangia (in July, at 14·5-h daylength) than tetrasporangia reflects the differences in the photoperiodic responses of the two phases. Both sporangial and gametangial blades form some reproductive organs under 15°C 16:$\overline{8}$ h conditions when they reach 10–20 mm length after about 2 months. The SD response may ensure that even if only very small plants are present in late summer, as a result of low temperature, these will then reproduce. The Roscoff isolate did not form tetrasporangia on fronds <20 mm long even under SD conditions, and the formation of reproductive organs by small plants of the Finavarra isolate may be an ecotypic response to the temperature conditions of Galway Bay. The west coast of Ireland is at the extreme northern distributional limit of *H. latifolia*, which extends southwards to Morocco and the Mediterranean Sea.

Temperature and daylength thus regulate the phenology of growth and reproduction in these three species by their effects on growth rate and by reproductive responses to long-day and short-day conditions, which are expressed over narrow temperature ranges.

ACKNOWLEDGMENTS

We are grateful for the technical assistance of W. Guiry, D. Burke, P. O'Rafferty (UCG) and D. Rogers (QUB), and to Dr M. J. Dring for help and advice.

REFERENCES

Boillot, A (1965). Sur l'alternance de générations hétéromorphes d'une Rhodophycée, *Halarachnion ligulatum* (Woodward) Kützing (Gigartinales, Furcellariacées). *Compte rendu hebdomadaire des Séances de l'Académie des Sciences, Paris*, **261**, 4191–4193.

Breeman, A. M. & ten Hoopen, A. (1981). Ecology and distribution of the subtidal red alga *Acrosymphyton purpuriferum* (J. Ag.) Sjost, (Rhodophyceae, Cryptonemiales). *Aquatic Botany*, **11**, 143–166.

Breeman, A. M., Bos, S., van Essan, S. & van Mulekon, L. L. (1984). Light–dark regimes in the intertidal zone and tetrasporangial periodicity in the red alga *Rhodochorton purpureum* (Lightf.) Rosenv. *Helgoländer Meeresuntersuchungen*, **38**, 365–387.

Chapman, A. R. O. & Craigie, J. S. (1977). Seasonal growth in *Laminaria longicruris*: relations with dissolved inorganic nutrients and internal reserves of nitrogen. *Marine Biology*, **40**, 197–205.

Dethier, M. N. (1981). Heteromorphic algal life histories: the seasonal pattern and response to herbivory of the brown crust, *Ralfsia californica*. *Oecologia (Berlin)*, **49**, 333–339.

Dethier, M. N. (1982). Pattern and process in tidepool algae: factors influencing seasonality and distribution. *Botanica Marina*, **25**, 55–66.

Dixon, P. S. & Irvine, L. M. (1977). *Seaweeds of the British Isles Volume 1 Rhodophyta Part 1 Introduction, Nemaliales, Gigartinales*. British Museum (Natural History), London.

Dring, M. J. (1984). Photoperiodism and phycology. *Progress in Phycological Research*, **3**, 159–192.

Dring, M. J. & West, J. A. (1983). Photoperiodic control of tetrasporangium formation in the red alga *Rhodochorton purpureum*. *Planta*, **159**, 143–150.

Farrow, G. E. & Clokie, J. J. P. (1979). Molluscan grazing of sublittoral algal-bored shells and the production of carbonate mud in the Firth of Clyde, Scotland. *Transactions of the Royal Society of Edinburgh*, **70**, 139–148.

Feldmann, J. (1954). Inventaire de la flore marine de Roscoff. Algues, champignons, lichens et spermatophytes. *Travaux de la Station Biologique de Roscoff, New Series*, **5** (*Supplément 6*), 1–152.

Guiry, M. D. & Cunningham, E. M. (1984). Photoperiodic and temperature responses in the reproduction of north-eastern Atlantic *Gigartina acicularis* (Rhodophyta: Gigartinales). *Phycologia*, **23**, 357–367.

Hunt, R. (1978). *Plant Growth Analysis*. Institute of Biology, Studies in Biology, No. 11.

Irvine, L. M. & Maggs, C. A. (1983). Peyssonneliaceae. *Seaweeds of the British Isles Volume 1 Part 2A Cryptonemiales* (sensu stricto), *Palmariales, Rhodymeniales* (Ed. by L. M. Irvine) pp. 52–61. British Museum (Natural History), London.

Kain, J. M. (1982). The reproductive phenology of nine species of Rhodophyta in the subtidal region of the Isle of Man. *British Phycological Journal*, **17**, 321–331.

Kapraun, D. F. (1978). Field and culture studies on growth and reproduction of *Callithamnion byssoides* (Rhodophyta, Ceramiales) in North Carolina. *Journal of Phycology*, **14**, 21–24.

Kornmann, P. & Sahling, P.-H. (1977). Meeresalgen von Helgoland. *Helgoländer Meeresuntersuchungen*, **29**, 1–289.

Littler, M. M. & Littler, D. S. (1983). Heteromorphic life-history strategies in the brown alga *Scytosiphon lomentaria* (Lyngb.) Link. *Journal of Phycology*, 19, 425–431.

Lubchenco, J. & Cubit, J. (1980). Heteromorphic life histories of certain marine algae as adaptations to variations in herbivory. *Ecology*, 61, 676–687.

Lüning, K. (1980). Control of algal life history by daylength and temperature. In *The Shore Environment, Vol. 2, Ecosystems* (Ed. by J. H. Price, D. E. G. Irvine & W. F. Farnham) pp. 915–945. Academic Press, London.

Lüning, K. & Dring, M. J. (1979). Continuous underwater light measurements near Helgoland (North Sea) and its significance for characteristic light limits in the sublittoral region. *Helgoländer Meeresuntersuchungen*, 32, 403–424.

Maggs, C. A. & Guiry, M. D. (1982). Morphology, phenology and photoperiodism in *Halymenia latifolia* Kütz. (Rhodophyta) from Ireland. *Botanica Marina*, 25, 589–599.

Maggs, C. A., Guiry, M. D. & Irvine, L. M. (1983). The life history in culture of an isolate of *Rhododiscus pulcherrimus* (Naccariaceae; Rhodophyta) from Ireland. *British Phycological Journal*, 18, 206.

Rietema, H. & Breeman, A. M. (1982). The regulation of the life history of *Dumontia contorta* in comparison to that of several other Dumontiaceae (Rhodophyta). *Botanica Marina*, 25, 569–576.

Slocum, C. J. (1980). Differential susceptibility to grazers in two phases of an intertidal alga: advantages of heteromorphic generations. *Journal of Experimental Marine Biology and Ecology*, 46, 99–110.

Steneck, R. S. (1982). A limpet-coralline alga association: adaptations and defences between a selective herbivore and its prey. *Ecology*, 63, 507–522.

van den Hoek, C. (1975). Phytogeographic provinces along the coasts of the northern Atlantic Ocean. *Phycologia*, 14, 317–330.

van den Hoek, C. (1982). Phytogeographic distribution groups of benthic marine algae in the North Atlantic Ocean. A review of experimental evidence from life history studies. *Helgoländer Meeresuntersuchungen*, 35, 153–214.

Enhancement of anoxia tolerance by removal of volatile products of anaerobiosis

R. M. M. CRAWFORD, LORNA S. MONK AND Z. M. ZOCHOWSKI

Department of Plant Biology and Ecology, The University, St Andrews, Scotland KY16 9TH

SUMMARY

1 Experiments with germinating seedlings show that circulating anaerobic atmospheres are less toxic than those that are kept static.

2 Circulating the anaerobic atmospheres lowers tissue ethanol accumulation and reduces exposure to high carbon dioxide concentrations.

3 Experiments with chickpea (*Cicer arietinum* L.) show that high carbon dioxide concentrations are only toxic in static anaerobic environments where they enhance ethanol production and accumulation.

4 Studies with rhizomes of anoxia-tolerant and anoxia-intolerant monocotyledonous species show a post-anoxic increase in catalase activity in the intolerant species. In *Glyceria maxima* this increase can be reduced by circulating the anaerobic atmosphere.

5 Exposure of anaerobic rhizomes of *Glyceria maxima* to ethanol vapour increases post-anoxic catalase activity.

6 It is suggested that the post-anoxic oxidation of anaerobically accumulated ethanol by catalase leads to the formation of toxic products such as acetaldehyde. The beneficial effect of circulating the anaerobic atmosphere could be due therefore to the prevention of ethanol accumulation during anaerobiosis and the consequent generation of acetaldehyde in the post-anoxic phase.

INTRODUCTION

Depriving higher plant tissues of oxygen will in all cases eventually lead to death. The length of time that higher plants can survive without oxygen varies with species and the conditions under which the anoxia is imposed. In cotton and soybean roots death can take place in as little as 3 and 5 h respectively, with a proportion of the roots being dead within 30 min (Huck 1970). By contrast, the rhizomes of some monocotyledonous species can be kept alive in an anaerobic incubator under strict anoxia for over 2 months (Crawford 1982). In some anoxia-sensitive species the damage caused by periods of anaerobiosis may not be immediately apparent, especially if the anoxic treatment is brief and the tissues are restored to air before any symptoms of injury are evident. Thus in pea seedlings a period of anaerobiosis induced by prolonged soaking does not cause a failure of germination when subsequently restored to air but is seen later as damage to the continued growth of

375

the apical bud (Berrie 1960). Many tissues, including rhizomes of *Iris germanica* and *Pteridium aquilinum*, emerge from several days of anaerobic incubation with apparently healthy rhizomes which then begin to degenerate within hours of being exposed to air. After a few hours in air these post-anoxic rhizomes show evidence of peroxidative damage to their cell membranes by the production of malondialdehyde (Hunter, Hetherington & Crawford 1983) and also of ethane. Both these products are considered to indicate peroxidative destruction of fatty acid chains.

Experimentally it is therefore possible to define two types of anaerobic injury, (i) anoxic and (ii) post-anoxic, depending on when the damage is caused to the plant tissues. When the injury is observed, however, does not necessarily determine when it was caused. The death of the apical bud in pea seedlings, although seen only on the extension of the shoot during the post-anoxic phase, may have taken place during the period of anaerobiosis. However, the peroxidative damage that has been reported in rhizomes on being restored to air after a period of anoxia (Hunter, Hetherington & Crawford 1983) is clearly a post-anoxic injury as it can take place only when there is access to a source of oxygen.

The lethal targets for both anoxic and post-anoxic injury are as yet unclear and some of the controversy that surrounds the possible causes of anaerobic death and flooding injury in plant tissues may be due to the fact that plants vary in the nature of the lethal target. Death that is caused by a few hours' anoxia, as in soybean and cotton roots (Huck 1970), may take place for entirely different reasons than those which are eventually responsible for the death of tissues that are capable of surviving several days or even weeks under anoxia.

The absence of oxygen impedes a number of aspects of cell function. Mitosis is prevented and as yet there is no clear evidence that any eukaryotic species is capable of carrying out cell division without oxygen. The rapid cessation of oxidative phosphorylation inhibits ATP-requiring processes such as H^+ transport from cytoplasm to vacuole (Bennett & Spanswick 1984). The removal of oxygen will also inhibit fatty acid desaturation and sterol biosynthesis. Thus alternations in energy metabolism and membrane biosynthesis may well be important factors in the pathology of anoxia and will be affected immediately when the oxygen supply is removed. Changes that take place as a consequence of anoxia will take place more gradually as conditions in the cell change, due to prolongation of anaerobic metabolism. Thus, reduction in cytoplasmic pH (glycosidic acidosis) is a consequence of hypoxia, and tissue acidification can begin to take place within 2 min of the onset of hypoxia, leading to root tip death in 12–24 h (Roberts *et al.* 1984a).

The accumulation of ethanol (Crawford & Zochowski 1984) has also been suggested as a possible mechanism whereby the continuation of anoxia will adversely affect the cell and could contribute to anaerobic injury. Ethanol accumulation is progressive and depends on the balance between the rate of production and removal by diffusion (Monk, Braendle & Crawford 1987), so it is unlikely to be the cause of sudden anaerobic death that is seen in some sensitive roots. The time-scale for the accumulation of potentially dangerous ethanol concentrations would suggest that it may play a role when anaerobic injury takes place over a matter of days or even weeks (Barclay & Crawford 1981, 1982).

Ethanol and anoxic injury in germinating seedlings

Considerable debate has taken place on the role of ethanol in anaerobic injury. A number of studies have raised doubts as to whether the amounts accumulated in plant tissues during periods of flooding or experimental anoxia are sufficient to have a directly toxic effect on plant cells (Jackson, Herman & Goodenough 1982; Alpi & Beevers 1983; Rumpho & Kennedy 1983). Recently it has been demonstrated that a circulating, as opposed to a static, anaerobic atmosphere increases the anaerobic life of chickpea seedlings (Crawford & Zochowski 1984). The experimental arrangement for these experiments is shown in Fig. 1, where both the static and moving anaerobic environments are maintained in an anaerobic incubator in order to ensure the same rigorous exclusion of oxygen.

Figure 2 shows the effect of circulating the anaerobic atmosphere around germinating seedlings on anoxic survival for a number of crop species. It can be seen that in lettuce, chickpea, cabbage, turnip, flax and wheat a circulating atmosphere produced significant prolongation of anoxic life but in rice seedlings viability declined rapidly with anoxic periods of 48 h or more irrespective of whether or not the atmosphere was circulated. In chickpea seedlings (Crawford & Zochowski 1984) it was observed that circulating the anaerobic atmosphere reduced the ethanol content of the anoxic tissues to 1/13th of that found in the static environment. These experiments therefore concluded that the removal of the volatile compounds generated under anoxia (e.g. carbon dioxide and ethanol) increased the survival of seedlings in the post-anoxic phase. Data so far published

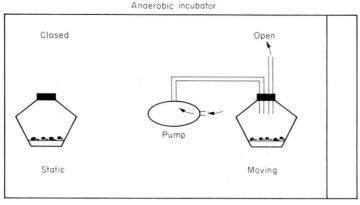

FIG.1. Diagram of arrangement in anaerobic incubator for maintaining a static and moving anaerobic atmosphere while still retaining rigorous anoxia. The atmosphere in the anaerobic incubator contains 85% nitrogen, 10% hydrogen and 5% carbon dioxide. Circulation of the atmosphere over a palladium catalyst ensures removal of any traces of oxygen. See Crawford & Zochowski (1984).

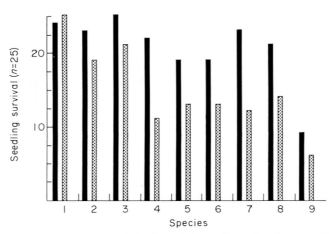

FIG. 2. Seedling survival after varying periods of anoxia in moving and static anaerobic environments. The anoxic treatment varied from 24 to 96 h depending on the sensitivity of the species. Species (1) maize, 96 h;(2) soya bean, 96 h; (3) chickpea, 96 h; (4) wheat, 96 h; (5) flax, 48 h; (6) cabbage var. Harbinger, 48 h; (7) turnip var. Milan purple top, 24 h (8) lettuce var. Webb's wonderful, 24 h; (9) rice, 96 h. (■) Moving atmosphere; (▦) static atmosphere.

does not distinguish whether it is the removal of carbon dioxide or ethanol or some other volatile product of anaerobiosis which contributes to the increased longevity of tissues that have been incubated in moving anaerobic atmospheres. However, we can now report recent factorial experiments on the interaction of carbon dioxide with moving and static environments. Using chickpea seedlings it was found that under static atmospheres the addition of high concentrations of carbon dioxide

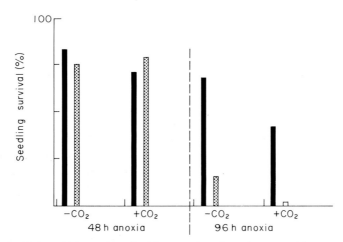

FIG. 3. Effect of addition of 25% carbon dioxide to static and moving anaerobic atmospheres on the anoxic survival of chickpea seedlings. Note that after 96 h anoxia viability is reduced most in the static atmosphere and that the greatest effect of carbon dioxide on viability is found in the static atmosphere. (■) Moving atmosphere; (▦) static atmosphere.

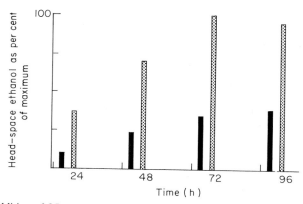

FIG. 4. Effect of addition of 25% carbon dioxide to a static anaerobic atmosphere on the head-space accumulation of ethanol above chickpea seedlings. The results are expressed as a percentage of the maximum concentration obtained over a 96 h anaerobic incubation. (■) Anoxia with no CO_2; (▨) anoxia with 25% CO_2.

(15–25%) will further reduce seedling viability. This effect cannot be attributed to carbon dioxide alone as under a moving anaerobic environment the addition of high concentrations of carbon dioxide has only a minimal effect on seedling viability (Fig. 3). Analysis of the composition of the anaerobic head-space shows that the presence of high concentrations of carbon dioxide on an anoxia-sensitive species such as chickpea results in a marked increase in ethanol production (Fig. 4). The above and the reported effects of glycosidic acidosis on root viability under anoxia (Roberts *et al.* 1984b) suggest that there is a need for attention to be paid to the possibility of interactive effects between carbon dioxide concentration and ethanol accumulation on anoxic survival.

Ethanol and post-anoxic injury in rhizomatous plants

The experiments described above using seedlings suggest that in the limited time-scale of 0–96 h minimizing ethanol accumulations with moving anaerobic atmospheres will result in increased survival after anaerobic incubation. Further experiments were undertaken to test this hypothesis using rhizomes in order to extend experimentation over several weeks. Particular attention was paid to the possibility that it is the oxidation of anaerobically accumulated ethanol in the post-anoxic phase that may contribute to the eventual death on return to air of anaerobically incubated plant tissues. It has already been shown that on restoration to air the deterioration of *Iris germanica* rhizomes is associated with the accumulation of products of lipid peroxidation (Hunter, Hetherington & Crawford 1983). The sensitivity of peroxidative injury that comes about as a result of anaerobic incubation may be a consequence of the inability of the plant to protect its tissues against such injury, as they may lack some of the necessary enzymes, catalase, peroxidase or superoxide dismutase. During anoxia the enzymes that protect tissue against the toxic effects of oxygen are sometimes reduced in activity (Morinaga 1925), thus rendering the plant sensitive to peroxidative damage on return to air. Alternatively, on returning to air, the oxidation of anaerobically accumulated metabolites such as ethanol may generate potentially damaging products such as acetaldehyde, by the peroxidatic activity of catalase (Oshino, Oshino & Chance 1973):

$$Catalase + H_2O \longrightarrow catalase\text{-}H_2O_2 \text{ (compound I)}$$
$$Catalase\text{-}H_2O_2 + CH_3CH_2OH \longrightarrow catalase + 2H_2O + CH_3CHO$$

Hydrogen peroxide generation occurs through the action of free ferrous iron and oxygen radicals:

$$Fe^{2+} + O_2 \longrightarrow Fe^{3+} + O_2^-$$

$$O_2^{.-} + O_2^{.-} + 2H^+ \xrightarrow{\text{superoxide dismutase}} H_2O_2 + O_2$$

(Hendry & Brocklebank 1985)

Within 30 min of return to air there can be observed in *Glyceria maxima* rhizomes a tenfold increase in acetaldeyde concentrations as compared with the anoxic phase (Table 1). Preliminary experiments had shown that in some tissues injured anoxically an increase in catalase activity takes place in the post-anoxic period. As catalase can mediate the oxidation of ethanol to acetaldehyde the experiments described below examine the effects of varying lengths of anaerobic treatment and circulation of the anaerobic atmosphere on the subsequent post-anoxic activity of catalase.

Catalase activity during the anoxic and post-anoxic phases was measured on freshly prepared 1 mm thick transverse rhizome slices. The slices were placed in an inverted oxygen electrode (Rank Brothers, Bottisham, Cambridge) in 3 ml of Sorensen's buffer pH 7·0 which had been previously saturated with air at 25°C. Conditions for the catalase estimation were based on the methods of del Rio *et al.*

TABLE 1. Ethanol and acetaldehyde in the head-space above *G. maxima* rhizomes after 4 days anoxia and during a post-anoxia phase. (Mean values from 4–9 rhizomes ± S.E.)

Treatment	Ethanol nmol g^{-1} fr wt (in gas phase)	Acetaldehyde nmol g^{-1} fr wt (in gas phase)	Ratio of ethanol: acetaldehyde
4 days anoxia	613·2 ± 114·0	9·5 ± 2·4	66
+ 30 min aerobic post-anoxia phase	385·7 ± 62·2	51·7 ± 8·0	8·3
+ 60 min	381·8 ± 71·2	49·8 ± 4·0	7·8
+ 120 min	356·4 ± 16·1	69·6 ± 12·3	5·8
+ 240 min	313·9 ± 27·3	66·4 ± 8·2	5·2
+ 360 min	327·8 ± 28·5	67·3 ± 4·8	5·0

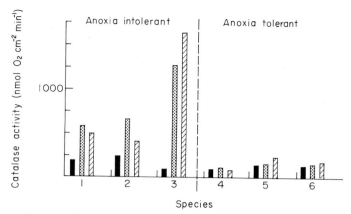

FIG. 5. Catalase activity developed after 3 days post-anoxic aerobic incubation in rhizomes of monocotyledonous species that vary in their tolerance of anoxia. The anoxic pre-treatments varied between 3 and 21 days. Intolerant species: (1) *Iris germanica*; (2) *Juncus effuses*; (3) *Glyceria maxima*. Tolerant species: (4) *Iris pseudacarus*; (5) *Schoenoplectus lacustris*; (6) *Acorus calamus*.

(1977). Respiration was measured for 15–20 min so that catalase estimations could be made below the oxygen saturation point. After this preliminary period 100 μl of hydrogen peroxide was injected into the electrode chamber giving a final concentration of 7 mM. For further experimental details see Monk, Braendle & Crawford (1987).

Figure 5 shows the difference in catalase activity between three anoxia-intolerant and three tolerant species after 3 days post-anoxic incubation in air. The intolerant species all show a rise in catalase activity when subjected to anaerobic incubation. The extent of this rise in catalase activity increased with the length of the anaerobic pre-treatment up to 14 days. After this period some species were beginning to suffer as seen above with *Glyceria maxima* and therefore a 28-day anaerobic incubation was not as effective as the 21-day treatment in stimulating the subsequent rise in catalase activity in air.

Earlier experiments (Crawford & Zochowski 1984) had shown that circulating the anaerobic atmosphere reduced anoxic injury. A similar experiment was carried out with *Glyceria maxima* to determine if circulating the anaerobic atmosphere would also have an effect on the post-anoxic increase in catalase activity. Rhizomes were therefore placed in the anaerobic incubator as shown in Fig. 1, with one set being kept in a static atmosphere and the other in a circulating anaerobic atmosphere. As both sets of rhizomes and the circulating pump were all inside the incubator the degree of anoxia was identical for all rhizomes. The result of this treatment on catalase activity in *Glyceria maxima* rhizomes is shown in Fig. 6 where it can be seen that the circulation of the anaerobic atmosphere markedly reduced the development of catalase activity.

The data in Fig. 7 illustrate the effect of exposing the rhizomes during anoxia to ethanol vapour. The rhizomes in this experiment were placed in two closed

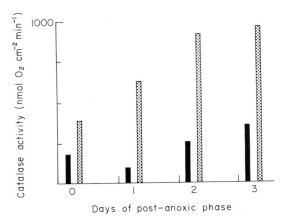

FIG. 6. Effect of moving and static anaerobic atmospheres on the subsequent activity of catalase in *Glyceria maxima* rhizomes measured over 3 days of post-anoxic aerobic incubation. Note the marked increase in rhizomes that had been kept for 5 days in the static anaerobic atmosphere as compared with other treatments. (■) Moving atmosphere; (▨) static atmosphere.

containers for a short 3-day anaerobic incubation. A small open beaker containing 500 ml of 1 M ethanol was placed in one of the closed vessels. It was known from the previous experiment that a 3-day anaerobic incubation would give rise to a relatively small increase in catalase activity. This was seen again in this experiment. However, the presence of the ethanol in the vapour phase produced a much greater increase in catalase activity (nineteenfold) when the tissues were returned to air.

A number of preliminary connections between anoxic treatment and post-anoxic metabolism are evident; notably, the connection between the length and type of anaerobic treatment on the subsequent development of catalase activity in air. Catalase can act to carry out a peroxidatic ethanol oxidation (Oshino,

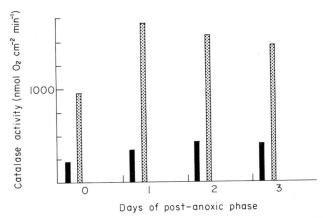

FIG. 7. Increase in catalase activity as measured during the post-anoxic incubation of *Glyceria maxima* rhizomes that had been exposed to ethanol during their previous incubation in a static anaerobic atmosphere. (■) 3 days anoxia with no ethanol vapour; (▨) 3 days anoxia with ethanol vapour.

& Chance 1973), and if hydrogen peroxide generation takes place it is possible that at high ethanol concentrations the reactions studied in rat liver catalase by the above authors could take place in rhizomes.

From previous experiments it would be expected that circulating the anaerobic atmosphere would reduce the ethanol concentration in the tissues, and this could explain the reduction in catalase activity in the post-anoxic period. Artificial addition of ethanol as a dilute vapour was quite remarkable for the rapid and large increase it produced in the catalase activity that was induced on returning the rhizomes to air.

CONCLUSIONS

The above research indicates that there is a close connection between anoxic and post-anoxic events in plant tissues and that these are related to the ability of certain species to survive a period of oxygen deprivation. It is also evident that the assessment of the potentially toxic role of any anaerobically produced metabolite, such as ethanol, is incomplete if its metabolism is not also studied during the subsequent post-anoxic phase. The alleviation of anoxic injury that is obtained by circulating the anaerobic atmosphere suggests that either ethanol or carbon dioxide or both these products of glycolysis, if allowed to accumulate, can prove toxic. Factorial experiments to establish if either ethanol or carbon dioxide on their own are toxic suggest that the removal of ethanol alone is sufficient to enhance anaerobic survival. The presence of high carbon dioxide concentrations in static atmospheres does increase ethanol accumulation and thus could contribute indirectly to anoxic injury as well as post-anoxic injury when ethanol is oxidized to acetaldehyde on return of the tissue to air.

ACKNOWLEDGMENTS

This research was made possible by a grant from the Natural Environment Research Council which is gratefully acknowledged. We are also indebted to Miss Susan Gibson for the data on seedling survival in static and moving anaerobic environments and to Dr Billy Good and Mr K. Fagerstedt for suggestions to improve the manuscript.

REFERENCES

Alpi, A. & Beevers, H. (1983). Effects of oxygen concentration on rice seedlings. *Plant Physiology*, 1, 30–34.
Barclay, A. M. & Crawford, R. M. M. (1981). Temperature and anoxic injury in pea seedlings. *Journal of Experimental Botany*, 32, 943–949.
Barclay, A. M. & Crawford, R. M. M. (1982). Plant growth and survival under strict anaerobiosis. *Journal of Experimental Botany*, 33, 541–549.
Bennett, A. B. & Spanswick, R. M. (1984). H$^+$-ATPase activity from storage tissue of *Beta vulgaris*. *Plant Physiology*, 74, 545–548.

Berrie, A. M. M. (1960). Abnormal growth and development in pea seedlings from exposure to adverse conditions during germination. *Nature*, **185**, 626–627.

Crawford, R. M. M. (1982). The anaerobic retreat as a survival strategy for aerobic plants and animals. *Transactions of the Botanical Society of Edinburgh*, **44**, 57–63.

Crawford, R. M. M. & Zochowski, Z. M. (1984). Tolerance of anoxia and ethanol toxicity in chickpea seedlings (*Cicer arietinum* L.). *Journal of Experimental Botany*, **35**, 1472–1480.

Hendry, G. A. F. & Brocklebank, K. J. (1985). Iron induced oxygen radical metabolism in water-logged plants. *New Phytologist*, **101**, 199–206.

Huck, M. G. (1970). Variation in taproot elongation rate as influenced by composition of the soil air. *Agronomics Journal*, **62**, 815–818.

Hunter, M. I. S., Hetherington, A. M. & Crawford, R. M. M. (1983). Lipid peroxidation—a factor in anoxia intolerance in *Iris* species. *Phytochemistry*, **22**, 1145–1147.

Jackson, M. B., Herman, B. & Goodenough, A. (1982). An examination of the importance of ethanol in causing injury to flooded plants. *Plant, Cell and Environment*, **5**, 163–72.

Monk, L. S., Braendle, R. & Crawford, R. M. M. (1987). Catalase activity and post-anoxic injury in monocotyledonous plants. *Journal of Experimental Botany* (In press).

Morinaga, T. (1925). Catalase activity and the aerobic and anaerobic germination of rice. *Botanical Gazette*, **79**, 73–84.

Oshino, O., Oshino, R. & Chance, B. (1973). The characteristics of the 'peroxidatic' reaction of catalase in ethanol oxidation. *Biochemical Journal*, **131**, 555–567.

del Rio, L. A., Ortega, M. G., Lopez, A. L. & Gorge, J. L. (1977). A more sensitive modification of the catalase assay with the Clarke O_2 electrode. Application of the kinetic theory of pea leaf enzymes. *Analytical Biochemistry*, **80**, 409–415.

Roberts, J. K. M., Callis, J., Wemmer, D., Walbot, V. & Jardetzky, D. (1984a). Mechanism of cytoplasmic pH regulation in hypoxic maize root tips and its role in survival under hypoxia. *Proceedings of the National Society of Sciences, USA*, **81**, 3379–3383.

Roberts, J. K. M., Callis, J., Jardetzky, O., Walbot, V. & Freeling, M. (1984b). Cytoplasmic acidosis as a determinant of flooding tolerance in plants. *Proceedings of the National Academy of Sciences, USA*, **81**, 6029–6033.

Rumpho, M. E. & Kennedy, R. A. (1983). Anaerobiosis in *Echinochloa crus-galli* (barnyard grass) seeds. *Plant Physiology*, **72**, 44–49.

Some rapid responses of sunflower to flooding

D. R. DRAKEFORD AND D. M. REID

Plant Physiology Research Group, Department of Biology, University of Calgary, Alberta, Canada T2N 1N4

SUMMARY

Anoxia, induced by flooding of roots of *Helianthus annuus*, inhibited acidification of the root bathing medium. Low acidity and flooding caused more rapid death of plants than either stress on its own. Flooding increased leakiness of roots to [^{3}H]-β-alanine, and in general, young roots were less damaged by anoxia than older tissues. Anoxia also promoted more rapid loss of potassium from older as compared to younger roots. The higher carbohydrate reserves in younger roots might be one of the reasons for their greater tolerance to flooding-induced anoxia.

INTRODUCTION

This paper deals with some short-term effects of flooding on sunflowers, *Helianthus annuus* L. cv. Russian. By short term, we mean physiological events that happen in 24 h or less, as contrasted with events that take days, weeks or months to manifest themselves. Sunflowers respond to flooding, in the long term, by reduction of stem elongation, leaf and petiole epinasty, premature senescence, chlorosis, production of adventitious roots and hypercotyl hypertrophy, and eventual death of the plant (Jackson & Drew 1984).

These effects can be divided into those that are anoxic in nature and others such as adventitious root formation and hypercotyl/hypertrophy that could be termed water effects, since they cannot be reversed by aeration of the flood water (Wample & Reid 1975).

The long-term survival of the plant may well depend on physiological and morphological adaptations that take days or even weeks to occur. For these adaptations to be effective, however, the plant has to be able to survive the short term. We were interested in this short-term survival and the biophysical triggers involved.

The species used was *Helianthus annuus* which could be regarded as a short-term tolerant plant since it can survive flooding for approximately 14 days (Wample & Reid 1975). We needed a set of symptoms of flooding that would appear within the time period in which we were interested and which could be easily measured *in vivo* without sacrificing the tissue. This led us to examine interactions between root and the environment since this organ has the first and closest contact with the stress of flooding.

Many plant species are able to alter the soil pH around their roots (Clarkson & Hanson 1980). Changes in H^+, HCO_3^-, or OH^- fluxes across the plasmalemma are known to take place to compensate for imbalances in cation and anion uptake (Clarkson & Hanson 1980). Since much of the metabolic control over ionic exchange occurs at the plasmalemma (Lüttge & Higinbotham 1979), in part because of the dependence of membrane bound ATPase activities on adequate oxygen supply; and since membranes have been postulated as general sites of stress injury (Levitt 1980), we decided to investigate membrane-related phenomena as mechanisms of short-term flood tolerance.

METHODS

In order to be able to examine such changes we developed growth systems that allowed us to control the precise conditions of flood stress as well as to measure changes in the external medium that might reflect alterations in ions' flux across membranes. To this end we designed two hydroponic systems which have been fully described elsewhere (Hubick, Drakeford & Reid 1981; Drakeford, Mukherjee & Reid 1985). Both systems gave us control over the oxygenation of the flood water as well as giving the flooded and nonflooded plants the same access to water and nutrients. The first of these two hydroponic systems (Hubick, Drakeford & Reid 1981) was designed to accommodate larger plants and in this system we could grow plants that were: (a) not flooded but adequately supplied with nutrients, oxygen and water (NF plants); (b) flooded-aerated plants (FA) in which the flood water was well aerated with oxygen; and (c) flooded plants (F) in which the flood water was depleted of oxygen.

The second system (Drakeford, Mukherjee & Reid 1985) again allowed us to produce NF, FA and F plants but was designed so that we could use much smaller amounts of tissue and also reduce the volume of water. Further, it allowed small changes in the ionic composition of the bathing medium to be easily detected. All other methods used are described in Drakeford, Mukherjee & Reid (1985).

RESULTS AND DISCUSSION

The pH of the Hoagland's solution was shown to be altered by the plants in different ways depending on whether the plants were flooded (F) or nonflooded (NF) (Fig. 1). The pH of the hydroponics system in which plants were not flooded dropped from over pH 5·0 to pH 3·9 over a period of 8 days, whereas that of the flooded system remained high. This difference in pH was observed over many experiments.

The acidification is probably an active event in this system and the high pH that correlates with flooding represents a switching off of active acidification by flooding. The system was run without plants in it (Fig. 2) but with a variety of inert media that have been used for root support. Under these conditions the pH rose rapidly from pH 4·8 to between pH 5·5 and 6·0 and remained virtually constant thereafter. This is analogous to the flooded (F) condition in Figs 1 and 3.

In Fig. 3a we see again the difference that is established between flooded (F)

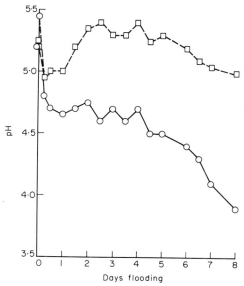

FIG. 1. The pH of the root bathing nutrient medium of plants 22 days old at O day of flooding. Not flooded (○); Flooded (□).

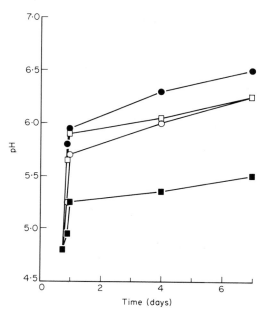

FIG. 2. The pH of the hydroponics system without plants, but with pots containing four different root-growing media. Terragreen (a crushed backed clay) (□); no medium (○); glass beads (■); granite grit (●).

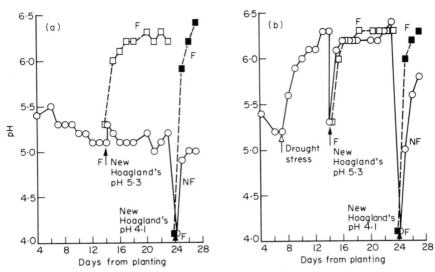

FIG. 3. The pH of the root bathing nutrient medium of plants that were 4 days old at the start of the experiment. The solution was replaced and flooding (F) started at times indicated by a solid arrow. Drought was started when indicated by an open arrow. (a) not flooded, continuous watering (○); flooded and underwent normal senescence (□); flooded and underwent 'catastrophic death' (■). (b) not flooded (NF), 1 h of watering every 24 h (○); flooded (F) and underwent normal senescence (□); flooded and underwent 'catastrophic death' (■).

and nonflooded (NF) roots. But there is an interesting difference in the manner of death of the plants flooded at pH 5·3 and those flooded at pH 4·1. The nonflooded controls at either pH appear healthy and continue to grow and acidify the nutrient medium after an initial adjustment of the pH to bring it back to the pH before the nutrient solution was changed. The plants that were flooded at pH 5·3 also behaved like flooded plants observed in many previous experiments. That is, they senesced slowly over a period of 14 days by which time they were dead. Senescence was a gradual process involving chlorisis, epinasty and death of pairs of leaves, starting with the oldest after 3–4 days and ending up with the apex. Plants that were flooded at pH 4·1, however, underwent a process that we have termed 'catastrophic death'. This involved loss of tensile strength in all the leaves simultaneously within 24 h of the onset of flooding so that they tore more easily, followed by a spread of dry lesions over the leaf surface. By the third day of flooding all expanded leaves were a pale green (not the yellow of chlorotic tissue) as opposed to the dark green of healthy tissue. They were so dry that they crumbled to the touch. Only the apex and the expanding leaves still had a healthy appearance and by the fifth day of flooding these too were dead.

Although these two pH treatments (Fig. 3) were carried out on plants differing in age, in another experiment, plants of identical age that had undergone flooding in nutrient media of pH 3·9 and pH 5·5 underwent catastrophic death or normal senescence respectively. Thus, it appears that a combination of low pH and flooding causes plants to die within a few days whereas each stress on its own causes no apparent immediate damage in the case of low pH or slow senescence in the case of flooding.

Droughted plants (Fig. 3b) also allow the pH of the medium to adopt a pH similar to that of flooded plants or the system without plants (Fig. 2), lending further support to the hypothesis that non-water-stressed plants actively acidify the nutrient medium but the onset of water stress shuts down this acidification.

Thus, we have established a symptom of flooding that appears within the time-frame of interest. This system is indicative of changes in the transport of ions which affect the pH of the external medium of which outward flux of H^+ is one, but by no means the only, possibility. This ion flux is shut down by water stress. This would indicate that the membranes are altered in some way and that anoxia is not the sole cause of the shut-down of acidification since drought-stressed plants, which were well aerated, had a similar effect as those which were flooded.

If membranes are being thus affected by flooding stress we hypothesized that they might show other measurable symptoms such as altered leakage of organic acids, amino acids and proteins, as well as various ions. We decided to test the leakiness of the membranes to β-alanine, an analogue of alanine, of similar size and charge configuration and which is little metabolized. Uptake and leakage were measured as in Fabijan *et al.* (1981). Uptake is achieved by incubating the tissue in $[^3H]$-β-alanine for 2 h. Leakage is measured by washing the labelled tissue in cold β-alanine and then incubating for 6 h in distilled water. Samples were taken at intervals and counted. Uptake was the sum of the DPM in the samples plus the incubating water plus those remaining in the tissue as measured on a sample oxidizer.

Figure 4 shows leakage of $[^3H]$-β-alanine from 2 cm root tips of 25-day-old

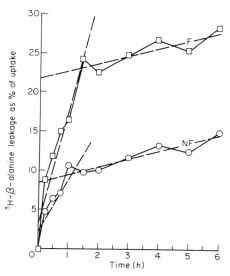

Fig. 4. Leakage of $[^3H]$ β-alanine from 20 mm root tips from 25-day-old plants expressed as a % of uptake. Analysis with a 2-way ANOV showed significant differences between all NF (nonflooded) as compared to F (flooded) points except at 1 h.

plants subjected either to flooding or nonflooding conditions. As can be seen, the leakage from the roots of plants occurs rapidly over the first hour and a half of incubation but slows down after that. Examination of the slopes of the lines shows that the large difference in leakiness between flooded (F) and nonflooded (NF) root tips is due to differences in the leakage rate in the first 1·5 h.

Washout experiments such as these can be divided into three phases. A rapid washout of a few minutes that is attributed to the apoplastic free space, a longer washout that is assigned to the cytoplasm and the remaining washout that is presumed to come from the vacuole (Lüttge & Higinbotham 1979).

In our case, the apoplastic free space washout was removed by washing with nonradioactive β-alanine. So the two phases apparent in Fig. 4 may be due to fluxes across the plasmalemma and the tonoplast in the first 1·5 h and the remaining incubation time respectively.

When the leakage of [³H]-β-alanine was examined in whole excised roots from two ages of plants of 7 and 14 days flooded for 24 h (Fig. 5), we see again that the leakage difference between flooded (F) and nonflooded plants (NF) is established during the period assigned to leakage from the cytoplasm. It is interesting to note that abscisic acid (ABA) apparently reduces the plasmalemma leakage slightly in aerated tissue and to control levels in flooded tissue for the younger roots but does not affect the older roots.

We therefore see that by 24 h, flooding causes the root tissue to become more leaky to [³H]-β-alanine, possibly at the plasmalemma. However, changes that result in the membrane becoming leaky to [³H]-β-alanine and that affect the plant's ability to acidify the nutrient medium are likely to affect the flux of other ions across the root membrane. Consequently, we examined the uptake and retention of K^+ by flooded and nonflooded roots.

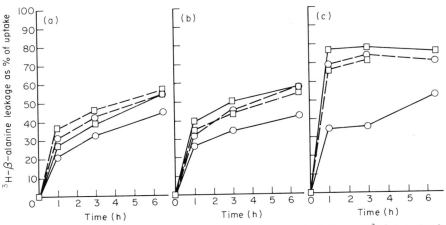

FIG. 5. The effect of age of tissue (7 days and 14 days) and ABA on the leakage of [³H] β-alanine from whole excised roots with (a) flooding; (b) flooding with aeration; and (c) not flooded. Solid lines, + ABA (1 mg dm⁻³); broken lines, no ABA; 14-day old tissue (□); 7-day old tissue (○).

Furthermore, since there have been hints in the literature that young plants may be more tolerant of flooding than older plants (Crawford 1972) we decided to look at two age groups of roots. Those that were designated 'young' plants were 5–7 days old and those designated 'old' plants were 11–14 days old with plants being selected to be at a similar stage of development from experiment to experiment. It was felt that by examining young roots versus slightly older roots it would be possible to study short-term tolerance in tissue of similar genetic make-up.

Figure 6 (top) shows changes in whole excised roots in the conductivity of a bathing medium of 1 mM KCl in 0·5 mM CaSO₄ over 24 h. It can be seen first of all that there is a difference in the way 'young' roots and 'old' roots affect the bathing medium. Five-day-old roots were able to actively take up ions from the medium and flooding stopped that uptake. Ten-day-old non-flooded roots, on the other

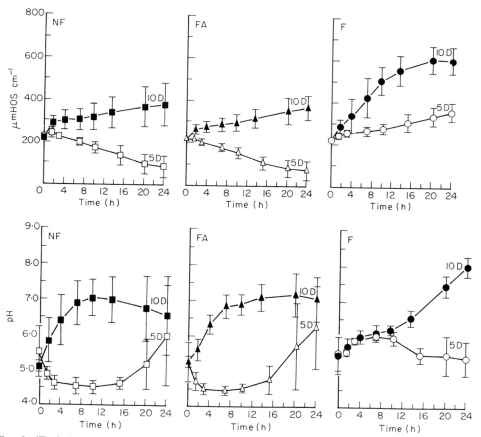

FIG. 6. (Top) changes in the conductivity and (bottom) changes in pH, of the bathing medium for 5-day-old (5D) and 10-day-old (10D) excised but whole roots, weighing 1 g fresh wt at time zero. NF = nonflooded; FA = flooded with aeration; F = flooded. Bathing medium was 1 mM KCl in 0·5 mM CaSO₄. Mean values of four replicates ± 95% confidence limits.

hand, were unable to take up ions from the medium, and flooded roots lost ions to the medium. These changes in conductivity correlated well with changes in K concentration in the medium (Fig. 9). We found that the pH of the medium undergoes active acidification when K^+ is being taken up but not when no uptake takes place or when K^+ is being leaked out (Fig. 6, bottom).

As well as differences in the fluxes of ions into or out of the roots there were visual indications of an age-dependent difference in the ability of roots to survive 24 h of flooding. The younger roots were white and healthy in appearance whereas the older roots had a brown necrotic look to them. These differences showed up quantitatively in the fresh weight (data not presented). Young roots showed no loss of fresh weight over 24 h for nonflooded controls and only a small loss (approximately 20%) of fresh weight when they were flooded. However, older roots lost fresh weight even when nonflooded and lost nearly 58% of their fresh weight when flooded.

We also examined the dry weight and found that the younger roots lost more dry weight than do older roots. These two sets of data combined as a fresh weight/dry weight ratio (Fig. 7) show that the younger plants sacrifice their greater reserves of dry weight in order to maintain their fresh weight.

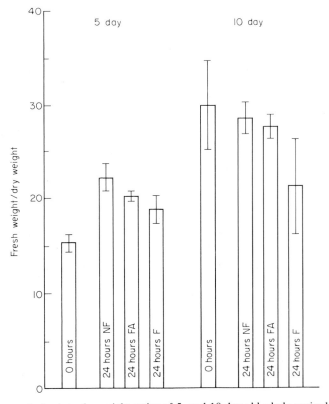

FIG. 7. Changes in fresh to dry weight ratios of 5- and 10-day old whole excised roots.

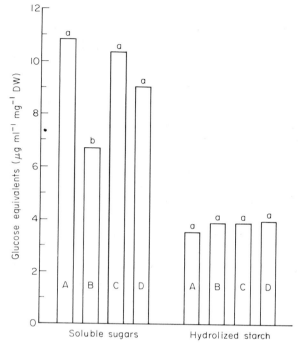

FIG. 8. Soluble sugar and hydrolized starch content of young (6 day) and older (13 day) roots. Plants grown under low (85 W m^{-2}) or higher irradiances (135 W m^{-2}). Different lower case letters above the histogram bars represent significant differences ($P = 0.5$). A, Low light, young roots; B, low light, old roots; C, high light, young roots; D, high light, old roots.

It seems likely that carbohydrate reserves might be sacrificed in this way especially during the anoxia induced by flooding. We examined the soluble sugars in the roots (Fig. 8). Under photosynthetically active radiation (PAR) levels of 85 Wm^{-2}, the soluble sugar levels were indeed lower for older roots than for younger. At higher PAR (135 Wm^{-2}), however, these differences disappear. This led us to suspect that at least part of the greater tolerance of young roots was due to carbohydrate reserves from the cotyledon irrespective of light levels.

If this hypothesis is true, then growing plants under high light levels (135 Wm^{-2}) might make the older plants less leaky to K$^+$ under flooded conditions. In Fig. 9 after 20 h there were relatively small differences between flooded roots and nonflooded controls but by 48 h the flooded roots were as leaky to K$^+$ as roots were at 24 h under low light levels (Fig. 6). Thus, growth of the plants under high PAR overcame the tendency for older roots to become more leaky to K$^+$. However, since these studies were carried out with excised roots even these greater reserves are eventually used up.

We also found that when whole plants were examined under low PAR conditions (85 Wm^{-2}) the older plants again showed no difference in leakiness between flooded roots and nonflooded controls. In separate experiments the

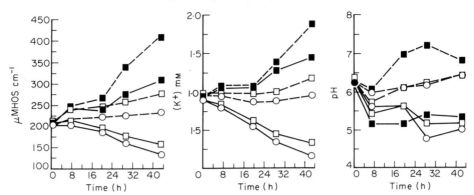

FIG. 9. Effects of flooding (nonflooded, NF□; flooded and aerated, FA○; flooded with nitrogenation, F,■) and age (6-day old (——) versus 13-day old (–––)) on conductivity (μM h cm^{-1}), K$^+$ concentration and pH of the bathing medium surrounding excised roots. Plants initially grown under the higher irradiances of 135 W m^{-2}. There was a significant effect of age on conductivity, K$^+$ and pH at 19·5 h in all cases ($P = 0\cdot0005$) and a significant effect on conductivity and K$^+$ at 42·5 h ($P = 0\cdot0005$).

putative stress hormones ethylene and abscisic acid did not either counteract or enhance the effects of flooding on the leakiness of excised root membranes.

CONCLUSIONS

We have demonstrated that flood-induced anoxia halts the acidification of a hydroponic nutrient medium by sunflowers and that a combination of low pH and flooding speeded up and altered the manner of death of plants compared with either stress acting alone. Membranes were shown to become more leaky to β-alanine with flooding. Young roots were shown to be more active than older roots at taking up potassium from the hydroponic medium. The younger roots seemed less affected by flood-induced anoxia than the older roots, ceasing active uptake, whereas the older roots became leaky and lost potassium to the medium.

These differences were shown to be partially due to greater carbohydrate reserves in the younger roots. By manipulation of the environmental conditions so as to provide higher reserves in the older roots, their resistance to leakiness can be prolonged. The effects of flooding on leakage were little affected by the stress hormones abscisic acid and ethylene.

REFERENCES

Clarkson, D. T. & Hanson, J. B. (1980). The mineral nutrition of higher plants. *Annual Review of Plant Physiology*, **31**, 239–298.
Crawford, R. M. M. (1972). Some metabolic aspects of ecology. *Botanical Society of Edinburgh*, **41**, 309–322.
Drakeford, D. R., Mukherjee I. & Reid, D. M. (1985). Some early responses of *Helianthus annuus* L. to flooding. I. The effects of flooding on the uptake and leakage of 'nonelectrolytes' by roots. *Journal of Experimental Botany*, **36**, 1705–1715.

Fabijan, D. M., Plumb, P., Dhindsa, D. & Reid, D. M. (1980). Effects of two growth retardants on tissue permeability in *Pisum sativium* and *Beta vulgaris*. *Planta*, **152**, 481–486.

Hoagland, D. R. & Arnon, D. I. (1938). The water culture method for growing plants without soil. Circular of the University of California Agriculture Experimental Station, **347**, 32 pp.

Hubick, K. T., Drakeford, D. R. & Reid, D. M. (1981). A comparison of two techniques for growing minimally water-stressed plants. *Canadian Journal of Botany*, **60**, 219–223.

Jackson, M. B. & Drew, M. C. (1984). Effects of flooding on growth and metabolism of herbaceous plants. *Flooding and Plant Growth* (Ed. by T. T. Kozlowski) pp. 47–128. Academic Press, Orlando.

Levitt, J. (1980). *Responses of Plants to Environmental Stress*, Vol. II. *Water, Radiation, Salt and Other Stresses*. Academic Press, New York.

Lüttge, U. & Higinbotham, N. (1979). *Transport in Plants*. Springer-Verlag, New York, 468 pp.

Wample, R. L. & Reid, D. M. (1975). Effect of aeration on the flood induced formation of adventitious roots and other changes in sunflower (*Helianthus annuus* L.) *Planta* **127**, 263–270.

Rhizome anoxia tolerance and habitat specialization in wetland plants

R. BRAENDLE

Institut für Pflanzenphysiologie, Universität Bern, Altenbergrain 21, Bern, CH 3013, Switzerland

R. M. M. CRAWFORD

Department of Plant Biology and Ecology, The University, St Andrews, Scotland KY16 9TH

SUMMARY

Variation in rhizome tolerance of anoxia, which can range from a few days to several months, is much greater than the range of anoxia tolerance found in roots. Because they act as regenerators for both roots and shoots, rhizomes are more important than roots for the survival of plants in anoxic conditions. This review explores the properties that confer on rhizomes their preferences for different wetland habitats, such as permanently submerged anaerobic muds, the interface zone between lake and free water, and soils that are flooded only seasonally.

INTRODUCTION

Roots and rhizomes are the two plant organs most likely to be subjected to periods of oxygen deficiency by flooding, although most studies on variation in flooding tolerance in higher plants have been concerned with roots. Armstrong (1979) has reviewed the positive role of root aerenchyma in facilitating root aeration in flooded soils. Numerous other studies have investigated the physiological reactions of roots to flooding to determine whether or not there are factors other than root aeration that confer a tolerance to flooding (Crawford 1982; ap Rees & Wilson 1984). Such physiological studies are frequently hindered by the need to use excised roots because the physiological responses to anoxia of intact plants can be confused with the differing abilities of the plants to ventilate their roots with oxygen diffusing downwards from the shoot. The limited life of excised roots prevents experiments from being prolonged for a sufficient length of time to determine if there is any long-term adaptive response of the tissues to extended periods of oxygen deficiency. As yet there are no reports that detached roots can survive more than a few days of anoxia. Under aerobic conditions excised *Glyceria maxima* roots survived satisfactorily for 3–4 days although the rate of respiration fell steadily (Jenkin & ap Rees 1983). After a nitrogen gassing treatment of as short as 24 h the capacity to respire aerobically when returned to air was drastically diminished. In this case carbohydrate supply does not appear to have been limiting; however, some authors consider that translocation of substrates to roots is a major limiting factor in determining the length of anaerobic life (Vartapetian, Andreeva & Nuritdinov 1978; Saglio & Pradet 1980; see also chapter by Setter *et al.* in this volume).

397

By contrast with roots, rhizomes are perennial organs with plentiful energy reserves. Furthermore, they have a capacity for regeneration and a prolonged and independent existence when deprived of both shoots and roots (Braendle 1980). This capacity for an extended and predictable existence as isolated organs makes rhizomes suitable for long-term physiological experimentation on the effects of anoxia. Ecologically, their survival capacity under anoxia is important because of their role as over-wintering perennating organs which are likely to be subjected to periods of oxygen shortage (Laing 1940, 1941; Boulter, Coult & Henshaw 1963). More recently, it has been shown that detached rhizomes differ widely in the length of time that they can be kept under anoxia and still regenerate healthy plants when brought out into air. Survival can vary from as little as 4 days in *Juncus effusus* to nearly 3 months in *Scirpus maritimus* (Table 1).

Wet and rhizomatous species also differ markedly in their degree of habitat specialization. In some species, e.g. *Schoenoplectus lacustris* and *Typha latifolia*, the rhizomes are normally found buried in permanently saturated mud. In other species, e.g. *Acorus calamus* and *Menyanthes trifoliata*, the rhizomes lie at or near the surface of submerged muds. *Phragmites australis*, however, shows a wider range from the completely submerged soil to land that is only periodically flooded. Some species occur where the soil is regularly flooded in winter but not in summer. Here too the species can be divided into those that have their rhizomes buried in the soil, such as *Iris pseudacorus* and *Phalaris arundinacea*, or else have their rhizomes nearer the surface, as with *Ranunculus flammula* and *Mentha aquatica*. The distinctly different habitats of these species make them suitable material for investigations of physiological adaptations to anoxia.

ANOXIA TOLERANCE IN RHIZOMES FROM DIFFERENT HABITATS

Table 1 lists the maximum length of time that it has been possible to keep detached rhizomes under anoxia and still achieve healthy shoot regeneration when returned to air. The table also records whether or not the rhizomes were observed to possess the ability to extend their shoots during a period of oxygen deprivation (see Fig. 1). It is important to note that both the condition of the rhizomes in terms of the carbohydrate reserves (Steinmann & Braendle 1984 a, b; Braendle 1985) and the type of anoxic treatment (Crawford & Zochowski 1984) are likely to affect the anaerobic vitality of the organs as well as their capacity for shoot elongation. In early spring, when they possess unexpanded buds and plentiful carbohydrate supplies, *Glyceria maxima* rhizomes can be kept alive for 7–14 days under anoxia and may even show some shoot elongation during the first few days of oxygen deprivation. In summer when the carbohydrate supplies in the rhizomes are reduced, they are usually killed by as little as 7 days' anoxia (Barclay & Crawford 1982). Similarly, in *Typha latifolia* there is prolonged survival under anoxia in spring but this is reduced to less than 3 weeks in summer when the carbohydrate reserves are low (Haldemann & Braendle 1986).

TABLE 1. Length of anaerobic incubation that can be endured in detached rhizomes without loss of regenerative power. For experimental methods see Barclay & Crawford (1982)

Species	Anoxia endurance* (days)	Shoot elongation
Carex rostrata	4	None
Juncus effusus	4–7	None
J. conglomeratus	4–7	None
Glyceria maxima	7–21	Occasional
Ranunculus lingua	7–9	None
R. repens	7–9	None
Mentha aquatica	4	None
Eleocharis palustris	7–12	None
Fililpendula ulmaria	7–14	None
Carex papyrus	7–14	None
C. alternifolius	7–14	None
Spartina anglica	>28	None
Iris pseudacorus	>28	None
Phragmites communis	>28	None
Typha latifolia	>28	Frequent
T. angustifolia	>28	Frequent
Scirpus americanus	>28	Frequent
S. maritimus	>90	Frequent
S. tabernaemontani	>90	Frequent
Schoenoplectus lacustris	>90	Frequent

*These figures represent the minimum time that the species were able to survive the anoxic treatment; longer periods of anoxia survival may be possible in those species that survived 90 days or more. It should be noted that tolerance can vary with season and the state of development of tissues. Normally the tolerance is greater in spring and less in summer when carbohydrate reserves are small. Similarly, well-formed terminal buds in Glyceria maxima can elongate under anoxia but this is not seen in smaller lateral buds.

Recently, it has been shown that circulating the anaerobic atmosphere as opposed to keeping it static can prolong the anaerobic life of chickpea seedlings (Crawford & Zochowski 1984). In Glyceria maxima circulating the anaerobic atmosphere can prevent the development of epidermal lesions in the distal regions of the rhizome bud (Fig. 2). These experiments show that for uniformity of response it is necessary to standardize experiments on anoxia tolerance both in relation to season and to whether the atmosphere is stagnant or circulating. In addition, the volume of the head-space above the anaerobically incubated tissues must be known if relevant comparisons are to be made. The data for relative anoxia tolerance recorded in Table 1 were obtained using a circulating anoxic atmosphere in an anaerobic incubator with a volume of $0 \cdot 613 \text{ m}^3$ (Barclay & Crawford 1982).

In addition to their ability to withstand prolonged periods of anoxia, Table 1 also records whether or not the species tested extended their shoots during the anaerobic treatment. Certain species regularly produced healthy, upright shoots from pre-formed buds, even although the atmosphere was strictly anoxic for several weeks. Anaerobic growth in plants is a rare occurrence; and apart from examples listed in Table 1 it has been observed only in relation to coleoptile extension in rice

FIG. 1. Examples of bud extension growth that took place during 14 days of total anoxia at 22°C in two wetland rhizomatous species: (a) *Schoenoplectus lacustris* and (b) *Scirpus maritimus*; the plant on the left is the aerobic control and the plant on the right the anaerobically incubated rhizome segment. (Reproduced with permission from the *Transactions of the Botanical Society of Edinburgh* **14**, 57–63).

and *Echinochloa* spp. (Kennedy *et al.* 1980; Rumpho *et al.* 1984). In rice, coleoptile survival is limited to a few days and is usually followed by epinasty and death if the period of anoxia is extended much beyond 48 h. By contrast, the wetland rhizomatous species capable of anaerobic shoot growth listed in Table 1 can sustain the periods of shoot extension over 1–2 weeks. These anaerobically produced shoots may eventually die within 2–3 weeks if not returned to air. However if restored to aerobic conditions within this period their growth will continue normally. Even after the first flush of anaerobic bud expansion, if the rhizomes are still kept under anoxia and have further non-expanded buds, these too can subsequently send out expanded shoots. *Scirpus maritimus* is capable of producing new shoot growth under anoxia after 2 months of continuous anaerobic incubation (Barclay & Crawford 1982, Fig. 1).

A comparison of the species listed in Table 1 with their habitat preferences (Table 2) shows that of all the species investigated so far, those from the most anaerobic habitats possess the highest degree of anoxia tolerance. Further, it is the species with rhizomes normally buried in anaerobic mud that exhibit the capacity

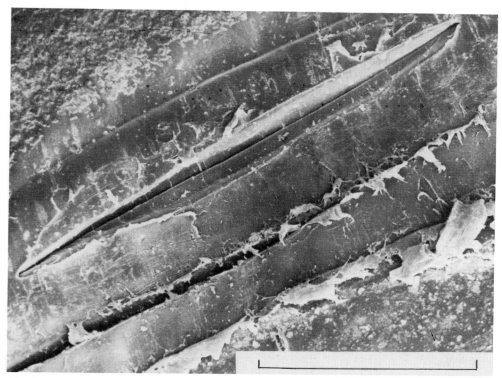

FIG. 2. Scanning electron microscope photograph of the epidermis of the tip of a *Glyceria maxima* rhizome bud that had been kept in a static anaerobic environment for 8 days. Note the 200 μm fissures. These did not appear in the rhizomes that were kept in a moving anaerobic atmosphere. The scale mark = 100 μm.

for extended shoot elongation under anoxia. However, in species where the rhizomes lie at the surface of the submerged mud, the phenomenon of prolonged anoxia tolerance does not frequently occur. The only species from this habitat that has so far been shown to be anoxia tolerant is *Acorus calamus*. This species is an interesting exception in that it also shows shoot extension under anoxia. Rhizomatous wetland species that die rapidly when placed under anoxia include *Juncus effusus* and *J. conglomeratus* as well as *Glyceria maxima* (note that this species has increased anoxia tolerance in early spring—see p. 398). It would appear that in contrast to the deeply buried species, the rhizomes living on or near the soil surface are more dependent on oxygen for survival.

Because rhizomes allow experimentation to be carried out over weeks instead of just a few days as with roots, it is possible to demonstrate more variation in their anoxia-tolerance than can be found in roots. It is important to note that all these species possess well developed aerenchyma and have a high degree of efficiency in ventilating their rhizomes when their shoots are in contact with air (Armstrong 1979; Studer & Braendle 1984). However, despite this universal capacity for rhizome ventilation, there are enormous differences in the tolerance of anoxia. This

TABLE 2. Habitat specialization in rhizomatous wetland plants

I. Buried in permanently saturated or water-covered muds

Schoenoplectus lacustris
Scirpus tabernaemontani
S. maritima
S. americanus
Spartina anglica
S. alterniflora
Phragmites australis
Typha latifolia
T. angustifolia

II. Buried in soils subject to seasonal flooding

Phragmites australis
Iris pseudacorus
Phalaris arundinacea
Glyceria maxima
Eleocharis palustris
Juncus conglomeratus
J. effusus

III. Rhizomes living at the interface between anaerobic mud and free water

Acorus calamus
Menyanthes trifoliata
Ranunculus flammula
R. repens
Mentha aquatica

difference does not occur at random and therefore suggests that physiological adaptation is closely related to habitat specificity.

CARBOHYDRATE RESERVES AND HABITAT SPECIALIZATION IN RHIZOMES

As shown above, it is the plants of the most extreme anaerobic habitats that show the greatest ability to survive under anaerobic conditions. Anaerobic metabolism is costly in terms of carbohydrate consumption as compared with normal respiration: those species that can survive extended periods of anoxia and live in the most anaerobic environments also have large carbohydrate reserves. These reserves show considerable fluctuations throughout the year with a typical maximum in autumn and a minimum in early summer after shoot extension (Steinmann & Braendle 1984a). In both *Typha latifolia* and *T. angustifolia* the highest dry weight values for the rhizomes are found in late autumn (Fiala 1978). Similarly tracer experiments with $^{14}CO_2$ on *Spartina alterniflora* show that the rhizome is the major carbohydrate sink in the growing season (Lyttle & Hull 1980). Over winter, a sharp fall in the carbohydrate content has been observed in the rhizomes of *Typha latifolia* from 45 to 27% of their total dry weight (Kausch, Seago & Marsh 1981). In *Spartina alterniflora* the high winter levels of carbohydrate reserves are most

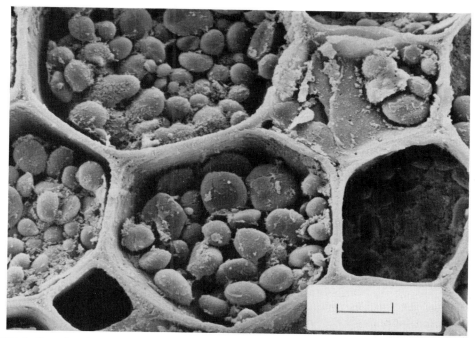

FIG. 3. Scanning electron microscope photograph of storage parenchyma cells in the vascular cylinder of *Schoenoplectus lacustris* rhizomes showing many large amyloplasts. The scale mark = 10 μm.

severely depleted in spring in zones of rapid development (Gallagher 1983; Gallagher, Wolf & Pfeiffer 1984). A similar picture is observed in *Scheonoplectus lacustris* (Steinmann & Braendle 1984b, Fig. 3). These observations show that rhizomes that live in anaerobic mud typically have high concentrations of carbohydrates in early autumn and minimal levels in early summer. These strongly marked fluctuations are evidence of the high metabolic cost to the plant of resuming growth in spring from rhizomes that are deeply submerged in an anaerobic habitat. In *A. calamus* it is noticeable that the carbohydrate reserves never sink in early summer to as low a level as that found in those of the deeper buried species and that the fluctuations are not so pronounced (Haldemann & Braendle 1986). It may be that surface-living rhizomes make less demands on their reserves by not having to extend shoots as far and by prolonging the photosynthetic season through the possession of green rhizome tissues in a way that is not possible in the species with buried rhizomes.

There also appears to be a difference between anoxia-tolerant and anoxia-intolerant species in their ability to conserve carbohydrate supplies when experimentally deprived of oxygen. As carbohydrate consumption by plant tissues is highly temperature dependent, it is striking that under the conditions of experimental anoxia (22°C) *Schoenoplectus lacustris* showed only a 2% reduction in total nonstructural carbohydrate reserves. By contrast under the same conditions of experimental anoxia *Glyceria maxima* (which is intolerant of prolonged periods of anaerobiosis) showed a 46% reduction in total nonstructural carbohydrate (Barclay & Crawford 1983). The soluble sugars also showed a similar pattern of behaviour with a marked reduction over 4 days in *Glyceria maxima* and little change in *Schoenoplectus lacustris*. Thus even although *Glyceria maxima* rhizomes may normally possess considerable carbohydrate reserves they are likely to suffer rapid depletion when exposed to low-oxygen stress. Such behaviour is maladaptive in habitats where the oxygen supplies may be limited. There appears therefore to be an ecological relationship between carbohydrate conservation and the degree of anoxia that is found in the preferred habitats of these species.

ANAEROBIC METABOLISM IN RHIZOMES

One of the first effects of anoxia on any tissue is a rapid change in energy charge values (Mocquot *et al.* 1981). This is also the case in rhizomes but in most species there is usually a recovery (Monk & Braendle 1982; McKee & Mendelssohn 1984). This recovery is also observed in maize roots (Saglio & Pradet 1980) where it is dependent on the provision of an external sugar supply which is not the case with rhizomes. These observations further support the important role of carbohydrate reserves in maintaining the anaerobic integrity of rhizome tissues.

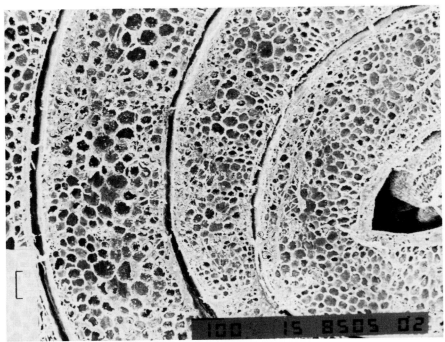

FIG. 4. Scanning electron microscope photograph of a transverse section of the basal region of the distal bud of a *Glyceria maxima* rhizome. Note that in common with most wetland plants there is a well-developed aerenchyma tissue. This species, however, by contrast with *Schoenoplectus lacustris* and *Iris pseudacorus* (see Figs 3 and 5) survives only a few days under static anaerobic atmospheres. The scale mark = 100 μm.

Rhizomes exhibit a natural pattern of anaerobiosis in the extent to which ethanol accumulates in relation to aeration. Even under well-aerated conditions *Schoenoplectus lacustris* rhizomes are evidently hypoxic, as ethanol can always be detected (Duss & Braendle 1982). When aeration is impeded in this species ethanol concentrations rise but reach a plateau. This stabilization can be achieved by a combination of reduced production, with removal by diffusion or translocation. This appears to be a feature of all wetland species so far examined (Monk, Crawford & Braendle 1984). In *Acorus calamus*, which is a surface-living rhizome (see Table 2), there is a high ADH activity in comparison with other wetland rhizomatous species. Although ADH is not a rate-limiting step in glycolysis its activity is regularly increased under anaerobic conditions (Smith & ap Rees 1979; Harberd & Edwards 1983). Thus on flooding or experimental anoxia there is an initial increase in ethanol concentration in the rhizomes but this reaches a plateau usually at concentrations less than 30–40 μmol g^{-1} fresh weight.

These results suggest that surface-living species show only minimal increases in ethanol concentration in their rhizomes when kept under anoxia, provided the conditions for ventilation of their tissues are maintained as they would be in their natural habitat. It appears that the porous nature of the rhizomes and their occurrence at the soil–water interface facilitates the rapid removal of ethanol (Figs. 4 & 5).

When a dryland rhizomatous species such as *Iris germanica* (Fig. 6) is subjected

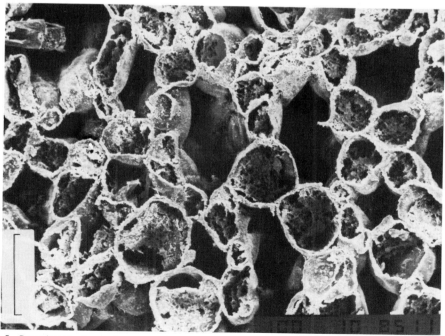

FIG. 5. Scanning electron microscope photograph of cortex of *Iris pseudacorus* rhizomes showing the development of extensive aerenchyma. Under anaerobic conditions rhizomes with these well-developed air-spaces reach a plateau in their ethanol concentrations which only rarely exceeds 40 μmol g^{-1} fresh weight. The scale mark = 100 μm.

to anoxia, the plateau effect that was common to the wetland species is not seen. Ethanol concentrations in *Iris germanica* rhizomes rise as long as the anoxic condition is imposed. Whether or not the concentrations of ethanol that accumulate as a result of anoxia (Monk, Crawford & Braendle 1984) are toxic to cells is a matter of some debate (Smith & ap Rees 1979; Jackson, Herman & Goodenough 1982). However, when rhizomes of *Glyceria maxima* are kept in a moving anaerobic environment their ethanol content is reduced to one sixth of that observed after the same length of time in a static environment. In addition the viability of the lateral buds on the rhizomes is significantly increased in the moving anaerobic atmosphere (Monk & Crawford, unpubl. results; see also chapter by Crawford *et al.* in this volume). Visible effects of a moving, as opposed to a static, anaerobic environment are also seen with scanning electron microscope studies (Fig. 2). In the moving anaerobic environment the epidermis of the rhizome tip is indistinguishable from the aerobic control. However, when the anaerobic environment is static longitudinal slits approximately 200 μM long appear in the epidermis.

As with carbohydrate levels, the rhizomes of wetland habitats show annual fluctuations in ADH activity. This activity is maximal in winter and minimal in summer (Haldemann & Braendle 1986). The regular nature of the change in activity with the seasons suggests that rhizomes are under a more anaerobic environment in winter than in summer. It appears that some dependence on anaerobic metabolism is a fact of life for buried rhizomes of wetland habitats and that tolerance of these conditions is greatest in those species that inhabit the most hypoxic habitats.

Clearly ethanol fermentation is the most important source of anaerobic energy supply in rhizomes. However, it is also possible that under certain conditions alternative products may be produced, e.g. shikimic acid shows annual fluctuation in *Iris pseudacorus* (Tyler & Crawford 1970). When intact plants are examined it is possible to observe increases in the concentration of malic acid in the roots of a number of species. Nevertheless, the quantitative importance of this in relation to total fermentation capacity is still doubtful (ap Rees & Wilson 1984). In the sea grass *Zostera* spp. alanine and amino butyric acid accumulate within a few hours of the imposition of root anoxia and this can account for 70% of the total amino acid pool. At the same time there is only a small increase in ethanol (Pregnall *et al.* 1984). For rhizomatous species, however, the re-oxidation of NADH appears to be achieved mainly through ethanol production, although some lactate production has been reported in *Schoenoplectus lacustris* (12 μmol g^{-1} fr. wt, Duss & Braendle 1982; 5 μmol g^{-2} fr. wt, Monk, Crawford & Braendle 1984). The role of alternative products to ethanol needs to be carefully quantified before any claims can be made for the substantial contribution of alternative products in glycolysis.

RHIZOME VENTILATION BY AERENCHYMA

The extensive development of aerenchymatous tissues in marsh and aquatic plants is well known and has been amply demonstrated as highly effective in transferring oxygen from the shoot to the root (Armstrong 1979; Dacey & Klug 1982a, b). In particular, monocotyledonous species of wetland habitats regularly show a well developed aerenchyma (Smirnoff & Crawford 1983). Among dicotyledons there is a more heterogeneous situation. Species such as *Filipendula ulmaria* inhabit wet marshes yet have relatively low amounts of aerenchymatous tissue. If we examine the species listed in Table 1, it is difficult to distinguish any relationship between the particular habitat characteristic of a wetland species and the extent of aerenchyma formation. The species from the most anaerobic muds, such as *Schoenoplectus lacustris*, are no different from those of the upper zones such as *Glyceria maxima* (Fig. 4) and *Carex rostrata*, in their reliance on aerenchyma for the downward diffusion of oxygen. Thus, the eco-physiological adaptation of these species must also depend on some other feature. As discussed above, these species differ in their tolerance of experimental anoxia. This does not mean that the development of aerenchyma is irrelevant. Oxygen is clearly required for root growth, and many aspects of cell metabolism are dependent on oxygen, e.g. fatty acid desaturation and sterol biosynthesis. Also there is no evidence, as yet, that cell division in roots or the development of new tissue can take place without oxygen.

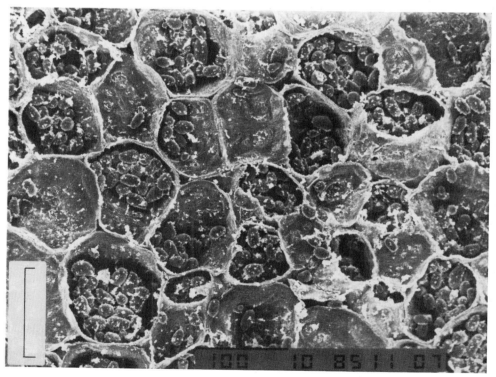

FIG. 6. Scanning electron microscope photograph of the cortex of *Iris germanica* rhizomes—a more solid tissue than that of *Iris pseudacorus* (Fig. 5) and where continuing anaerobic conditions cause ethanol concentrations to rise without reaching any plateau. The scale mark = 100 μm.

Ethylene has been clearly shown to stimulate the production of aerenchyma in maize (Jackson, Fenning & Jenkins 1985) but not in rice (Jackson *et al.* 1985). However, in those cases where ethylene is the stimulant for aerenchyma formation, its synthesis from its precursor (ACC) is an oxygen-dependent reaction (Young & Hoffman 1984) and will therefore necessitate an adequate level of tissue aeration. Shoot extension is also generally assumed to be ethylene controlled (Osborne 1984). Nevertheless, the most anoxia-tolerant species described above which are capable of shoot extension in complete absence of oxygen must presumably receive a stimulus for bud extension that is not dependent on ethylene production. For a more detailed discussion of this problem see Ridge (in this volume).

The extent to which oxygen diffuses downwards may play an important role in the resistance of plants to soil and water toxins such as ferrous iron. In the Mississippi Delta, Mendelssohn & Postek (1982) have shown that in different regions there is a marked variation in the production of ferric deposits on root surfaces (see also Taylor, Crowder & Rodden 1984). Aerenchyma will also permit the upward diffusion of the volatile products of glycolysis from the rhizomes. It has been shown that plant survival under anoxia can be prolonged by ventilation without the supply of oxygen (Crawford & Zochowski 1984). Thus, removing the volatile products of anoxia may also be an important function of aerenchyma in addition to the supply of oxygen to the roots and rhizomes.

CONCLUSIONS

The object of this review has been to highlight the role of the rhizome in relation to survival mechanisms and habitat specificity of a range of wetland species. Within wetland habitats there are many micro-sites in which specialized survival strategies have evolved. The variation in anoxia tolerance among rhizomes of different wetland species and their ability to succeed in different sites has made it possible to examine some of the properties that are associated with survival under different environmental conditions. Many authors in the past have looked at roots as the organs which determine the success or failure of plants in flood-prone soils. This study shows that rhizomes exhibit a much greater range of tolerance to anoxia than roots. It is suggested that rhizome properties play a major role in determining the success of such species in various wetland conditions.

ACKNOWLEDGMENTS

The co-operation between Bern and St Andrews which produced the results on which this paper is based, was made possible by financial support from the British Council which is gratefully acknowledged. We are also indebted to Dr B. Good, Miss Lorna Monk and Mr K. Fagerstedt for helpful comments on the preparation of the manuscript.

REFERENCES

Armstrong, W. (1979). Aeration in higher plants. *Advances in Botanical Research*, 7, 225–332.

Barclay, A. M. & Crawford, R. M. M. (1982). Plant growth and survival under strict anaerobiosis. *Journal of Experimental Botany*, 33, 541–549.

Barclay, A. M. & Crawford, R. M. M. (1983). The effect of anaerobiosis on carbohydrate levels in storage tissues of wetland plants. *Annals of Botany*, 51, 255–259.

Boulter, D., Coult, D. A. & Henshaw, G. G. (1963). Some effects of gas concentrations on metabolism of the rhizome of *Iris pseudacorus* (L.). *Physiologia Plantarum*, 16, 541–548.

Braendle, R. (1980). Die Ueberflutingstoleranz der Seebinse (*Schoenoplectus lacustris* L. Palla). Uebersicht ueber die verschiedenen Anpassungsstrategien. *Vierteljahrschrift Naturforschende Gesellschaft Zurich*, 125/2, 177–185.

Braendle, R. (1985). Kohlenhydratgehalt und Vitalitaet isolierter Rhizome von *Phragmites australis*, *Schoenoplectus lacustris* und *Typha latifolia* nach mehrwöchigem O_2-Mangelstress. *Flora*, 177, 317–321.

Crawford, R. M. M. (1982). Physiological responses to flooding. *Encyclopedia of Plant Physiology* (Ed. by O. L. Lange, P. S. Nobel, C. B. Osomomd, & H. Ziegler) Vol. 12B, 453–477.

Crawford, R. M. M. & Zochowski, Z. M. (1984). Tolerance of anoxia and ethanol toxicity in chickpea seedlings (*Cicer arietinum* L.). *Journal of Experimental Botany*, 35, 1472–1480.

Dacey, J. W. H. & Klug, M. J. (1982a). Tracer studies of gas circulation in *Nuphar*: $^{18}O_2$ and $^{14}CO_2$ transport. *Physiologia Plantarum*, 56, 361–366.

Dacey, J. W. H. & Klug, M. J. (1982b). Ventilation by floating leaves in *Nuphar*. *American Journal of Botany*, 69, 999–1003.

Duss, F. & Braendle, R. (1982). Die Ueberflutungstoleranz der Teichbinse (*Schoenoplectus lacustris* L. Palla). V. Die Bildung von verschiedenen Gaerungsprodukten und Transportsubstanzen im Rhizomegewebe bei Sauerstoffmangel. *Flora*, 172, 217–222.

Fiala, K. (1978). Underground organs in *Typha angustifolia* and *Typha latifolia*, their growth propagation and production. *Acta Sc. Nat. Brno.*, 12 (8), 1–43.

Gallagher, J. L. (1983). Seasonal patterns in recoverable underground reserves in *Spartina alterniflora* Loisel. *American Journal of Botany*, 70, 212–215.

Gallagher, J. L., Wolf, P. L. & Pfeiffer, W. J. (1984). Rhizome and root growth rates and cycles in protein and carbohydrate concentrations in Georgia *Spartina alterniflora* Loisel. plants. *American Journal of Botany*, 71, 165–169.

Haldemann, C. & Braendle, R. (1986). Jahrzeitliche Unterschiede im Reservstoffgehalt und von Gaerungsprozessen in Rhizomen von Sumpfund Roehrichpflanzen aus dem Freiland. *Flora*, 178, 307–313.

Harberd, N. P. & Edwards, K. J. R. (1983). Further studies on the alcohol dehydrogenases in barley: evidence for a third alcohol dehydrogenase locus and data on the effect of an alcohol—1 null mutation in homozygous and in heterozygous condition. *Genetical Research in Cambridge*, 41, 109–116.

Jackson, M. B., Fenning, T. M. & Jenkins, W. (1985). Aerenchyma (gas-space) formation in adventitious roots of rice (*Oryza sativa* L.) is not controlled by ethylene or small partial pressures of oxygen. *Journal of Experimental Botany*, 36, 1566–1572.

Jackson, M. B., Fenning, T. M., Drew, M. C. & Saker, L. R. (1985). Stimulation of ethylene production and gas-space (aerenchymas) formation in adventitious roots of *Zea mays* L. by small partial pressures of oxygen. *Planta*, 165, 486–492.

Jackson, M. B., Herman, B. & Goodenough, A. (1982). An examination of the importance of ethanol in causing injury to flooded plants. *Plant, Cell and Environment*, 5, 163–172.

Jenkin, L. E. T. & ap Rees, T. (1983). Effects of anoxia and flooding on alcohol dehydrogenase in roots of *Glyceria maxima* and *Pisum sativum*. *Phytochemistry*, 22, 2389–2393.

Kausch, A. P., Seago, J. L. & Marsh, L. C. (1981). Changes in starch distribution in the overwintering organs of *Typha latifolia* (Typhaceae). *American Journal of Botany*, 68, 877–880.

Kennedy, R. A., Barrett, S. C. H., Zee, D. V. & Rumpho, M. E. (1980). Germination and seedling growth under anaerobic conditions in *Echinochloa crus-galli* (Barnyard grass). *Plant, Cell and Environment*, 3, 243–248.

Laing, H. E. (1940). Respiration of the rhizomes of *Nuphar advenum* and other water plants. *American Journal of Botany* 27, 547–581.

Laing, H. E. (1941). Effect of concentration of oxygen upon growth of rhizomes and semi-submerged water plants. *Botanical Gazette*, 102, 712–724.

Lyttle, R. W. & Hull, R. J. (1980). Photoassimilate distribution in *Spartina alterniflora* Loisel. I. Vegetative and floral development. *Agronomy Journal*, 72, 933–946.

McKee, K. L. & Mendelssohn, I. A. (1984). The influence of season on adenine nucleotide concentration and energy charge in four marsh plant species. *Physiologia Plantarum*, 62, 1–7.

Mendelssohn, I. A. & Postek, M. T. (1982). Elemental analysis of deposits on the roots of *Spartina alterniflora* Loisel. *American Journal of Botany*, 69, 904–912.

Mocquot, B., Prat, C., Mouches, C. & Pradet, A. (1981). Effect of anoxia on energy charge and protein synthesis in rice embryo. *Plant Physiology*, 68, 636–640.

Monk, L. S. & Braendle, R. (1982). Adaptation of respiration and fermentation to changing levels of O_2 in rhizomes of *Scheonoplectus lacustris* (L.) Palla and its significance to flooding tolerance. *Zeitschrift für Pflanzenphysiologie*, 105, 369–374.

Monk, L. S., Crawford, R. M. M. & Braendle, R. (1984). Fermentation rates and ethanol accumulation in rhizomes of *Schoenoplectus lacustris* (L.) Palla and its significance to flooding tolerance. *Botany*, 35, 738–745.

Osborne, D. J. (1984). Ethylene of plants of aquatic and semi-aquatic environments: a review. *Plant Growth Regulation*, 2, 167–185.

Pregnall, A. M., Smith, R. D., Kursar, T. A. & Alberte, R. S. (1984). Metabolic adaptation of *Zostera marina* (eelgrass) to diurnal periods of root anoxia. *Marine Biology*, 83, 141–147.

ap Rees, T. & Wilson, P. M. (1984). Effects of reduced supply of oxygen on the metabolism of roots of *Glyceria maxima* and *Pisum sativum*. *Zeitschrift für Pflanzenphysiologie*, 114, 493–503.

Rumpho, M. E., Pradet, A., Khalik, A. & Kennedy, R. A. (1984). Energy charge and emergence of the coleoptile and radicle at varying oxygen levels in *Echinochloa crus-galli*. *Physiologica Plantarum*, 62, 133–138.

Saglio, P. H. & Pradet, A. (1980). Soluble sugars, respiration and energy charge during ageing of excised maize root tips. *Plant Physiology*, 66, 516–519.

Smirnoff, N. & Crawford, R. M. M. (1983). Variation in the structure and response to flooding of root aerenchyma in some wetland plants. *Annals of Botany*, 51, 237–249.

Smith, A. M. & ap Rees, T. (1979). Pathways of carbohydrate fermentation in the roots of marsh plants. *Planta*, 146, 327–334.

Steinmann, F. & Braendle, R. (1984a). Carbohydrate and protein metabolism in the rhizomes of the bulrush (*Schoenoplectus lacustris* L. Palla) in relation to natural development of the whole plants. *Aquatic Botany*, 19, 53–63.

Steinmann, F. & Braendle, R. (1984b). Auswirkung von Halmverlusten auf der Kohlen-hydratstoffwechsel ueberfluteter Seebinsenrhizome (*Schoenoplectus lacustris* L. Palla). *Flora*, 175, 295–299.

Studer, C. & Braendle, R. (1984). Sauerstoffkonsum und Versorgung der Rhizome von *Acorus calamus* L. *Glyceria maxima* (Hartm.) Holmberg, *Menyanthes trifoliata* L. *Phalaris arundinacea* L. *Phragmites communis* Trin. und *Typha latifolia* L. *Botanica Helvetica*, 94, 25–31.

Taylor, V. J., Crowder, A. A. & Rodden, R. (1984). Formation and morphology of an iron plaque on the roots of *Typha latifolia* L. grown in solution culture. *American Journal of Botany*, 71, 666–675.

Tyler, P. D. & Crawford, R. M. M. (1970). The role of shikimic acid in waterlogged roots and rhizomes of *Iris pseudacorus* L. *Journal of Experimental Botany*, 21, 677–682.

Young, S. F. & Hoffman, N. E. (1984). Ethylene biosynthesis and its regulation in higher plants. *Annual Review of Plant Physiology*, 35, 155–189.

Vartapetian, B. B., Andreeva, I. N. & Nuritdinov, N. (1978). Plant cells under oxygen stress. *Plant Life in Anaerobic Environments* (Ed. by D. D. Hook, & R. M. M. Crawford) pp. 12–88. Ann Arbor Michigan.

Carbohydrate status of terrestrial plants during flooding

T. L. SETTER, IRENE WATERS AND H. GREENWAY

School of Agriculture, University of Western Australia, Nedlands, W.A. 6009, Australia

B. J. ATWELL

Dryland Crops and Soil Research Program, CSIRO, Wembley, W.A. 6014, Australia

T. KUPKANCHANAKUL

Huntra Rice Experiment Station, Ayutthaya, New Zealand

SUMMARY

1 Some views are presented on the effects of flooding on the carbohydrate status of terrestrial vascular plants.

2 Plants with shoots in air and roots subjected to waterlogging or to O_2 deficiency usually have higher concentrations of carbohydrates than plants grown in totally aerobic conditions. Higher carbohydrate levels are probably a consequence of reduced growth of the roots. That is, the increase in carbohydrate levels resulting from reduced utilization in growth generally exceeds the depletion of carbohydrates which result from (i) decreases in photosynthesis due to increased stomatal resistance, and (ii) accelerated breakdown of carbohydrates in anaerobic conditions.

3 Rice is often subjected to complete or partial submergence of the shoots (see List of Definitions). Complete submergence drastically reduces the rate of growth and in contrast to the above cases usually reduces the levels of carbohydrates in tissues. The few data available indicate that low light intensities and low CO_2 concentrations in the water reduce growth and carbohydrate concentrations. Preliminary studies on partial submergence of lowland rice varieties also showed large decreases in carbohydrate levels, the magnitude of these changes depending on the proportion of the leaves below the water. These results contrast with the response of a deepwater rice variety in which carbohydrates remain high even when 70% of the shoot is below the water.

4 In the concluding section we briefly discuss two issues: (i) regulation of growth and maintenance processes in O_2 deficient tissue and (ii) possible interactions between flooding and other environmental factors.

LIST OF DEFINITIONS

Flooding: The inundation of land resulting in either (i) waterlogging of the soil or (ii) partial or complete submergence of the shoots of plants.

Lowland, deepwater and floating rice: Rice cultivars which are grown in areas where, during much of the growing season, water depths range between 0–0·5, 0·5–1·0 and 1·0–5·0 m respectively.

411

Partial submergence: The partial covering of plant shoots with water during flooding.

Relative carbon loss: The relative carbon loss was approximated by the weight of carbon dioxide and ethanol lost per unit weight of tissue and expressed as day^{-1}.

Relative growth rates: All relative growth rates are calculated on a dry weight basis and expressed as day^{-1}.

INTRODUCTION

Importance of carbohydrates in plants subjected to flooding

Many of the effects of flooding emanate from the slower diffusion of gases such as O_2, CO_2 and ethylene in the liquid than in the gas phase. These have been thoroughly reviewed by Jackson & Drew (1984) and Jackson (1985). This paper focuses on the consequences of these changes on the carbohydrate status of terrestrial vascular plants during flooding. Carbohydrate levels are determined by the balance between consumption in respiration and growth, access to stored reserves and production of carbohydrates in photosynthesis.

This review is not comprehensive, and hence only selected examples are used to illustrate the changes in carbohydrates in plants during flooding. As is well known, flooding also leads to widespread changes in soil chemistry, biology and structure (Ponnamperuma 1984). These will not be discussed here.

Two cases of flooding are considered in this paper:
(i) Plants with shoots in air and roots either in waterlogged soil or in nutrient solution at low O_2 concentrations.
(ii) Rice subjected to partial or complete submergence of shoots. This is a common occurrence in southeast Asia, and is in addition to the normal waterlogged condition of the soil (HilleRisLambers & Seshu 1982; Khush 1984).

It has been demonstrated in several studies that an adequate supply of carbohydrates can be critical to the survival of plant tissues at low O_2 concentrations. In excised coleoptiles and leaves of rice, addition of glucose prevented mitochondrial disintegration, which commenced otherwise within 1–2 days after imposing anoxia. No exogenous glucose was needed to retain mitochondrial integrity in these organs when intact rice seedlings were exposed for 3–5 days to anoxia (Vartapetian, Andreeva & Kozlova 1976). Exogenous glucose also prolonged the structural integrity of mitochondria in excised roots for 2–3 days; disintegration otherwise occurred within 1 day of commencing anoxia (Vartapetian, Andreeva & Nuritdinov 1978).

Glucose addition also considerably delayed the death of root tips when anoxia was imposed on either intact plants or excised roots of rice, peas and pumpkin (Webb & Armstrong 1983). These authors concluded that anoxia reduced the available energy below a critical level required for survival of the cells. It is relevant to note that the plants they used were grown at a light intensity of only

100 μmol m^{-2}s^{-1} or in diffuse light, i.e. the plants probably had low initial sugar concentrations. We suggest that the effects of anoxia may be less severe for plants grown in the field at high light intensities (at approximately 500–2000 μmol m^{-2} s^{-1}.

Apart from the role of soluble carbohydrates in preservation of ultrastructure and survival of root tips, storage carbohydrates also contribute to plant survival during flooding. Crawford (1978) has emphasized that species tolerant to waterlogging have large rhizomes and tubers. Presumably these organs have adaptive value in spring because the carbohydrate reserves could sustain anaerobic fermentation until new shoots develop (Crawford 1978).

Possible factors determining carbohydrate status

Carbohydrate levels in vascular plants subjected to O_2 deficiency could increase due to a direct effect of low ATP production on synthetic reactions involved in growth. However, any consequent increase in levels of carbohydrate due to decreased utilization in some plant parts may be counteracted in at least three ways.

Firstly, changes in concentrations of hormones can modulate patterns of growth (reviewed by Jackson & Drew 1984; Jackson 1985). Alternatively there may be changes in levels of receptors of hormones. The changes in growth pattern may be associated with large changes in carbohydrate levels in different parts of the plant. One example is the action of ethylene in the rapid elongation of isolated internodes of deepwater rice (Raskin & Kende 1984a). Rapid elongation ensures that in the intact plant a substantial proportion of the foliage is raised above the water. Such developmental changes may affect the carbohydrate status of the plant as a whole or its individual organs.

Secondly, O_2 deficiency may occur in roots in waterlogged soils and presumably also in submerged shoots in the dark. The resulting decrease in ATP regeneration from ADP will result in low ATP and high ADP levels and consequently may lead to an accelerated breakdown of sugars ('Pasteur Effect', Beevers 1961).

Thirdly, CO_2 supply for photosynthesis may become limiting. For plants with partially or completely submerged shoots this limitation may arise even at ($CO_2 + HCO_3^-$) concentrations as high as $1 \cdot 5$ mol m^{-3} due to the slow diffusion of CO_2 and HCO_3^- in the floodwaters (Smith & Walker 1980, for aquatic species; own observations for rice).

On the other hand, for plants with shoots in air, CO_2 uptake could become restricted due to the increase in stomatal resistance which has often been observed for plants with their roots in waterlogged soil (Reid & Bradford 1984; Kozlowski & Pallardy 1984). Some discussion on the causes for this increase in stomatal resistance is relevant to the main theme of this paper. It may arise from water deficits in the leaves induced by a reduction in the water conductivity of the root system as a whole (Kozlowski & Pallardy 1984). This in turn could result from a lower root mass or from a decrease in water permeability per unit surface area of root. Alternatively, an increase in stomatal resistance may arise from the accumulation of abscisic acid or reduction in levels of gibberellins and cytokinins which has been observed in waterlogged plants (reviewed by Reid & Bradford 1984). Such changes in hormones may explain the response of peas in waterlogged soil (Jackson & Kowalewska 1983). Firstly, there was a decrease in stomatal conductance which occurred within 24 h of flooding without any indication of water deficits in the leaves. Secondly, after 4–5 days there was a decline in the chlorophyll content of leaves and in the ability of the leaf cells to retain water. However, Jackson & Kowalewska (1983) felt these latter effects were probably due to phosphate toxicity rather than to changes in hormones. In the case of peas it is likely that reduced photosynthesis, which would result from damage to the leaves and increased stomatal resistance, may contribute to the adverse effects of flooding. However, only a knowledge of the carbohydrate status of the plants could provide the evidence required. This applies both to the experiment by Jackson & Kowalewska (1983) and to cases of reductions in stomatal opening and photosynthesis referred to by Kozlowski & Pallardy (1984).

A completely different explanation for the increase in stomatal resistance and/or reductions in photosynthesis is a feedback effect due to a reduction in utilization of carbohydrates associated with inhibition of growth. An elegant demonstration of such a feedback effect was made by Hall & Milthorpe (1978) where they showed the importance of sinks for carbohydrates in determining the rate of photosynthesis. In these studies the fruits were removed from *Capsicum* plants grown in well-drained potting mixtures. The removal of such a large sink resulted, within 1 day, in substantial increases in leaf diffusive resistance and in intracellular resistance (Hall & Milthorpe 1978). The leaf diffusive resistance would presumably have been mainly stomatal resistance.

Results consistent with this concept were found in a waterlogging study on maize. In these experiments Wenkert, Fausey & Watters (1981) showed that decreases in root and leaf extension rates occurred within 1–12 h of waterlogging, while transpiration and stomatal conductance either remained the same or decreased slightly after 2–3 days.

In summary, the accumulation of carbohydrates due to impaired growth would oppose the tendency for carbohydrates to be depleted by reduced photosynthesis or increased glycolytic rates. It is therefore likely that carbohydrate levels are affected by several environmental factors during flooding including light intensity, temperature and, in the case of submerged rice, O_2, CO_2 and other gas concentrations in the water. These factors will be further considered in the final section of this paper.

In the following sections changes in carbohydrate status in different flooding situations will be discussed in detail. Experiments in the next section describe the effects of low O_2 concentration on plants with an adequate carbohydrate source (see Tables 1 and 2). Cases discussed are either (1) seedlings which have access to large carbohydrate reserves in the endosperm or (2) mature photosynthesizing plants which are subjected to O_2 deficiency in the root environment but which have their shoots in the air. The section on Submergence of Rice deals with partial or

TABLE 1. Effects of low O_2 concentrations on sugar and starch concentrations of intact 3–5 day-old seedlings with access to large reserves of carbohydrates in the remaining endosperm

Plant	Environment and treatment	Relative growth rate* (day^{-1})		Sampling time (days)	Tissues	Sugars† (percent of dry weight)		Starch (percent of dry weight)		Reference
		High O_2	Low O_2			High O_2	Low O_2	High O_2	Low O_2	
Rice (cv. Calrose)	Comletely submerged with O_2 at 0·25 or 0·00 mol m^{-3}. Treatment started at time of germination.	0·58	0·26	3–4	Coleoptile	6·0	7·9	—	—	Atwell (1982)
Rice (cv. Arborio)	Completely submerged. Treatment started at 3 days after germination.	0·57	0·0	1	Roots	Not available	65% of initial level	—	—	For growth Bertani et al. (1980) For sugars Bertani et al. (1981)
Maize	Roots with O_2 at 0·25 or 0·04 mol m^{-3} and shoots in air. Treatment started at 5 days after imbibition.	0·29	0·12	2	Roots (mm)§ 0–2 2–15 15–55	14·2 10·8 3·4	31·5 13·0 7·7	1·4 1·6 1·3‡	1·8 0·7 1·8‡	Atwell et al. (1985)

All experiments reported in this table were in continuous darkness. High O_2 and Low O_2 refer to concentrations described in the column Environment and Treatment.

* Rates calculated between 3–3·5 days after start of treatment for rice coleoptiles and between 0–2 days after treatment for rice roots and maize.

† For maize, assuming fresh weight/dry weight of 10; for rice based on observed fresh weight/dry weight ratios.

‡ For 15–200 mm zone of roots.

§ Distance from root tip.

TABLE 2. Sugar concentrations in photosynthesizing plants with roots exposed to low O_2 concentrations, stagnant conditions, or to waterlogged soil

Plant	Light* (μmol m^{-2} s^{-1})	Environment and treatment	Relative growth rate‡ (day^{-1}) High O_2	Low O_2	Sampling time (days)	Tissues	Sugars (% of dry weight) High O_2	Low O_2	Reference
Wheat	500–1500	Field; with waterlogged and drained soil. High mineral nutrition	Not available		17	Shoots	8·8	16·9	E. Barrett-Lennard (pers. comm.)
Barley	200–700	Glasshouse; in nutrient solution with O_2 at >0·22 or <0·02 mol m^{-3} (Nutrient solution)	0·17 / 0·13 / 0·25	0·14 / 0·04 / 0·17	10	Shoots / Seminal roots / Nodal roots	5·5 / 1·0 / 1·8	10·4 / 8·8 / 8·5	Barley: Limpinuntana & Greenway (1979); similar results with wheat (F. Buwalda & P. Leighton, pers. comm.).
Barley	174	Growth room; in nutrient solution flushed with air or N$_2$. Low mineral nutrition / High mineral nutrition	Not available		7 / 7	Roots / Roots	9·6 / 1·3	14·2 / 1·5	De Witt (1974, in Table 1.) Similar results for low mineral nutrition: van de Heide et al. (1963)
Beans (P. vulgaris)	200	Growth room; in aerated or stagnant solution O_2 about 0·25 and 0·06 mol m^{-3} respectively (Nutrient solution)	0·03 / 0·20	0·14 / 0·08	3	Hypocotyl / Roots	2·8 / 2·3	9·9 / 5·0	Papenhuijzen (1983)
Rice (cv. Calrose)	600–1200	Glasshouse; summer O_2 >0·22 or <0·02 mol m^{-3} (Nutrient solution)	‡		12	Roots	4·4	5·6	Limpinuntana & Greenway (1979)
	200–700	Glasshouse; autumn; O_2 >0·22 or <0·02 mol m^{-3}.	‡		18	Root tips (5–7·5 mm long) / Rest of root	3·8 / 1·1	4·1 / 0·9	Limpinuntana & Greenway (1979)

Sugar concentrations are based on anthrone reaction or sum of glucose, fructose and sucrose determined enzymatically. High O_2 or Low O_2 refer to concentrations described in the column Environment and treatment.

*Photosynthetically active radiation (400–700 nm) for glasshouse and field experiments are measured at midday.

‡ Growth rates: for barley between 0 and 10 days of start of treatment for beans between 0 and 3 days of start of treatment.

‡ After 12–18 days treatment the shoot and root dry weight at low O_2 were 10–15% lower than at high O_2.

complete submergence when photosynthesis may be reduced due to limited access to inorganic carbon or light (Table 4).

THE OCCURRENCE AND CAUSES OF HIGH SUGAR CONCENTRATIONS IN PLANTS EXPOSED TO LOW O_2 CONCENTRATIONS IN THE ROOT ENVIRONMENT

Reduced utilization of carbohydrates in growth processes

Carbohydrate concentrations remained the same or increased in all cases reported in Tables 1 and 2 except in one experiment where there was a decrease of sugars in roots of young rice seedlings suddenly exposed to anoxia (Bertani, Brambilla & Menegus 1981). Access to reserve carbohydrates in the remaining endosperm is probably a dominant factor in the maintenance of high sugar concentrations in coleoptiles of submerged, intact rice seedlings and roots of maize seedlings. The sugar levels increased in all parts of root systems of maize at low O_2 concentrations, i.e. root tips, rapidly expanding and fully mature tissues (0–2, 2–15 and 15–55 mm from the tip, respectively; Table 1). High sugar concentrations in the root tips conflict with the observations of Saglio (1985). In his experiments, strict anoxia reduced transport of a sugar analogue from the scutellum of maize seedlings to the root tips by inhibiting unloading from the phloem rather than reducing the sink size. In the experiment reported in Table 1 sugars accumulated in the tip, presumably because sugar utilization was reduced by direct effects of low ATP regeneration on processes such as cell division and protein synthesis (see also the General Discussion). This view receives further support from the observation that soluble amino acids in root tips and other root tissues of maize were 2–3·5 times higher at 0·04 than at 0·25 mol m^{-3} O_2 (Atwell *et al.* 1985). Similar results were observed for rice roots which contained up to twice the concentration of soluble amino acids at 0·03 mol m^{-3} as at 0·25 mol m^{-3} O_2 (Bertani & Brambilla 1982).

Nutrient solution culture at low O_2 concentrations rather than waterlogged soil has generally been used to evaluate the effect of flooding on carbohydrate concentrations in photosynthesizing plants with shoots in air (Table 2). It is therefore reassuring that wheat, which had grown for 2·5 weeks in waterlogged soil in the field, also had double the sugar concentrations in the shoots compared with plants growing in adjacent drained plots (Table 2).

The hypothesis that the increase in carbohydrates occurs due to reduced utilization by the plant as a whole is supported by the observed carbohydrate accumulation even in plant parts which are not exposed to low O_2 concentrations (Table 2). For example, in barley and wheat increases occurred in the shoots as well as the roots. Furthermore, in both nodal and seminal roots the sugar concentrations were higher at 0·02 than at 0·22 mol m^{-3} O_2 (barley, Table 2; wheat, P. Leighton, pers. comm.). This increase of sugar concentration in both root types is of particular interest in view of their different responses to low O_2: seminal root growth is almost completely inhibited, while nodal root growth remains substantial (Limpinuntana

& Greenway 1979; Trought & Drew 1980) or is even faster (Benjamin & Greenway 1979) at low than at high O_2. Carbohydrate levels increased in the hypocotyl of beans even though the growth rate was 4·5 times higher at low than at high O_2 concentrations (Table 2). We suggest that this accumulation of sugars in the hypocotyl is due to reduced growth of the roots and to a lesser extent to the shoot (Papenhuijzen 1983).

Higher sugar concentrations at low than at high O_2 concentrations were also found in roots of rice grown in summer (Table 2). In contrast to other cereals, low O_2 concentrations for 12–18 days in the root medium of rice reduced root and shoot growth by only 10–15% (Limpinuntana & Greenway 1979). For rice the accumulation of sugars at low O_2 concentrations was observed even when plants were sampled just before sunrise (Limpinuntana & Greenway 1979). Different results were found for rice grown in autumn, where there was no appreciable effect of low O_2 on sugar concentration, even in the root tips (Table 2). This discrepancy probably resulted from low light intensities and /or short days during autumn.

Before discussing the carbon budget, we would like to emphasize that an accumulation of carbohydrates in plants with roots at low O_2 may not always occur. In particular there may be interactions between flooding and other environmental factors and these will be considered in the General Discussion.

Construction of a carbon budget

In order to construct a comprehensive carbon budget one requires values for (i) the import of carbohydrates into plant organs, (ii) the partitioning of carbon loss between respiration and leakage from the tissue and (iii) the utilization of carbon as both soluble and insoluble cell constituents in the growing tissues. Such data should be obtained from the same experiment, and as far as we are aware, this has not been done for roots of plants exposed to low O_2 concentrations.

A preliminary carbon budget

A carbon budget has been constructed for the simple system of slices of carrot roots (James & Ritchie 1955). Here we have constructed a preliminary carbon budget for intact plants based on carbon utilization in aerobic and anaerobic tissues using several assumptions discussed in the following section. This budget is based on two ratios which are derived as follows:

Ratio 1 is the quotient of (carbon lost as CO_2 and ethanol under anaerobic conditions, C anaerobic) to (carbon lost as CO_2 in aerobic conditions, C aerobic); therefore:

$$\text{Ratio 1} = \frac{\text{C anaerobic}}{\text{C aerobic}}$$

Ratio 2 is the quotient of the sum of (carbon lost as CO_2 under aerobic conditions, C aerobic) plus (the difference between carbon incorporated into growth under aerobic and under anaerobic conditions, ΔC growth) to (the carbon lost as CO_2 under aerobic conditions, C aerobic), therefore:

$$\text{Ratio 2} = \frac{\text{C aerobic} + \Delta \text{C growth}}{\text{C aerobic}}$$

In Ratio 2, C aerobic + ΔC growth gives the amount of carbon which would become available for fermentation under anaerobic conditions. This assumes that the import of sugar into the organ is not affected by anoxia (see assumption 1, following section). Therefore, Ratio 2 indicates the rate of fermentation at which sugar concentrations under anaerobic conditions would be maintained at the same level as under aerobic conditions. During anoxia carbohydrates will accumulate in tissue if Ratio 2 > Ratio 1, because the consumption of carbon by fermentation and any growth will be less than that supplied for respiration and growth in aerobic tissue. Conversely carbohydrates will be depleted if Ratio 2 < Ratio 1.

It should be emphasized that Ratio 1 is used here to describe carbon loss in aerobic and anaerobic tissue. It is not a valid method for identifying a Pasteur Effect (Beevers 1961) and in this analysis is not intended as an assessment of glycolytic flux.

The Ratios 1 and 2 for four different cereals including rice are shown in Table 3. The carbon losses in maize, wheat and barley at low O_2 concentrations usually exceed those at high O_2, i.e. Ratio 1 > 1·0. Nevertheless this ratio is less than Ratio 2, and sugars therefore accumulate. This is borne out by carbohydrate measurements (Table 2). Ratio 2 would of course have been even bigger if growth had ceased altogether under low O_2 concentrations. In the most pronounced case maize roots still grew at 0·12 day^{-1} at 0·04 mol m^{-3} O_2, and complete cessation of growth would have increased Ratio 2 from 2·1 to 3·0 (Table 3).

In contrast to the above species, the carbon loss in 3–6 day old rice seedlings in anoxia was at least three times that in aerobic conditions (Ratio 1, Table 3). These estimated carbon losses are consistent with most earlier studies on rice seedlings in which the carbon losses can be estimated from decreases in dry weight of the seedling as a whole: Tsuji (1968) quotes a doubling in the rate of carbon loss. However, conflicting results were found for rice seedlings after being soaked in solution for 72 h (Crawford 1977). Ratio 1 was only 1·4, being the lowest of five species under investigation. The reason for this discrepancy with the experiments shown in Table 3 is unknown.

For rice Ratio 2 equalled Ratio 1 in one experiment by Bertani, Brambilla & Menegus (1980; Table 3). However, Ratio 2 was substantially smaller than Ratio 1 in the experiment by Atwell (1982; Table 3), yet the coleoptiles grown under anoxia contained the same concentrations of sugars as coleoptiles grown under aerobic conditions. So there must have been an increased import of sugars via the phloem, and this can be further shown by the following calculations. Seedlings were germinated in anoxic and aerobic solution (Atwell 1982), and relative growth rates of seedlings (excluding endosperm) were about 0·26 day^{-1} and 0·58 day^{-1}, while

TABLE 3. Carbon budgets for various species exposed to anaerobic conditions

Plant	Ratio 1		Ratio 2		References		
							For Ratio 2
	Tissue	$\dfrac{C\ anaerobic*}{C\ aerobic}$	Tissue	$\dfrac{C\ aerobic+\Delta C\ growth\dagger}{C\ aerobic}$	For Ratio 1	Growth	Respiration
Maize	Root tips	1·4	Whole roots	2:1	Neal & Girton (1955)	Atwell et al. (1985)	Atwell et al. (1985)
Wheat	Whole germinating seedlings	0·6–1·5	Seminal roots of intact plants	2:1	Taylor (1942)	P. Leighton (pers. comm.)	C. Thomson (pers. comm.)
Barley	Whole germinating seedlings	1·5	Seminal roots of intact plants	2·0	Phillips (1947)‖	Limpinuntana & Greenway (1979)	de Witt (1974)
Rice cv. Calrose	Coleoptiles of germinating seedlings	3·3	Coleoptiles of germinating seedlings	2·6	Atwell (1982)	Atwell (1982)	Atwell (1982)
Rice (cv. Arborio)	Whole germinating Seedlings	3·0	Whole germinating seedlings	3:1	Bertani et al. (1980)	Bertani et al. (1980)	Bertani et al. (1980)
Rice (cv. Early Prolific)	Whole germinating Seedlings	4·5–6	Whole germinating seedlings	not available	Taylor (1942)§	Taylor (1942)§	

The carbon budgets are assessed by the use of Ratio 1 and Ratio 2 which are explained in detail in the text (pp. 418–419). If Ratio 2 > Ratio 1, carbohydrates accumulate. If Ratio 2 < Ratio 1, carbohydrates deplete. C anaerobic = carbon loss under anaerobic conditions. C aerobic = carbon loss under aerobic conditions. ΔC growth = the difference in growth between aerobic and anaerobic conditions, i.e. carbon incorporated into new growth.

*Ratio 1 is the quotient of (CO_2 plus ethanol production at low O_2) to (CO_2 production at high O_2); the CO_2 evolution at high O_2 is sometimes calculated from O_2 uptake assuming a respiratory quotient of 1·0.

†Based on relative growth rates and either CO_2 loss, or O_2 uptake assuming a respiratory quotient of 1·0.

‡Growth rates for seminal roots of intact plants; respiration for excised roots. The respiration was adjusted to temperature at which growth was measured assuming a Q_{10} of 2·0. Fresh weight/dry weight ratios of 9 and 10 were used for shoots and roots respectively.

§Assuming all CO_2 evolution at low O_2 came from alcoholic fermentation; if this was not true Ratio 1 would have been smaller.

‖Established for times later than 6 h after start of anoxia treatment.

the relative carbon losses were 0.58 day^{-1} and 0.19 day^{-1}, respectively. That is, growing tissue in the aerated treatment imported carbon at a relative rate of 0.77 day^{-1}, whereas coleoptiles in anoxic solution imported carbon at 0.84 day^{-1}. This shows that carbon loss due to high fermentative activity increased by 0.39 day^{-1} and carbon import from the endosperm increased by 0.07 day^{-1} in anoxia. Therefore, about 18% of the increase in carbon loss due to high fermentative activity in anoxic rice seedlings was due to greater carbohydrate import from the endosperm, the remainder being yielded by impairment of growth due to O_2 deficiency. Sugar levels were therefore maintained (Table 1). It is well established that the growth of rice seedlings observed in anoxia is due to coleoptile and not root growth (Kordan 1974).

Overall, the rate of carbon loss is increased in anaerobic rice seedlings, whether growth ceases (Bertani, Brambilla & Menegus 1980) or continues (e.g. Atwell 1982). On the other hand, carbon losses in beans, barley and wheat, though generally stimulated by anoxia, remain less than the sum of the amount of carbon which would be invested into new growth and lost as CO_2 in aerobic conditions. No doubt, not all plant species or tissues behave the same in anoxia. Turner (1956) listed four species where Ratio 1 was two or less and as many where Ratio 1 exceeded three. Therefore, carbon depletion may occur in some species subject to anoxia.

The increase in carbohydrates in the plant as a whole (Table 2) needs to be considered in relation to the carbon budget of the roots. There are two possible causes for carbohydrate accumulation in the shoots.

(i) A decrease in export to the roots, due either to a reduced sink for carbohydrates or to inhibition of phloem transport. The first proposition is consistent with the higher values for Ratio 2 than for Ratio 1 (Table 3). The second is also possible since in the stolon of *Saxifraga sarmentosa* large inhibitions of phloem transport occur during 5 h of anoxia (Qureshi & Spanner 1973).

(ii) A reduction in growth of the shoots due to changes in messages, e.g. hormonal or nutritional, from the anaerobic roots (Jackson & Kowalewska 1983). This is presumably the case in barley (Limpinuntana & Greenway 1979). Shoot dry weights 5 days after imposing treatments were slightly lower at 0.02 than 0.22 mol m^{-3} O_2, and more importantly the relative growth rates of the shoots on the basis of total dry weight minus weight of sugars were 0.12 and 0.15 day^{-1} at 0.02 and 0.22 mol m^{-3} O_2 respectively.

Assumptions required for the construction of a carbon budget
The carbon budget discussed in the previous section was based on the following three assumptions which provide the criteria for future work on construction of carbon budgets.

(1) Anaerobiosis of the roots does not interfere with sugar import from the shoots. This is by no means certain; for example, Saglio (1985) showed that phloem transport of a labelled glucose analogue in roots of maize was inhibited by anoxia.

(2) Any leakage of carbon excluding ethanol from roots is not exacerbated by anoxia. This has never been thoroughly checked in the steady state. The leakage of amino acids and organic acids from excised roots was greatly increased during the first 2–5 h after transfer from high to low O_2 or anoxia; this occurred for barley (Hiatt & Lowe 1967) and peas (Smith & ap Rees 1979a). Similarly, loss of potassium from excised roots also occurred at low O_2 concentrations in maize (Marschner, Handley & Overstreet 1966) and barley (Marschner & Mengel 1966; Hiatt & Lowe 1967). There was no evidence for substantial leakage of substances other than ethanol in carrot roots exposed to anoxia for 115 h (James & Ritchie 1955). In addition, the permeability of excised wheat roots to sorbitol at 0·01 mol m^{-3} was as low as in roots at 0·22 mol m^{-3} O_2 (F. Buwalda, pers. comm.). These results with sorbitol suggest that solute retention depends on the physical characteristics of the solute. The alternative possibility that the retention of a solute is peculiar to the plant species seems less likely. It is important to know whether roots of intact plants which have been grown for several days under anaerobic conditions lose more solutes than plants grown under aerobic conditions.

(3) Alcoholic fermentation is the sole pathway of anaerobic metabolism. This consideration is important because a conversion of sugars to end-products, other than CO_2 and ethanol, would affect the carbohydrate status without changing either Ratio 1 or 2. This assumption 3 is not strictly true; nevertheless in the few species studied ethanol is probably the dominant end-product of glycolysis. Clear evidence for this notion is available (James & Ritchie 1955; Davies 1980), and further support is given in the following section.

Relationship between end-products of anaerobic pathways and carbohydrate status
As stated above, the use of Ratios 1 and 2 in the carbon budget depends on the assumption that ethanol is the dominant end-product of anaerobic metabolism. Evidence for this is obtained from the ethanol:CO_2 ratio which is 1·0 if pure alcoholic fermentation prevails. This ratio is 0·9 in intact rice seedlings (Phillips 1947) and 0·83 in excised maize roots supplied with glucose (Neal & Girton 1955). Curiously, the ratio in barley seedlings only increased to close to 1 about 6 h after they became anaerobic (Phillips 1947).

In other experiments with peas and three flood-tolerant marsh species, 75–85% of the determined end-products of anaerobic metabolism consisted of ethanol (the total did not include CO_2; Smith & ap Rees 1979a, b). In excised root tips of maize the loss of endogenous sugars during the first hour of anoxia could be accounted for by ethanol (60%), CO_2 (30%, calculated by us), lactate (10%) and alanine (5%), with no additional malate produced (Saglio, Raymond & Pradet 1980). Similar results were obtained in rice seedlings where ethanol and CO_2 accounted for more than 75% of the starch consumed in anaerobic metabolism (Atwell 1982).

Supporting evidence for the view that ethanol is the main product of anaerobic metabolism comes from a comparison between rates of alcoholic fermentation and long-term accumulation of products of other pathways, for example, maize produced about 11 μmol CO_2 and ethanol per gram fresh weight per hour (Neal & Girton 1955). So, if all carbon flowed to a three-carbon compound instead of ethanol and CO_2, some 264 μmol per gram fresh wt would accumulate each day. This figure contrasts with the low levels of anaerobic end-products which have been reported in the literature. Crawford & Tyler (1969) measured the concentrations of malate, lactate and succinate in roots of three waterlogging tolerant species 4 days after being waterlogged. Malate accumulated to 2–4 μmol per gram fresh wt and lactate and succinate to less than 0·5 μmol per gram fresh wt. Malate concentrations in rice seedlings subjected to anoxia for 4 days were 7–24 μmol per gram fresh wt compared with 4–6 μmol per gram fresh wt in aerobic seedlings (Avadhani *et al.* 1978). These values contrast with the 1056 μmol ethanol which would be produced by maize over 4 days under anaerobic conditions (calculated from above).

It should further be noted that substantial synthesis of end-products other than ethanol would raise the problem of their disposal. Retention of the bulk of these products in the tissues is implausible simply on osmotic grounds. For example, in maize roots, the production of a three-carbon end-product at 264 μmol per gram fresh wt would generate an additional 0·7 to 1·5 MPa per day, depending on whether the solute was non-dissociating or an organic acid balanced by a monovalent cation respectively; alternatively a four-carbon organic acid balanced by two monovalent cations would generate 2·2 MPa per day. (All these figures were calculated assuming 90% water content and 0·1 MPa osmotic pressure generated by 40 μmol non-dissociating solute per ml.) This rate of accumulation is difficult to contemplate even if other osmotica were to be eliminated from the cell simultaneously. Any anaerobic end-products would therefore have to leak into the surrounding medium as does ethanol or be exported in the xylem to the shoots (Davies 1980).

There is little evidence that transport of end-products from anaerobic roots to the shoot is an adaptive feature of vascular plants, but transport of ethanol from the roots to the shoots has been shown in several studies (reviewed by Chirkova 1978). Fulton & Erickson (1964) showed that ethanol moved to the shoots of tomato plants when the roots were waterlogged. It is unknown what proportion of the ethanol produced this represented, or whether ethanol was directed into the xylem sap under some control. ap Rees & Wilson (1984) found that only 1% of the ^{14}C-sucrose metabolized in anaerobic roots of *Glyceria maxima* ended up in the shoots after 6 h; however the plants were in the dark. This, therefore, needs to be repeated in the light while transpiration rates are high.

SUBMERGENCE OF RICE

The response of rice to complete submergence is less well documented than the response of plants to an anaerobic root environment. Hence a brief description of complete submergence is given before discussing changes in the carbohydrate status of the rice plants.

Complete submergence occurs frequently in southeast Asia (HilleRisLambers & Seshu 1982). The main problem of complete submergence occurs during short-term flash-floods in lowland rice areas which may last up to approximately 10 days (Khush 1984). Little is known about the characteristics of the floodwater and there may be large variations between locations in light intensities under water, turbulence, and concentrations of dissolved gases such as O_2, CO_2 and ethylene.

Partial submergence of rice is of even greater agronomic importance than complete submergence. Short-term partial submergence occurs frequently in many regions with rain-fed lowland rice. This type of flooding reduced grain yield proportionally to the percent of plant height which was submerged (experiment by Pande as cited by Yoshida 1981). For example, flooding to 50% of plant height between the establishment of seedlings and maximum tillering reduced grain yield by 30%. Partial submergence is particularly important in the deepwater and floating rice areas where there are long-term floods with water depths of $0 \cdot 5 - 5 \cdot 0$ m above the soil surface, which occur in many southeast Asian flood plains. During the initial fast rise of the floodwaters the rice may occasionally become completely submerged (HilleRisLambers & Seshu 1982). However, this problem is usually remedied by the rapid elongation of the internodes of deepwater and floating rice cultivars, which are adapted to these plains. Nevertheless even these cultivars have a substantial percentage of their shoots submerged for up to 3–5 months of the flooding (HilleRisLambers & Seshu 1982).

We have recently measured some characteristics of floodwaters in deepwater and floating rice areas of Thailand (unpubl. data). Oxygen concentrations vary with depth and over diurnal cycles. At 1500 h, O_2 in the top 20 cm of water sometimes reaches supersaturation, while it usually drops to 10–25% of saturation overnight.

Response of rice during complete submergence

The extreme changes in the environment of the shoots during submergence elicit a multitude of responses by the rice plants. During complete submergence these include a severe curtailment of dry weight increments (Palada & Vergara 1972), increased elongation of sheaths (Hanada 1983) and development of chlorosis in some intolerant varieties (unpubl. data). Furthermore, death of many plants has been observed following desubmergence (Palada & Vergara 1972). Whether death occurs during or after submergence is unknown, and the possible interrelations between the various responses to submergence are not yet clear.

The complex responses described above should be kept in mind even though this paper deals only with the severe curtailment of dry weight increments (Table 4), and the associated decreases in concentration of carbohydrates. Relative growth rates during 10 days of complete submergence were decreased somewhat more in turbid than in clear water (Table 4). In both sets of experiments on complete submergence (Table 4) the submerged plants had much lower concentrations of carbohydrates than the non-submerged plants, so there are strong arguments that this decrease in carbohydrates was the explanation for the growth reduction. The

TABLE 4. Effect of complete or partial submergence of shoots on sugar concentrations in shoots or leaves of rice (all plants were grown in soil)

Type of rice	Light* (μmol m^{-2} s^{-1})	Environment and treatment	Relative growth rate (day^{-1}) Not submerged	Submerged	Sampling time (days)	Tissues	Sugars (percent of dry weight) Not submerged	Submerged	Reference
Lowland rice (cv. hort peta)	Unavailable	Glasshouse; complete submergence clear water	0·17	0·06	10	Shoots	4·9	1·7	Palada & Vergara (1972)
		turbid water†	0·17	0·03	10	Shoots	4·9	1·0	Unpubl. data
Lowland rice (cv. Kurkaruppan)	600–1200	Glasshouse; complete submergence	0·21	0·03	6	Shoots	14·2	3·6	Unpubl. data
Lowland rice (cv. IR 42)	600–1200	Glasshouse; 50% submerged‡	66%		5	Leaf blades	15·5	7·4§	A. Polvatana (pers. comm.)
		85% submerged§	50%		5		15·5	5·0§	
Deepwater rice (cv. Leuang Pratew 123)	800–2400	Field. Water level increased at 5 m day^{-1} for 16 days. Final water depth 1·0 m. 70% submerged.			16	Emerging leaf	5·5	5·3	Unpubl. data
						Growing point	16·8	22·9	
						Youngest internode	23·4	25·7	

*Photosynthetically active radiation (400–700 nm) for glasshouse and field experiments are measured at midday.

†Light intensity at base of plant in turbid water was 55% of light at water surface.

‡As measured on length of longest leaf, i.e. from tip of leaf to soil surface.

§Growth data expressed as dry weights at end of treatment period. Similar responses for sugars found for sheaths which were all under water in the 85% submerged treatment.

low levels of carbohydrates could be due to any of the following responses or factors:

(i) Leakage from the tissues due to partial loss of membrane integrity.

(ii) Increased consumption of carbohydrates due to low O_2 concentrations in either the roots or shoots, particularly during the night. This is conceivable in view of the high carbon loss in anoxia/carbon loss in air found for germinating rice (Table 3).

(iii) Decreased photosynthesis due to either low light intensity (Palada & Vergara 1972), low CO_2 during the day, and/or high O_2 during the day (unpubl. data). The latter may result in increased carbon loss in photorespiration. The effects related to CO_2 and O_2 depend on the turbulence of the water. For example, turbulence greatly increased photosynthesis of aquatic species in water at low concentrations of HCO_3^- and/or CO_2 as the diffusion limitation was removed (Smith & Walker 1980).

Which of the above responses or factors have dominant roles in the curtailment of growth during complete submergence presumably depends on the characteristics of the floodwaters, incoming radiation and, perhaps, on the particular rice cultivar. In one of the few definitive studies available there were strong indications that light was an important factor (Table 4). Another response was found in our glasshouse studies at high light intensities where the supply of CO_2 played a dominant role (Table 5). Flushing the floodwater with 3% CO_2 in air greatly increased both dry weight accumulation and sugar concentrations in submerged plants (Table 5). Nevertheless, neither growth nor sugar concentrations reached the same values as in plants which were not submerged (Table 5). It is unknown whether these remaining discrepancies can be overcome by increasing the CO_2 concentration

TABLE 5. Effects of CO_2 enrichment on growth and sugar concentration in completely submerged rice (cv. Kurkaruppan, a submergent tolerant variety)

| Non-submerged | | Submerged | |
	Stagnant solution	Solution flushed with air containing 0·03% CO_2	Solution flushed with air containing 3% CO_2
Relative growth rate (day^{-1})			
0·21	0·03	0·03	0·16
Sugar concentrations in shoots (percent of dry weight)			
14·2	3·6	3·2	9·9
CO_2 concentrations (μmol m^{-3}) in floodwater			
Not applicable	0·01–0·19	0·01	0·45

Plants were grown for 14 days in soil and completely submerged for 6 days in floodwater with a pH 5·3 (unpubl. data). A 12 h day with 600–1200 μmol m^{-2} s^{-1} of photosynthetically active radiation (400–700 nm) was used.

further, or whether they are associated with other factors, e.g. by high ethylene concentrations or by low O_2 concentrations at night.

Response of rice during partial submergence

Recent studies with a lowland rice cultivar have indicated that the carbohydrate level in the plants depends on the percentage of the foliage which remains above water (Table 4), but the data are still too scanty for a detailed interpretation. In deepwater and floating rice cultivars partial or complete submergence also results in changes in growth pattern, consisting of a shift from leaf development to rapid elongation of sheaths and/or internodes. Such changes in growth may therefore affect the carbohydrate status of the plant or its organs.

Some data on growth and sugar concentrations in a recent experiment on partial submergence of a deepwater rice cultivar are shown in Table 4. The experiment was conducted in deepwater tanks at Huntra Rice Experiment Station, 100 km north of Bangkok, Thailand. Water level was raised 5 cm day^{-1}, and the data presented in Table 4 were collected when the water level had reached 1·0 m and 70% of the shoots were under water. This flooding did not change the carbohydrate levels of the emerging leaf, which remained above water, nor did it decrease the levels in plant parts under water, such as the internodes and the growing point (Table 4).

Similar maintenance of high sugar concentrations was obtained for a floating rice cultivar, Pin Gaew 56, which elongated faster than the deepwater rice cultivar as the water rose, and hence only had 50% of shoots submerged at 1·0 m. In this floating rice cultivar, flooding of part of the shoots resulted in a drastic change in plant development with a shift from leaf growth to rapid elongation of internodes. It is particularly interesting that these changes did not materially affect the carbohydrate status of the plants.

The above observation of steady sugar levels appears to contradict recent observations of Raskin & Kende (1984b) on large decreases in starch for submerged excised internodes from the floating rice cultivar Habiganj Aman II. However this is not so, because we also found substantial starch decreases in the flooded plants. In the floating rice cultivar Pin Gaew 56 starch levels in the submerged plants as a percentage of non-submerged plants were 80% for the sheath of the youngest leaf, 60% for the growing point and 70% for the youngest internode. Nevertheless, the sum of sugars plus starch was usually the same or higher in the flooded than in the control plants. For example in the young internode the dry weight contained 21% carbohydrates in non-flooded plants and 27% in flooded plants (unpubl. data). Starch in the deepwater rice cultivar Leuang Pratew 123 declined even more during flooding; nevertheless as in the floating rice cultivar flooding did not cause an appreciable decrease in the sum of sugars and starch in any of the plant parts described above (unpubl. data).

GENERAL DISCUSSION

Regulation of energy generation in hypoxia and anoxia

The data we have discussed show a range of responses to hypoxia in plant tissue: carbohydrate levels may increase or decrease depending on the efficacy of the source of carbohydrates (e.g. whether the shoots are submerged or in air), and the performance of the sink (e.g. rice coleoptiles grow in anoxia, pea roots die). The paradox of the metabolism of the anaerobic sink is that sugars often accumulate (Tables 1 and 2), yet the growth is either impaired or stopped. Such data indicate that at low O_2 concentrations (hypoxia) the plant may direct the limited amount of ATP available to maintenance processes by switching off growth. This balance between carbohydrate supply, energy for survival (maintenance respiration) and energy for growth (growth respiration) is therefore paramount to an assessment of flooding tolerance. It should be emphasized that these metabolic changes occur in concert with other mechanisms of flood tolerance such as increased aerenchyma formation. In classical studies on pea roots, Armstrong, Healy & Lythe (1983) have demonstrated quantitatively the boundaries to which porosity may benefit roots, a point also recently emphasized by Saglio, Raymond & Pradet (1983). Beyond the limits of these physical adaptations, the role of carbohydrate supply and regulation of carbohydrate and energy metabolism may become most important.

An interesting example of the balance between maintenance and growth is found in seminal roots of intact wheat plants. In solution culture at 0.01 mol m^{-3} O_2, only the apical 2–4 mm of the seminal axis showed evidence of decay (P. Leighton, pers. comm.). In contrast, the bulk of the root system survived for 10 days at 0.01 mol m^{-3} O_2, as shown by both its low permeability to sorbitol (F. Buwalda, pers. comm.) and its rapid recovery after return to aerated solution; relative growth rates increased to 1.4 times those in aerated controls (P. Leighton, pers. comm.). Thus, lateral root primordia were able to survive even if the apical meristem succumbed.

Just how limited ATP supply in anoxic roots and coleoptiles is apportioned between growth and maintenance is still debated. Conceivably, small variations in the O_2 concentration of the surrounding medium may have considerable influence on the intracellular ATP/ADP ratio and thus profoundly influence metabolism, possibly via 'pacemaker' reactions (Faiz-ur-Rahman, Trewavas & Davies 1974). For example, Saglio (1985) found that phloem unloading in maize root tips was inhibited by strict anoxia, whereas at 0.04 mol m^{-3} O_2 Atwell *et al.* (1985) showed that accumulation of sugars occurred in root tips of maize. The slow growth of these maize roots was not decreased further when the endosperm and roots of decapitated plants were completely submerged. This suggests that there was no supply of atmospheric O_2 via internal gas spaces in the roots of plants which were not completely submerged. The dramatic difference in response to O_2 concentrations of zero and only 0.04 mol m^{-3} (Saglio 1985; Atwell *et al.* 1985, respectively) strongly suggests that sugar transport and possibly other cell maintenance processes are less sensitive to O_2 deficiency than growth. This type of regulation is consistent with the operation of a Pasteur Effect as a survival mechanism in non-growing anoxic tissue. The absolute necessity of sugar supply for survival of root meristems in anoxia has been demonstrated by Webb & Armstrong (1983).

Finally, it is interesting and comforting to note that the ATP/ADP ratio in maize roots is very sensitive to the O_2 concentration in the incubation medium of maize roots (Saglio, Raymond & Pradet 1983). This relationship, coupled with curves of O_2 uptake versus concentration for rice coleoptiles (Atwell 1982) and maize roots (Atwell *et al.* 1985) suggests that O_2 concentrations below $0.05–0.10$ mol m^{-3} would cause dramatic shifts in energy consumption. These shifts would be in the direction of maintenance at the expense of energy consumption in growth (Atwell 1982). Presumably such shifts would occur at much higher O_2 concentrations in bulky tissues such as seeds. Ironically, the attempts of Saglio, Raymond & Pradet (1983) to evaluate the effects of internal O_2 concentrations on ATP/ADP ratios were frustrated by ATP derived from alcoholic fermentation, a process to which the maize root tip owes its survival in severe O_2 deficiency. It would be interesting to measure the amount of ATP generated from both alcoholic fermentation and aerobic respiration over a range of O_2 concentrations. This could be done by relating the energy states of the cell to the intracellular O_2 concentration, possibly measured using micro-electrodes.

Interactions between flooding and other environmental factors

The complexity of both the flooded environment and the diverse response of plant species to this environment is impressive. This paper has dealt with only one aspect of the response of plants to flooding and even then not comprehensively. In this conclusion we have tried to orient the changes in the carbohydrate status into the general problem of flooding response by speculating on the conditions under which flooded plants may show an accumulation or depletion of carbohydrates. These speculations are schematically presented in Table 6, which gives suggestions of how carbohydrate concentrations may be altered in the plant either before or during flooding. It should be emphasized that definitive evidence for several of these suggestions is very scarce.

Both duration of light and light intensity may be important to the response of plants to flooding, be it with shoots remaining in air or with submerged shoots (Table 6A). As suggested earlier, low O_2 concentrations may exert adverse effects particularly when the carbohydrate status of the plant is low (see Introduction). Furthermore, when shoots are submerged the maintenance of high light intensities over long periods of the day may mitigate the adverse effects of the slow diffusion of CO_2 in water (Table 6C). In contrast, high temperatures may be detrimental since these would accelerate respiration without a concomitant increase in photosynthesis when limited by CO_2 diffusion of low light.

TABLE 6. Some characteristics of plants and environment which may determine whether flooding causes carbohydrate depletion or accumulation (it should be emphasized that several of these suggestions are highly speculative)

Carbohydrate depletion	Carbohydrate accumulation
A. Factors in common to plants in flooded soil with shoots in air and to plants which are partially or completely submerged	
Short days	Long days
Low light	High light
High temperature	Low temperature
Factors inducing rapid growth before flooding	Factors leading to slow growth during flooding
B. Factors applicable only to plants with shoots in air	
High vapour pressure deficit	Low vapour pressure deficit
Stomatal closure due to increased abscisic acid levels	Decreased growth due to decreased cytokinin levels
C. Factors applicable only to partially and completely submerged plants	
Low light in water (turbidity)	High light in water (clear water)
Low CO_2 in water	High CO_2 in water
Low turbulence	High turbulence

The suggestion that a surplus of carbohydrates due to slow growth before flooding might increase the tolerance of the plant (Table 6A) is supported experimentally: rice plants grown at low nitrogen concentrations had higher initial carbohydrate levels and a greater percent survival after submergence than plants grown at high concentrations (Palada & Vergara 1972). Furthermore, as shown in Table 2, accumulation of a carbohydrate surplus in barley grown at low light intensities occurred only at low mineral nutrition. Low concentrations of nutrients may also enhance development of aerenchyma before sudden flooding, possibly conferring a greater tolerance of plants to inundation. This increased development of the aerenchyma has been demonstrated in maize at low nitrogen (Konings & Verschuren 1980) and low nitrogen or phosphate concentrations (Drew & Saker 1983).

Other environmental conditions are also likely to interact with flooding. For example, when shoots are in air, a high vapour pressure deficit may well aggravate the plant response to flooding. It is well known that low root/shoot ratios occur in flooded plants, and it is likely that these may result in water deficits in the shoots.

In summary, the ecophysiology of plant adaptation to flooding cannot be evaluated properly without studying the possible interactions with other environmental factors. Such studies would benefit from using some representative genotypes with different tolerances to flooding.

ACKNOWLEDGMENTS

The authors acknowledge financial support from the Australian Center for International Agricultural Research and the Australian Federal Wheat Council. For critical comments which led to this paper we thank E. G. Barrett-Lennard, F. Buwalda and M. B. Jackson. Details of unpublished data were kindly made available by E. G. Barrett-Lennard, F. Buwalda, Kalaya Kupkanchanakul, Penny Leighton, I. MacPharlin, Ladda Pakinnaka and C. Thompson. For typing of the manuscript and tables we thank Angela Fudge and Nancy Scade.

REFERENCES

Armstrong, W., Healey, M. T. & Lythe, S. (1983). Oxygen diffusion in pea. II. Oxygen concentrations in the primary pea root apex as affected by growth, the production of laterals and radial oxygen loss. *New Phytologist*, **94**, 549–555.

Atwell, B. J. (1982). *Growth and metabolism of rice seedlings at low oxygen concentrations.* Ph.D. thesis. University of Western Australia, Nedlands, 6009.

Atwell, B. J., Thomson, C. J., Greenway, H., Ward, G. & Waters, I. (1985). A study of the impaired growth of roots of *Zea mays* seedlings at low oxygen concentrations. *Plant, Cell & Environment*, **8**, 179–188.

Avadhani, P. N., Greenway, H., Lefroy, R. & Prior, L. (1978). Alcoholic fermentation and malate metabolism in rice germinating at low oxygen concentrations. *Australian Journal of Plant Physiology*, **5**, 15–25.

Beevers, H. (1961). The internal regulation of respiratory rates. *Respiratory Metabolism in Plants* Chapter 7. pp. 147–160. Row, Peterson and Company, New York.

Benjamin, L. R. & Greenway, H. (1979). Effects of a range of O_2 concentrations on porosity of barley roots and on their sugar and protein concentrations. *Annals of Botany*, **43**, 383–391.

Bertani, A. & Brambilla, I. (1982). Effect of decreasing oxygen concentration on some aspects of protein and amino-acid metabolism in rice roots. *Zeitschrift für Pflanzenphysiologie*, **107**, 193–200.

Bertani, A., Brambilla, I. & Menegus, F. (1980). Effect of anaerobics on rice seedlings: growth, metabolic rate, and fate of fermentation products. *Journal of Experimental Botany*, **31**, 325–331.

Bertani, A., Brambilla, I. & Menegus, F. (1981). Effect of anaerobiosis on carbohydrate content in rice roots. *Biochemie und Physiologie de Pflanzen*, **176**, 835–840.

Chirkova, T. V. (1978). Some regulatory mechanisms of plant adaptation to temporal anaerobiosis. *Plant Life in Anaerobic Environments.* (Ed. by D. D. Hook & R. M. M. Crawford) pp. 137–154. Ann Arbor Science, Michigan.

Crawford, R. M. M. (1977). Tolerance of anoxia and ethanol metabolism in germinating seeds. *New Phytologist*, **79**, 511–517.

Crawford, R. M. M. (1978). Metabolic adaptation to anoxia. *Plant Life in Anaerobic Environments.* (Ed. by D. D. Hook & R. M. M. Crawford) pp. 119–136. Ann Arbor Science, Michigan.

Crawford, R. M. M. & Tyler, P. D. (1969). Organic acid metabolism in relation to flooding tolerance in roots. *Journal of Ecology*, **57**, 235–244.

Davies, D. D. (1980). Anaerobic metabolism and the production of organic acids. *The Biochemistry of Plants, Vol. 2. Metabolism and Respiration.* (Ed. by D. Davies) pp. 581–611. Academic Press, London.

Drew, M. C. & Saker, L. R. (1983). *Agricultural Research Council Letcombe Laboratory Annual Report, 1982*, pp. 43–44. Agricultural Research Council: London.

Faiz-ur-Rahman, A. T. M., Trewavas, A. J. & Davies, D. D. (1974). The Pasteur effect in carrot root tissue. *Planta (Berlin)*, **118**, 195–210.

Fulton, J. M. & Erickson, A. E. (1964). Relation between soil aeration and ethyl alcohol accumulation in xylem exudates of tomatoes. *Soil Science Society of America Proceedings*, **28**, 610–614.

Hall, A. J. & Milthorpe, F. L. (1978). Assimilate source-sink relationships in *Capsicum annuum* L. III. The effects of fruit excision on photosynthesis and leaf and stem carbohydrates. *Australian Journal of Plant Physiology*, **5**, 1–13.

Hanada, K. (1983). Growth of internodes, leaves and tillers of floating and non-floating rice varieties under deepwater conditions. *Japan Journal of Tropical Agronomy*, **27**, 221–236.

van de Heide, H., Berendina, M. de Boer-Bolt & Van Raalte, M. H. (1963). The effect of a low oxygen content of the medium on the roots of barley seedlings. *Acta Botanica Neerlandica*, **12**, 231–247.

Hiatt, A. J. & Lowe, R. H. (1967). Loss of organic acids, amino-acids, K, and Cl from barley roots treated anaerobically and with metabolic inhibitors. *Plant Physiology*, **42**, 1731–1736.

HilleRisLambers, D. & Seshu, D. V. (1982). Some ideas on breeding procedures and requirements for deepwater rice improvement. *Proceedings of the 1981 International Deepwater Rice Workshop*, pp. 29–44. International Rice Research Institute, Phillipines.

Jackson, M. B. (1985). Ethylene and the responses of plants to excess water in their environment—a review. *Ethylene and Plant Development* (Ed. by J. A. Roberts & G. A. Tucker) pp. 241–265 Butterworths, London.

Jackson, M. B. & Drew, M. C. (1984). Effects of flooding on growth and metabolism of herbaceous Plants. In *Flooding and Plant Growth* (Ed. by T. T. Kozlowski) pp. 47–128. Academic Press, London.

Jackson, M. B. & Kowalewska, A. K. B. (1983). Positive and negative message from roots induce foliar desiccation and stomatal closure in flooded pea plants. *Journal of Experimental Botany*, **34**, 493–506.

James, W. O. & Ritchie, A. F. (1955). The anaerobic respiration of carrot tissue. *Proceedings of the Royal Society of London, Series B*, **143**, 302–310.

Khush, G. S. (1984). *Terminology for Rice Growing Environments*. International Rice Research Institute, Phillipines.

Konings, H. & Verschuren, G. (1980). Formation of aerenchyma in roots of *Zea mays* in aerated solutions, and its relation to nutrient supply. *Physiologia Plantarum*, **49**, 265–270.

Kordan, H. A. (1974). Patterns of shoot and root growth in rice seedlings germinated under water. *Journal of Applied Ecology*, **11**, 685–690.

Kozlowski, T. T. & Pallardy, S. G. (1984). Effect of flooding on water, carbohydrate, and mineral relations. In *Flooding and Plant Growth* (Ed. by T. T. Kozlowski) pp. 165–193. Academic Press, London.

Lao Tsu (300–500BC). *Tao Te Ching* (Translated by Gia Fu Feng & Jane English, 1973). Wildwood House Ltd., London.

Limpinuntana, V. & Greenway, H. (1979). Sugar accumulation in barley and rice grown in solutions with low concentrations of oxygen. *Annals of Botany*, **43**, 373–381.

Marschner, H., Handley, R. & Overstreet, R. (1966). Potassium loss and changes in the fine structure of corn root tips induced by H-ions. *Plant Physiology*, **41**, 1725–1735.

Marschner, H. & Mengel, K. (1966). Der Einfluss von Ca- und H- Ionen bei unterschiedlichen Stoffwechselbedingungen auf die membranpermeabilitat junger Gerstenwurzeln. *Zeitschrift feur Pflanzenernaehrung Duengung Bodenkun De*, **112**, 39–49.

Neal, M. J. & Girton, R. E. (1955). The Pasteur effect in maize. *American Journal of Botany*, **42**, 733–737.

Palada, M. C. & Vergara, B. S. (1972). Environmental effects on the resistance of rice seedlings to complete submergence. *Crop Science*, **12**, 209–212.

Papenhuijzen, C. (1983) Effect of interruption of aeration of the root medium on distribution of dry matter, sugar and starch in young plants of *Phaseolus vulgaris*. *Acta Botanica Neerlandica*, **32**, 63–67.

Phillips, J. W. (1947). Studies on fermentation in rice and barley. *American Journal of Botany*, **34**, 62–72.

Ponnamperuma, F. N. (1984). Effects of flooding on soils. In *Flooding and Plant Growth*. (Ed. by T. T. Kozlowski pp. 9–45. Academic Press, London.

Qureshi, F. A. & Spanner, D. C. (1973). The effect of nitrogen on the movement of tracers down the stolon of *Saxifraga sarmentosa* with some observations on the influence of light. *Planta (Berlin)*, **110**, 131–144.

Raskin, I. & Kende, H. (1984a). Regulation of growth in stem sections of deepwater rice. *Planta (Berlin)*, **160**, 66–72.

Raskin, I. & Kende, H. (1984b). Effect of submergence on translocation, starch content and amylolytic activity in deepwater rice. *Planta (Berlin)*, **162**, 556–559.

ap Rees, T. & Wilson, P. M. (1984). Effects of reduced supply of oxygen on the metabolism of roots of *Glyceria maxima* and *Pisum sativum*. *Zeitschrift für Pflanzenphysiologie*, **114**, 493–503.

Reid, D. M. & Bradford, K. J. (1984). Effects of flooding on hormone relations. In *Flooding and Plant Growth* (Ed. by T. T. Kozlowski) pp. 195–219. Academic Press, London.

Saglio, P. H. (1985). Effect of path or sink anoxia on sugar translocation in roots of maize seedlings. *Plant Physiology*, 77, 285–290.

Saglio, P. H., Raymond, P. & Pradet, A. (1980). Metabolic activity and energy charge of excised maize root tips under anoxia. *Plant Physiology*, 66, 1053–1057.

Saglio, P. H., Raymond, P. & Pradet, A. (1983). Oxygen transport and root respiration of maize seedlings. *Plant Physiology*, 72, 1035–1039.

Smith, A. M. & ap Rees, T. (1979a). Effects of anaerobiosis on carbohydrate oxidation by roots of *Pisum sativum*. *Phytochemistry*, 18, 1453–1458.

Smith, A. M. & ap Rees, T. (1979b). Pathways of carbohydrate fermentation in the roots of marsh plants. *Planta (Berlin)*, 146, 327–334.

Smith, F. A. & Walker, N. A. (1980). Photosynthesis by aquatic plants: Effects of unstirred layers in relation to assimilation of CO_2 and HCO_3 and to carbon isotopic discrimination. *New Phytologist*, 86, 245–259.

Taylor, D. L. (1942). Influence of oxygen tension on respiration, fermentation, and growth in wheat and rice. *American Journal of Botany*, 29, 721–738.

Trought, M. C. T. & Drew, M. C. (1980). The development of waterlogging damage in wheat seedlings (*Triticum aestivum* L.). I. Shoot and root growth in relation to changes in the concentrations of dissolved gases and solutes in the soil solution. *Plant and Soil*, 54, 77–94.

Tsuji, H. (1968). Effect of anaerobic condition on the dry weight decrease of germinating rice seeds. *Botanical Magazine of Tokyo*, 81, 233–242.

Tsuji, H. (1969). Rise in oxygen uptake during air-adaptation of anaerobically treated rice seedlings. *Botanical Magazine of Tokyo*, 82, 226–238.

Turner, J. S. (1956). Respiration: The Pasteur effect in plants. *Annual Review of Plant Physiology*, 2, 145–168.

Vartapetian, B. B., Andreeva, I. N. & Kozlova, G. I. (1976). The resistance to anoxia and the mitochondrial fine structure of rice seedlings. *Protoplasma*, 88, 215–224.

Vartapetian, B. B., Andreeva, I. N. & Nuritdinov, N. (1978). Plant cells under oxygen stress. In *Plant Life in Anaerobic Environments* (Ed. by D. D. Hook & R. M. M. Crawford) pp. 13–88. Ann Arbor Science, Michigan.

Webb, T. & Armstrong, W. (1983). The effects of anoxia and carbohydrates on the growth and viability of rice, pea and pumpkin roots. *Journal of Experimental Botany*, 34, 579–603.

Wenkert, W., Fausey, N. R. & Watters, H. D. (1981). Flooding responses in *Zea mays* L. *Plant and Soil*, 62, 351–366.

de Witt, M. C. J. (1974). Reacties van gerstwortels op een tekort aan zuurstot. Publ. V. R. B. Offsetdrukkerij, Groningen.

Yoshida, S. (1981). *Fundamentals of Rice Crop Science*. International Rice Research Institute, Los Banos, Lagguna, Philippines. 269pp.

Author index

Figures in italics refer to pages where full references appear; figures in bold to articles in this volume.

Abbey, K., 335, *343*
Abeles, F. B., 70, *74*
Adams, D. O., 294, *300*
Adams, M. S., 91, 93, *96*:126, *127*:138, *146*
Ahmad, I., 337, 338, 339, 341, *345*
Ahmadjian, V., 355, *357*
Albert, L. S., 87, *96*, *98*
Allaway, G. W., 341, *343*
Allen, E. D., 83, *95*:101, 106, *111*:136, *146*:158, *166*
Allen, S., 132, *148*
Alpi, A., 249, *252*:377, *383*
Amarsinghe, I., 55, 57, 58, 64, 68, *76*
Amrhein, N., 294, 299, *300*
Andersen, J. M., 100, *112*
Anderson, B., 338, *343*
Anderson, L. W. T., 87, *95*
Andreeva, I. N., 206, 221, *223*:270, *277*:412, *433*
Andrews, T. J., 334, 340, 341, 342, *343*, *344*
Anthony, D. S., 144, *149*
App, A. A., 255, *263*
Arber, A., 54, 55, 56, *74*
Aris, J. J. A. M., 55, *75*
Armstrong, W., **303–20**:55, *74*:206, 223, 227, *237*:266, *276*:293, *300*:303, 304, 305, 308, 311, 312, *320*:325, *330*:397, 401, 407, *409*:412, 428, 429, *431*, *433*
Arnon, D. I., 255, *263*
Asahi, T., 217, *222*, *223*
Asker, H., 268, *276*
Atkinson, M. R., 339, 341, *343*
Attiwill, P. M., 334, *343*
Atwell, B. J., 417, 419, 420, 421, 422, 428, 429, *431*
Avadhani, P. N., 423, *431*

Baardseth, E., 36, *49*:141, *146*
Badger, M. R., 81, *95*
Bain, J. T., 84, *95*:105, *111*:136, *146*
Baker, I., 246, *252*
Ball, M. C., 338, 341, *343*
Barber, D. A., 55, *76*
Barclay, A. M., 159, *166*:206, *222*:294, *300*:322, 330:377, *383*:398, 399, 400, 404, *409*
Barret, S. C. H., 193, *202*
Barron, J. A., 42, 43, 45, *51*
Bartels, P. M., 56, 59, 60, 66, 69, *75*
Barth, H., 333, *343*
Bartley, M. R., **153–65**:**167–76**:155, 156, 161, *165*:170, 174, 175, 176, *177*

Basiouny, F. M., 188, *189*
Bazier, R., 271, *276*
Beardall, J., 81, *95*:130, 131, 133, 134, 135, 136, 140, 144, *146*, *148*
Beer, S., 90, *95*:101, 102, *111*:122, *127*
Beevers, H., 249, *252*:377, *383*:413, 419, *431*
Beggs, C. J., 5, *19*
Belford, R. K., 322, *330*
DeBell, D. S., 282, 289, 290, *291*, *292*
Benedict, C. R., *146*
Benjamin, L. R., 416, *431*
Bennert, H. W., 341, *343*
Bennett, A. B., 376, *383*
Berard-Therriault, L., 37, *49*
Berrie, A. M. M., 376, *384*
Berry, J. A., 81, *95*:132, *147*
Bertani, A., **255–63**:249, *252*:256, 257, 260, *263*, *264*:417, 419, 420, 421, *431*
Bertness, M. D., *372*
Bewley, J. D., 167, *177*
Biebl, R., 42, 44, *49*:340, *343*
Birch, W. R., 185, *189*
Birchler, J. A., 231, *238*
Bisson, M. A., 36, *50*
Björn, L. O., 14, 15, 16, 17, *21*
Black, M., 167, *177*
Black, M. A., 82, 85, *95*
Blackman, F. F., 265, *276*
Bligny, B., 270, *277*
Bobylev, Y. S., 207, *223*
Bodkin, P. C., 13, *19*:87, 88, *95*:156, *165*
Boillot, A., 362, *372*
Bolton, J. J., 44, 46, *49*, *51*
Bommerson, J. C., 130, 135, *147*
Boney, A. D., 36, *49*
Boon, P. I., 341, *343*
Borthwick, H. A., 17, *19*, *21*
Bosler, M. E., 262, *264*
Boston, H. L., 91, 93, *96*:126, *127*:138, *146*
Bostrack, J. M., *19*
Boulter, D., 398, *409*
Bowes, G., **79–95**:81, 82, 83, 84, 85, 86, 88, 89, 90, 91, 92, 93, 94, 95, *96*, 97, *98*:99, 100, 108, *111*:115, 119, 122, *127*, *128*:146, *148*
Boylen, C. W., 83, *96*:125, *127*
Bradford, K. J., 299, *301*:413, 414, *433*
Bradford, M. M., 243, *252*
Bradshaw, A. D., 70, *74*
Braendle, R., **233–9**:**398–408**:246, *253*:293, 294, 295, 299, *300*, *301*:305, *320*:381, *384*:398, 401, 402, 403, 404, 405, 406, *409*, *410*

434

Brambilla, I., **255–63**:249, *252*:256, 257, 260, *263*, *264*:417, 421, *431*
Breeman, A. M., 360, 371, *372*, *373*
Bridgwater, F., 281, *292*
Briggs, W. R., 4, 5, *21*, *22*
Bristow, J. M., 82, 84, 86, 88, 94, *96*, *97*
Brocklebank, K. J., 380, *384*
Brove, P., 282, *292*
Brown, A. D., 282, *291*
Brown, A. D. H., 231, *238*
Brown, A. F., *372*
Brown, C. L., 284, *291*
Brown, F. A., 37, 42, *49*:348, *357*
Brown, J. M. A., 83, 85, 86, *96*
Brown, L. M., 36, *49*
Browse, J. A., 83, 86, *96*:106, *111*
Bufler, G., 295, *300*
Burns, J. A., 265, *276*
Burrows, E. M., 48, *49*:348, *357*
Busch, G., 115, 116, 119, 123, *128*:139, 143, *147*
Bushby, C. H., 144, *147*
Butler, W. L., 4, *19*:155, *165*
Butter, M. E., 55, *75*

Callow, M. E., 41, *49*
Campbell, R., 323, *330*
Cannell, R. Q., 321, 322, *330*
Canoy, M. J., 337, *343*
Canvin, D. T., 82, 86, 94, *97*
Card, K. A., 132, *149*
Cardinal, A., 37, *49*
Carlson, P. R., 240, *252*
Carver, K. A., 257, *263*
Casey, S., 44, 48, *52*
ten Cate, H. J., 44, *49*
Catling, P. M., 186, *189*
Chalmers, A. G., 239, *253*
Chambers, P. A., 17, 18, *19*:156, 157, *165*:176, *177*
Chandler, G., 335, *343*
Chang, L. A., 289, 290, *291*
Chapman, A. R. O., 360, *372*
Chester, R., 36, *51*
Child, R., **153–65**:158, *165*
Chirkova, T. V., 423, *431*
Chollet, R., 133, *147*
Chrysler, M. A., 19, *20*
Chudek, J. A., 36, 41, *49*
Clark, L. H., 325, 329, *330*
Clarkson, D. T., 329, *331*:385, 386, *394*
Clements, H. F., 17, *20*
Clokie, J. J. P., 371, *372*
Clough, B. F., 334, 339, 340, 342, *343*, *344*
Cockburn, W., 114, *127*
Colijn, F., 28, 32, *34*
Collins, J. C., 36, 44, *51*, *52*
Combes, R., 19, *20*
Cook, C. D. K., **179–88**:181, 182, 183, 185, 186, 188, *189*
Cooke, R. J., 268, *276*

Cookson, E. C., 57, 59, 60, 63, *74*, *75*
Cossins, E. A., 299, *300*
Costes, C., 271, *276*
Coughlin, S. J., 41, *49*
Coult, D. A., 398, *409*
Cowan, I. R., 340, 342, *344*
Craigie, J. S., 360, *372*
Crawford, R. M. M., **375–83**:**398–408**:206, 222:227, 228, 230, 234, *237*, *238*:246, 248, *252*, *253*:262, *264*:266, 273, *276*:282, 289, 290, *291*, *330*, *331*:375, 376, 377, 379, 381, *383*, *384*:391, *394*:397, 398, 399, 400, 404, 405, 406, 407, 408, *409*, *410*:413, 419, 423, *431*
Critchley, C., 338, *344*
Cronan, J. E., 270, *276*
Crothers, J. H., 28, 32, *34*
Crowder, A. A., 408, *410*
Cubit, J., 360, *373*
Cunningham, E. H., 360, 361, *372*
Curcio, J. A., 8, *20*
Czopek, M., 176, *177*

Dacey, J. W. H., 60, *75*:100, *111*:240, *253*:305, *320*:407, *409*
Dahl, E., 48, *49*
Dale, H. M., 87, *97*:158, *166*
van Dam, L., 337, *345*
Davenport, J., 36, 41, *49*
Davies, D. D., **265–76**:234, *238*:255, *263*:266, 267, 268, *276*:422, 423, 428, *431*
Davis, B. J., 194, *202*
Davis, G. J., 87, *96*
Dawes, C. J., 239, *252*
Day, D. A., 208, *222*
Deitzer, G. F., 17, *20*
Delistraty, D. A., 242, *252*
Denslow, S., **281–91**:285, *291*
Dethier, M. N., 360, 371, *372*
Dey, B., 260, *264*
Deysher, L., 348, *357*
Dickinson, H. G., 161, *166*
Dickson, D. M. J., 36, 41, *49*
Dixon, P. S., 38, 44, *49*, *50*:364, 370, *372*
Dore, W. G., 186, *189*
Dorgelo, J., 37, *49*
Douce, R., 270, *277*
Dow, E. J., 174, *177*
Dowling, R. M., 235, *238*
Drakeford, D. R., **385–94**:385, 386, *394*, *395*
Drew, M. C., **321–30**:255, *263*, *264*:294, *300*:322, 323, 324, 325, 326, 327, 328, *330*, *331*:385, *394*:412, 413, 418, 430, *431*, *432*, *433*
Dring, M. J., **23–34**:4, 5, 12, 13, 17, 19, *20*, *21*:23, 26, 33, *34*:37, 42, *49*:348, *357*:360, 361, 370, *372* *373*
Dromgoole, F. I., 37, 38, 39, 45, *49*:83, 86, *96*:107, *111*
Dunn, E. L., 240, *252*
Duss, F., 405, 406, *409*

Edwards, G. E., 80, *96*, *97*
Edwards, K. J. R., 231, *238*:405, *409*
Edwards, P., 44, 48, *52*
Effer, W. R., 266, *276*
Ehleringer, J., 80, 81, *97*
Epstein, E., 36, *49*
Epstein, S., 144, *148*
Erickson, A. E., 423, *431*
Eshel, A., 101, 102, *111*
Evans, L. V., 41, *49*:355, 356, *357*
Everett, M., 5, *20*

Fabijan, D. M., 389, *395*
Faiz-ur-Rahman, A. T. M., 428, *431*
Farquahar, G. D., 132, *147*
Farrar, J. F., 355, *357*
Farrow, G. E., 371, *372*
Fassett, N. C., 181, *189*
Fausy, N. R., 414, *433*
Fauth, A., 182, *189*
Federer, C. A., 4, *20*
Feldmann, J., 369, *372*
Fenning, T. M., 408, *409*
Ferrari, T. E., 255, 258, 262, *263*
Fiala, K., 402, *409*
Field, C. D., 339, *344*
Filner, P., 258, 262, *263*
Fisher, E., 351, *357*
Flesher, D., 255, 262, *264*
Flowers, T. J., 338, *344*
Fox, T. C., **193–202**
Francke, J. A., 44, *49*
Frankland, B., **167–76**:155, 161, *165*:168, 169, 170, 175, *177*
Frederick, S. E., 90, *96*
Freeling, M., 231, *238*:268, *276*, *277*
Fries, N., 355, *357*
Fritsch, F. E., 41, *49*
Fuhrer, J., 294, *300*
Fuhrer-Fries, C. B., 294, *300*
Fukshansky, L., 8, 14, 16, 17, *20*, *21*:161, *165*:169, *177*
Fulcher, R. G., 351, *357*
Fulton, J. M., 423, *431*
Funke, G. L., 56, 59, 60, 66, 69, *75*

Gabrielsen, E. K., 4, *20*
Gallagher, J. L., 402, *409*
Gambrell, R. P., 321, 322, *331*
Garcia-Novo, F., 262, *264*
Garrad, L. A., 176, *177*:188, *189*
Gates, D. M., 342, *344*
Gates, J. E., 144, *147*
Gaudet, J. J., 87, *96*
Gaynard, T. J., **303–20**:304, 305, 306, 308, 309, 312, *320*
Generozova, I. P., **205–22**
Gessner, F., 36, 42, *50*
Giffard, S., 324, *330*
Gill, A. M., 335, *344*
Gill, C. J., 55, *75*

Girton, R. E., 420, 422, 423, *432*
Gleason, M. L., 240, *252*:305, 316, *320*
Glidewell, S. M., 82, *97*
Glime, N. M., 136, *147*
Goldberg, B., 6, *20*
Gomes, A. R. Sena, 55, *75*
Good, R., 265, *276*
Goodenough, A., 377, *384*:406, *409*
Goodman, D., 325, *331*
Gopal, B., 174, *177*
Gordon, D. M., 102, 103, 105, 106, 107, 108, 109, *112*:141, *148*
Gorham, P. R., 176, *177*
Grace, J. B., 118, *128*
Greenway, H., **411–31**:246, *252*, *253*:416, 418, 420, 421, *431*, *432*
Greenwood, D. J., 304, *320*:321, 325, *331*
Grego, S., 266, 267, *276*
Griffiths, H., **129–46**:130, 131, 132, 133, 134, 135, 136, 140, 144, *146*, *147*, *148*
Grime, J. P., 56, 76:171, *177*
DeGroote, D., 85, *96*
Grosselink, J. G., 251, *253*
Gruber, P. J., 90, *96*
Gude, H., 341, *345*
Guiry, M. D., **359–71**:360, 361, 367, 368, *372*, *373*
Gunning, P. E. S., 144, *147*
Gutmann, I., 284, *291*

Hageman, R. H., 255, 256, 262, *264*
Hahn, H., 262, *264*
Haines, F. M., 46, *50*
Haldemann, C., 297, *300*:398, 403, 406, *409*
Hall, A. J., 414, *431*
Haller, W. T., 81, 82, 83, 84, 85, 86, 88, 89, 90, 92, 93, 95, *96*, *98*:176, *177*:188, *189*
Hamelin, J., 342, *344*
Hammett, L. K., 289, 290, *291*
Hanada, K., 424, *432*
Handley, R., 422, *432*
Hanson, A. D., 231, *238*:266, *276*
Hanson, J. B., 208, *222*:385, 386, *394*
Harberd, N. P., 231, *238*:405, *409*
Harris, G. P., 355, *357*
Harris, P., 270, *276*
Harris, W. H., 325, 329, *330*
Hartmann, E., 5, *20*
den Hartog, C., 36, 37, *50*
Harvey, D. M. R., 338, *344*
Haslam, J. M., 220, *223*
Hastorf, C. A., 132, *147*
Hayes, R., 17, *20*
Healy, M. T., 304, *320*:428, *431*
Hedgpeth, J. W., 36, *50*
van der Heide, H., 416, *432*
van der Heijden, L. A., 181, *190*
Hellebust, J. A., 36, *49*, *50*
Hendricks, S. B., 17, *19*, *21*:155, *165*
Hendry, G. A. F., 380, *384*
Hendy, C. H., 132, *149*

Henshaw, G. G., 398, *409*
Herman, A. I., 144, 145, *147*
Herman, B., 377, *384*:406, *409*
Hershner, C., 242, *252*
Hetherington, A. M., 273, *276*:376, 379, *384*
Hiatt, A. J., 248, *252*:422, *432*
Hickey, R. J., 125, *127*
Higinbotham, N., 386, 390, *395*
Hill, D. J., 355, *357*
HilleRisLambers, D., 412, 424, *432*
Hirel, B., 335, *344*
Hiscock, K., 28, 32, *34*
van den Hock, C., 370, 371, *373*
Hoffman, N. E., 324, *331*:408, *410*
Holaday, A. S., 81, 82, 83, 84, 85, 88, 89, 90,
    92, 93, *96*:108, *111*:122, *127*:133, *147*
Holmes, M. G., **3–19**:4, 5, 6, 7, 8, 15, 16, 17, *19*,
    20, *21*:154, 155, 156, *165*
Holt, J. A., 340, *345*
Hook, D. D., **281–91**:282, 284, 285, 290, *291*,
    292
ten Hoopen, A., 360, 371, *372*
Hopkinson, C. S., 251, *253*
Horton, R. F., 59, 63, 75, *76*
Hough, R. A., 85, *96*
Howes, B. L., 239, *252*
Huber, S. C., 80, *96*
Hubick, K. T., 386, *395*
Huck, M. G., 375, 376, *384*
Hucklesby, D. P., 256, *264*
Hull, R. J., 402, *410*
Hunt, R., 361, *372*
Hunter, M. I. S., 273, *276*:376, 379, *384*
Hutchinson, G. E., 10, *20*:84, *96*

Ikeda, K., 17, *21*
Inada, K., 4, *20*:66, *76*
Irvine, L. M., 360, 364, 370, *372*, *373*

Jabben, M., 17, *19*, *20*
Jackson, M. B., 54, 55, 57, 60, 61, 62, 63, 64,
    66, 75:87, 97, 294, *300*:322, 324, *330*,
    331:377, *384*:385, *394*:406, 408, *409*:412,
    413, 414, 421, *432*
Jackson, W. T., 322, *331*
Jacobsen, J. V., 231, *238*:266, *276*
Jahromi, S., 282, *292*
James, A. T., 270, *276*
James, W. O., 418, 422, *432*
Jazwozski, E. G., 262, *264*
Jenkin, L. E. T., **227–37**:230, 236, *238*:246,
    252:397, *409*
Jenkins, W., 408, *409*
Jennings, D. H., 340, *344*
Jerlov, N. G., 10, 13, *20*:23, 24, 32, *34*
John, C. D., 246, *252*, *253*
Johnson, C. B., 161, *165*
Johnson, M. P., 19, *20*:87, *96*
Johnston, A. M., 133, 134, 136, 138, 140, 141,
    142, 144, *147*, *148*
Jolivet, Y., 342, *344*
Joly, C. A., 330, *331*

Jones, H. G., 37, *50*:132, *147*
Joshi, G. V., 334, 335, 337, *344*

Kacser, H., 265, *276*
Kadono, Y., 176, *177*
Kain, J. M., 4, *20*:360, *372*
Kalsi, G., 234, *238*:290, *292*
Kane, M. E., 87, *96*
Kanwisher, J. W., 240, *253*
Kaplan, A., 81, *95*
Kapraun, D. F., 370, *372*
Karnovsky, M. J., 207, *222*
Kasha, M., 9, *20*
Katsura, N., 66, *76*
Kausch, A. P., 402, *409*
Kauss, H., 36, *50*
Kawase, M., 324, *331*
Kays, S. E., 262, *264*
Keeley, J., 290, *292*
Keeley, J. E., **113–27**:83, 85, 89, 91, 92, 95, *96*,
    97:114, 115, 116, 118, 119, 120, 121, 122,
    123, 125, 126, *127*, *128*:136, 138, 139,
    143, 144, *147*, *148*:246, 248, *252*:305, *320*
Kellerman, G. M., 220, *223*
Kelly, P. M., 268, *276*
Kende, H., 58, 59, 61, 63, 64, 66, 67, 74, *75*,
    76:87, *97*:262, *264*:266, 277:294, *300*:305,
    320:413, 427, *432*
Kendrick, R. E., 8, *21*:168, *177*
Kennedy, R. A., **193–202**:85, *96*:193, 194, *202*,
    203:206, 217, *222*:266, *276*:377, *384*:400,
    409
Kenworthy, P., **265–76**:266, 267, *276*
Kerby, N. W., 141, 142, 143, *147*
Kershaw, K. A., 355, *357*
Keys, A. J., 335, *344*
Khfaji, A. K., 48, *50*
Khush, G. S., 412, 424, *432*
Kimmerer, T. W., 294, 295, *300*
King, G., 239, *252*
King, R. J., 33, *34*
Kingham, D. L., 355, 356, *357*
Kinzel, H., 340, *343*
Kionka, C., 294, 299, *300*
Kirk, J. T. O., 23, 28, *34*
Kirst, G. O., 36, *50*
Klaine, S. J., 158, *165*
Klein, W. H., **3–19**:5, 6, *20*, *21*:154, 156
Klepper, L., 262, *264*
Klug, M. J., 305, *320*:407, *409*
Kluge, M., 80, *97*:114, 118, *128*
Knoth, A., 48, *50*
Kohlmeyer, J., 355, *357*
Konings, H., 324, *331*:430, *432*
Kopecko, K. J., 121, *128*
Kordan, H. A., 206, 217, *222*:421, *432*
Kornmann, P., 362, *372*
Kowalewska, A. K. B., 414, 421, *432*
Kozlova, G. I., 221, *223*:270, *277*:412, *433*
Kozlowski, T. T., 55, *75*:290, *292*:294, 295,
    300:413, 414, *432*
Kramer, P. J., 322, *331*

Krattinger, K., 179, 182, *189*
Kremner, B. P., 37, 41, 44, *50*:356, *357*
Kristensen, I., 37, 38, 40, *50*
Ku, H. S., 59, 74, *75*
Ku, M. S. B., 80, *97*
Kopanchanakul, T., **411–31**
Kusanagi, T., 66, 74, 76:87, *98*

Laing, H. E., 294, *300*:398, *410*
Lambert, J. M., 233, *238*
Lance, C., 271, *277*
Larher, F., 339, 342, *344*
Larkum, A. W. D., 36, *51*
Lathrop, E. W., 116, 121, *128*
Lauchli, A., 48, *51*:337, *345*
Lazerte, B. D., 130, 133, 135, *147*
Lea, P. J., 335, *344*
Leach, W., 266, *276*
Lear, R., 337, *344*
Lee, K. W., 133, *147*
Lee, R. B., 255, *264*
Leggett, J. E., 248, *252*
Leibig, J., 265, *276*
Leopold, A. C., 70, *75*
de Lestang, G., 38, *50*
Levitt, J., 386, *395*
Lewis, D. H., 355, *357*
Lewis, J. R., 37, 43, *50*
Limpinuntana, V., 416, 418, 420, 421, *432*
Lindberg, B., 356, *357*
Ling, E., 57, 60, 62, 63, 64, 75:87, *97*
Linhart, Y. B., 246, *252*
Linnane, A. W., 220, *223*
Little, C., 33, *34*
Littler, D. S., 360, *373*
Littler, M. M., 360, *373*
Liu, Y., 294, 295, *300*
Lizada, M. C. C., 295, *300*
Lloyd, N. D. H., 82, 86, 94, *97*
Lodge, S. M., 348, *357*
Loftus, M. E., 11, *21*
Looi, A. S., 88, *96*
Lowe, R. H., 422, *432*
Lubchenco, J., 360, *373*
Lucas, W. J., 106, *111*
Lüning, K., 4, 5, 12, 13, 17, 19, *20*, *21*:26, 33, *34*:360, 371, *373*
Lüönd, R., 182, 185, *189*
Luttge, U., 386, 390, *395*
Luzikov, V. N., 220, *222*
Lynch, J. M., 322, *331*
Lythe, S., 428, *431*
Lyttle, R. W., 402, *410*

Maberly, S. C., 82, 84, 85, 86, 90, 94, *95*, *97*, *98*:99, 101, 103, 104, 105, 106, 107, *111*, *112*
McCallum, W. B., 87, 88, *97*
McCartney, H. A., 8, 16, *20*
McComb, A. J., 59, 66, *75*
McCree, K. J., 4, *21*
McCully, M. E., 87, *97*:351, *357*

McDonald, T. J., 235, *238*
MacFarlane, J. J., **129–46**:134, 136, *147*
McKee, K. L., **239–52**:234, *238*:239, 240, 242, 243, 244, 247, 248, *252*, *253*:404, *410*
McKee, W. H. Jr., 283, *292*
Mackereth, F. J. H., 316, *320*
McManmon, M., 228, 230, *238*:248, *252*
McNally, S. F., 335, *344*
McPherson, D. C., 323, *331*
Madsen, T. V., 108, 110, *111*
Maggs, C. A., **359–72**:360, 361, 363, 367, 368, 372, 373
Malavolta, E., 255, *264*
Malone, M., 60, 61, 63, 64, 66, 69, 75:294, *300*
Mann, K. H., 142, *148*
Marschner, H., 422, *432*
Marsh, L. C., 402, *409*
Marshall, D. R., 282, *291*
Martin, A. C., 181, *190*
Maslova, I. P., 206, *223*
Mason, G., 171, *177*
Mathews, R. P., 125, 126, *128*
Mazliak, P., 271, 272, *276*, *277*
Meiss, A. N., 255, *263*
Mendelssohn, I. A., **239–52**:234, *238*:239, 240, 242, 243, 244, 247, 248, *252*, *253*:404, 408, *410*
de Mendoza, D., 270, *276*
Menegus, F., 249, *252*:256, 257, *263*:417, 421, 431
Mengel, K., 422, *432*
Metcalfe, C. R., 305, *320*
Métraux, J-P., 58, 59, 61, 75:87, *97*
Meysel, M. N., 220, *222*
Michael, P. W., *203*
Miflin, B. J., 335, *344*
Miller, J. L., 176, *177*
Miller, P. C., 341, *344*
Millington, W. F., *19*, *95*
Milthorpe, F. L., 414, *431*
Mitchell, C. A., 246, 247, *253*
Mitchell, R. S., 58, 59, *75*
Miyachi, S., 81, 92, *98*
Mocquot, B., **265–76**:249, *253*:268, *276*:404, *410*
Monk, L. S., **375–83**:246, *253*:294, *300*:381, 384:404, 405, 406, *410*
Monson, R. K., 80, *97*
Mook, W. G., 130, 135, *147*
Moore, P. D., 89, 95, *97*
Moore, P. G., 354, *357*
Moore, R. T., 334, 341, *344*
Morgan, D. C., 8, *21*:158, *165*
Morgan, J. J., 101, *112*
Morinaga, T., 380, *384*
Morris, J. T., 240, *253*
Morrisset, C., 270, *277*
Morton, B. A., 97:121, 122, 126, *128*:136, *147*
Moss, B. L., 39, 42, *50*
Mukherjee, I., 386, *394*
Mukherji, S., 260, *264*
Muller, G. J., 334, *343*

Munda, I. M., 37, 41, *50*
Munday, J., 282, *291*
Munz, P. A., 121, *128*
Musgrave, A., 57, 59, 60, 61, 62, 63, 64, 67, 68, 69, 75:87, *97*

Nance, J. F., 255, *264*
Nash, D., 342, *344*
Nawa, Y., 217, *222, 223*
Naylor, A. W., 83, 90, 92, *98*
Neal, M. J., 420, 422, 423, *432*
Nelson, R. B., 282, *292*
Newton, J. W., 144, 145, *147*
Newton, R. J., 176, *177*
Nichiporovich, A. A., 4, *21*
Niedhardt, F. C., 268, *277*
De Niro, M. J., 132, 143, *147, 148*
Nooden, L. D., 19, *21*, 176, *177*
Northcote, D. H., 272, *277*
Norton, T. A., **347–56**:37, 39, 42, 48, *50*, *51*:348, 349, 350, 351, 352, 353, 355, *357*, *358*

Oda, Y., 176, *177*
Ogata, E., 36, *50*
Ogren, W. L., 81, 92, *98*
Okimoto, R., 268, *277*
O'Leary, M. H., 130, 131, 132, *147, 148*
Opik, H., 206, *223*
Orebamjo, T. O., 335, *345*
Osborne, B. A., 133, 134, 136, 138, 140, 144, *148*
Osborne, D. J., 54, 57, 59, 62, 63, 70, *74, 75*, *76*:299, *300*:408, *410*
Oshina, O., 382, *384*
Osmond, B. B., 335, *345*
Osmond, C. B., 83, *97*:127, *128*:133, 135, 138, 144, *147*:305, *320*
Overstreet, R., 422, *432*
Owens, R. J., 272, *277*

Page, C. N., 143, *147*
Palada, M. C., 424, 426, 430, *432*
Paleg, L. G., 342, *344*
Pallardy, S. G., 413, 414, *432*
Palmer, J. H., *75*
Papenhuijzen, C., 416, 418, *432*
Parke, M., 38, *50*
Parker, M. W., 17, *19, 21*
Parronda, R. T., 251, *253*
Pasteur, L., 266, *277*
Paton, J. A., 137, *148*
Patrick, W. H., 234, *238*:243, 244, 247, 248, *253*:321, *331*
Pearce, J. V., 137, *148*
Pearcy, R. W., 80, 81, *97*
Penuelas, J., 105, *111*:136, *148*
Percival, E., 36, 38, *50*
Petty, C. C., 8, *20*
Pfeiffer, N. E., 114, *128*
Pfeiffer, W. J., 403, *409*
Pfund, A. H., 9, *21*

Pharr, P. M., 289, 290, *291*
Phillips, J. W., 420, 422, *432*
Phillips, M. E., 305, *320*
Phillips, W. L., 186, 187, *190*
Pieterse, A. H., 55, *75*
Pieterse, A. J., 189, *190*
Pollak, G., 341, *345*
Ponnamperuma, F. N., 321, 322, *331*:412, *432*
Poo, W. K., 175, *177*
Poovaiah, B. W., 70, *75*
Popp, M., **333–43**:337, 338, 339, 340, 341, 342, *344*
Postek, M. T., 239, *253*:408, *410*
Pradet, A., 218, *223*:268, 272, 274, 277:325, *331*:397, 404, *410*:422, 428, 429, *433*
Prahl, C., 100, *112*:138, *148*:305, *320*
Pratt, L. H., 4, *21*
Pregnall, A. M., 406, *410*
Price, L., 9, *21, 22*
Pringsheim, E. G., 39, *50*
Prins, H. B. A., 83, *97*
Proctor, M. C. F., *95*:105, *111*:136, 137, *146*, *148*
Pryor, A. J., 282, *292*

Quail, P., 4, *21*
Quebedeaux, B., 260, *264*
Quillet, M., 38, *50*:356, *357*
Qureshi, F. A., 421, *432*

Rainina, E. J., 220, *222*
Randerson, P. F., 239, *253*
Ranson, S. L., 266, *276*
Raskin, I., 61, 63, 64, 66, 67, 74, *75*:266, *277*:305, *320*:413, 427, *432*
Raven, J., 36, *50*
Raven, J. A., **129–46**:81, 82, 83, 84, 90, 95, *97*:101, *111*:127, *128*:130, 131, 132, 133, 134, 135, 136, 138, 140, 141, 142, 143, 144, *146, 147, 148*:305, *320*
Ray, T. B., 144, *148*
Raymond, B., 218, *223*
Raymond, P., 268, 274, *277*:325, *331*:422, 428, 429, *433*
Rebeille, F., 270, *277*
Rebsdorf, Aa., 101, *111*
Reed, R. H., 36, 41, 42, 43, 44, 45, 46, 48, *51*
ap Rees, T., **227–37**:206, *222, 223*:228, 229, 230, 231, 232, 233, 235, *237, 238*:246, 247, *252, 253*:397, 405, 406, *409, 410*:422, 423, *432, 433*
Reggiani, R., **255–63**:260, *264*
Reid, D. M., **385–94**:299, *301*:386, *395*:413, 414, *433*
Reynolds, E. S., 207, *223*
Rhebergen, L. J., 44, *49*
Richardson, K., 126, *128*:138, 140, 142, 143, 144, *148*
Ridge, I., **53–74**:55, 57, 58, 59, 62, 63, 64, 68, 69, 75, *76*:294, *300*
Ried, A., 37, 42, 44, *51*
Rietema, H., 360, *373*

Riley, J. P., 36, *51*
del Rio, L. A., 380, *384*
Ritchie, A. F., 418, 422, *432*
Ritchie, R. J., 36, *51*
Robards, A. W., 329, *331*
Roberts, J. K. M., 231, 234, 237, *238*:255,
    264:266, 267, *277*:323, *331*:376, 379, *384*
Roberts, K., **265–76**:268, *276*
Rodden, R., 408, *410*
Roelofs, J. G. M., 89, *97*
Roeske, C. A., 131, *148*
Rose-John, S., 61, *76*
Rosenberg, R. M., 131, *148*
Rozema, J., 341, *345*
Rozenberg, G. V., 9, *21*
Rugg, D. A., **347–56**
Rumpho, M. E., **193–202**:193, 194, *202*,
    203:266, *276*:377, *384*:400, *410*
Rusanowski, P. C., 351, *357*
Russell, G., **35–48**:36, 38, 40, 42, 44, 45, 47, 48,
    51, *52*

Sachs, M. M., 268, *277*
Saenger, P., 335, 337, 341, 342, *345*
Saglio, P. H., 218, *223*:268, 271, *277*:325,
    331:*397*:404, *410*:417, 421, 422, 428, 429,
    433
Sahling, P.-H., 362, *372*
Saker, L. R., 255, *263*:327, 328, *331*:430, *431*
Sale, P. J. M., 294, *301*
Salisbury, F. B., 17, *21*
Salvucci, M. E., 81, 82, 84, 85, 86, 88, 90, 91,
    92, 93, 94, *96*, *97*:99, 100, 108, *111*:146,
    148
Samarakoon, A. B., 59, 63, *75*, *76*
Sanderson, J., 329, *331*
Sand-Jensen, K., **99–111**:84, 89, 93, 95, *97*,
    98:99, 100, 102, 103, 105, 106, 107, 108,
    109, 110, *111*, *112*:138, 141, *148*:305, *320*
Schäfer, E., *19*:161, *165*:169, *177*
Schmidt, B., 341, *343*
Schmidt, B. L., *95*
Schmidt, K., 5, *20*
Schneller, J., 182, *190*
Scholander, P. F., 337, 339, 341, *345*
Scholander, S. I., 337, *345*
Schonbeck, M., 37, 39, 42, *51*
Schonbeck, M. W., 348, 349, 350, 351, 352,
    353, 355, *357*, *358*
Schramm, W., 33, *34*:36, 37, 42, *50*, *51*
Schuurkes, J. A. A. R., 89, *97*
Scott, G. D., 355, *358*
Sculthorpe, C. D., 54, *76*:81, 85, 88, *97*:115,
    128:138, 140, 143, *148*:180, 185, *190*
Seago, J. L., 402, *409*
Seamon, D. E., 193, *202*
Segestrale, S. G., 36, *51*
Seliger, H. H., 11, *21*
Seneca, E. D., 239, *253*
Senger, H., 5, *21*
Seshu, D. V., 412, 424, *432*

Setter, T. L., **411–31**
Seyfried, M., 14, *21*
Sheath, R. G., 134, *148*
Sheldon, R. B., 83, *96*:125, *127*
Shropshire, W., Jr., 6, *21*
Sibasaki, T., 176, *177*
Siegelman, H. W., *19*:155, *165*
Sifton, H. B., 172, *177*
Silk, W. K., 65, *76*
Simpson, G. M., 173, *177*
Sircar, S. M., 260, *264*
Sisworo, E. J., 255, *263*
Sharma, K. P., 174, *177*
Slocum, C. J., 371, *373*
Smirnoff, N., 342, *345*:407, *410*
Smith, A. J. E., 136, 137, *148*
Smith, A. M., **227–37**:206, *223*:228, 229, 234,
    238:246, 247, *253*:290, *292*:405, *410*:422,
    433
Smith, B. N., 144, *148*
Smith, F. A., 36, *50*:85, *97*:131, 133, *148*:413,
    426, *433*
Smith, H., 4, 6, 7, 8, 9, 12, 15, 17, *20*, *21*:153,
    154, 155, 158, 161, *165*
Smith, J. A. C., 132, *147*
Smith, L. P., 33, *34*
Smith, M. W., 268, *277*
Smith, R. C., 9, *21*:158, *166*
Smith, S. E., 36, *50*
Smits, A. J. M., 89, *97*
Snkhchian, H. H., **205–22**
Søndergaard, M., 93, *98*:138, *148*
Spain, A. V., 340, *345*
Spalding, M. H., 81, 92, *98*
Spanner, D. C., 421, *432*
Spanswick, R. M., 376, *383*
Spence, D. H. N., **153–65**:167**–76**:9, 10, 11, 12,
    13, 17, 18, *20*, *21*:82, 83, 84, 85, 86, 87, 88,
    90, 94, *95*, *97*, *98*:101, 103, 104, 106, 107,
    111, *112*:136, *146*:154, 155, 156, 157, 158,
    159, *165*, *166*:172, 174, 176, *177*
Spencer, W. E., 82, 83, 88, 90, 94, *98*
Spreitzer, R. J., 81, 92, *98*
Sprent, J. I., 144, *148*
Spruit, C. J. P., 8, *21*
Stanley, R. A., 83, 90, 92, *98*
Staverman, W. H., 130, 135, *147*
Steeman-Nielsen, E., 83, 84, 86, 92, *98*:101,
    106, *112*
Steinmann, F., 294, 295, 299, *301*:398, 402,
    403, *410*
Stelzer, R., 337, *345*
Steneck, R. S., 371, *373*
Stephenson, R. L., 142, *148*
Sternberg, L., 143, *148*
Stewart, G. R., **333–43**:335, 337, 338, 339, 342,
    345
Stewart, W. D. P., 36, *51*
Stokholm, H., 305, *320*
Storey, R., 338, *345*
Streeter, J. G., 262, *264*
Stuckey, R. L., 186, 187, 188, *190*

Studer, C., **293–9**:293, 294, 295, *301*:305, *320*:401, *410*
Stumm, W., 101, *112*
Subrahmanyan, R., 348, 350, *358*
Suge, H., 59, 66, 74, *75*, 76:87, *98*
Sutherland, G. K., 354, *358*
Szalados, J. E., 130, 133, 135, *147*

Taiz, L., 63, *76*
Takada, H., 36, *50*
Takimoto, A., 17, *21*
Tan, F. C., 142, *148*
Tanner, C. B., 4, *20*
Tasker, R., 161, *165*
Taylor, D. L., 266, 277:420, *433*
Taylor, S. E., 338, *343*
Taylor, V. J., 408, *410*
Taylorson, R., 168, *177*
Teal, J. M., 240, *253*
Teeri, J. A., 115, *128*
Terry, N., 338, *343*
Thomas, B., 161, *166*
Thompson, K., 56, *76*:171, *177*
Thorne, R. F., 116, *128*
Ting, I. P., 80, *97*:114, 118, *128*
Tolbert, N. E., 90, *96*
Tomlinson, P. B., 335, *344*
Topa, M. A., 281, *292*
Townsend, L. R., 262, *264*
Trewavas, A. J., 428, *431*
Tripepi, R. R., 246, 247, *253*
Troke, P. F., 338, *344*
Trought, M. C. T., 255, 263, *264*:417, *433*
Troughton, J. H., 132, *149*
Tsuji, H., 206, 217, *223*:419, *433*
Tsuzuki, M., 81, 92, *98*
Turner, J. S., 421, *433*
Turner, R. E., 239, *253*
Turner, T., 337, *344*
Tyler, J. E., 9, 158, *166*
Tyler, M. J., 9, *21*
Tyler, P. D., 234, *238*:406, *410*:423, *431*
Tyron, A. F., 114, *128*
Tyron, R. M., 114, *128*

Ueda, K., 206, 217, *223*
Uhler, F. M., 181, *190*
Ultsch, G. R., 144, *149*
Urmi-König, K., 182, 183, 188, *189*

Vadas, R. L., 351, *357*
Van, T. K., 82, 83, 84, 86, 88, 89, 90, 92, 95, *96*, *98*:106, *112*
Varner, J. E., 255, 258, *263*
Vartapetian, B. B., **205–22**:206, 207, 217, 220, 221, *223*:270, 271, *276*, *277*:397, *410*:412, *433*
van der Velde, G., 181, *190*
Vergara, B. S., 424, 426, 430, *432*
Veschuren, G., 324, *331*:430, *432*
Vierstra, R. D., 4, *21*
Vierssen, W. van, 173, *177*

Vince-Prue, D., 17, *21*
Vogelmann, T. C., 14, 15, 16, 17, *21*

Wagner, E., 5, *20*
Wahlefed, A. W., 284, *291*
Waisel, Y., 101, 102, *111*
Walker, C. M., 125, 126, *128*
Walker, D. A., 257, *263*
Walker, N. A., 85, *97*:131, 133, *148*:413, 426, *433*
Wallace, P. G., 220, *223*
Wallenstein, A., 87, *98*
Walters, J., 59, 60, 61, 63, 67, 68, 69, *75*, *76*
Wample, R. L., 385, *395*
Warburg, O., 266, *277*
Ward, C. H., 158, *165*
Waters, I., **411–31**
Watson, K., 220, *223*
Watters, H. D., 414, *433*
Webb, T., 206, *223*:303, 304, *320*:412, 429, *433*
Webber, F. C., 354, 355, *358*
Weber, J. A., 19, *21*:176, *177*
Weeks, D. C., *19*:87, 88, *95*:156, 157, *165*:176, *177*
Weigel, P., 339, *344*
Weinberg, S., 9, *22*
Wenkert, W., 414, *433*
West, J. A., 44, *51*:360, *372*
Westlake, D. F., 88, *98*
Wetzel, R. G., 85, 90, *95*, *96*:118, 122, *127*, *128*:294, *301*
Wiegert, R. G., 239, *253*
Wiencke, C., 48, *50*, *51*
Wignarajah, K., 246, *253*
Wilkinson, M., 42, 45, *52*
Williams, C., 281, *292*
Williams, W. T., 55, *76*
Wilson, P. M., **227–37**:206, *222*:231, 232, 233, 235, *237*:397, 406, *410*:423, *432*
Winston, R. D., 176, *177*
Winter, E., 340, *345*
Wiskich, J. T., 342, *344*
van Wisselingh, C., 183, *190*
Withrow, R. B., 5, 9, *21*, *22*
de Witt, M. C. J., 416, 420, *433*
Wium-Andersen, S., 93, *98*:100, *112*:138, 149:315, *320*
de Wolf, A., 324, *331*
Wolf, P. L., 403, *409*
Woo, K. C., 335, *345*
Woolhouse, H. W., 234, *238*:290, *292*
Wyn Jones, R. G., 36, 41, *49*, *52*:338, *345*

Yang, S. F., 294, 295, *300*, *301*
Yarish, C., 44, 48, *52*
Yellow, D., 220, *223*
Yeo, A. R., 338, *344*
Yeo, R. R., 174, *177*
Yoder, O. C., 258, 262, *263*
Yoshida, S., 315, *320*:424, *433*
Young, A. J., 44, 46, *52*

Young, M., 356, *358*
Young, S. F., 324, *331*:408, *410*
Yund, P. O., *372*

de Zacks, R., 63, 64, *75*
Zanefeld, J. S., 37, 38, 42, *52*
Zedler, P. H., 116, 122, *128*
van der Zee, D., 193, 194, *202*:266, *276*

Zenkevitch, L., 36, *52*
Zieman, J. C., 240, *252*:305, *320*
Zimmerman, B. K., 5, *22*
Zimmermann, U., 36, *52*
Zochowski, Z. M., **375–83**:376, 377, 381, *384*:398, 399, 408, *409*
Zubatov, A. S., 220, *222*
Zwar, J. A., 231, *238*

Subject index

Figures in italics refer to tables or figures.

Abscisic acid, 87, 89, 390, 394, 414
Acanthus, 341
  A. ilicifolius, *338*, *339*
ACC *see* 1-aminocyclopropane- 1- carboxylic
  acid
Acer pseudoplatanus, 270–1
Acetaldehyde production, 380
  in *Glyceria maxima*, anoxic and post-anoxic,
    380
  in marsh plant rhizomes, 293, 294, 296, 297
Acorus calamus, 188
  rhizomes, 293, 398, 401, *402*
    ACC concentration, 297–8, 299
    acetaldehyde release, *296*, 297
    ADH activity, 405
    carbohydrate reserves, 403
    ethanol release, *295*, 297, 299
    ethylene release, *297*, 299
    malonyl-ACC content, *298*, 299
Acorus clanus, 381
Acrostichum aureum, 333, 339
Acrosymphyton purpuriferum, 360, 371
ADA, 61
Adenosine diphosphate, (ADP)
  ATP/ADP ratio
    and carbohydrate levels, 413, 428–9
    in maize roots, *326*, 327
    in pea roots, *274*
Adenosine disphosphate (ADP) (*cont.*)
    levels, pea roots, *274*
    in *Spartina alterniflora*, hypoxic roots, 250
Adenosine triphosphate (ATP)
  and carbohydrate levels, 413, 428–9
  in fermentation, 227, 228
  in glycolysis, 268
  levels, anoxic conditions, 227–8
    in maize, *326*
    in pea roots, *273*, 274, 276
  levels, hypoxic conditions
    in *Spartina alterniflora*, 250
Adenylate energy charge (AEC)
  of anoxic pea roots, 273–4
  of anoxic rhizomes, 404
  in *Echinochloa* sp., 198, *199*
  of hypoxic *Spartina alterniflora* root, 244,
    249–51
  of maize roots, 325–7
Adenylic energy charge, 257
  and CO$_2$ evolution, nitrate affected, 261–2,
    263
ADP *see* Adenosine diphosphate

Adventitious roots, and flooding acclimation,
    322–30
Aegialitis, 335, 337, 339
  A. annulata, *338*, *339*, *340*, *341*
Aegiceras, 335, 339, 342
  A. coriculatum, *338*, *339*, *340*, *341*
Aerenchyma, 55, 182, 305
  in adventitious shoots, 322–4
  in *Eriophorum angustifolia*, 303, 306–8, 318,
    319
  in *Glyceria maxima*, root oxygen supply, 233,
    234, 397
  internal oxygen transport, 306–8, 325–7
  of marsh plants, 293–4
  rhizome, ventilation by, 397, 401, 406–8
  of roots, ion transport function, 327–9
  in *Spartina alterniflora*, 239, 240, 247, 305
Aerial leaves, 79, 80, 81, 82, 87
  in heterophyllous plants, 86–8, *89*
  photosynthetic function, 89, 90, 94–5
  production, phytochrome controlled, 156
Alanine production, anoxic roots, 229, 234, 406
β-alanine, uptake and leakage, flooded
    sunflowers, 389–90, 394
Alcohol dehydrogenase (ADH), 255
  and flooding tolerance, 228, 230–1, 282
    in loblolly pines, 286–7, *288*, 289
    in mangroves, 336, 337
    in rice and rice weeds, 193, 196–8
    in *Spartina alterniflora*, *244*, 245–7, 248,
      249, 251
    in wetland rhizomes, 405, 406
Alcoholic fermentation, 228–31, 234, 422, 429
  in loblolly pine, 289
  in marsh plants, 293–9
  in *Spartina alterniflora*, 246, 251, 252
  see also under products of fermentation
Aldrovanda vesiculosa, 182
Algae
  benthic, light climate model, 23–34
  bicarbonate utilization, 92–3, 106, 107
  brown *see* Brown algae
  desiccation tolerances, 37–42
  foliose, 23, 26, 28, 32
  green, 81, 92
  photoperiodism in, 17, 19, 360, *362*, 364,
    368, 370–2
  red, 13, 26, 358
    phenology, environmental controls, 361–72
Alisma plantago-aquatica, 59
Alopecurus geniculatus, 58

443

Amaranthus caudatus, 170
Amino acid release, anaerobic conditions, *257*
Amino butyric acid, 406
1-aminocyclopropane-1-carboxylic acid, 61,
    324, 408
    in marsh plant rhizomes, 293, 294, 297–8,
    299
Amphibolis, 183
Anabaena, 129
    A. azollae, carbon isotope values, 144, *145*
Angiosperms, 17, 81, 88, 90, 106, 239
    freshwater, photomorphogenesis, 13, 153–65
    and temperate environment, 19, 156, 159–60
Animals, diaspore dispersal, 181–2
Apium nodiflorum, *59*
Aponogeton, 180, 183
Aquatic acid metabolism, 114
Armeria maritima, *341*
Arundo, 179
Aschynomene, 181
Ascophyllum nodosum, 107
    carbon isotope values, 141–3
    fungal relationship, 355–6
Asteraceae, 179
ATP *see* Adenosine triphosphate
Atractophora hypnoides, reproduction, 359,
    363–6
    photoperiodic response, *364*, 370–1
    seasonality, *366*
    temperature affected, 365–6, 370–1
Auxin, 63–4, 70, 72
Avicennia, 333, 335, 337, 339, 342
    A. germinans, 335, 337
    A. marina, 227, *235*, 237, *338*, *339*, *340*, *341*
    A. nitida (*germinas*), 335, *336*, 338, *341*
Azolla, 129
    A. caroliniana, carbon isotope values, 144–5

Baldellia ranunculoides, 180
Baltic Sea, algae of, 47–8
Barley *see* Hordeum
Batrachian Ranunculi, 129, 140–1, 146
Betula erecta, *59*
Betula nigra, 246
Betula pendula, 246
Bicarbonate utilization, 99, 101
    carbon isotope discrimination, 131, 133–4,
    135, 140–3
    concentrations, and water type, 101
    measuring techniques, 101–3
Bicarbonate utilization (*cont.*)
    and oxygen supply, *Eriophorum
    angustifolium*, 315
    plasticity of use, *Elodea canadensis*, 107–10
    in submerged macrophytes, 79, 80, 81, 83–5,
    89, 90, 92–3, 106–11
Bidens tripartia, 171
Birds, and plant dispersal, 181–2
Blyxa, 181
    B. aubertii, *185*

Boundary layer effects, carbon acquisition, 85,
    86, 133, 134, 136, 142
Brackish waters, 36
Brasenia schreberi, 19
Brassica, 377, *378*
    B. napus, 322
Brown algae, 13, 26
    desiccation studies, 37–41
    evolution, and salinity, 47–8
    gels, role of, 40–2
    gene flow, 44–5
    sexual reproduction, and salinity, 41–2
    thalli changes, 35, 36, 38, 39–41
    zonation, and salt tolerance, 42–4
Brugiera, 333
    B. exaristata, *340*
    B. gymnorhiza, *338*, *339*
Bryophytes, CO_2 sources, *104*
Bud extension, with anoxia, 399–400, 408
Buoyant tension, and elongation, 67–70
Butomus umbellatus, 188

C_3 photosynthesis, 80, 90
    carbon isotope discrimination
        aquatic plant, 134–43, 144–5
        terrestrial plant, 131–2
    of *Isoetes*, 114, 118
    of mangroves, 334
C_3–C_4 intermediates, 80, 133
C_4 acid system, 79, 80–1, 89, 91–2
    photosynthesis, 132, 334
    terrestrial plant, carbon isotope values, 132
Cabbage, 377, *378*
Cadiscus, 179
Callithamnion byssoides, 370
Callitriche, 72, *121*, 181
    C. platycarpa, 57, *59*, 60, 62, 63, 87
    C. stagnalis, *59*, 66, *100*, *102*, *103*, *104*, *106*
Caltha palustris, *59*, 71
Calvin cycle *see* Photosynthetic carbon
    reduction cycle
Capsicum, 414
Carbohydrate levels, 411
    and anaerobic metabolism products, 262, 263,
    412–13, 422–3
    and anoxia endurance, *Glyceria maxima*, 398
    and diel fluctuation, in *Isoetes*, 113, *120*
    factors affecting. 413–17
    and irradiance levels, 393, *416*
    of maize, 218
    and rhizome habitat specialization, 402–4
    of rice plants, 424–7
    and sugar concentrations, with hypoxia,
    417–23
Carbon
    budget construction, 418–22
    pathways, 79–80
        and isotope discrimination, 131–2
Carbon dioxide
    and anoxic injury, 377–9, 383

anoxic root production, 229, 234, 235, 288
  in loblolly pine, 286–7, *288*, 289, 290
  with nitrate, 261–2
bicarbonate conversion to *see* Bicarbonate utilization
and carbohydrate status, 413, 414, 426, 429, *430*
compensation point, 5
  in submerged aquatic macrophytes, 79, 81, 90, 92, 94, *104*
dark fixation, 91–2, 93, 115, 118
diel related availability, 115–20
diffusion resistance, 79, 85
and heterophylly, 88
hydrosoil, utilization, 93–4
and internal aeration, 303, 314–17
and isotope discrimination, 130–4
  plant studies, 134–45
K_m values, 79, *86*, 92–3, 94
-oxygen equilibrium, in water, 85–6
and photosynthesis, 4–5, 80–2, 89
sources, submerged macrophytes, 103–6
submerged *Eriophorum angustifolium*, radial oxygen loss, 313–17
submerged shoot elongation, 65–6, 73, 74
uptake, CAM species, 114–20, 123, 126
utilization, non HCO_3-users, 99, 103–6, 113
vascular plants, as users, *104*
Carex alternifolius, 399
Carex arenaria, 329
Carex papyrus, 399
Carex rostrata, *399*, 407
Carpophyllum, 107
Catabolic reducing charge, in anaerobic metabolism, 255, 257, *260*, 261, 262, 263
Catalase activity, anoxic and post-anoxic phases, 380–3
Cell division, elongation responses, 64–5, 71
Cell expansion, elongation responses, 62–4, 69–70, 72–3
Centella, 181
Ceramium rubrum, 43–4, *102*, *106*
Ceratophyllum, 183, *185*
  C. demersum, *106*
Ceratopteris, 182
Cereal crops, acclimation to flooding, 322–30
Ceriops, 333, 335, 342
  C. tagal, *338*, *339*, *340*
Chaetomorpha linum, *102*
Chickpea, *see Cicer arietinum*
Chilling requirement, in germination, 171, 174
Chlorine accumulation, in mangroves, 337–41
Chlorophyll
  and bicarbonate utilization, 99, 108, 110, 111
  levels, *Isoetes howellii*, *123*
  and radiation absorption, 5, 6, 8
Cicer arietinum, 375, 399
  anoxic seedlings, ethanol affected, 377–9
Circulating anaerobic atmosphere, 375–83, 399
Cladophora rupestris, 43–4

Conocarpus, 333
Cotton, 375, 376
Cotton grass, 305
Cotula, 179
  C. coronopifolia, 188
Crassula, 89, 91, 126
  C. aquatica, *121*, 122
  C. erecta, 122, 123
  C. falcata, 14, *15*, *16*, *17*
Crassulacean acid metabolism (CAM), 80, 91–2, 113
  in *Isoetes*, 114–26, 143–4
  photosynthesis, carbon isotope values, 132, 143–4
Cruoria rosea, 362
Crustose corallines, 26
Cryptochrome, and shoot elongation, 161–4
Cucurbita, 412
  C. pepo, 14
Cucumis sativus, 170
Cyanobacteria, 81
  of *Azolla caroliniana*, 144–5
Cyperaceae, 179, 305
Cytokinins, 414

Damasonium, 181
Dark CO_2 fixation, 91–2, 93
  in *Isoetes*, 115, 118, *120*
Dark reactions of phytochrome, *17*, 161
Dark respiration, *Eriophorum angustifolium*, 311, 312, 313
Depth accommodation, 55–60
  and fluence rate, 153–65
Diaspores
  floating, 180–1
  polymorphic, 185–6
  sunken, 180, 182–3
  of *Typha*, 179–80
Dispersal mechanisms
  animal, 179, 181–2
  diaspore, polymorphic, 185–6
    sunken, 180, 182–3
  and range extensions, 186–8
  seed specialization, 183–5
  by water, 180–1
  by wind, 179–80
Dissolved inorganic carbon (DIC), 79, 133
  and bicarbonate measurement, 101–3
  concentrating mechanism, 81, 92–3
  and carbon isotope values, 129
  and photosynthetic rates, 82, 83, 86, 88–9, 95, 99
Dormancy, 167, 172, 176
  photoperiodic introduction, 19
Downingia cuspidata, *121*
Drepanocladus aduncus, *104*
Dumontia, 360

Echinochloa, 182, 200–2, 400
  adenylate content, 198, 199, 201
  ADH activity, 193, 194, 197–8, 201

Echinochloa (cont.)
  crus-galli, 193–201, 206, 217
  crus-pavonis, 196, 197, 198, *200*, 201
    ethanol production, 193, 194, 196
    germination requirements, 194–5, 201
    metabolic inhibitors, 193, 198, 200
  muricata, 194, 196, 197, 198, *200*, 201
  oryzicola, 197
  phyllopogon, 193
  utilis, 197
Echinodorus, 182
Eclipta, 179, 181
Egeria, 183
Egeria densa, 91, *106*, 188
Eichornia crassipes, 55, 144
Elatine, *121*, *183*
Eleocharis, 182
  E. acicularis, *121*
  E. coloradoensis, 174
  E. macrostachya, *121*
  E. palustris, *399*, *402*
Elodea, 183
  E. callitrichoides, 188
  E. canadensis, 91, 99, 105, *106*
    bicarbonate use, 107–10, 111
  E. nuttallii, 188
Elongation responses, 55–60, 70–2
  and buoyant tension, 67–70
  CO_2 and O_2 induced, 65–6
  ethylene-induced, 53, 60–5, 72–4, 87, 413
  fluence rate-dependent, 153, 158–65
  light-induced, 66–7
Enteromorpha, 107
  E. intestinalis, *142*, 143
    salinity tolerance, 35, 44, *45*
Epilobium hirsutum, *59*
Erigeron, 179
Eriphorum, 179
  E. angustifolium, 303, 305–8
    radial oxygen loss, 308, 309, *311*, *312*,
      313–17
    root aeration, partial submergence, 308–17
    total submergence, 317–19
Ethanol
  and anoxic injury, 375, 376–9
  in anoxic roots, 228, 229, 234, 235
  in hypoxic roots, *Spartina alterniflora*, *244*,
    248–9, 252
  leakage, and carbohydrate status, 418, 422–3
  in loblolly pine, 282, 284, 287, *288*, 289
  in marsh plant rhizomes, 293, 295, 297, 298
  and pH, in maize roots, 267
    in pea seedlings, 267
  and post-anoxic injury, rhizomatous plants,
    379–83
  in rhizomes, 405–6
  in rice and rice weed germination, 195–6
Ethylene
  and aerenchyma formation, 324, 408
  and cell lysis, 323, 324
  gas entrapment, 60–1, 62
  –gibberellin interactions, 63, 64

  leaf age, and sensitivity to, 71
  production, marsh plant rhizomes, 293, 294,
    297
  and shoot elongation, 53, 60–5, 72–4, 87, 413
    mechanism of action, 62–5
Eurhynchium rusciforme, *104*
Excoecaria, 337
  E. agallocha, *339*

Fatty acid synthesis, anaerobic conditions,
  270–3
FCCP, and energy charge, 273, *274*
Fermentation pathways, 228–31
Filipendula ulmaria, *399*, 407
Flavoprotein photoreceptor, 168, 174
Flax, 377, *378*
Flood-tolerant plant categories, 206–7
Fontinalis, 129
  F. antipyretica, *104*, 105
    carbon isotope discrimination, *135*, 136–7
Freshwater, carbon isotope values, 130
Freshwater macrophytes, bicarbonate use, 99,
  106–11
Fucus, 347, 348
  F. serratus, light climate, 33–4
    salinity, 38–41
  F. spiralis, competition with *Pelvetia
    canaliculata*, 348–51
    salinity, 38–41
  F. versiculosus, 38–41, *103*, *106*
Fungi, 354–6

Gelbstoff (yellow substance), 10, 11, 14
Gene activation, and anaerobic response, 268,
  276
Germination
  and anoxia, *Zea mays* mitochondria, 205–22
  and hydration, 168
  and light, 168–70
  and oxygen, 172–3
  of rice and rice weeds, 193–202
  and temperature, 170–1, 173
  under water, 172–6
Geum rivale, *59*
Gibberellins, 414
  and aerial leaf production, 87
  and ethylene interaction, 63, 64, 72
Gigartina acicularis, 360
Glaux maritima, *341*
Glucose, exogenous, in anoxia, 412–13
  metabolism, mangrove roots, 235
Glyceria maxima
  ACC concentration, 297–8, 299
  acetaldehyde, production, *296*, 297, *380*
  ADH activity, 230–1, 246
  anoxia endurance, 398, *399*, 401
  carbohydrate reserves, 398, 404
  catalase activity, anoxic and post-anoxic, 380–3
  ethanol production, *295*, 297, 299, *380*
  ethylene production, *297*
  fermentation pathways, 228–9

habitat, *402*
malonyl-ACC content, *298*
sucrose metabolism, with anoxia, 227, 231–4
Glycerol with anoxia, 228, 234
Glycine max, 375, 376, *378*
Glycolysis, 228, 229
and carbohydrate levels, 414, 422
and mitochondrial membrane formation, 205, 220–1
pH affected, 266–8
rhizome metabolism, 405, 408
Glycosidic acidosis, 376, 379
Gynerium, 179

Habitat, 80
algal, and salt tolerance, 42–4
of *Isoetes*, *124*
rhizome specialization, with anoxia, 397–402
Halarachnion ligulatum, reproduction, 362–3
photoperiod affected, *362*, 369–70
seasonality, *363*
temperature affected, 362, 369–70
Halophila, 185
Halymenia latifolia, reproduction, 359, 361, 367–9
mechanical disturbance, 367, 371
photoperiod affected, *368*, 371
seasonality, *369*
temperature affected, *367*, 371
Heat stability, of enzymes, 342
Helianthus annus, 322, 324
responses to flooding, 385–94
β-alanine leakage, 389–90, 394
pH affected, 386–9
potassium uptake and leakage, 390–2, *394*
Heritiera littoratis, *339*
Heterophylly, 80, 81, 82, 86–8, *89*, 90
in *Hippuris vulgaris*, 156
Hibiscus tiliaceus, *339*
High irradiance reaction, 170
Hippuris, *89*, 183
H. vulgaris, 19, 87, 88, *104*
aerial leaf, change to, 153, 156
shoot elongation in, 159–65
Hordeum, 231, 255, 258, 263, 322
carbon budget, 419, *420*, 421, 422
sugar concentrations, *416*, 417
Hormones, 413
and cell expansion, 62–4, 70
and heterophylly, 87
Hyale nilssoni, 354
Hydration, and germination, 168
Hydrilla, 188
H. verticillata, 146
dispersal, 1791
photosynthesis, 79, 83, *86*, 88, *89*, 90, 91, 92
and bicarbonate use, *106*
stem tubers, 156, 158, 176
Hydrocharis, 57, 182, 183, 189
H. morsus-ranae, *59*, 60, 186

Hydrogen ion concentration *see* pH
Hydropectis, 179
Hydrophytes, 206
Hydrosoil carbon dioxide, 93–4
Hygrophila polysperma, 83, 88, *89*, 94
bicarbonate utilization, *84*
Hygrophytes, 206

Ion transport, aerenchymous roots, 327–9
Iris germanica, 273, 376, 379, *381*, 405–6
Iris pseudacorus, 273, *381*, 398, 399, *402*, 404, 406
Isoetes, *89*, 91, 113–14, *121*, 139
amphibious, adaptive radiation in, 123–5
I. andicola, 126, *144*
carbon capture, 100, 104, 129
I. durieu, 126, *135*, 144
I. howellii, 79, 92, 95, 113, 125
CAM in, 114–23
lacustrine, adaptive radiation in, 125–6
I. lacustris, *104*
carbon isotope values, *139*, 143–4
terrestrial, adaptive radiation in, 126–7
Isotope discrimination, and carbon acquisition, 129
aquatic and amphibious plants, 133–45
in mangroves, 334
photosynthetic, component reactions of, 130–1
terrestrial vascular plants, 131–2

Jaegeria, 179
Jerlov water classification, 23, 24
Juncus bulbosus, *104*, *105*
Juncus conglomeratus, *399*, 401, *402*
Juncus effusus, *381*, 398, 399, 401, *402*

Kalanchoe blossfeldiana, *340*
Kandelia, 333
Kelp, 23, 32, 33, 360
Kranz anatomy, 79, 85, 86, 91

Lactate production, anoxic roots, 229, 234, 255, 266, 422, 423
of pea seeds, 267, 268
of *Schoenoplectus lacustris*, 406
Lactuca cativa, 377, *378*
Lagarosiphon major, 188
Laguncularis racema, *340*
Languncularis racemosa, *340*
Lagarosiphon, 181
L. major, 91
Laminaria digitata, 38, *40*, *142*, 143
Laminaria hyperborea, *142*, 143
Laminaria longicruris, *142*, 143
Laminarian zones, 26, 27–8, 32, 33
Leaves
aerial *see* Aerial leaves
green and etiolated, radiation, 5
phytochrome function, and radiation, 14–17
submersed *see* Submersed leaves

Lemanea, 129, 146
  L. mammilosa, carbon isotope values, 134–6
Lemna minor, 182
Lettuce, 377, *378*
Lichen symbiosis, 355
Light environments
  and algal growth, 33–4
  and carbohydrate levels, 393, 411, 412–13,
      414, *425*, 426, 429
  diurnal changes in, 5–6, 28–32
  and germination, 168–72, 174
  and heterophylly, 87, 94
  intertidal and subtidal areas, 23–34
  macroalgal phenology, 360, 361, 364–5
  and photomorphogenesis, 153–65
  and photosynthesis, 12–13, 80
  and radial oxygen loss, *Eriophorum
      angustifolium*, 313–17
  and solar elevation, 6, 9
  and submerged shoot elongation, 66–7, 73
  and temperature, 18–19
  see also Photoperiod; Radiation
Lilaea, 181
  L. scilloides, *121*
Limnobium, 182, 183
Limnocharis, *183*
Limnophila sessiliflora, 83, 94
  bicarbonate utilization, *84*
  standing crop values, 88, *89*
Limonium vulgare, *341*
Linum, 377, *378*
Littorella, *89*, 91, 126, 183
  L. uniflora, 80, 93, *104*, *105*
Lobelia, 89
  L. dortmanna, 80, 93, *104*, *105*
    carbon isotope values, 138–40, 143
Loblolly pine *see Pinus taeda*
Ludwigia, 181
Lumnitzera, 333, 339, 342
  L. littorea, *339*
Lycopersicon, 270, 322, 423
Lycopus europaeus, 171, 186, *187*
Lythrum salicaria, 171, 188

Macrophytes *see* Freshwater macrophytes;
      Marine macrophytes; Submerged aquatic
      macrophytes
Maize *see Zea mays*
Malic acid, 80
  in fermentation, 228, 229, 423
    of *Glyceria maxima*, 233, 234
    of loblolly pine, 282, 284, 287, *288*, 289
    of rhizomes, 406
    of *Spartina alterniflora*, 240, *244*, 247–8
    of *Typha* shoots, 236
  in photosynthesis, 90, 91, 92, 93
    of *Isoetes*, 113, 115, 118, 119–22, 123, 144
Malonyl-ACC, marsh plant rhizomes, 293, 294,
      298, 299
Mangroves
  ammonia assimilation, 335

glucose metabolism, 235
glutamine synthetase, 335, 336, 337, 342
nitrogen assimilation, 334–5
photosynthesis, 334
salt regulation, 337–41
succulence, 340
temperature adaptation, 341–2
Maple, fatty acid composition, *271*
Marine macrophytes, bicarbonate use, 99,
      106–7, 110–11
Marine waters *see* Seawater
Marsh plants, 171, 172, 206, 246, 247
  fermentation products, and ACC
      concentration, 293–9
Marsilea vestita, 87, 88
Mean internode length (MIL), and depth,
      159–60
Megalodonta, 179, 181
Mentha aquatica, 58, 398, *399*, *402*
Menyanthes trifoliata, 59, 398, *402*
Mesophytes, 206
Mitochondria, anoxic responses in *Zea mays*,
      207–22
Mosses, aquatic, 84, 104–5
Mycorrhizal fungi, 100
Mycosphaerella ascophyllii, 354–6
Myriophyllum, 89, 183
  M. brasiliense, 86, 88, 92, 93
  M. spicatum, 80, *84*, 91, 92, *106*
  M. verticilliatus, 19

NAD^+ and NADH levels, anaerobic root
      metabolism, 255, 258–61
Najas marina, *173*
Nasturtium officinale, 59, 60
Navarretia, 182
Neostapfia, 181
Neptunia, 181
Nicotiana, 258
Nitella flexilis, *106*
Nitrate, in anaerobic metabolism, 255–63
  and nitrite production, 258
  and root survival, 257–8
  shortage, and aerenchyma formation, 324
  utilization, in mangroves, 335
Nitrogen fixing cyanobacteria, 144–5
Nuphar, 56, 305
  N. lutea, *104*, *174*
Nymphaea, 56
  N. alba, 59
Nymphoides, 181, 182
  N. peltata, 56, 57, 59, 60–1, 63, 64, 65, 66,
      67, 68–9, 70, 71, 72
  dispersal and spread, 181, 188

Oenanthe crocota, 59
Oenanthe fistulosa, 59
Oilseed rape, 322
Orontium, 180
Oryza sativa
  adenylate content, with anoxia, *199*, 201

adenylic energy charge, with anoxia, 261–2
  with hypoxia, 246
ADH activity, with anoxia, 246
carbohydrate levels, in flooding, 411–30
carbon budget, 419–21
carbon dioxide evolution, with anoxia, 261–2, 263
coleoptile, anaerobic growth, 399–400
  ethanol production, 196
and ethylene, 61–2, 66–7, 72, 413
fatty acid synthesis, with anoxia, *271*
gene activation, 268
germination requirements, 194–5, 201, 206
ion transport in, 328–9
metabolic inhibitors, 198, *200*
nitrate reduction, 258–61, 263
roots, and nitrate supply, 256–63
  and oxygen status, *304*, 305
seedling survival, with anoxia, *378*
shoot elongation, 56, 57–8, *59*, 63, 66, 74, 413
starch levels, *415*, 427
submergence responses, 66–7, 74, 423–7
sugar levels, anaerobic conditions, *415*, *416*, 417–18, *425*, 426, 427
Osbornea octodents, *339*
Ottelia, 183
Oxydative phosphorylation, 205, 220, 227, 228, 376
  uncoupler, and pea root energy charge, 265, 273–4
Oxygen
  –carbon dioxide equilibrium, in water, 85–6
  concentrations, and carbohydrate levels, 412, 413
  and sucrose metabolism, 231–4, 236
  and germination, 172–3
  internal transport of, aerenchyma function, 325–7
  and submerged shoot elongation, 65–6, 73

Pasteur effect, 266, 413
Pea *see Pisum sativum*
Pellia epiphylla, *104*
Peltandra, 180
Pelvetia canaliculata, 38–9, *40*, 42, 347
  competition with *Fucus spiralis*, 348–51
  decay, submersion-induced, 351–4
  fungal relationships, 354–6
  narrow zone of 348–51
  volemitol synthesis, 356
Pelvetia fastigiata, 356
PEP carboxylase activity, and carbon dioxide fixation, 115, 130, 131, 132
PEPCase, and carbon dioxide fixation, 80, 91
pH
  effects on glycolysis, 266–8
  and photosynthesis, 83–5, 100–3
  effects on radial oxygen loss, 315
  of soil, plant altered, in flooding, 385, 386–9
Phalaris arundinacea, 58, *402*

Phaseolus vulgaris, 8, 16, 17, 282
  sugar levels, with anoxia, *416*, 418
Phosphate, 324, 414
Photomorphogenesis, 3, 153–4
  photoreceptors, 3–5
    see also under names of photoreceptors
  and radiation environment, 5, 13–17
  and shade, 154–5, 158–9
  and shoot elongation, 67, 158–65
  and stem tuber production, 156, 158
  and turion production, 156–8
Photomorphogenetically active radiation, 4–5
  and macroalgal growth, 364–5
  and underwater depth, 13–17
  see also Cryptochrome; Phytochrome
Photoperiod
  and dormancy, 19
  and germination, 168, 176
  and growth responses, 156–8
  and heterophylly, 87
  and macroalgal growth, 19, 359, *362*, 364, *368*, 370–2
  and radiation environment. 5–6
    underwater, 17–18, 19
  and temperature, 18, 19
Photorespiratory carbon oxidation (PCO) cycle, 79, 80, 90, 91, 92
Photorespiratory (PR) states
  bicarbonate affected, *83*, 99, 109
  and mangrove glutamine synthesis, 335
  submerged aquatic macrophytes, 76, 78, 80, 81, *84*, *86*, 90, 91, 92–3
Photosynthesis
  in benthic algae, tide affected, 23, 33–4
  and bicarbonate utilization, 99–111
  and carbohydrate levels, 410, 413, 414, 426
  and carbon isotope ratios, 130–1, 133–4
  and carbon pathways, 79–80
    constraints, aquatic habitats, 82–9
    strategies of aquatic plants, 89–95
    see also names of individual pathways
  emersed and submersed, carbon isotope discrimination, 134–46
  and light wavelengths, 4–5
  and oxygen supply, 303, 312, 313–14
  and sugar concentrations, 416
  terrestrial plants, in flooding, 413, 414, 417, *425*, 426, 429
  under water, 12–13
    constraints on, 13–17, 82–9
Photosynthetically active radiation (PAR), 4–5
  availability, canopy shade, 6, 7
  and growth responses, *157*, 159, 161, *162*, 164
  levels, and carbohydrate reserves, 393, *414*
  in turbid water, 10–11
  underwater limitations, 12–13, *155*, 156, 159
Photosynthetic carbon reduction (PCR) cycle 79, 80, *89*, 90, 91, 95, 115
Phragmites, 179
  P. australis, 293, 398, *402*

Phragmites (*cont.*)
    ACC concentrations, 297–8, 299
    acetaldehyde release, *296, 297*
    ethanol release, *295, 297, 298*
    ethylene release, *297*
malonyl-ACC content, *298*
P. communis, 399
Phycodrys rubens, 43–4
Phytochrome
    cycling rate, 161–2, 164, 168–70
    dark reactions, 161
    function, and external radiation variation,
        14–17
    and germination, 168–70, 174, *175*
    growth responses mediated by, 156–65
    and heterophylly, 87
    photoequilibrium, 6, 8, 16, *17*, 154, 161, 162,
        164, 169
        as depth sensor, *155*, 156, 158
    and photoperiodism, 17–18
    quantum requirements, 5
    and R : FR ratio, 4, 6–8, 14–17, 18, 87, 94,
        154–6
    and shading, 6–8, 154, 164, *175*
    and shoot elongation, *Hippuris vulgaris*, 161–5
    spectral requirements, 4
Phytoplankton, 10, 11, 13, 17
Pilayella littoralis, 35
    cell mortality, and salinity, *43*
    esturine and marine, salinity tolerance, *46*
    populations, gene flow, 44–5
Pilularia americana, 121
Pilularia globulifera, 104
Pinus elliotti, 282
Pinus taeda, flooding regimes
    ADH activity, 286–7, 288, 289
    CO_2 production, 286–7
    dry weight, 285–6
    effects of phosphorus, 282, 283, *286*
    ethanol production, 282, 287
    growth rates, 285, 289, 290
    malate production, 282, 287
    root morphology, 285–6
Pistia, 182
Pisum sativum, anaerobic responses, 265–76,
        414
    ADH activity, 230–1, 246
    endoplasmic reticulum (ER), *269, 270, 275,*
    energy charge, 273–4
    ethanol and lactate synthesis, 267, 268
    fatty acid synthesis, 271–3
    fermentation, 228–9
    gene activation, 268, 276
    glucose addition, 412
    root sucrose metabolism, 227, 231–4
    ultrastructure changes, 268–70
Plagiobothrys undulatus, 121
Plantago major, 53, 54, *59*, 71, 170
Plumaria elegans, 43–4
Poaceae, 179
Poa pratensis, 70

Podostemonaceae, 183
Pollen, wind dispersal, 179
Polygonum, 58
    P. amphibium, 59
Polymer synthesis, flood-tolerant plants, 227,
        231–7
Polysiphonia violaceae, 102
Potamogeton, 179
    P. amphifolius, 86
    P. crispus, 17, 18, *106*
        turion formation, 19, 153, 156–8
    P. distinctus, 59, 66, 87
    P. lucens, 106
    P. natans, 104
    P. nodosus, 87
    P. obtusifolius, 19
    P. pectinatus, 100, 103, 106
    P. polygonifolius, 104
    P. richardsonii, 153, 158–9
    P. schweinfurthii, 174
Potassium
    transport, and aerenchyma, 327–9
    uptake and leakage, flooded and non-flooded
        roots, 390–4, 422
Prasiola stipitata, 43
Promitochondria, 208, *209*, 220, 221
Proserpinaca, 89
    P. palustris, 87, 88
Pteridium aquilinum, 376
Pteridophytes, 84
Puccinella peisonis, 337
Pumpkin, 14, 412
Pyridine nucleotide pools, in anaerobic
        metabolism, 255, 258–61

Racomitrium aciculare, 104
Radiation, 3–5
    aerial, 5–6
    attenuation, in water, 8–12
        and depth, equation for, 13
    diurnal changes to, 6, 8, 17–18, 24
    measurements, interpretation, 5
    and shading, 6–8, 82, 154–6, 164, 169, 174,
        175
    and underwater photosynthesis, 12–13
    see also photomorphogenetically active
        radiation; Photosynthetically active
        radiation
Ranunculus aquatilis, 121
Ranunculus baudotii, 135, 141
Ranunculus flabellaris, 87, 88
Ranunculus flammula, 59, 398, *402*
Ranunculus fluitans, 74
Ranunculus lingua, 399
Ranunculus lingus, 59
Ranunculus penicillatus, 135, 140–1
Ranunculus repens, 53, 58, *59*, 64, 71, *399*, 402
Ranunculus rionii, 182
Ranunculus sceleratus, 54, 57, *59*, 60, 63, 67, 71
    fermentation pathway, 227, 228–9, 246
Ranunculus trichophyllous, 135, 141

Regnellidium diphyllum, 57, *58*, *59*, 60–1, 62, 63, 64, 65, 67, *68*, 69, 71, 72, 73
Reussia, 181
Rhizomes
    acetaldehyde levels, 293, *296*
    aerenchyma, ventilation by, 404–8
    carbohydrate reserves, 398, 402–4, 413
    ethanol levels, 293, 294, *295*
        and post-anoxic injury, 379–83
    ethylene levels, 293, 294, *297*, 299
    habitat, and anoxia tolerance, 397, 402–4, 408
Rhizophora, 333, 335, 337, 342
    R. apiculata, 334, *338*, *339*
    R. lamarckii, *338*
    R. mucronata, 334, 339
    R. stylosa, 227, *235*, 237, *338*, *340*
Rhizophoraceae, 339, 340, 341
Rhodochorton, 360
Rhododiscus pulcherrimus, 364
Rhynchostegium aciculare, *104*
Rice, *see Oryza sativa*
Rice, wild *see Zizania*
Rice weeds *see Echinochloa*
Rotala, 183
RUBISCO carboxylase, and CO_2 fixation, 130, 131–2, 134, 136, 137, 138, 141, 142, 143, 145
RuBPCase, and CO_2 fixation, 80, 86, 90, 91, 92, 94

Saccharum, 179
Sagittaria, 181
    S. pygmaea, *59*, 66, 87
    S. sagittifolia, *59*
Salt tolerance
    in brown algae, 35–48
    in mangroves, 337–41
Salvinia, 182
    S. molesta, 188
Saxifrage sarmentosa, 421
Schoenoplectus lacustris, 247, 293, 299, 398, *402*
    acetaldehyde release, *296*, *297*
    aerenchyma formation, 407
    anoxia endurance, *399*
    bud extension, 400
    carbohydrate reserves, 403, 404
    catalase activity, post-anoxic, *381*
    ethanol release, *295*, *297*, 298, 405
    ethylene release, *297*
Scirpus americanus, *399*, *402*
Scirpus lacustris, 400
Scirpus maritimus, 398, *399*, 400, *402*
Scirpus tabernaemontani, *399*, *402*
Sclerolepis, 179
Scyphiphora hydrophylacea, *339*
Scytosiphon, 360
Seawater
    carbon isotope values, 130
    HCO_3^- concentration, 101
    salinity, 36

Seaweed, *see* Algae
Sediment, as CO_2 source, 104–5, 126, 138
Seed bank, 169
    underwater, 182
Seeds and seedlings
    anchorage, 182–3
    germination, 161, 167–76
        underwater, 172–5
    photoreceptor pigments, 161, 168
    specialization, dispersal mechanisms, 183–5
Senecio, 282
    S. aquaticus, 227, 228–9, 246
Shading, 154–5
    and elongation response, 158–65
    and radiation environment, 6–8, 82, 154, 166, 169, 174, 175
    and temperature, 18
Shikimic acid 228, 406
Shinnersia, 179
Sinapis arvensis, 170, *171*
Sodium concentrations, mangroves, 337–41
Soil drainage, 321
Solar elevation
    and radiation attenuation, 6, 9
    and temperature fluctuation, 18
Sonneratia, 333, 335, 339
    S. alba, *338*, *339*
Soybean, 375, 376, *378*
Sparganium, 183
    S. erectum, 185–6
    S. simplex, *104*, *106*
Spartina alterniflora, 305, 316, 402
    adenylate activity, 242, 244, 249–51
    ADH activity, 243, 244, 245–7, 248, 249–51, 252
    AEC ratio, 244, 249–51
    aerenchyma, 239, 240, 247, 251–2
    ethanol accumulation, 243, 244, 248–9
    hypoxic root responses, 239–52
    malate accumulation, 243, 244, 248–9
Spartina anglica, *341*, *399*, *402*
Spartina patens, 305
Sphagnum cuspidatum, *104*
Stigeoclonium, 44
Stomatal resistance, with anoxia, 411, 413–14
Stratiotes, 183
Suaeda maritima, 338
Submersed aquatic macrophytes (SAM)
    bicarbonate use, controls of, 99–111
    CO_2 compensation point, 79, 81, 90, 92, 94
    CO_2 sources, 103–6
    leaf morphology, 101
    light utilization, 82
    photosynthesis, 79, 80–2
        constraints, 82–9
        strategies, 89–95
Submersed leaves, 80, 82, 85, 87
    and diffusion resistance, 85
    in heterophyllous plants, 81, 86–8
    photosynthesis, 79, 82, 83, *84*, 86
Subularia aquatica, *104*
Succinate production, anoxic mitochondria, 211
    anoxic roots, 423

Sucrose metabolism, and oxygen concentration
  in *Glyceria maxima* and *Pisum sativum*,
    231–4
  in *Typha angustifolia* shoots, 236
Sugar concentrations, with anoxia, 417–23
Sunflower *see Helianthus annus*
Suspended particles, and radiation attenuation,
    3, 9, 10–12, 154, 174
Sweet potato, 282, 289

Temperature, 3
  adaptation, in mangroves, 341–2
  and germination, 170–1, 173
  and heterophylly, 87
  and macroalgal reproduction, 359, 360,
    362–72
  and photoperiod interactions, 18–19
  and photosynthesis, 82–3
    in *Isoetes*, *116*, 117
  and solar elevation, 18
Tides, light climate model
  diel cycle, surface irradiance, 28–32
  extreme spring tides, 27–8
  low water spring tide, 31–2
  lowest astronomical tide, 23, 27–8
  spring neap tide, 27–8
  water depth, 24–7
Tobacco, 258
Tomato, 270, 322, 423
Toxin elimination, anaerobic atmosphere,
    375–83, 399
Trailliella, 360
Trapa, 181, 183, *184*, 185
Triticum
  aerenchyma formation, 322
  amino acid release, nitrate affected, *257*
  carbon budget, 419, *420*
  CO_2 evolution, and adenylic energy charge,
    261–2, 263
  NADH levels, anaerobic conditions, 255,
    258–61
  nitrate reduction, 258, 263
  radiation, within canopy, 7
  seedling, survival, with anoxia, 377, *378*
  sugar concentrations, with hypoxia, *416*, 417,
    418
  viability, and nitrate, 257–8
  yield losses, in flooding, 322
Turbud water, 10–11, 23, 28, 161
Turion production, 19, 153, 156–8, 176
Turnip, 377, *378*
Typha
  dispersal, 179–80, 182
  T. angustata, 174
  T. angustifolia, 174, *399*, *402*
    sucrose metabolism, 227, *236*
  T. latifolia, 172–3, 174, 293–9, *402*
    acetaldehyde release, *296*, 297

anoxia endurance, 398–9
ethanol release, *295*, 297, 298
ethylene release, *297*

Ulva lactuca, 41, *102*, *103*, *106*
Utricularia, 183
  U. australis, 182
  U. intermedia, 104

Vallisneria, 180, *183*
Veronica, 181
Victoria, 180
  V. amazonica, *59*
Volemitol production, and cyclical flooding, 356

Water
  density, and photosynthesis, 88–9
  depth changes, and light climate, 24–7
  and diaspore dispersal, 180–1
  radiation attenuation, 8–9, 174, 175
    freshwater, 174
    in turbid water, 10–12
Websteria, 182
Wind dispersal, 179–80
Wolffia, 182

Xylocarpus, 335
  X. mekongensis, *339*

Yeast, 266
  responses to anoxia, 205, 206, 220–1

Zea mays
  ADH activity, 282
  aerenchyma formation, 322–3, 324
    and internal oxygen transport, 325–7
    and ion transport, 327–9
  amino acid levels, with anoxia, 417
  ATP utilization, with anoxia, 205, 220, 221
Zea mays (*cont.*)
  ATD/ADP ratio, 428–9
  carbon budget, 428–9
  cristae formation, anoxic mitochondria, 208,
    209, *210*, 212, *213*, *214*, *215*, 216, 220,
    222
  ethanol production, 267, 422–3
  in primary anaerobiosis, 207, 208–11, 217,
    378
  respiratory function, seeds and seedlings, 207,
    211, 217
  in secondary anaerobiosis, 207, 211–16,
    217–20
  starch levels, with anoxia, *415*
  sugar levels, with anoxia, 404, *415*, 417, 421
Zizania, 180
  Z. aquatica, 173
Zostera, 406
  Z. marina, *103*, *106*